Donor–Acceptor Cyclopropanes in Organic Synthesis

Donor–Acceptor Cyclopropanes in Organic Synthesis

Edited by
Prabal Banerjee and Akkattu T. Biju

WILEY-VCH

Editors

Prof. Prabal Banerjee
Indian Institute of Technology Ropar
Department of Chemistry
Rupnagar-140001 Punjab
India

Prof. Akkattu T. Biju
Indian Institute of Science
Department of Organic Chemistry
Bangalore 560 012
India

Cover Image: Courtesy of the Editors

Library of Congress Card No.: applied for

British Library Cataloguing-in-Publication Data
A catalogue record for this book is available from the British Library.

Bibliographic information published by the Deutsche Nationalbibliothek
The Deutsche Nationalbibliothek lists this publication in the Deutsche Nationalbibliografie; detailed bibliographic data are available on the Internet at <http://dnb.d-nb.de>.

Print ISBN: 978-3-527-34987-6
ePDF ISBN: 978-3-527-83563-8
ePub ISBN: 978-3-527-83564-5
oBook ISBN: 978-3-527-83565-2

Typesetting: Straive, Chennai, India
Printing and Binding: CPI Group (UK) Ltd, Croydon, CR0 4YY

C9783527349876_290124

Contents

Preface

Donor–Acceptor Cyclopropanes (DACs) constitute a preeminent class of building blocks in organic synthesis featuring a strain-induced incredibly wide variety of reactivity. Besides, the cyclopropanes in general are also decorated with a plethora of structurally and biologically important scaffolds. The book is aimed at focusing on the chemistry involved with DACs and their utilization toward the construction of various carbo- and heterocycles of biological and industrial importance. Continuous progress made in the field of DACs, involving $(3+n)$ annulation, ring-opening, molecular rearrangements, and the asymmetric versions of many of these reactions have captured the tremendous interest of organic chemists. Signing up of DACs toward the synthesis of an array of carbo- and heterocycles of pharmaceutical importance have proved very beneficial. However, the lack of a book in the area of DACs has prevented chemists from appreciating the complete developments in DAC chemistry. The focus of the book is on recent developments in cycloaddition, ring-opening, and molecular rearrangements involving DACs. Examples from the literature that have demonstrated the synthetic application of these processes as well as unprecedented contributions that have directed research in the area are included in the book. In addition, the simpler and milder reaction conditions in DAC chemistry will inspire a broad range of organic and medicinal chemists to explore the applications of DACs in their respective fields.

The first chapter is documented by *Hans-Ulrich Reissig*, who significantly contributed to the early phase of this chemistry and coined the term Donor–Acceptor Cyclopropanes. This chapter provides a detailed introduction with a historical and personal perspective on this field. An inward understanding of the structure, bonding, and reactivity plays a central facilitating role in engendering molecules of choice. From this perspective, Chapter 2, authored by *Daniel B. Werz* and coworkers, demonstrated the bonding and rationale of strain-induced reactivity of these three carbon synthons, considering the structural and electronic insights. Chapters 3–12 focus on the chemistry of diverse DACs and their reactivity, including the mechanistic studies and utilization toward the construction of various carbo- and heterocycles of biological and industrial importance. Chapter 3, written by *Roman A. Novikov* and coworkers, covers the cycloaddition and annulation reactions of DACs. Moreover, it integrates a number of 1,3-zwitterionic synthons that could be generated from DACs and their responsiveness toward the formation of carbon–carbon and carbon–heteroatom multiple bonds. The advent of organocatalysis has created new opportunities in organic synthesis. In Chapter 4, *Jose L. Vicario* and *Efraim Reyes* focus on the organocatalytic *in-situ* generation of the DACs to undergo a variety of interesting transformations.

In continuation, Chapter 5 by *Akkattu T. Biju* and coworkers deals with the ring-opening 1,3-bisfunctionalization of DACs. The exercise of bisfunctionalization works through the introduction of two functional groups at the same time to the reacting partner. Rearranging the carbon skeleton is a recognized aspect of organic chemistry and demonstrates an alluring strategy to modify existing structures to construct important molecular frameworks. Chapter 6, authored by *Igor V. Trushkov*, demonstrates the detailed molecular rearrangements of DACs to produce a broad diversity of products. Because of their structural and pharmaceutical importance, the construction of nitrogenous compounds has always been an attractive area of research for organic chemists. Chapter 7, documented by *Jerome Waser* and coworkers, introduces DA aminocyclopropanes as a reliable building block and provides an in-depth description of their synthetic applications in ring-opening reactions and formal cycloadditions. DACs with one carbonyl group have emerged as interesting molecules with diverse reactivity lines as compared to the classical DACs with diester group as an acceptor. Chapter 8, written by *Prabal Banerjee* demonstrates the synthetic strategies and reactivity profile of these cyclopropyl carbonyls, followed by a discussion on their exploitation in the total synthesis of complex biologically important molecules.

The aroyl- and nitro-substituted DACs are relatively emerging entities in the area of DAC chemistry, and Chapter 9, documented by *Thangavel Selvi* and *Kannupal Srinivasan*, focuses on the synthesis of these DACs and their further utilization toward the facile access to various acyclic, carbocyclic, and heterocyclic compounds. Fascinated by the metal-free activation of DACs, *Yong Tang* and *Lijia Wang* have written Chapter 10. In the following chapter, they referred to various strategies for the C—C bond cleavage of the DACs by means of protic acids, non-metal Lewis acids, bases, thermal reactions, and so on. Asymmetric catalysis has emerged as a robust platform with a huge potential in organic synthesis, despite the tremendous challenges associated. Chapter 11, documented by *Xiaoming Feng* and coworkers, summarizes the stereospecific transformations or enantioselective reactions of racemic, prochiral, or optically enriched cyclopropanes. Cyclopropanation followed by ring-opening or cycloaddition and cyclopropanation ring-opening cyclization are attractive strategies widely used in the total synthesis of complex natural products. Chapter 12, authored by *Manas K. Ghorai* and coworkers, demonstrates the exploitation of DACs in the total synthesis of bioactive natural products over the last two decades.

The crafting of this book was possible only because of the outstanding contributions of all the colleagues engaged in DAC chemistry, and we are very thankful to them. In addition, we thank all the authors for their invaluable contributions, and we appreciate their time and patience. We believe this book will be a source of inspiration for youngsters as well as senior chemists practicing synthetic organic chemistry. We are thankful to Katherine Wong and Dr. Sakeena Quraishi at Wiley-VCH for their strong support and constructive suggestions in preparing this book.

Prabal Banerjee
Ropar / Bangalore, November 2023

Akkattu T. Biju
31 May 2023

1

Introduction to the Chemistry of Donor–Acceptor Cyclopropanes: A Historical and Personal Perspective

Hans-Ulrich Reissig

Institut für Chemie und Biochemie, Freie Universität Berlin, D-14195 Berlin, Germany

CHAPTER MENU

1.1 Introduction

During the past 15 years, we have seen tremendous progress in new applications of donor–acceptor cyclopropanes (DACs). Between 1980 and 2005, only a handful of papers per year were published mentioning this term; however, starting in 2006, a constant increase of interest could be observed, and recently, 80–100 articles dealing with this type of cyclopropanes as key compounds were released annually (Figure 1.1). This increasing number of contributions and the growing importance of this field are confirmed by the high number of recent review articles and, of course, by the fact that this book will collect articles from many of the key players in this research area. We introduced the term "donor-acceptor-substituted cyclopropane" in 1980 [1] and contributed to this field in its early phase. However, we did not use the term regularly; sometimes, we preferred the more specific name "siloxy-substituted cyclopropanecarboxylate," assuming that it is more precise.

Donor–Acceptor Cyclopropanes in Organic Synthesis, First Edition.
Edited by Prabal Banerjee and Akkattu T. Biju.

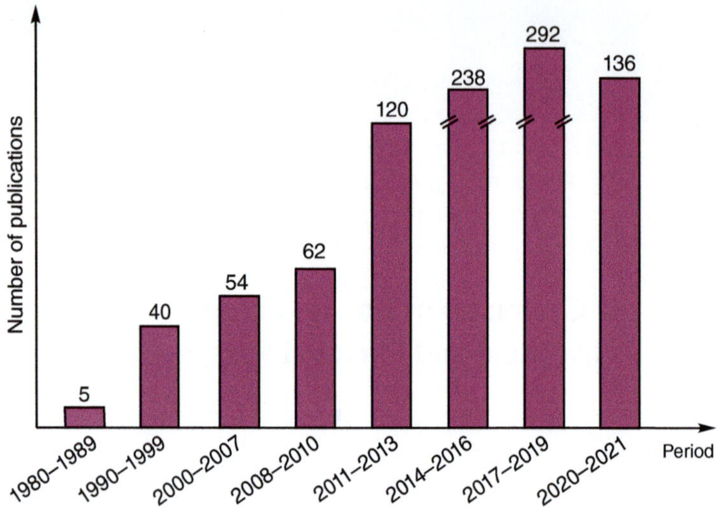

Figure 1.1 Number of publications dealing with the topic "donor–acceptor cyclopropane" or synonyma (according to a search in Web of Knowledge on 26 September 2021).

Also, several of the important contributions of Ernest Wenkert do not name their substrates DACs [2]. Therefore, the statistics in Figure 1.1 are not fully representative of the early period of 1980–2005.

Why did DACs receive this importance in organic synthesis? For a long time, cyclopropanes were regarded as exotic laboratory curiosa. In 1882, August Freund prepared the parent compound in Lemberg [3]; shortly after, in 1884, William Henry Perkin Jr. synthesized the first functionalized cyclopropane (diethyl cyclopropanedicarboxylate) [4] in the Munich laboratory of Adolf von Baeyer, who recognized the special properties of this type of hydrocarbons and formulated his famous concept of ring strain [5]. Over the years and decades, cyclopropane derivatives with different substituents and functional groups were prepared and investigated; however, in general, the reaction mechanisms involved were at the center of interest. The development of efficient methods for their synthesis was essential for this progress, in particular, the use of carbenes and carbenoids allowed simple and selective approaches to various classes of cyclopropanes. It was only in the 1960s and 1970s that it became evident that cyclopropanes can also serve as building blocks in organic synthesis, and very famous chemists were involved in exploring these possibilities. A systematic treatment of "Methods of Reactivity Umpolung" by Dieter Seebach [6] also included certain aspects of cyclopropane chemistry in this seminal review. Here the phrase "cyclopropane trick" was mentioned and connected with reactivity umpolung. A second early key player in this period was Armin de Meijere, who entered the field as a physical organic chemist but subsequently also provided important synthetic contributions in the cyclopropane field [7]. Very important contributors to the use of cyclopropanes in organic synthesis, in particular, in natural product synthesis, were Samuel Danishefsky, Robert V. Stevens, and Ernest Wenkert. Danishesky et al. exploited cyclopropanes activated by two acceptor substituents that can be smoothly ring-opened (homo-Michael addition), especially in an intramolecular fashion, to give skeletons suitable for further synthetic elaboration [8]. The known Cloke rearrangement of cyclopropyl imines to dihydropyrrole derivatives was further developed by Stevens and applied to natural product synthesis [9]. On the other hand, Wenkert et al. explored the chemistry of oxycyclopropanes for the synthesis of terpenes and alkaloids. His publications also contained a few examples of alkoxy-substituted cyclopropyl ketones or esters; however, these DACs were semantically not distinguished from the other oxycyclopropanes [2]. Nevertheless, his group should receive the credit for being the first to use DACs in natural product synthesis.

1.2 My Personal Entry to Donor–Acceptor Cyclopropanes

After my doctoral studies with Rolf Huisgen [10] at Ludwig-Maximilians University in Munich, I started a postdoctoral stint in the laboratory of Edward Piers at the University of British Columbia in Vancouver, Canada, in the fall of 1978. In Munich, I worked with diazoalkanes and studied kinetics, as well as the mechanistic aspects of their 1,3-dipolar cycloadditions. In the group of Piers, I was trained as a synthetic chemist, with a research project dealing with cuprate chemistry, the generation of divinylcyclopropanes, and their Cope rearrangements to cycloheptadiene derivatives [11]. My project and the contemporary literature taught me that cyclopropanes are very suitable compounds to achieve synthetic processes, which are not easily possible by alternative methods. Afterward, I had the chance to start my independent academic career as an associate of the group of Siegfried Hünig [12] in Würzburg, and as my first research project, I suggested to use donor-acceptor-substituted cyclopropanes. This idea originated when reading the publications of Danishefsky [8]: instead of an external nucleophile, a directly connected nucleophilic center (donor center) should open the acceptor-activated cyclopropane ring by a strain-driven retro-aldol reaction. For this type of process, only a few related examples could be found in the literature [2]. The original drawing of my grant application to the Fonds der Chemischen Industrie, a very supportive institution in Germany for young scientists, is shown as a copy in Figure 1.2. My proposal was apparently considered to be reasonable, and equipped with a Liebig fellowship, I could start with my project at the end of 1979.

In Vancouver, I had learned that silyl enol ethers are very useful starting materials for many synthetic operations, whereas during my doctoral work in Munich, methyl diazoacetate was one of the key compounds. It was, therefore, a nearby idea to combine this knowledge for the synthesis of siloxy-substituted cyclopropanecarboxylate **2** (Scheme 1.1). They were efficiently available by copper-catalyzed addition of the carbenoid derived from methyl diazoacetate to the silyl enol ethers **1**. As the simplest subsequent reaction, we first studied the ring-opening with fluoride sources to give 1,4-dicarbonyl compounds **3** under very mild conditions. In my very first independent paper published in 1980, we used the term "donor-acceptor-substituted

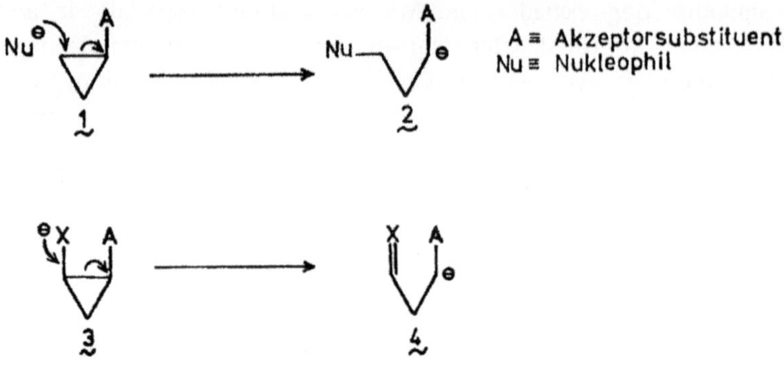

Figure 1.2 Copy of a hand-drawn scheme in a grant proposal submitted by the author to the Fonds der Chemischen Industrie in the summer of 1979.

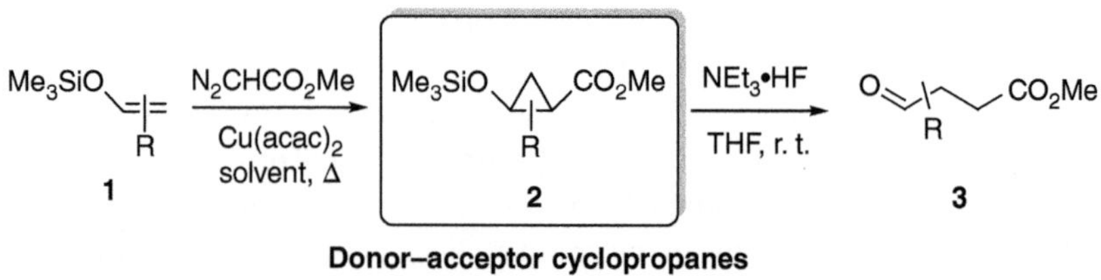

Scheme 1.1 Synthesis and ring-opening of siloxy-substituted cyclopropanecarboxylate **2**, the first cyclopropanes named DACs.

cyclopropanes" for this type of compound [1], which was later shortened to donor–acceptor cyclopropanes (DACs). I am not entirely sure why I had chosen this name, but my thoughts were probably influenced by the review of Seebach, who classified compounds by donor and acceptor centers [6].

One of the initial ideas of this project – the ring-opening with fluoride under aprotic conditions and the trapping of the resulting ester enolate with electrophiles – did not work satisfactorily [13]. However, as an excellent alternative, we found a step-wise method for forming new C—C bonds at the acceptor-substituted cyclopropane carbon atom. Methyl cyclopropanecarboxylate **2** could be smoothly deprotonated with lithium diisopropylamide (LDA) and subsequently trapped with a broad range of electrophiles (Scheme 1.2). This clean deprotonation reaction was not self-evident, since enolates incorporating a cyclopropane ring were essentially unknown around 1980. The reaction with alkyl halides R′-X occurred with surprisingly high stereoselectivity [14], leading to C-1 substituted cyclopropanes **4**, whose ring-opening led to higher substituted 1,4-dicarbonyl compounds. The trapping of the enolates with aldehydes or ketones furnished highly substituted tetrahydrofuran derivatives **5** (synthetically very useful γ-lactols) after treatment with fluoride [15]. The reaction of the enolates with carbon disulfide or aryl isothiocyanates, followed by the addition of methyl iodide, provided a nice route to interestingly functionalized thiophene or pyrrole derivatives **6** [16].

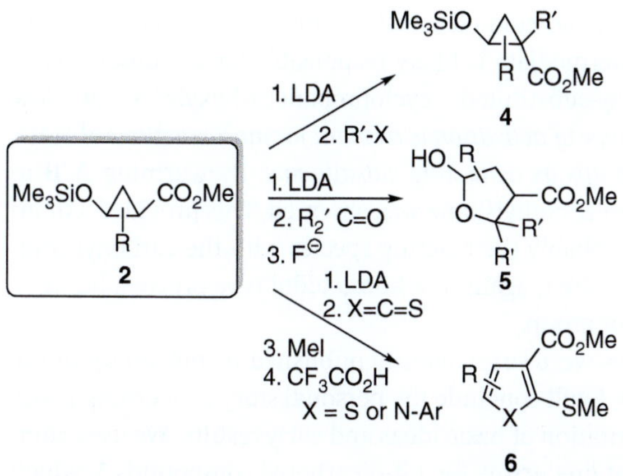

Scheme 1.2 Deprotonation of siloxy-substituted cyclopropanecarboxylate **2** with LDA and subsequent reactions with electrophiles leading to products such as **4–6**.

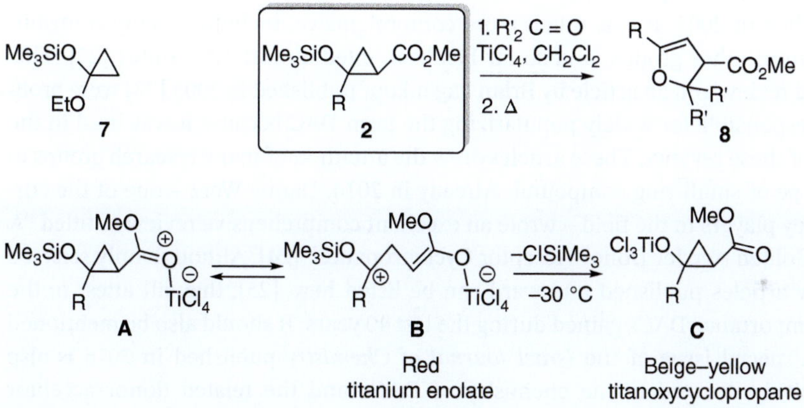

Scheme 1.3 Titanium tetrachloride-promoted ring-opening of siloxy-substituted cyclopropanecarboxylate **2** and reactions with carbonyl compounds leading to dihydrofuran derivatives **8**; titanium intermediates **A–C** involved in this process.

An alternative mode of activation and ring-opening of the siloxycyclopropanes **2** employed strong Lewis acids such as titanium tetrachloride (Scheme 1.3). This idea was deduced from the results of Kuwajima and Nakamura published in 1977 [17]. They demonstrated that 1-ethoxy-1-(trimethylsiloxy)cyclopropane **7** (an oxycyclopropane bearing **no** acceptor group) reacts with aldehydes and ketones under $TiCl_4$-promotion in a ring-opening homo-aldol process to yield γ-hydroxy carbonyl compounds. Under similar conditions, siloxy-substituted cyclopropanecarboxylate **2** furnished acyclic products or γ-lactols in good yield. Through elimination, the intermediate γ-lactols could be converted into dihydrofuran derivatives **8** [18]. After these preliminary results, a subsequent mechanistic study revealed an equilibrium between a $TiCl_4$-complex **A** and a red-colored acyclic titanium enolate **B** at low temperatures [19]. This type of intermediate is likely responsible for the observed *cis-trans*-isomerization of siloxy-substituted cyclopropanecarboxylates at low temperatures [20]. *Today, this type of activation is relevant for most reactions of DACs bearing two alkoxycarbonyl groups as activating substituents.* By warming **A/B** to −30 °C, an elimination of chlorotrimethylsilane was observed. This provided a titanoxycyclopropane **C**, which is probably the reacting species with the carbonyl compound to form the new C–C product, again in a homo-aldol type process, but now with the support of the acceptor group.

Although many more studies were subsequently published by our group in the period between 1980 and 1995, I will conclude my personal story concerning DACs with this rather detailed presentation of basic ideas and early results. We used compounds such as **2** as convenient precursors for 1,4-dicarbonyl compounds **3**, which were *in situ* trapped by subsequent reactions; for instance, intramolecular Diels–Alder reactions, and in addition, we also studied Ugi and Gewald multicomponent reactions. The results of the first years were already summarized in a review article published in 1988 [21], and later, Reinhold Zimmer and I wrote a second review published in 2003. It was much more comprehensive, including many contributions from other groups, and so far has been cited almost 1200 times [22]. This second review and an article by Brian Pagenkopf published in 2005 [23] were probably responsible for widely popularizing the term DAC because it was used in the titles of these reviews. These articles drew the attention of many research groups to this type of small-ring compound. Already in 2014, Daniel Werz – one of the current key players in the field – wrote an excellent comprehensive review entitled "A New Golden Age for Donor-Acceptor Cyclopropanes" [24]. Although only selected review articles published afterward can be listed here [25], they all attest to the high importance DACs gained during the last 40 years. It should also be mentioned that a special issue of the *Israel Journal of Chemistry* published in 2016 is also exclusively devoted to the chemistry of DACs and the related donor–acceptor cyclobutanes [26].

1.3 A Few Principles of the Chemistry of Donor–Acceptor Cyclopropanes

We must start with an unanswerable question: What is a donor, and what is an acceptor substituent? When the author of this chapter introduced the term DAC in 1980, his intention was to apply this name to cyclopropane derivatives bearing relatively strong electron-donating substituents such as alkoxy, siloxy, amido, or amino groups, in combination with a vicinally positioned electron-withdrawing substituent, and in most cases this substituent was a carbonyl or a cyano group. Later, the term DAC was less strictly employed, and cyclopropanes with substituents that are able to stabilize a positive or a negative charge can now be generally regarded.

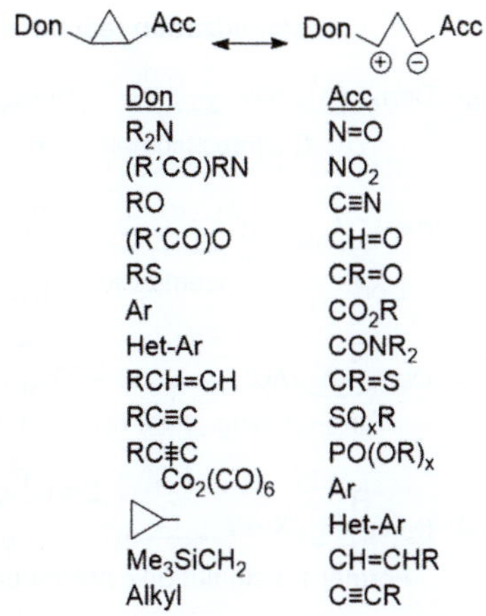

Figure 1.3 DACs polarized by different donor and acceptor substituents, respectively, and the respective zwitterionic mesomeric formula.

These substituents polarize the bond between the carbon atoms bearing the activating groups, a situation that can be characterized by a *zwitterionic mesomeric formula*, as depicted in a simplified manner in Figure 1.3. Many substituents can now be included in this (incomplete) list.

On the donor side, many more substituents can be considered, in particular, aryl, hetaryl, or alkenyl groups. Even alkyl groups are able to stabilize a positive charge better than a hydrogen atom. The acceptor substituents are generally X=Y double bond or X≡Y triple bond systems containing one or two heteroatoms. However, aryl, hetaryl, alkenyl, and alkynyl groups are also able to stabilize a negative charge and should be added as acceptor substituents in a systematic survey. It is obvious that two donor or two acceptor substituents should have a stronger influence on the reactivity of the cyclopropane than just one of these substituents. A very useful scale of the activating property of many of these substituents has been provided by Werz et al. who performed DFT calculations of the rearrangement of DACs to the corresponding five-membered ring systems [27]. The effect of substituents on the reactivity of DACs in the presence of Lewis acids was also studied by Werz et al. [28], whereas Ofial et al. examined the reactivity of several acceptor-substituted cyclopropanes, including a few DACs, in nucleophilic ring-opening reactions by thiols in the absence of Lewis acids [29].

The reactions of DACs can be roughly classified into the categories of isomerization reactions, ring-opening reactions, and cycloadditions. In Scheme 1.4, isomerizations are subdivided into reactions under the maintenance of the cyclopropane ring [case a)] *cis*/*trans*-isomerizations and [case b)] racemizations), reactions under ring-opening [case c)], and rearrangements under ring-enlargement [case d)]. The latter case can be regarded as a (formal) 1,3-sigmatropic rearrangement; the 3,3-sigmatropic Cope-rearrangements of cyclopropane derivatives bearing two alkenyl groups or their heteroanalogs, which lead to seven-membered carbo- or heterocycles, are not listed here.

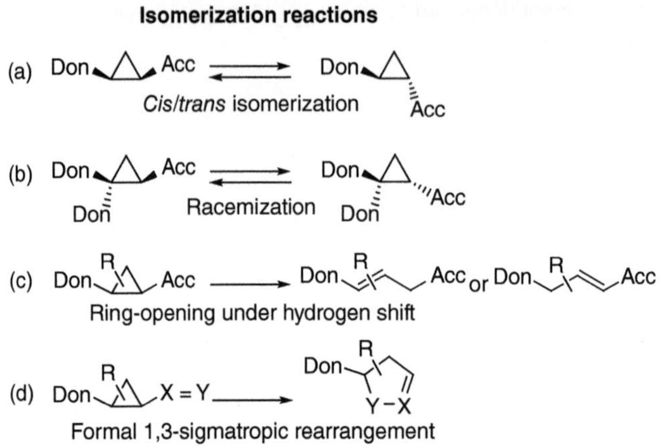

Scheme 1.4 Isomerization reactions of DACs (presentation in part under disregard of configurational aspects).

Ring-opening reactions under the incorporation of reagents (Scheme 1.5) can occur through a primary nucleophilic attack at the donor-substituted carbon [case a)], an electrophilic attack at the acceptor-substituted carbon [case b)], or the attack of a radical, often observed as the addition of the radical to an alkenyl group as donor substituent [case c)]. Donors such as siloxy groups allow a reaction with fluoride as a nucleophile at the silicon center, delivering a cyclopropoxy anion whose ring opening directly generates a carbonyl group in a retro-aldol process [case a), also see Scheme 1.1]. Similarly, the reactions of electrophiles can also lead to the formation of a carbonyl group if their counter ion X^- reacts with the siloxy group [case b), also see Scheme 1.3]. The ring-opening of acceptor-substituted vinylcyclopropanes by palladium or other metal complexes, which affords π-allyl complexes ready for further reactions with external or internal nucleophiles, is not presented here.

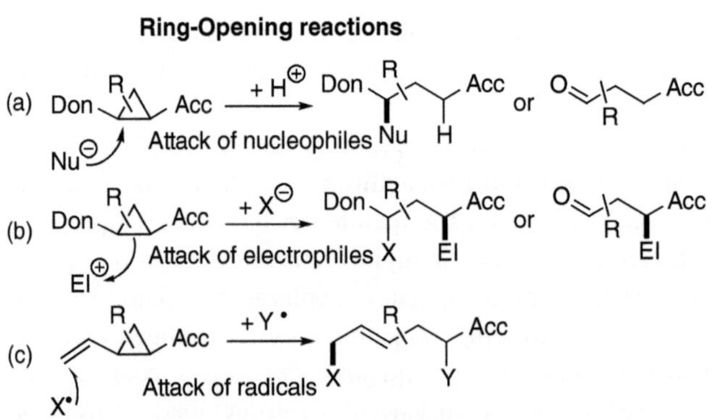

Scheme 1.5 Ring-opening reactions of DACs.

Cycloadditions constitute a particularly important class of reactions of DACs. Only a few can occur without external promoters; however, Lewis acids are generally employed in stoichiometric or catalytic amounts in order to activate the cyclopropanes. Most frequently, the combination of two alkoxycarbonyl groups on the acceptor side and an (electron-rich) aryl group on the donor side was used. In the simplified Scheme 1.6, only the overall processes are illustrated, showing that (3+2)-cycloadditions give rise to five-membered ring systems [case a)], and that several 1,3-dipoles have been used in (3+3)-cycloadditions, which furnish six-membered heterocycles [case b)]. A few examples of dimerizations of DACs have been reported, which also belong to the category of (3+3)-cycloadditions. To complete the picture, (3+4)-cycloadditions of DACs with 1,3-(hetero)dienes are listed here [case c)], although this process is relatively rare. Higher-order cycloadditions and reactions where a rearrangement of the DAC occurs before a cycloaddition proceeds have also been studied.

Scheme 1.6 Schematic presentation of (3+*n*)-cycloaddition reactions of DACs (simplified presentation without Lewis acids frequently required in these processes).

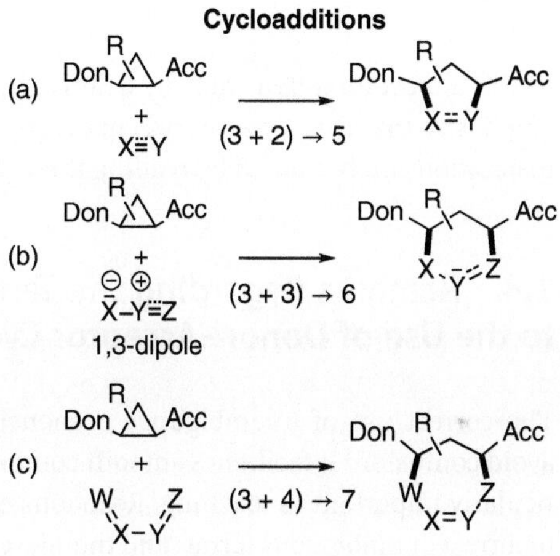

All these basic reactions have been described in detail in the published reviews, and certainly, they will be discussed again in the following chapters. Here, I just want to draw attention to the very interesting early studies reported by Cram and coworkers, starting in 1970. As physical–organic chemists, they carefully studied the isomerizations of specifically substituted cyclopropanes, which are clearly DACs according to our current definition [30]. Scheme 1.7 shows a typical example: the thermal racemization of compound **9** involves zwitterionic intermediate **D** as a crucial species. Cram et al. called these systems “carbanion-carbonium ion intermediates,” discussed the character of the intermediates (1,3-zwitterion vs. singlet 1,3-diradical), and determined the activation parameters. As expected, the rate of racemization is dependent on the polarity of the solvent, with a relative rate of 1 in benzene and 75 in dimethylformamide in the presence of lithium bromide. The lithium cation acts as a Lewis acid in these reactions, which are more complex due to subsequent ring-opening and ring-enlargement steps. In the case depicted in Scheme 1.7, the zwitterion **D** is a real existing intermediate with a certain lifetime, which is formed by the ring-opening of **9** or *ent*-**9**. It should not be mixed up with the zwitterionic mesomeric formula presented in Figure 1.3, which expresses the charge distribution of DACs.

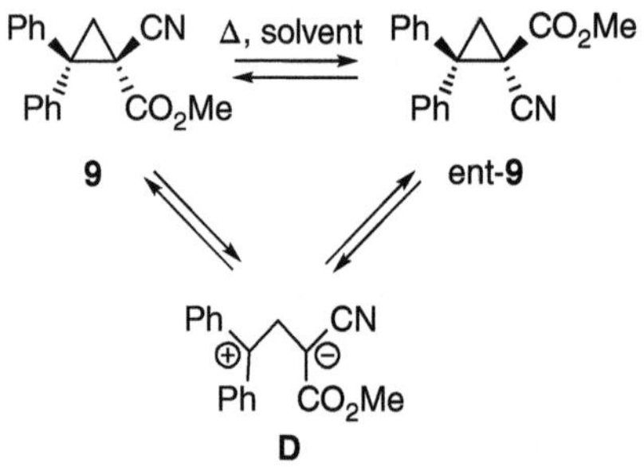

Scheme 1.7 Racemization of DAC **9** via zwitterionic intermediate **D** as studied by Cram et al.

This almost forgotten study by Cram et al., published 50 years ago, is presented to emphasize that DAC chemistry is not entirely new and that a lot of information and inspiration can be gained by reading these detailed reports.

1.4 Remarks Regarding the Terminology Applied to the Use of Donor–Acceptor Cyclopropanes

The correct use of unambiguous nomenclature and terminology is inevitable to avoid confusion. It facilitates smooth communication between scientists and is particularly important in teaching. Reactions of DACs are often described by applying incorrect or ambiguous terms, and therefore a few of these issues are discussed here.

(a) The term "1,3-dipole" or "1,3-dipolar synthon" for DACs is misleading and incorrect. Huisgen clearly defined 1,3-dipoles [31] as conjugated 4π-system as depicted in Figure 1.4. These species always contain sp^2- or sp-hybridized heteroatoms Y in their center, which can bear a positive charge in an electron octet formula.

The bonding situation of cyclopropanes can be described by the MO-model by Walsh, which suggests that the C—C "single" bonds have considerable π-character (below, σ/π is used to describe this type of bonding). The interaction of cyclopropane bonds with adjacent substituents is therefore stronger than expected [7a, 27]. The zwitterionic mesomeric formula of DACs reflects the polarization of the bond between donor- and acceptor-substituted carbons and the dominating interaction with the substituents, but it has no common feature with 1,3-dipoles. The central atom is – by definition – a carbon atom and therefore not in strong interaction with the two adjacent carbons.

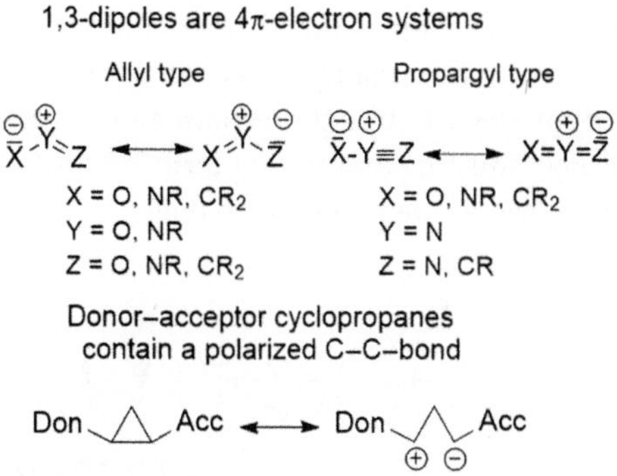

Figure 1.4 1,3-Dipoles according to Huisgen's systematic classification and polarization of DACs.

(b) Very often, DACs are named "1,3-dipolar synthons," which is wrong in two aspects. For the use of "1,3-dipole," see the the previous discussion under (a). The term "synthon" was initially introduced by Corey to characterize a hypothetical (charged) unit within a target molecule that represents a potential precursor reagent [32]. However, Corey noted in 1988 that "synthon" has now come to be used to mean "synthetic building block" rather than a retrosynthetic fragment. Since the original meaning of "synthon" is still useful in retrosynthetic analysis, I suggest calling DACs "1,3-zwitterionic building blocks" (or synthetic equivalents of a "1,3-zwitterionic synthon").

(c) Surprisingly, the term "(3+n)-cycloaddition" is inconsistently used when the reactions of DACs are discussed. It should be recalled first that parentheses, for instance, in (3+2)-cycloaddition, should be used to define the number of centers involved in a cycloaddition, whereas brackets, for instance, in [4+2]-cycloaddition, denote the number of (π)-electrons involved in a cycloaddition. For reactions of DACs, several authors prefer the more general term "(3+n)-annulation" or the even less specific "(3+n)-cyclizations".

Indeed, the definition of cycloaddition reactions is ambiguous if cyclopropanes are involved. According to the criteria, as collected long ago by Huisgen [33], cycloadditions are ring-forming reactions with an increase in the number of σ bonds. They are not associated with the elimination of small compounds or with the shift of atoms – at least in the ring-forming step. The reaction mechanism involved (thermally or photochemically, concerted as pericyclic reactions or stepwise via intermediates, uncatalyzed or catalyzed) is irrelevant. Later, IUPAC recommends similar criteria but notes that two or more unsaturated molecules should participate in the formation of a cyclic adduct in which a net reduction of bond multiplicity can be observed [34]. Cyclopropanes are usually not regarded as unsaturated molecules, but their partial π-character (see the previous discussion under (b)) justifies treating them as ethene homologs and calling the reactions summarized in Scheme 1.6 "real" cycloadditions if all other criteria are met. The prototype of a (3+2)-cycloaddition of a cyclopropane to a five-membered ring (Scheme 1.8) is electronically characterized as a [2σ/π+2π]-process, in analogy to a (2+2)-cycloaddition of two alkenes, which is a [2π+2π]-process. I strongly recommend to act pragmatic and stay with the well-established and frequently used term "(3+*n*)"-cycloadditions for cyclopropanes if the above-mentioned criteria are met.

Scheme 1.8 (3+2)-Cycloadditions of cyclopropanes to a double bond system X = Y.

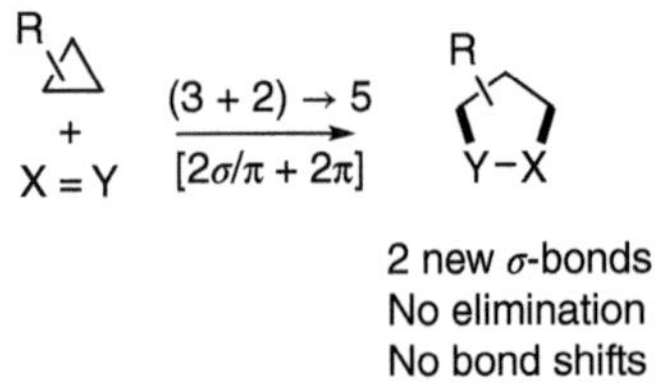

1.5 Conclusions

This introductory chapter should illustrate how the author was guided to introduce the term DACs and which types of reactions are possible with DACs. Studies conducted in the 1980s already revealed some of the important features of reactivity, for instance, the activation of DACs by Lewis acids. Later, the definition of DACs was expanded to include many new cyclopropane derivatives, particularly compounds with aryl groups as donor substituents. A tremendous development could be observed with many synthetically very useful transformations employing DACs as crucial C3-building blocks. Impressive examples already exist on catalytic enantioselective processes [25h, 25i], and it can be expected that these will be further advanced. The currently observed increase in electron-transfer-promoted reactions may also influence the chemistry of DACs. Recent examples employing electrochemical methods can already be found in the literature [35]. Some functionalized bicyclo[1.1.0]butanes can also be classified as DACs, and a few ring-opening reactions of these very strained compounds were reported [36]. This compound class is not particularly difficult to access, and therefore, more applications can be expected in the future. Finally, and most importantly, surprising and entirely new reactions of DACs are certainly still possible. The golden age of DACs is not finished.

Abbreviations

Acc	acceptor substituent
DAC	donor–acceptor cyclopropane
Don	donor substituent
LDA	lithium diisopropylamide

References

1 Reissig, H.-U. and Hirsch, E. (1980). *Angew. Chem. Int. Ed. Engl.* 22: 813–814.

2 (a) Wenkert, E. (1980). *Acc. Chem. Res.* 3: 27–31.
(b) Wenkert, E. (1980). *Heterocycles* 14: 1703–1708.
(c) For an excellent summary of Wenkert's contributions to cyclopropane chemistry, see:Hudlicky, T. (2016). *Isr. J. Chem.* 56: 540–552.

3 Freund, A. (1882). *Monatsh. Chem.* 3: 625–635.

4 Perkin, W.H. Jr. (1884). *Ber. Dtsch. Chem. Ges.* 17: 54–59.

5 Baeyer, A. (1885). *Ber. Dtsch. Chem. Ges.* 18: 2269–2281. The important chapter on strain theory is presented in an appendix on page 2278–2281 of this article about polyacetylenes.

6 Seebach, D. (1979). *Angew. Chem. Int. Ed. Engl.* 18: 239–258.

7 (a) De Meijere, A. (1979). *Angew. Chem. Int. Ed. Engl.* 18: 809–826.
(b) De Meijere, A. (1987). *Chem. Ber. 23*: 865–879.

8 Danishefsky, S. (1979). *Acc. Chem. Res.* 12: 66–72.

9 Stevens, R.V. (1977). *Acc. Chem. Res.* 10: 193–198.

10 (a) Houk, K.N. and Reissig, H.-U. (2019). *Chem.* 5: 2499–2505.
(b) Giese, B., Mayr, H., and Reissig, H.-U. (2020). *Angew. Chem. Int. Ed.* 59: 12228–12232.

11 Piers, E. and Reissig, H.-U. (1979). *Angew. Chem. Int. Ed. Engl.* 18: 791–792.

12 Reissig, H.-U. (2021). *Angew. Chem. Int. Ed.* 60: 9180–9191.

13 Kunkel, E., Reichelt, I., and Reissig, H.-U. (1984). *Liebigs Ann. Chem.* 802–819.

14 (a) Reissig, H.-U. and Böhm, I. (1982). *J. Am. Chem. Soc.* 104: 1735–1737.
(b) Reichelt, I. and Reissig, H.-U. (1984). *Liebigs Ann. Chem.* 531–551.

15 (a) Brückner, C. and Reissig, H.-U. (1985). *J. Chem. Soc. Chem. Commun.* 1512–1513.
(b) Brückner, C. and Reissig, H.-U. (1988). *J. Org. Chem.* 53: 2440–2450.

16 (a) Brückner, C. and Reissig, H.-U. (1985). *Angew. Chem. Int. Ed. Engl.* 24: 588–589.
(b) Brückner, C. and Reissig, H.-U. (1988). *Liebigs Ann. Chem.* 465–470.
(c) Brückner, C., Suchland, B., and Reissig, H.-U. (1988). *Liebigs Ann. Chem.* 471–473.

17 Nakamura, E. and Kuwajima, I. (1977). *J. Am. Chem. Soc.* 22: 7360–7362.

18 (a) Reissig, H.-U. (1981). *Tetrahedron Lett.* 22: 2981–2984.
(b) Reissig, H.-U., Reichelt, I., and Lorey, H. (1986). *Liebigs Ann. Chem.* 1924–1939.

19 Reissig, H.-U., Holzinger, H., and Glomsda, G. (1989). *Tetrahedron* 45: 3139–3150.

20 Reissig, H.-U. and Böhm, I. (1983). *Tetrahedron Lett.* 24: 715–718.

21 Reissig, H.-U. (1988). *Top. Curr. Chem.* 144: 73–135.

22 Reissig, H.-U. and Zimmer, R. (2003). *Chem. Rev.* 103: 1151–1196.

23 Yu, M. and Pagenkopf, B.L. (2005). *Tetrahedron* 61: 321–347.

24 Schneider, T.F., Kaschel, J., and Werz, D.B. (2014). *Angew. Chem. Int. Ed.* 53: 5504–5523.

25 (a) de Nanteuil, F., De Simone, F., Frei, R. et al. (2014). *Chem. Commun.* 50: 10912–10928.
(b) Cavitt, M.A., Phun, L.H., and France, S. (2014). *Chem. Soc. Rev.* 43: 804–818.

(c) Grover, H.K., Emmett, M.R., and Kerr, M.A. (2015). *Org. Biomol. Chem.* 13: 655–671.
(d) Pagenkopf, B.L. and Vemula, N. (2017). *Eur. J. Org. Chem.* 2561–2567.
(e) Budynina, E.M., Ivanov, K.I., Sorokin, I.D., and Melnikov, M.Y. (2017). *Synthesis* 49: 3035–3068.
(f) Ivanova, O.A. and Trushkov, I.V. (2019). *Chem. Rec.* 19: 1–21.
(g) Singh, P., Varshnaya, R.K., Dey, R., and Banerjee, P. (2020). *Adv. Synth. Catal.* 362: 1447–1484.
(h) Pirenne, V., Muriel, B., and Waser, J. (2021). *Chem. Rev.* 121: 227–263.
(i) Xia, Y., Liu, X., and Feng, X. (2021). *Angew. Chem. Int. Ed.* 60: 9192–9204.
(j) Augustin, A.U. and Werz, D.B. (2021). *Acc. Chem. Res.* 54: 1528–1541.
(k) Gosh, K. and Das, S. (2021). *Org. Biomol. Chem.* 19: 965–982.
(l) Gosh, A., Dey, R., and Banerjee, P. (2021). *Chem. Commun.* 57: 5359–5373.

26 Introductory editorial for 16 articles: Reissig, H.-U. and Werz, D.B. (2016). *Isr. J. Chem.* 56: 367–368.

27 Schneider, T.F. and Werz, D.B. (2011). *Org. Lett.* 13: 1848–1851.

28 Kreft, A., Lücht, A., Grunenberg, J. et al. (2019). *Angew. Chem. Int. Ed.* 58: 1955–1959.

29 Jüstel, P.M., Stan, A., Pignot, C.D., and Ofial, A.R. (2021). *Chem. Eur. J.* 27: 15928–15935.

30 (a) Yankee, W.E. and Cram, D.J. (1970). *J. Am. Chem. Soc.* 92: 6328–6329.
(b) Yankee, W.E. and Cram, D.J. (1970). *J. Am. Chem. Soc.* 92: 6329–6331.
(c) Yankee, W.E. and Cram, D.J. (1970). *J. Am. Chem. Soc.* 92: 6331–6333.
(d) Yankee, W.E., Badea, F.D., Howe, N.E., and Cram, D.J. (1973). *J. Am. Chem. Soc.* 95: 4210–4219.
(e) Yankee, W.E., Spencer, B., Howe, N.E., and Cram, D.J. (1973). *J. Am. Chem. Soc.* 95: 4220–4230.
(f) Howe, N.E., Yankee, W.E., and Cram, D.J. (1973). *J. Am. Chem. Soc.* 95: 4230–4237.
(g) Chmurny, A.B. and Cram, D.J. (1973). *J. Am. Chem. Soc.* 95: 4237–4244.

31 (a) Huisgen, R. (1963). *Angew. Chem. Int. Ed. Engl.* 2: 565–598.
(b) Huisgen, R. (1963). *Angew. Chem. Int. Ed. Engl.* 2: 633–645.
(c) Breugst, M. and Reissig, H.-U. (2020). *Angew. Chem. Int. Ed.* 59: 12293–12307.

32 Corey, E.J. (1967). *Pure Appl. Chem.* 14: 19–37.

33 Huisgen, R. (1968). *Angew. Chem. Int. Ed. Engl.* 7: 321–328.

34 McNaught, A.D. and Wilkinson, A. (ed.) (1997). *IUPAC-Compendium of Chemical Terminology (the "Gold Book")*, 2e; compiled by, 6. Oxford: Blackwell Scientific Publications.

35 (a) Kolb, S., Ahlburg, N.L., and Werz, D.B. (2021). *Org. Lett.* 23: 5549–5553.
(b) Kolb, S., Petzold, M., Brandt, F. et al. (2021). *Angew. Chem. Int. Ed.* 60: 15928–15934.
(c) Saha, D., Taily, I.M., and Banerjee, P. (2021). *Eur. J. Org. Chem.* 5053–5057.

36 McNamee, R.E., Haugland, M.M., Nugent, J. et al. (2021). *Chem. Sci.* 12: 7480–7485.

2

Understanding the Reactivity of Donor–Acceptor Cyclopropanes: Structural and Electronic Analysis

Anu Jacob[1]*, Gwyndaf A. Oliver*[2]*, and Daniel B. Werz*[2]

[1]*Institut für Organische Chemie, Technische Universität Braunschweig, Hagenring 30, D-38106 Braunschweig, Germany*
[2]*Institut für Organische Chemie, Albert-Ludwigs-Universität Freiburg, Albertstraße 21, D-79104 Freiburg, Germany*

CHAPTER MENU

2.1 Introduction

Cyclopropanes with their strained bonds show great potential for simple and rapid transformations, leading to complexity in very few steps [1]. Although cyclopropanes are highly energetic molecules with a ring strain of about 115 kJ mol^{-1} [2], the C—C bonds of these molecules are kinetically rather inert and maintain the integrity of the ring. Unlike other saturated carbocycles, cyclopropanes undergo electrophilic reactions in the presence of mineral acids [3, 4] and add oxidatively to transition metals [5, 6].

The bonding and corresponding unique reactivity of cyclopropanes can be rationalized by various models. The strain model provides a simple explanation of how the highly strained system lowers its energy through a ring-opening reaction. For cyclopropane, the internuclear angle is 60°, which is far from the ideal sp^3 geometry and minimizes electrostatic repulsion requirements. The geometry of cyclopropane

Donor–Acceptor Cyclopropanes in Organic Synthesis, First Edition.
Edited by Prabal Banerjee and Akkattu T. Biju.

Figure 2.1 Orbitals by Förster, Coulson and Moffitt.

requires co-planarity of the three carbon atoms. This fact leads to an eclipsed arrangement of the C—H bonds, causing, in addition to the Baeyer strain, a torsional strain (Pitzer strain). Therefore, many explanations for the increased reactivity of cyclopropanes in comparison with other cyclic alkanes are based on the relief of ring strain. However, this model is insufficient to entirely account for the reactivity of cyclopropanes.

Another rationale using elaborate bonding models was suggested by Förster [7] and refined by Coulson and Moffitt [8, 9]. They regard the σ-bonds in cyclopropanes as being bent (Figure 2.1). There is about a 20% reduction in the overlap between two carbon atoms compared to that in ethane. This adds to the increased angular strain, thereby an increased reactivity is observed. Their theory advocates that the C—C bonds are built from sp^5-hybridized orbitals and C—H bonds are built from $sp^{2.3}$-hybridized orbitals. This assumption explains the high π character of the ring and leads to an understanding of why similar reactivity to that of alkenes is observed. X-ray crystallographic data for cyclopropane derivatives validated the concept of bent bonds. The maximum electron density is found outside the conventional bond axes [10]. However, this model was unsuccessful in rationalizing the effect of electron-donating and withdrawing substituents on the reactivity and stability of cyclopropanes.

The Walsh model gives a detailed description of bonding in cyclopropanes [11]. It was assumed that the bonding in cyclopropanes is represented more precisely by the combination of sp^2 hybrid orbitals and p_z-orbitals. The in-phase combination of the three sp^2 orbitals produces a molecular orbital (MO) with a large electron density in the center of the ring, which makes up the lowest energy orbital ψ_1. The three p_z orbitals directed outside to the corners of a hypothetical triangle form a degenerate pair of highest-occupied molecular orbitals (HOMOs) ψ_2 and ψ_3. These orbitals are called the Walsh orbitals of cyclopropane (Figure 2.2). Their electron density does not lie along the internuclear axis; they mirror the kind of bent bonds in the model

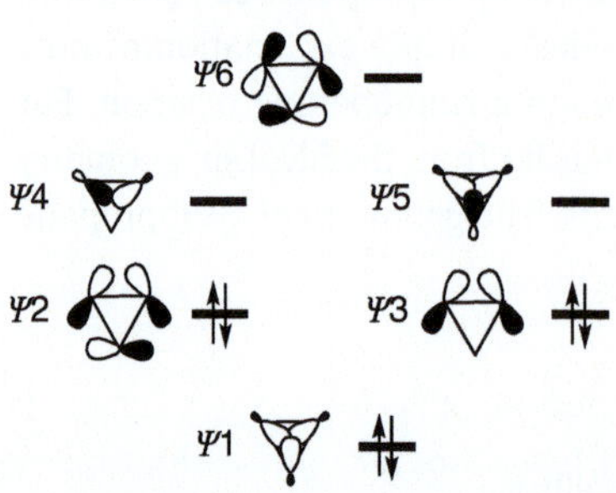

Figure 2.2 Walsh molecular orbitals of cyclopropane.

of Förster, Coulson, and Moffitt. These orbitals account for the π-character of the ring; energetically, they are relatively high-lying orbitals for a saturated hydrocarbon. The lowest unoccupied molecular orbitals (LUMOs) ψ_4 and ψ_5 are formed similarly to the ψ_1 orbital of sp^2 orbitals, which are directed to the center of the ring, but in a destructive fashion. The highest-energy unoccupied orbital ψ_6 is formed by the destructive overlap of p-orbitals. The Walsh orbitals are excellent π-donors, interacting well with low-lying vacant orbitals [12].

The presence of electron density at the center of the ring, explained by ψ_1, brings forth the concept of σ-aromaticity of cyclopropanes. The stabilization comes from the aromaticity due to σ-bonds. The ring consists of six electrons positioned across the three atoms, obeying the $(4n+2)$ rule of Hückel for aromaticity. This stabilization might explain its reactivity toward electrophiles while also explaining the strong localized diatropic ring current due to which a significant shielding of its ^1H nuclei (signals at $\delta = 0.22$) is observed in comparison to larger cycloalkanes (signals at $\delta = 1.44$–1.54) [13]. The current understanding of the structure and reactivity of cyclopropanes can be drawn from all the concepts, including ring strain, distorted π-like bonds, and σ-aromaticity.

2.2 Activated Cyclopropanes

Although cyclopropanes show a significant amount of ring strain, they are kinetically rather inert. Thus, additional functional groups need to be attached to the ring to allow mild and synthetically efficient transformations. Only a few reactions of cyclopropanes – such as some special ring-opening, ring-enlargement and pericyclic reactions – take place without the aid of activating substituents; however, harsh conditions are required. Examples of this reactivity are the vinylcyclopropane–cyclopentene or divinylcyclopropane–cycloheptadiene reactions [14]. Cyclopropanes substituted with electron-withdrawing groups are termed acceptor cyclopropanes. They behave analogously to γ-carbonyl cation equivalents/homologous Michael acceptors. Cyclopropanes substituted with electron-donating groups are called donor cyclopropanes, and they act as homologous enolate equivalents [15]. The first example of a ring-opening reaction of an activated cyclopropane was published by Bone and Perkin in 1895 [16]. They showed the ring-opening of cyclopropane **1** using an enolate generated from diethyl malonate **2** to obtain **3** in 50% yield (Scheme 2.1).

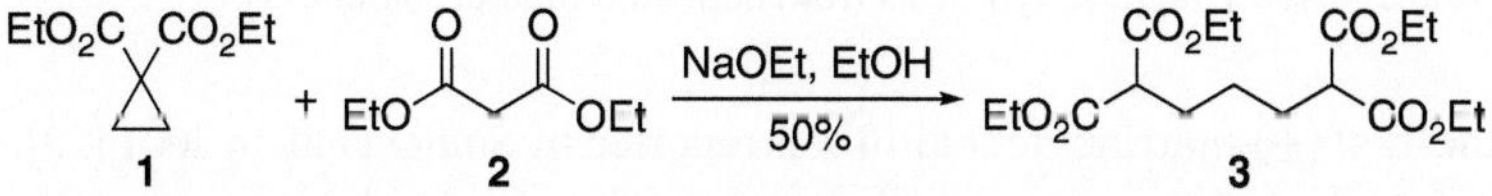

Scheme 2.1 Nucleophilic ring-opening of acceptor cyclopropanes.

Homologous Michael reactions of activated cyclopropanes employing a range of nucleophiles such as amines [17], mercaptans [17], enamines [18], and cuprates [19]

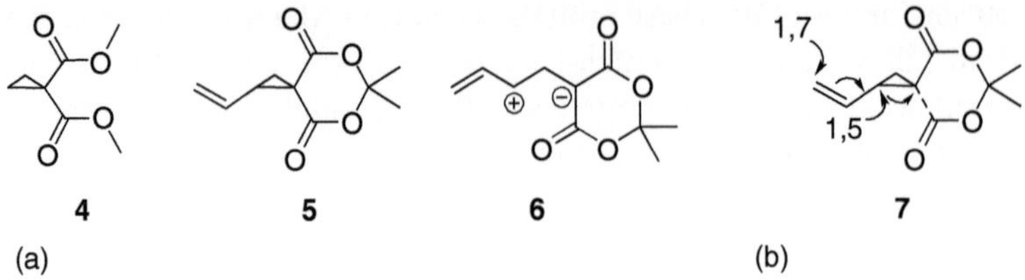

Figure 2.3 (a) Spiroacylal mode of activation in **5** and **6**, (b) modes of attack on a vinyl cyclopropane.

were later explored. However, the harsh reaction conditions and mediocre yields dimmed the utility of these intermolecular reactions. This was however tackled by Danishefsky and coworkers by choosing a spiroacylal for diester activation, instead of **4**, where the carbonyl groups are orthogonal to the plane of the cyclopropane (**5**) [20]. This enables the emerging carbanion to delocalize over the alkoxycarbonyl groups (Figure 2.3).

A spiroacylal linkage may allow greater charge separation in the transition state for ring-opening, reflecting the stabilization of the transition state leading to the presumed intermediate **6** (Figure 2.3a). This fact is easily understood from the relative acidity of Meldrum's acid ($pK_a \sim 5$) compared to acyclic malonic esters ($pK_a \sim 14$). Investigations into nucleophilic attacks on vinylcyclopropane by Singh and Danishefsky [21] showed that an exclusive 1,5-attack was observed when spiroacylal was chosen as the acceptor motif as in **7** (Figure 2.3a), whereas a competing 1,5- and 1,7-mode of nucleophilic addition took place in case of **4** (Figure 2.3b). This transformation under mild conditions underscores the value of spiroactivated vinylcyclopropanes in various synthetic strategies.

The remarkable electrophilic properties of spiroactivated cyclopropanes suggested a simple synthesis of *trans*-fused γ-lactones **10**, improving upon the conventional synthetic route involving epoxidation of olefins followed by the nucleophilic attack by an acetic acid ester carbanion equivalent. This *trans*-diaxial solvolysis reaction underlined the effectiveness of this activation mode (Scheme 2.2) [22].

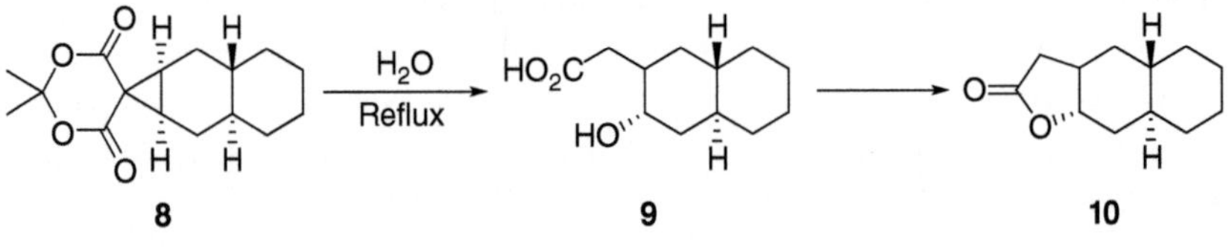

Scheme 2.2 *trans*-Fused γ-lactone synthesis from activated cyclopropanes.

Total synthesis of (±)-martinellic acid **14** was reported by Snider et al. in 2001 [23]. Derivatives of **14** have been recognized as antagonists for bradykinin receptors (B_1, B_2). Activated vinyl cyclopropane **5** was utilized in this total synthesis route. The synthetic strategy included ring-opening of **5** by substituted aniline **11**, followed by lactamization–oxidation to yield vinylpyrrolidone **12**. Reaction with *N*-benzylglycine and a subsequent intramolecular (3+2)-cycloaddition deliver **13**, a precursor to (±)-martinellic acid **14** (Scheme 2.3).

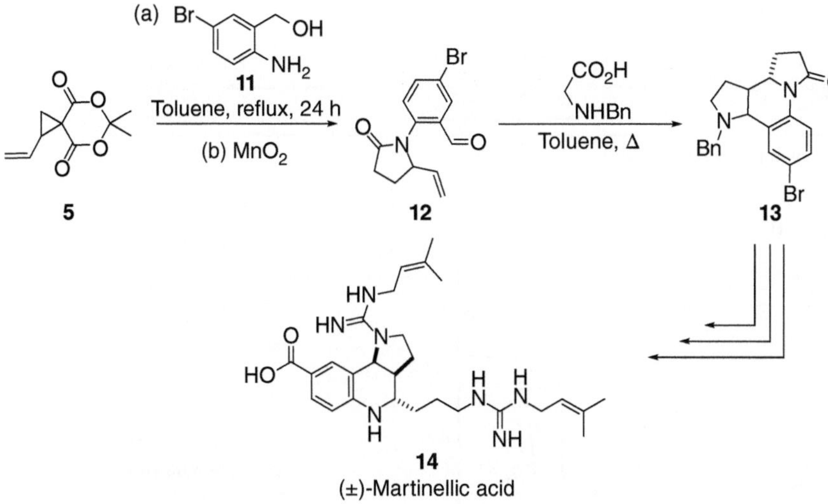

Scheme 2.3 Total synthesis of (±)-martinellic acid **14**.

2.3 Donor–Acceptor Cyclopropanes (DACs)

A significant breakthrough in the realm of cyclopropane chemistry was realized by substituting the three-membered ring with donor and acceptor moieties in a vicinal fashion. Donor and acceptor substituents act as a push–pull trigger and activate the ring, thereby enabling the molecule to be vulnerable to even mild reaction conditions. The transformations of this molecule were extensively researched by Wenkert and Reißig in the 1980s [24–31]. All the reactions could be rationalized by taking a 1,3-zwitterionic relationship into account, wherein the negative charge is stabilized by the electron-withdrawing group and the positive charge by an electron-donating group (Figure 2.4).

A variety of activation methods are known for donor–acceptor cyclopropanes (DACs), as shown in Scheme 2.4. The most widely studied of these is Lewis acid catalysis [32–36]. However activation by organocatalysis [37, 38], radicals [39], and electrochemical single electron transfer (SET) oxidation [40] are also known and give access to a variety of different intermediates for interesting chemical processes.

DACs can be easily activated by Lewis acids, which coordinate with the acceptor motifs and enhance their electron-withdrawing ability. This polarization weakens the σ-bond between the donor and acceptor carbon atoms, facilitating various reactions. DACs have been employed for ring-opening reactions [41–63], cycloadditions [64–85], and rearrangements [86–91]. Scheme 2.5 depicts the general reactions of DACs under Lewis acid catalysis.

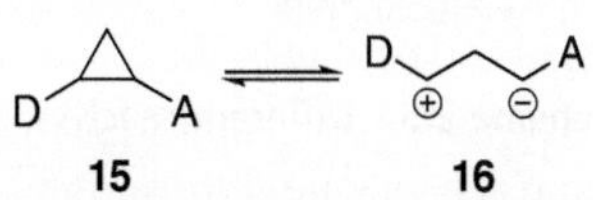

Figure 2.4 Zwitterionic relationship in vicinally substituted donor–acceptor cyclopropanes **15**.

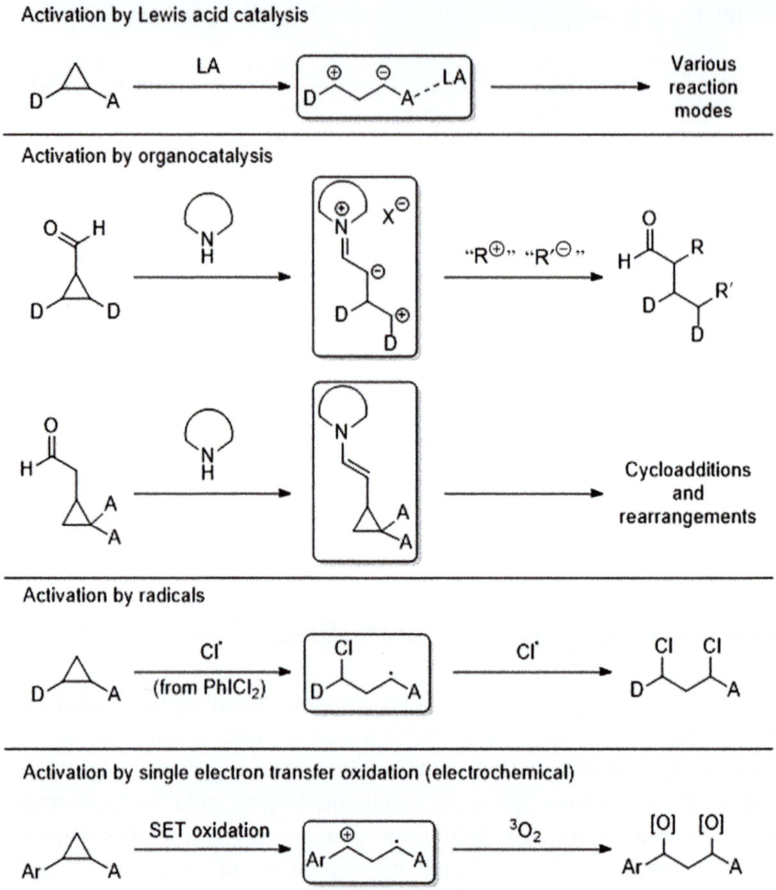

Scheme 2.4 Known activation strategies for DACs.

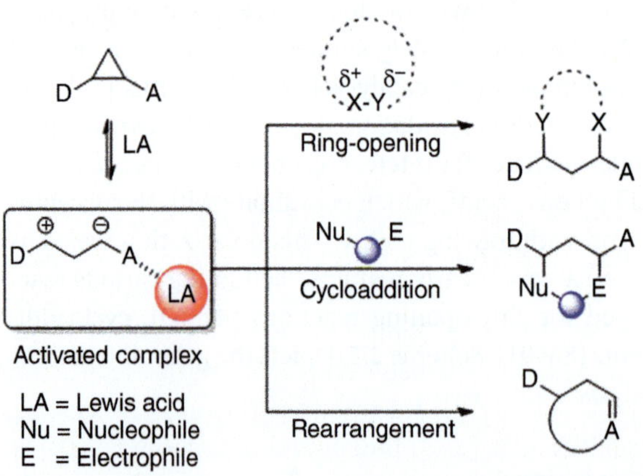

Scheme 2.5 Different reactivities of DACs by Lewis acid catalysis.

Ring-opening reactions in DACs are among the most straightforward and efficient ways to construct 1,3-functionalized structures. Nucleophilic ring-opening reactions of activated cyclopropanes could be viewed as homologous Michael additions. In contrast, for DACs, such ring-opening reactions can be considered as bimolecular nucleophilic substitution reactions (S_N2), taking into consideration the inversion of configuration at the reactive center. Ring-opening reactions by electrophiles require umpolung, which is generally achieved by transition metal catalysis of DACs with a vinyl donor [92–95]. Recently, multicomponent reactions where a nucleophile opens the ring and a suitable electrophile is trapped by the malonate anion have attracted the attention of cyclopropane chemists [53–63].

Cycloaddition reactions in DACs deliver highly functionalized four-, five-, six-, or seven-membered rings. Saturated or partially unsaturated rings are obtained by these (3+*n*) reactions [64–85]. These reactions are highly regioselective, as the carbon atom bearing the donor motif is preferentially attacked by the more nucleophilic part of the incoming reaction partner. Diastereo- and enantioselective versions have been explored with the aid of chiral substrates, chiral Lewis acids, or dynamic kinetic resolutions [96–108].

Rearrangements are another major class of transformation undergone by DACs. Thereby, the acceptor moiety is incorporated into the three-membered ring. Delocalization of the negative charge from the carbon atom to a heteroatom of the acceptor allows the heteroatom to attack the emerging positive charge next to the donor. The classical rearrangements involving cyclopropanes are cyclopropylimine-pyrroline and cyclopropylcarbaldehyde-dihydrofuran rearrangements investigated originally by Cloke [109] and Wilson [110] in 1929 and 1947, respectively. However, the first rearrangement involving a pure hydrocarbon backbone dates back to 1959, when Neureiter transformed 1,1-dichloro-2,2-dimethylcyclopropane to 4,4-dichlorocyclopentene under pyrolysis conditions [111]. The fact that such harsh conditions are required for the hydrocarbon system can be traced back to the non-polarized nature of this cyclopropane derivative, which impressively demonstrates how polarization lowers the activation barrier. In contrast, with DACs, numerous rearrangements were realized under very mild reaction conditions. Activation of the acceptor motifs by Lewis or Brønsted acid further lowers the activation barrier for the intramolecular attack on the donor-substituted carbon of the cyclopropane. Thus, DACs have been subject to a multitude of explorations as versatile synthons [112–115].

Cyclopropanes bearing an aldehyde were first activated by organocatalysis in 2011 by Sparr and Gilmour [37]. In this system, the electron-withdrawing aldehyde group becomes a donor group upon the formation of iminium/enamine [37, 116]. The initial work with an aldehyde attached directly to the three-membered ring showed concomitant electrophilic and nucleophilic chlorination. Later works employing a system with an extra CH_2 linkage between the cyclopropane and the aldehyde led to cycloadditions [38, 117–119], and in one particularly interesting case by Jørgensen et al. the formation of a [2+2] product rather than the classical 1,3-dipolar chemistry (Scheme 2.6).

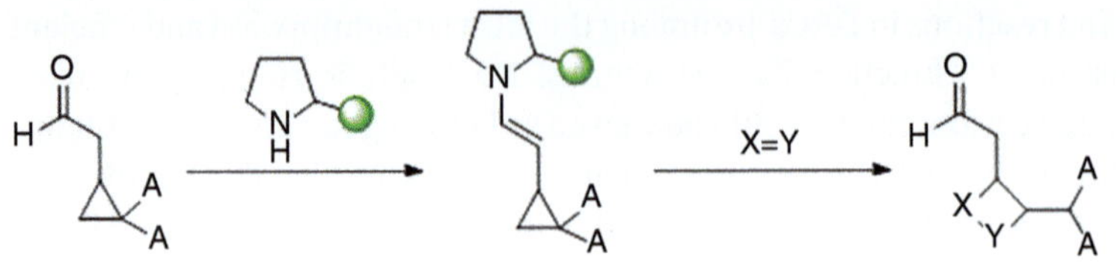

Scheme 2.6 Unexpected [2+2] products formed by organocatalytic activation.

In contrast to the organocatalytic method for the formation of 1,3-dichlorinated products, the Werz group developed a radical approach for this same reaction [39]. Homolytic cleavage of the polarized C—C bond, leading to a radical intermediate, allowed a reaction with further chlorine radicals. Following this, they went further and developed electrochemical methods for the activation of DACs [40, 120]. Direct anodic SET oxidation of the arene donor group is followed by a homolytic C—C bond cleavage. The resulting radical cation intermediate was then shown to react with triplet oxygen forming β-hydroxyketones, or with arenes in Friedel–Crafts-type reactions (Scheme 2.7, please see Chapter 10, Scheme 10.46, 10.47, for more details).

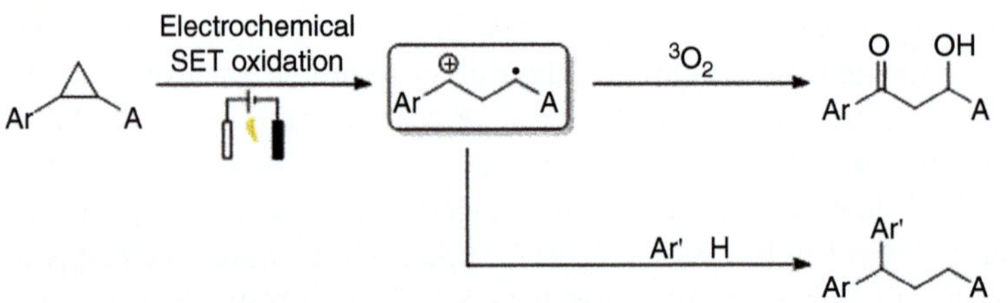

Scheme 2.7 Electrochemical reactions of DACs.

Investigations on the structural and kinetic properties of different DACs provided quantitative data for their strongly enhanced reactivity. This book chapter compiles the published information about the influence of structure and electronics on the reactivity of these strained, commonly used building blocks in organic synthesis.

2.4 Computational and Kinetic Investigations

An astonishing renaissance has been seen in the field of DAC chemistry in the last few decades. Nevertheless, investigations into the physical organic details were not satisfactorily carried out until 2019. In 1974, Danishefsky and Rovnyak investigated the effects of alkyl substituents on the nucleophilic ring-opening of activated cyclopropanes [121]. They chose compounds **17** and **22** as model substrates and investigated their reactions with pyrrolidine.

The nucleophile was observed to attack at the C2 and C3 positions of cyclopropane **17**, delivering four products **18–21** (Scheme 2.8). The reaction is rather low-yielding and only poorly chemoselective (3:2 ratio of attack at C2 vs. C3). A slow and low-yielding reaction with pyrrolidine was observed for **22**. The emergence of carbonium ion character in the ring-opening transition state is consistent with the preferential attack at the highly substituted carbon. However, the steric hindrance and the Thorpe–Ingold effect caused by the geminal alkyl groups retard the attack of the nucleophile on **22** compared to **17**.

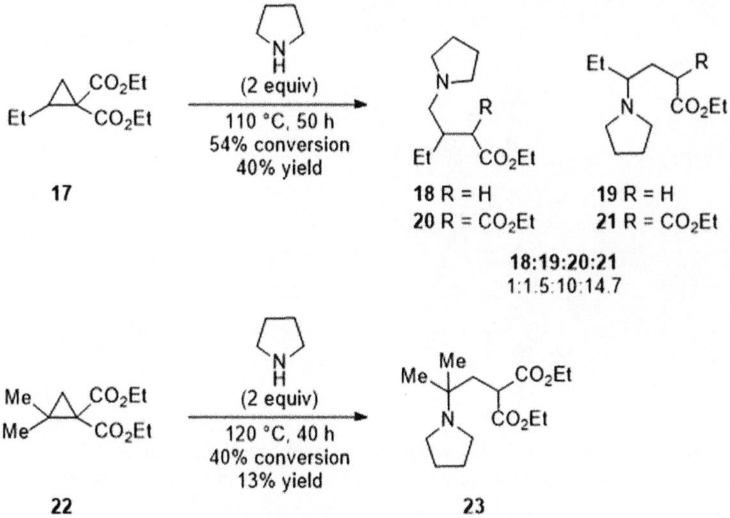

Scheme 2.8 Chemoselectivity in reactions of DACs substituted with an alkyl group.

In 2011, the influence of different push–pull combinations of donors and acceptors on ring-enlargement reactions of DACs was studied by Werz and Schneider using density functional theory (DFT) computations [122]. Figure 2.5 shows the model systems chosen for the study. Eight donor and nine acceptor moieties gave 72

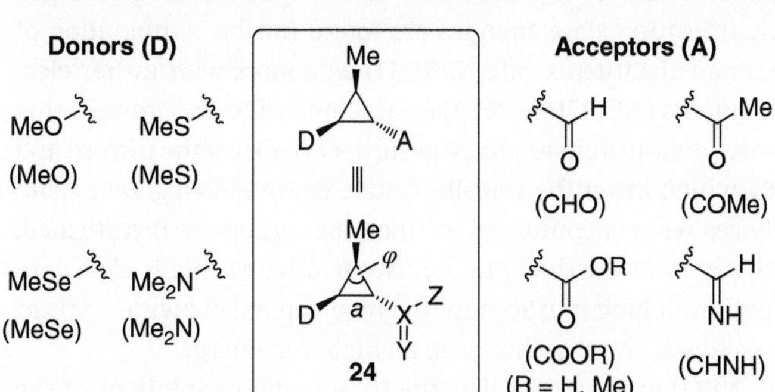

Figure 2.5 Donor (D) and acceptor (A) combinations investigated in a rearrangement study by Werz and Schneider.

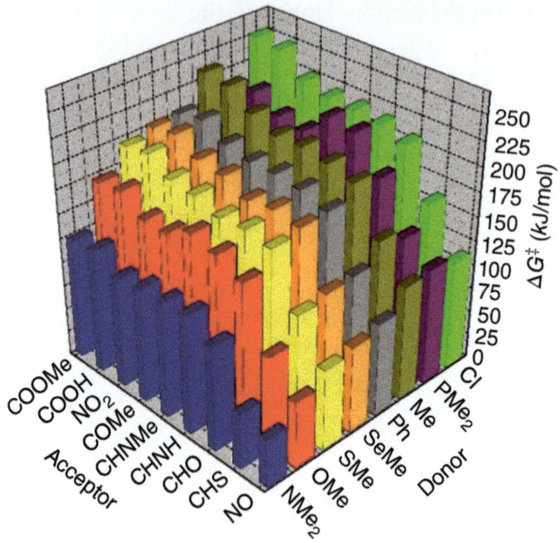

Figure 2.6 Transition state energies for the ring-enlargement reactions from **24** to **25** derived by B3LYP/6-311G(d) calculations for different combinations of donor and acceptor substituents.

D–A combinations with the cyclopropanes **24** as starting materials and the respective products **25** connected via transition states $\mathbf{TS_{24\text{–}25}}$ (Figure 2.5).

Figure 2.6 depicts the combinations arranged from the most efficient (Me_2N/NO) to the least efficient (Cl/COOMe) system for ring-enlargement. The activation barrier for the 1-amino-2-nitroso-substituted cyclopropane is so low that this compound should not exist; rather, it would immediately rearrange to the ring-enlarged system. In general, low transition state energies are found for the combination of strongly electron-donating substituents (Me_2N, MeO) as donors with rather electron-withdrawing substituents (NO, CHS, CHO) as acceptors. The data reveals that aromatic groups are better donors than aliphatic groups. Other than the nitroso and thioaldehyde acceptors, which lower the transition state energies to a great extent, relatively small differences were encountered for the other acceptors investigated. The lowest energy pathway from the three- to the five-membered ring is *via* a concerted process; computations taking into account the involvement of zwitterionic or diradical intermediates showed these pathways to be higher in energy.

Natural bond orbital (NBO) analyses translate the computational solutions of the Schrödinger wave equation into a simple and familiar chemical bonding concept [123]. The Wiberg bond index (WBI) [124] obtained from NBO analyses provides a quantification of electron population overlap between two atoms. The WBI of the formal single bond between the donor and the acceptor substituent ranges from 0.86 (**24(MeS/CHS)**) to 0.94 (**24(Cl/NO$_2$)**). In general, a value below 1 indicates a weakened bond. A small WBI for the bond, a (the bond between donor and acceptor) and a large angle ϕ (see Figure 2.5) in the late transition states are observed in combinations such as (**Me$_2$N/CHS**) and (**MeO/NO**). Figure 2.7 depicts the facile ring-enlargement of **24(Me$_2$N/CHS)** to **25(Me$_2$N/CHS)** with a late transition state $\mathbf{TS_{24\text{–}25}}$.

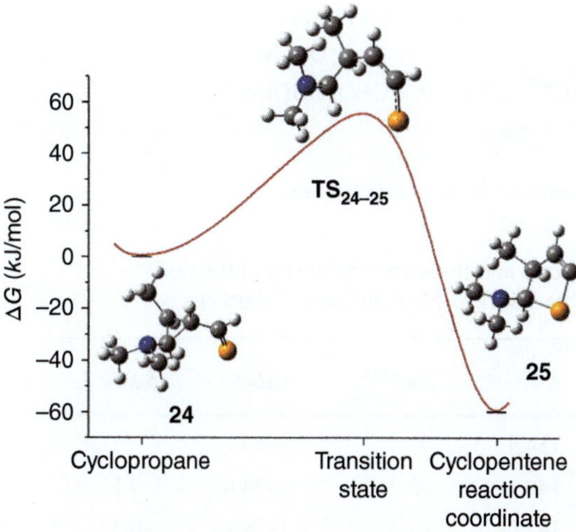

Figure 2.7 Optimized structures of cyclopropane **24(Me2N/CHS)**, transition state **TS24–25**, and dihydrothiophene **25(Me2N/CHS)** as calculated at the B3LYP/6-311G(d) level.

Table 2.1 Comparison of transition state energies ($\Delta G^{\dagger}$), reaction enthalpies (ΔG_R) *in vacuo*, and different solvents based on the PCM model (calculated at the B3LYP/6-311G(d) level of theory).

24 (D/A)	Solvent	$\Delta G^{\ddagger a}$	$\Delta\Delta G^{\ddagger a,b}$	$\Delta G_R{}^a$	$\Delta\Delta G_R{}^{a,b}$
(MeO/CHO)	*In vacuo*	154.4	—	−25.7	—
—	PhMe	146.4	−8.0	−22.1	3.6
—	CH_2Cl_2	137.8	−16.6	−19.3	6.4
	DMSO	134.2	−20.2	−18.5	7.2
(MeO/COOMe)	*In vacuo*	171.6	—	8.0	—
—	PhMe	164.2	−7.4	14.2	5.9
—	CH_2Cl_2	154.5	−17.1	18.7	9.8
—	DMSO	148.8	−22.8	18.9	10.9
(Me_2N/CHO)	*In vacuo*	108.2	—	−4.2	—
—	PhMe	92.0	−16.2	8.0	12.2
—	CH_2Cl_2	74.4	−33.8	10.6	14.7
—	DMSO	67.2	−41.1	11.4	15.5

a) Values are given in $kJ\,mol^{-1}$.
b) Differences of $\Delta G^{\ddagger}$ and ΔG_R, respectively, to the corresponding values *in vacuo*.

The role of solvents in the reaction was studied using computational methods. Table 2.1 compares activation barriers ($\Delta G^{\ddagger a}$) and reaction enthalpies ($\Delta G_R{}^a$) *in vacuo* and in three different solvents (toluene, dichloromethane, and dimethyl sulfoxide (DMSO)) for three selected examples. DMSO was found to tremendously

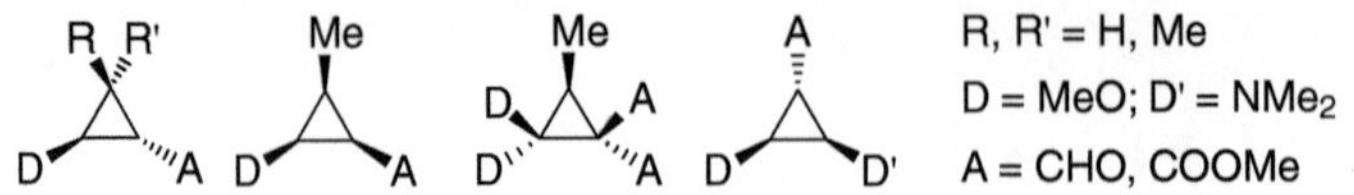

26 27 28 29

Figure 2.8 Further substitution patterns of D–A cyclopropanes.

Table 2.2 Influence of the methyl group and the stereochemistry of the donor–acceptor-substitution (calculated at the B3LYP/6-311G(d) level of theory).

26 (D/A)	R/R′	$\Delta G^{\ddagger a}$	$\Delta\Delta G^{\ddagger a,b}$	$\Delta G_R{}^a$	$\Delta\Delta G_R{}^{a,b}$
(MeO/CHO)	Me/H	154.4	—	−25.7	—
—	H/H	149.1	−5.3	−34.8	−9.1
—	Me/Me	147.0	−7.4	−26.1	−0.4
(MeO/COOMe)	Me/H	171.6	—	1.8	—
	H/H	168.6	−3.0	7.4	−6.2
	Me/Me	163.3	−8.4	8.0	−0.6
27 (D/A)	**R/R′**	$\Delta G^{\ddagger a}$	$\Delta\Delta G^{\ddagger a,b}$	$\Delta G_R{}^a$	$\Delta\Delta G_R{}^{a,b}$
(MeO/CHO)	Me/H	132.8	−21.6	−40.7	15.0
(MeO/COOMe)	Me/H	164.8	−6.8	30.6	22.6

a) Values are given in $kJ\,mol^{-1}$.
b) Differences of $\Delta G^{\ddagger}$ and ΔG_R, respectively, to the corresponding values for the system with one methyl substituent.

lower the transition state energy. This behavior was attributed to the stabilization of highly polarized transition states in polar aprotic solvents.

Studies on the influence of methyl groups as further substituents on the ring and the stereochemistry of the donor–acceptor substitution were conducted (Figure 2.8). It was found that the presence of methyl groups (**26**) has no significant influence on the energetics of the ring-enlargement (Table 2.2). *cis*-Substituted D–A cyclopropane **27** tends to have a smaller activation barrier than the corresponding *trans*-isomer.

Furthermore, the effect of additional donors and acceptors on **28** was investigated (Table 2.3). A decrease in transition state energies ($\Delta G^{\ddagger}$) was observed as expected with an increase in substitution, although no significant influence was observed for the reaction enthalpies (ΔG_R).

In order to access seven-membered ring **31** starting from a DAC, ring-enlargement of a cyclopropane with a vinylogous aldehyde acceptor **30** might be a favorable route (Scheme 2.9). An energy gain of 57 $kJ\,mol^{-1}$ was computed when a vinylogous aldehyde was employed as the acceptor because of the much less strained transition state $\mathbf{TS_{30-31}}$, compared to the five-membered transition state that would form when employing a MeO/CHO system.

Table 2.3 Effects of substitution with an additional acceptor and a further donor, respectively (calculated at the B3LYP/6-311G(d) level of theory).

28 (DD/AA)	$\Delta G^{\ddagger a}$	$\Delta\Delta G^{\ddagger a,b}$	$\Delta G_R{}^{a}$	$\Delta\Delta G_R{}^{a,b}$
(MeO/CHO)	154.4	—	−25.7	—
(MeO/(CHO)$_2$)	94.9	−59.5	−60.0	−34.3
((MeO)$_2$/CHO)	85.0	−69.4	−41.9	−16.2
(MeO/COOMe)	171.6	—	1.8	—
(MeO/(COOMe)$_2$)	147.1	−24.5	9.4	1.4

a) Values are given in kJ mol^{-1}.
b) Differences of $\Delta G^{\ddagger}$ and ΔG_R, respectively, to the corresponding values in system **24**.

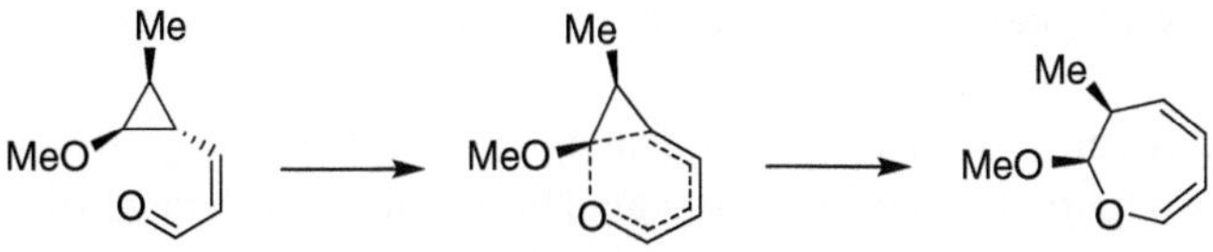

30 **TS$_{30–31}$** **31**

Scheme 2.9 Ring-enlargement of cyclopropane **30** to seven-membered ring **31**.

The requirement for a further, detailed kinetic study of several DACs was realized by the same group in 2019 [125]. Werz and co-workers shed light on the influences of structural and electronic properties on the reactivity of DACs. Physical-organic data of these molecules, such as bond lengths obtained from X-ray studies or DFT calculations, and relaxed force constants (RFC), were compared with kinetic rate constants. The (3+2)-cycloaddition reaction of an aldehyde to a DAC in the presence of $SnCl_4$ to form tetrahydrofuran was chosen as the model reaction for the kinetic investigation of different D–A systems. It was found that there was a first-order dependence on reaction rate with respect to the aldehyde. Because the Lewis acid loading is very low, the cyclopropane is found to have a zero-order dependence on rate at higher concentrations. A pseudo-first-order rate law was specified for the above system,

$$\frac{d[product]}{dt} = k \times [cyclopropane]^0 \times [aldehyde]^1$$

The rate constants of 18 different D–A cyclopropanes were determined by examining the consumption of aldehyde (Figure 2.9a). This data was logarithmically evaluated by employing the least-squares fitting method (Figure 2.9b). The rate constants were determined from the slope of this fitting curve. To allow for comparison, the value for the system with a phenyl group as a donor and two geminal methyl esters as acceptors was arbitrarily designated as 1 ($k_{rel} = 1$).

Table 2.4 provides the relative rate constants for the reaction of 4-fluorobenzaldehyde with different D–A cyclopropanes. It was observed that electron-rich donors on the

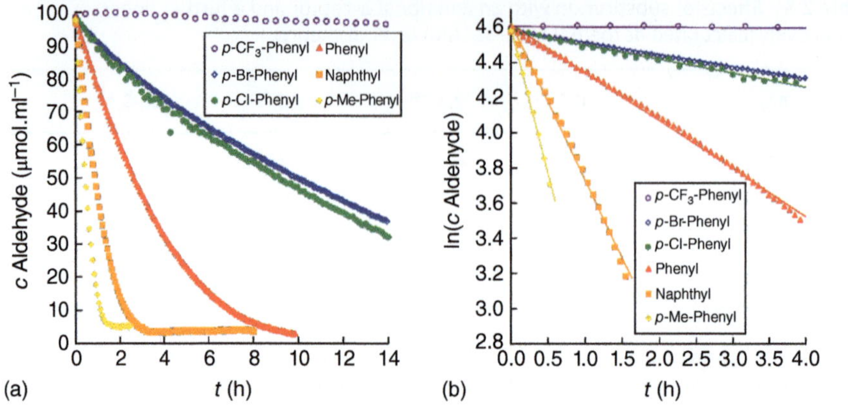

Figure 2.9 (a) Plot of consumption of aldehyde over time; (b) logarithmic plot of consumption of aldehyde over time.

cyclopropane such as *p*-methoxyphenyl ($k_{\text{rel}} = 49.7$) and *p*-methylphenyl ($k_{\text{rel}} = 6.32$) led to fast reactions. A naphthyl donor with its extended π-system accelerated the reaction by a factor of 3.12, the same holds true for the cyclopropyl donor with a factor close to 2. It has been observed that halogen-substituted donors lowered the rate slightly, whereas succinimide reduced the rate to 0.12. A sluggish reaction has been observed when electron-poor groups such as *p*-$F_3CC_6H_4$ or *p*-$O_2NC_6H_4$ were chosen as donors ($k_{\text{rel}} = 0.0077$ and $k_{\text{rel}} = 0.0015$, respectively). It should be noted that the cyclopropane with a phthalimide donor reacts about 350 times faster than with a succinimide donor. Investigating the influence of different esters as acceptors revealed that replacing methyl with ethyl or benzyl groups decreased the rate of reaction. Most probably, this is due to the inability of these more sterically encumbered ester groups to efficiently coordinate with Lewis acids.

An excellent correlation was observed when the measured logarithmic rate constants ($\log(k_{\text{rel}})$) for *para*-substituted D–A cyclopropanes were tabulated against Hammett substituent parameters (σ) (Figure 2.10). The slope of the Hammett plot, known as the reaction parameter ρ, provided information about the associated reaction mechanism [126]. The negative slope indicates that there is a build-up of positive charge during the course of the reaction, and electron-rich donors influence the rate-determining step, resulting in a faster reaction.

To get a clearer picture of the influence substituents have on the donor moiety, not only electronic but also physical parameters were systematically evaluated. Bond lengths are sometimes considered indicators of bond strengths and are correlated with the reactivity of compounds. Although longer bonds for better donors were expected in this case, the calculated bond lengths in the gas phase or the solid phase obtained from X-ray analyses showed no correlation with the kinetic trends. Similarly, the ^{1}H- and ^{13}C-NMR shifts of the hydrogen and carbon atoms next to the donor moiety did not provide any correlation with the corresponding rate constants (Table 2.5).

Table 2.4 The relative reaction rate constants for the reaction of 4-fluorobenzaldehyde with different D–A cyclopropanes.

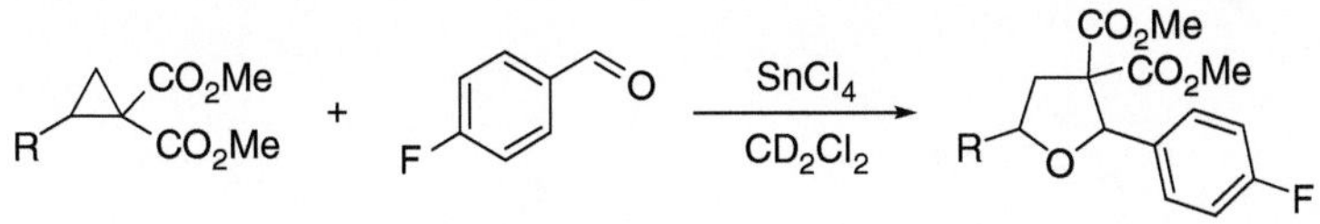

31 32 33

	Residue	k_{rel} [a]
a	*p*-MeO-Ph[b]	49.7
b	Phthalimide	42.0
c	*p*-Me-Ph	6.32
d	*o*-MeO-Ph	3.20
e	Naphthyl	3.12
f	Cyclopropyl	1.92
g	Ph	1.00
h	*p*-F-Ph	0.62
i	*o*-Me-Ph	0.50
j	Ph[c]	0.32
k	*p*-Cl-Ph	0.27
l	*p*-Br-Ph	0.25
m	Ph[d]	0.22
n	Succinimide	0.12
o	H_2C=CH-	0.093
p	*p*-F_3C-Ph	0.0077
q	*m*-O_2N-Ph	0.0017
r	*p*-O_2N-Ph	0.0015

Reaction conditions: D–A cyclopropane (1.00 equiv.), 4-fluorobenzaldehyde (1.00 equiv.), $SnCl_4$ (0.10 M in CD_2Cl_2, 0.015 equiv.), CD_2Cl_2 (0.10 M), 298 K.

a) $k_{rel} = k_{Residue}/k_{Phenyl}$.
b) Fast formation of a side product was observed.
c) Benzyl ester instead of methyl ester.
d) Ethyl ester instead of methyl ester.

Another physical parameter investigated was the RFC. RFCs measure the force required to distort a coordinate by a unit amount while allowing all other coordinates to relax [132]. The RFCs were calculated for the cyclopropane bond broken during the reaction. This was based on the notion that the second derivative of energy is a more sensitive kinetic descriptor than equilibrium bond lengths (Table 2.5). These data were found to be in congruence with the kinetic trends (Figure 2.11). A higher rate constant generally corresponds to a lower RFC and a weaker bond.

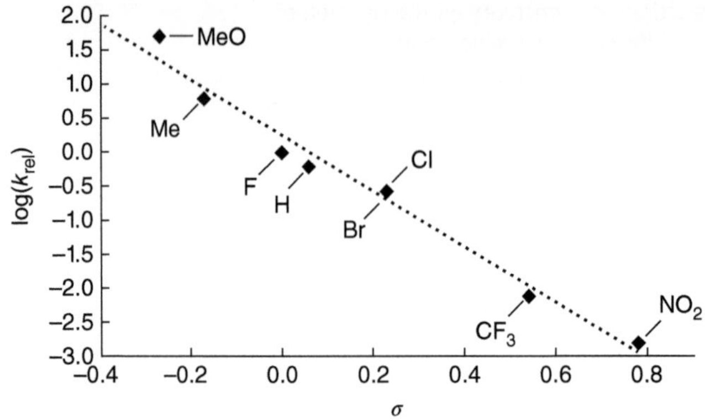

Figure 2.10 Hammett plot with *p*-substituted aryl residues.

Table 2.5 Comparison of physical parameters for *para*-substituted D–A cyclopropanes.

Donor	Bond length gas phase[a] (Å)	NMR shift C-2-H (ppm)	NMR shift C-2 (ppm)	RFC[b] (mdyn Å^{-1})
p-MeO-Ph	1.561	3.17[c]	32.1[c]	2.584
p-Me-Ph	1.559	3.19[c]	32.4[c]	2.646
Naphthyl	1.559	3.38[d]	32.9[d]	2.660
Ph	1.558 (1.539)	3.22[c]	32.4[c]	2.688
p-F-Ph	1.558	3.19[e]	31.7[e]	2.674
p-Cl-Ph	1.558 (1.537)	3.16[f]	30.7[f]	2.688
p-Br-Ph	1.558 (1.536)	3.21[f]	32.6[f]	2.688
p-CF_3-Ph	1.556 (1.549)	3.25[g]	29.7[g]	2.725
p-NO_2-Ph	1.556 (1.542)	3.28[e]	31.5[e]	2.732

a) B3LYP/6-311G(d,p) level of theory. Bond lengths in the solid state from X-ray analysis are given in parentheses.
b) RFC = relaxed force constant.
c) Müller and Fernandez [127].
d) Talukdar et al. [128].
e) Perreault et al. [129].
f) Ghanem et al. [130].
g) Doyle et al. [131].

Further investigations on different cycloaddition reactions of D–A cyclopropanes, even with different catalytic systems, revealed the reproducibility and generality of the observed kinetic trends. (3+3)- and (4+3)-cycloadditions of DACs with either nitrone **35** [97] or isobenzofuran **36** [133] as starting material were investigated (Scheme 2.10). The reactions were followed using an *in-operando* IR spectrometer, where the progress of the reaction was analyzed by the decrease in the intensity of characteristic IR bands of the starting materials.

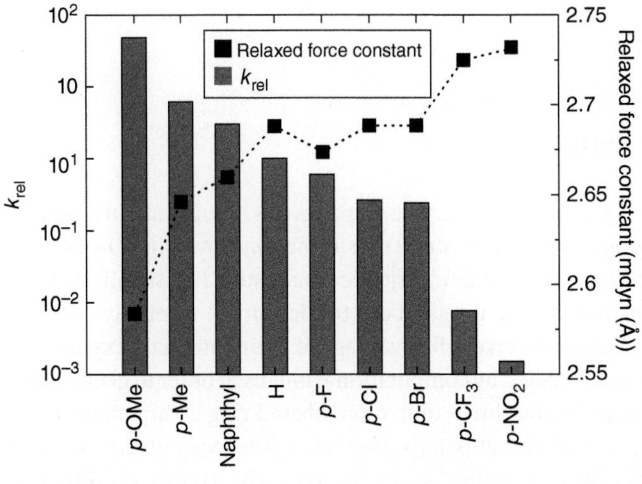

Figure 2.11 Reaction rate constants (logarithmic scale) vs. relaxed force constants of the bond to be broken.

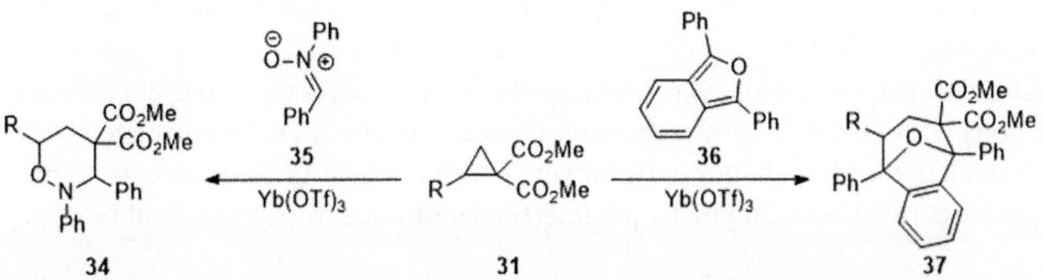

Scheme 2.10 (3+3)- and (4+3)-Cycloaddition reactions of DACs with nitrone and isobenzofuran.

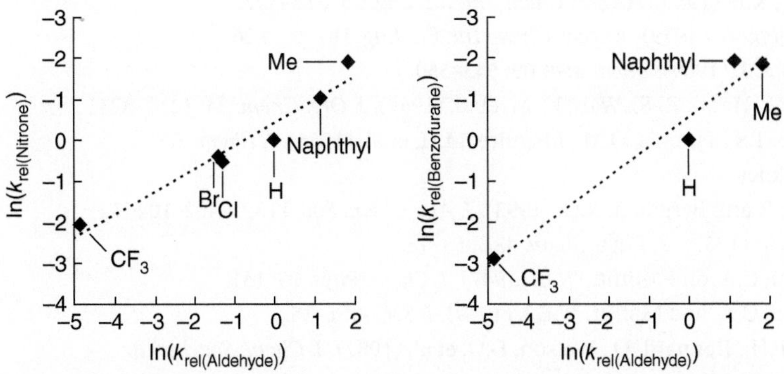

Figure 2.12 Plotting of the logarithmic specific rate constants for the cycloaddition reactions of DACs with aldehydes against those of nitrone and isobenzofuran.

Figure 2.12 was obtained by logarithmic plotting of specific rate constants for the reactions of the aldehyde against the ones with nitrone and isobenzofuran. The linear dependence reflects comparable kinetics. A good correlation was observed in the reaction of DACs with nitrones, whereas a slight deviation in the case of DACs with phenyl and naphthyl donors was noted.

2.5 Concluding Remarks

The development of DACs from activated cyclopropanes has brought about a renaissance in the field of cyclopropane chemistry. Transformations involving DACs have found valuable applications in synthetic organic chemistry. Investigations into understanding the influence of structure and electronics on the reactivity of these moieties have enabled chemists to take full advantage of their potential as synthons.

Different DACs have been studied by computational means in order to get insights into the activation barriers for the three- to five-membered ring enlargement reactions. These studies also revealed that polar solvents are able to significantly lower the energy of the highly polar transition states. Comparative studies on different DACs gave insights into the strong influence of the type of donor in cycloaddition reaction of DACs. (3+2)-Cycloadditions of DACs with aldehydes were found to be about 30 000 times faster when *p*-methoxyphenyl was chosen as donor instead of the *p*-nitrophenyl derivative. This observation was found to correlate with the Hammett substitution parameters and was similarly observed for other types of cycloaddition reactions with DACs. Interestingly, such kinetic data could not be correlated with the bond lengths of the bonds to be broken; however, RFCs of the respective bonds allowed a correlation with kinetic parameters determined by experimental means.

References

1 Wiberg, K.B. (1986). *Angew. Chem. Int. Ed. Eng.* 25: 312–322.
2 de Meijere, A. (1979). *Angew. Chem. Int. Ed. Eng.* 18: 809–826.
3 Collins, C.J. (1969). *Chem. Rev.* 69: 543–550.
4 Lee, C.C., Hahn, B.-S., Wan, K.-M. et al. (1969). *J. Org. Chem.* 34: 3210–3211.
5 Osdene, T.S., Timmis, G.M., Maguire, M.H. et al. (1955). *J. Chem. Soc.* 2038. Notes.
6 Burger, P. and Bergman, R.G. (1993). *J. Am. Chem. Soc.* 115: 10462–10463.
7 Förster, T. (1939). *Z. Phys. Chem.* 43B: 58–78.
8 Coulson, C.A. and Moffitt, W.E. (1947). *J. Chem. Phys.* 15: 151.
9 Coulson, C.A. and Moffitt, W.E.I. (1949). *J. Sci.* 40: 1–35.
10 Allen, F.H., Kennard, O., Watson, D.G. et al. (1987). *J. Chem. Soc. Perkin Trans. 2*: S1.
11 Walsh, A.D. (1947). *Nature* 159: 165.
12 Walsh, A.D. (1949). *Trans. Faraday Soc.* 45: 179.
13 Baranac-Stojanović, M. and Stojanović, M. (2013). *J. Org. Chem.* 78: 1504–1507.
14 Goldschmidt, Z. and Crammer, B. (1988). *Chem. Soc. Rev.* 17: 229.
15 Danishefsky, S. (1979). *Acc. Chem. Res.* 12: 66–72.
16 Bone, W.A. and Perkin, W.H. (1895). *J. Chem. Soc. Trans.* 67: 108–119.

17 Stewart, J.M. and Westberg, H.H. (1965). *J. Org. Chem.* 30: 1951–1955.
18 Dolfini, J.E., Menich, K., Corliss, P. et al. (1966). *Tetrahedron Lett.* 7: 4421–4426.
19 Corey, E.J. and Fuchs, P.L. (1972). *J. Am. Chem. Soc.* 94: 4014–4015.
20 Danishefsky, S. and Singh, R.K. (1975). *J. Am. Chem. Soc.* 97: 3239–3241.
21 Danishefsky, S. and Singh, R.K. (1975). *J. Org. Chem.* 40: 3807–3808.
22 Singh, R.K. and Danishefsky, S. (1976). *J. Org. Chem.* 41: 1668–1669.
23 Snider, B.B., Ahn, Y., and O'Hare, S.M. (2001). *Org. Lett.* 3: 4217–4220.
24 Wenkert, E., Alonso, M.E., Buckwalter, B.L. et al. (1977). *J. Am. Chem. Soc.* 99: 4778–4782.
25 Reissig, H.-U. and Hirsch, E. (1980). *Angew. Chem. Int. Ed. Eng.* 19: 813–814.
26 Wenkert, E. (1980). *Acc. Chem. Res.* 13: 27–31.
27 Reissig, H.-U. (1981). *Tetrahedron Lett.* 22: 2981–2984.
28 Brückner, C. and Reissig, H.-U. (1985). *Angew. Chem. Int. Ed. Eng.* 24: 588–589.
29 Grimm, E.L., Zschiesche, R., and Reissig, H.U. (1985). *J. Org. Chem.* 50: 5543–5545.
30 Brueckner, C. and Reissig, H.U. (1988). *J. Org. Chem.* 53: 2440–2450.
31 Hofmann, B. and Reissig, H.-U. (1993). *Synlett* 1993: 27–29.
32 Yu, M. and Pagenkopf, B.L. (2005). *Tetrahedron* 61: 321–347.
33 Cavitt, M.A., Phun, L.H., and France, S. (2014). *Chem. Soc. Rev.* 43: 804–818.
34 Schneider, T.F., Kaschel, J., and Werz, D.B. (2014). *Angew. Chem. Int. Ed.* 53: 5504–5523.
35 O'Connor, N.R., Wood, J.L., and Stoltz, B.M. (2016). *Isr. J. Chem.* 56: 431–444.
36 Werz, D.B. and Biju, A.T. (2020). *Angew. Chem. Int. Ed.* 59: 3385–3398.
37 Sparr, C. and Gilmour, R. (2011). *Angew. Chem. Int. Ed.* 50: 8391–8395.
38 Halskov, K.S., Kniep, F., Lauridsen, V.H. et al. (2015). *J. Am. Chem. Soc.* 137: 1685–1691.
39 Garve, L.K.B., Barkawitz, P., Jones, P.G. et al. (2014). *Org. Lett.* 16: 5804–5807.
40 Kolb, S., Petzold, M., Brandt, F. et al. (2021). *Angew. Chem. Int. Ed.* 60: 15928–15934.
41 Shimada, S., Hashimoto, Y., Sudo, A. et al. (1992). *J. Org. Chem.* 57: 7126–7133.
42 Yu, M. and Pagenkopf, B.L. (2003). *Tetrahedron* 59: 2765–2771.
43 Yu, M. and Pagenkopf, B.L. (2003). *Org. Lett.* 5: 4639–4640.
44 Tanaka, M., Ubukata, M., Matsuo, T. et al. (2007). *Org. Lett.* 9: 3331–3334.
45 Lifchits, O., Alberico, D., Zakharian, I. et al. (2008). *J. Org. Chem.* 73: 6838–6840.
46 Lifchits, O. and Charette, A.B. (2008). *Org. Lett.* 10: 2809–2812.
47 Cérat, P., Gritsch, P.J., Goudreau, S.R. et al. (2010). *Org. Lett.* 12: 564–567.
48 Roy, S. and Reiser, O. (2012). *Angew. Chem. Int. Ed.* 51: 4722–4725.
49 Sridhar, P.R. and Venukumar, P. (2012). *Org. Lett.* 14: 5558–5561.
50 de Nanteuil, F., Loup, J., and Waser, J. (2013). *Org. Lett.* 15: 3738–3741.
51 Wales, S.M., Walker, M.M., and Johnson, J.S. (2013). *Org. Lett.* 15: 2558–2561.
52 Xia, Y., Lin, L., Chang, F. et al. (2015). *Angew. Chem. Int. Ed.* 54: 13748–13752.
53 Das, S., Daniliuc, C.G., and Studer, A. (2016). *Org. Lett.* 18: 5576–5579.
54 Karmakar, R., Suneja, A., and Singh, V.K. (2016). *Org. Lett.* 18: 2636–2639.
55 Das, S., Daniliuc, C.G., and Studer, A. (2017). *Angew. Chem. Int. Ed.* 56: 11554–11558.

56 Lücht, A., Patalag, L.J., Augustin, A.U. et al. (2017). *Angew. Chem. Int. Ed.* 56: 10587–10591.
57 Wallbaum, J., Garve, L.K.B., Jones, P.G. et al. (2017). *Org. Lett.* 19: 98–101.
58 Das, S., Daniliuc, C.G., and Studer, A. (2018). *Angew. Chem. Int. Ed.* 57: 4053–4057.
59 Richmond, E., Vuković, V.D., and Moran, J. (2018). *Org. Lett.* 20: 574–577.
60 Augustin, A.U., Jones, P.G., and Werz, D.B. (2019). *Chem. Eur. J.* 25: 11620–11624.
61 Lücht, A., Sobottka, S., Patalag, L.J. et al. (2019). *Chem. Eur. J.* 25: 10359–10365.
62 Singh, K., Bera, T., Jaiswal, V. et al. (2019). *J. Org. Chem.* 84: 710–725.
63 Guin, A., Rathod, T., Gaykar, R.N. et al. (2020). *Org. Lett.* 22: 2276–2280.
64 Benfatti, F., de Nanteuil, F., and Waser, J. (2012). *Chem. Eur. J.* 18: 4844–4849.
65 Goldberg, A.F.G., O'Connor, N.R., Craig, R.A. et al. (2012). *Org. Lett.* 14: 5314–5317.
66 Zhu, W., Fang, J., Liu, Y. et al. (2013). *Angew. Chem. Int. Ed.* 52: 2032–2037.
67 Chakrabarty, S., Chatterjee, I., Wibbeling, B. et al. (2014). *Angew. Chem. Int. Ed.* 53: 5964–5968.
68 Mackay, W.D., Fistikci, M., Carris, R.M. et al. (2014). *Org. Lett.* 16: 1626–1629.
69 Xu, H., Hu, J.-L., Wang, L. et al. (2015). *J. Am. Chem. Soc.* 137: 8006–8009.
70 Chidley, T., Vemula, N., Carson, C.A. et al. (2016). *Org. Lett.* 18: 2922–2925.
71 Garve, L.K.B., Petzold, M., Jones, P.G. et al. (2016). *Org. Lett.* 18: 564–567.
72 Kaicharla, T., Roy, T., Thangaraj, M. et al. (2016). *Angew. Chem. Int. Ed.* 55: 10061–10064.
73 Racine, S., Hegedüs, B., Scopelliti, R. et al. (2016). *Chem. Eur. J.* 22: 11997–12001.
74 Sabbatani, J. and Maulide, N. (2016). *Angew. Chem. Int. Ed.* 55: 6780–6783.
75 Augustin, A.U., Sensse, M., Jones, P.G. et al. (2017). *Angew. Chem. Int. Ed.* 56: 14293–14296.
76 Dey, R. and Banerjee, P. (2017). *Org. Lett.* 19: 304–307.
77 Augustin, A.U., Busse, M., Jones, P.G. et al. (2018). *Org. Lett.* 20: 820–823.
78 Chagarovskiy, A.O., Vasin, V.S., Kuznetsov, V.V. et al. (2018). *Angew. Chem. Int. Ed.* 57: 10338–10342.
79 Kreft, A., Jones, P.G., and Werz, D.B. (2018). *Org. Lett.* 20: 2059–2062.
80 Augustin, A.U., Merz, J.L., Jones, P.G. et al. (2019). *Org. Lett.* 21: 9405–9409.
81 Petzold, M., Jones, P.G., and Werz, D.B. (2019). *Angew. Chem. Int. Ed.* 58: 6225–6229.
82 Ahlburg, N.L., Jones, P.G., and Werz, D.B. (2020). *Org. Lett.* 22: 6404–6408.
83 Jacob, A., Jones, P.G., and Werz, D.B. (2020). *Org. Lett.* 22: 8720–8724.
84 Lücht, A., Kreft, A., Jones, P.G. et al. (2020). *Eur. J. Org. Chem.* 2560–2564.
85 Oliver, G.A., Loch, M.N., Augustin, A.U. et al. (2021). *Angew. Chem. Int. Ed.* 49: 25825–25831.
86 Schneider, T.F., Kaschel, J., Dittrich, B. et al. (2009). *Org. Lett.* 11: 2317–2320.
87 Schneider, T.F., Kaschel, J., Awan, S.I. et al. (2010). *Chem. Eur. J.* 16: 11276–11288.
88 Kaschel, J., Schneider, T.F., Kratzert, D. et al. (2012). *Angew. Chem. Int. Ed.* 51: 11153–11156.
89 Kaschel, J., Schmidt, C.D., Mumby, M. et al. (2013). *Chem. Commun.* 49: 4403–4405.

90 Ivanova, O.A., Chagarovskiy, A.O., Shumsky, A.N. et al. (2018). *J. Org. Chem.* 83: 543–560.
91 Shim, S.Y., Choi, Y., and Ryu, D.H. (2018). *J. Am. Chem. Soc.* 140: 11184–11188.
92 Sebelius, S., Olsson, V.J., and Szabó, K.J. (2005). *J. Am. Chem. Soc.* 127: 10478–10479.
93 Selander, N. and Szabó, K.J. (2008). *Chem. Commun.* 3420–3422.
94 Sumida, Y., Yorimitsu, H., and Oshima, K. (2008). *Org. Lett.* 10: 4677–4679.
95 Moran, J., Smith, A.G., Carris, R.M. et al. (2011). *J. Am. Chem. Soc.* 133: 18618–18621.
96 Meyers, C. and Carreira, E.M. (2003). *Angew. Chem. Int. Ed.* 42: 694–696.
97 Young, I.S. and Kerr, M.A. (2003). *Angew. Chem. Int. Ed.* 42: 3023–3026.
98 Pohlhaus, P.D. and Johnson, J.S. (2005). *J. Am. Chem. Soc.* 127: 16014–16015.
99 Parsons, A.T. and Johnson, J.S. (2009). *J. Am. Chem. Soc.* 131: 3122–3123.
100 Sapeta, K. and Kerr, M.A. (2009). *Org. Lett.* 11: 2081–2084.
101 de Nanteuil, F. and Waser, J. (2011). *Angew. Chem. Int. Ed.* 50: 12075–12079.
102 Sathishkannan, G. and Srinivasan, K. (2011). *Org. Lett.* 13: 6002–6005.
103 Smith, A.G., Slade, M.C., and Johnson, J.S. (2011). *Org. Lett.* 13: 1996–1999.
104 Yang, G., Shen, Y., Li, K. et al. (2011). *J. Org. Chem.* 76: 229–233.
105 Benfatti, F., de Nanteuil, F., and Waser, J. (2012). *Org. Lett.* 14: 386–389.
106 Miyake, Y., Endo, S., Moriyama, T. et al. (2013). *Angew. Chem. Int. Ed.* 52: 1758–1762.
107 Rivero, A.R., Fernández, I., and Sierra, M.Á. (2013). *Org. Lett.* 15: 4928–4931.
108 Volkova, Y.A., Budynina, E.M., Kaplun, A.E. et al. (2013). *Chem. Eur. J.* 19: 6586–6590.
109 Cloke, J.B. (1929). *J. Am. Chem. Soc.* 51: 1174–1187.
110 Wilson, C.L. (1947). *J. Am. Chem. Soc.* 69: 3002–3004.
111 Neureiter, N. (1959). *J. Org. Chem.* 24: 2044–2046.
112 Reissig, H.-U. and Zimmer, R. (2003). *Chem. Rev.* 103: 1151–1196.
113 Carson, C.A. and Kerr, M.A. (2009). *Chem. Soc. Rev.* 38: 3051–3060.
114 Singh, P., Varshnaya, R.K., Dey, R. et al. (2020). *Adv. Synth. Catal.* 362: 1447–1484.
115 Augustin, A.U. and Werz, D.B. (2021). *Chem. Res.* 54: 1528–1541.
116 Wallbaum, J., Garve, L.K.B., Jones, P.G. et al. (2016). *Chem. Eur. J.* 22: 18756–18759.
117 Halskov, K.S., Næsborg, L., Tur, F. et al. (2016). *Org. Lett.* 18: 2220–2223.
118 Sanchez-Diez, E., Vesga, D.L., Reyes, E. et al. (2016). *Org. Lett.* 18: 1270–1273.
119 Blom, J., Vidal-Albalat, A., Jørgensen, J. et al. (2017). *Angew. Chem. Int. Ed.* 56: 11831–11835.
120 Kolb, S., Ahlburg, N.L., and Werz, D.B. (2021). *Org. Lett.* 23: 5549–5553.
121 Danishefsky, S. and Rovnyak, G. (1975). *J. Org. Chem.* 40: 114–115.
122 Schneider, T.F. and Werz, D.B. (2011). *Org. Lett.* 13: 1848–1851.
123 Foster, J.P. and Weinhold, F. (1980). *J. Am. Chem. Soc.* 102: 7211–7218.
124 Wiberg, K.B. (1968). *Tetrahedron* 24: 1083–1096.
125 Kreft, A., Lücht, A., Grunenberg, J. et al. (2019). *Angew. Chem. Int. Ed.* 58: 1955–1959.

126 Hansch, C., Leo, A., and Taft, R.W. (1991). *Chem. Rev.* 91: 165–195.

127 Müller, P. and Fernandez, D. (1995). *Helv. Chim. Acta* 78: 947–958.

128 Talukdar, R., Tiwari, D.P., Saha, A. et al. (2014). *Org. Lett.* 16: 3954–3957.

129 Perreault, C., Goudreau, S.R., Zimmer, L.E. et al. (2008). *Org. Lett.* 10: 689–692.

130 Ghanem, A., Lacrampe, F., and Schurig, V. (2005). *Helv. Chim. Acta* 88: 216–239.

131 Doyle, M.P., Davies, S.B., and Hu, W. (2000). *Org. Lett.* 2: 1145–1147.

132 Markopoulos, G. and Grunenberg, J. (2013). *Angew. Chem. Int. Ed.* 52: 10648–10651.

133 Ivanova, O.A., Budynina, E.M., Grishin, Y.K. et al. (2008). *Angew. Chem. Int. Ed.* 47: 1107–1110.

3

Cycloaddition and Annulation Reactions of Donor–Acceptor Cyclopropanes

Roman A. Novikov, Denis D. Borisov, and Yury V. Tomilov

N. D. Zelinsky Institute of Organic Chemistry, Russian Academy of Sciences, 119991, Moscow, Russian Federation

CHAPTER MENU

3.1 Introduction

The cycloaddition and annulation reactions of donor–acceptor (D–A) cyclopropanes are one of the most widespread, important, and most recognizable processes in their chemistry. And often D–A cyclopropanes are associated precisely with cycloaddition reactions. Indeed, the great ability of D–A cyclopropanes to cleavage the three-membered ring to form 1,3-zwitterionic intermediates allows access to an exceptionally wide variety of different cycloaddition processes. Cycloaddition/annulation reactions are discussed in many review articles on the chemistry of activated cyclopropanes from various points of view and types of transformations [1–8]. The main direction of cycloaddition/annulation processes is the reactivity of D–A cyclopropanes as the 1,3-zwitterionic synthones (**I**) (Figure 3.1), usually realized

Donor–Acceptor Cyclopropanes in Organic Synthesis, First Edition.
Edited by Prabal Banerjee and Akkattu T. Biju.

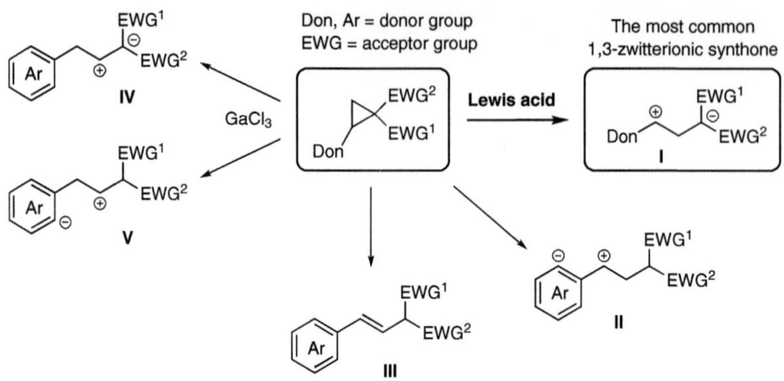

Figure 3.1 Main types of zwitterionic synthones in cycloaddition/annulation processes of the D–A cyclopropanes.

under the action of Lewis acids, or catalysis with Pd(0) for vinyl cyclopropanes. Generated 1,3-zwitterionic synthones react easily with various types of substrates, of which the most common are formal [3+2]-cycloaddition to multiple bonds, as well as formal [3+3]-cycloaddition with 1,3-dipoles or their synthetic equivalents, and also formal [4+3]-cycloaddition with (hetero)diene systems (Figure 3.2). There are also known a large number of other types of cyclization processes, in which D–A cyclopropanes act as various zwitterionic synthons. Of these, it can be noted the [3+2]-annulation at the aromatic donor substituent in D–A cyclopropane (synthon **II**), reactions proceeding through the in situ-generation of β-styrylmalonates (synthon **III**), cyclodimerization reactions of D–A cyclopropanes, and the annulation reactions under the action of $GaCl_3$, in which D–A cyclopropanes act as 1,2- and 1,4-zwitterionic intermediate synthons (**IV** and **V**) (Figure 3.1). All this leads to a very significant variety of implemented processes. All these aspects of the D–A cyclopropane chemistry are discussed in detail and comprehensively in the current book chapter.

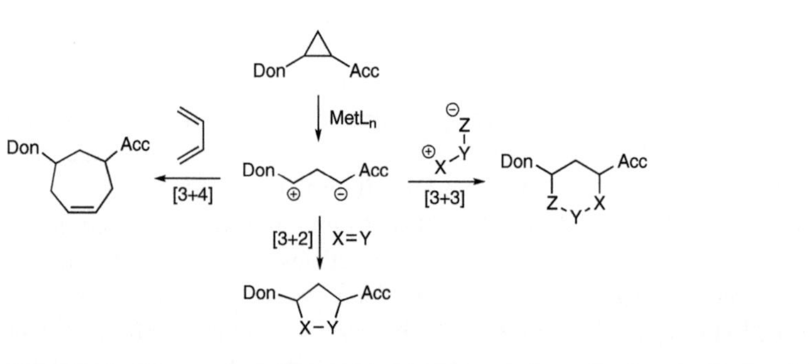

Figure 3.2 Three main formal cycloaddition processes of D–A cyclopropanes as 1,3-zwitterionic synthones.

3.2 Formal [3+2]-Cycloaddition with Carbon–Carbon Multiple Bonds

3.2.1 General Aspects

The formal [3+2]-cycloaddition of D–A cyclopropanes **1** to multiple C–C bonds is one of the main and most common processes in the chemistry of D–A cyclopropanes, and is rightfully considered one of the effective modern methods for creating five-membered carbocycles, including various polysubstituted and polycyclic derivatives, and products with multiple bonds in the cycle. The concept of this method is quite simple and lies in the fact that D–A cyclopropanes, after opening the three-membered ring, act as 1,3-zwitterionic synthons, which are added to multiple bonds via the [3+2]-cycloaddition pathway to form five-membered rings **2,3** (Figure 3.3). There are several different approaches to activating D–A cyclopropanes. The most common is the classical catalysis by Lewis acids, which are coordinated over one or two acceptor groups, followed by polarization of the σ-bond. In this case, the attack of the substrate is usually carried out through the carbocation center formed during the opening of the cyclopropane ring. Separately, we can single out the use of Lewis acids such as MgI_2, CaI_2, and ZnI_2 for activation, in which the iodide anion is initially temporarily attached to cyclopropane **1** and the substrate is further nucleophilically attacked, which allows the use of electron-deficient double bonds (Figure 3.4). Another approach is to use palladium catalysis Pd(0) for vinylcyclopropanes **4** with acceptor groups, or similar Ni(0) catalysis for cyclopropyl ketones. In this case, the activation and the opening of the three-membered ring occur upon coordination of Pd(0) with the vinyl group to form analogous 1,3-zwitterionic synthons and similar types of reactions occur (Figure 3.5). In addition, other approaches

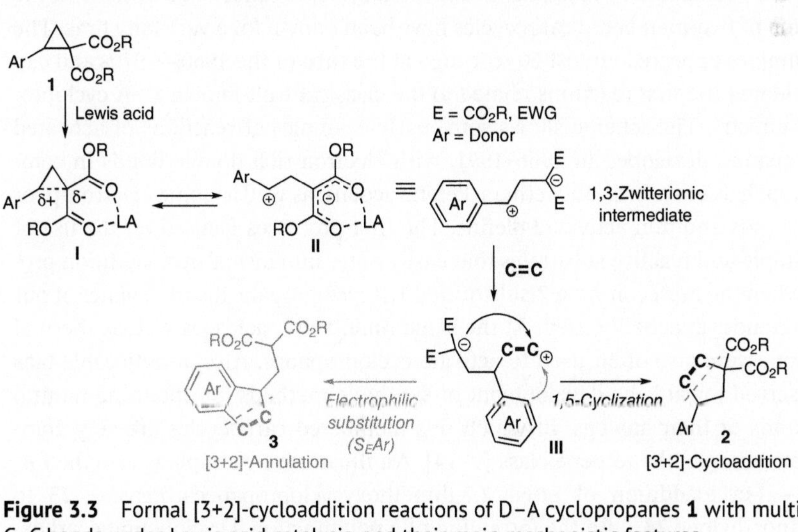

Figure 3.3 Formal [3+2]-cycloaddition reactions of D–A cyclopropanes **1** with multiple C–C bonds under Lewis acid catalysis, and their main mechanistic features.

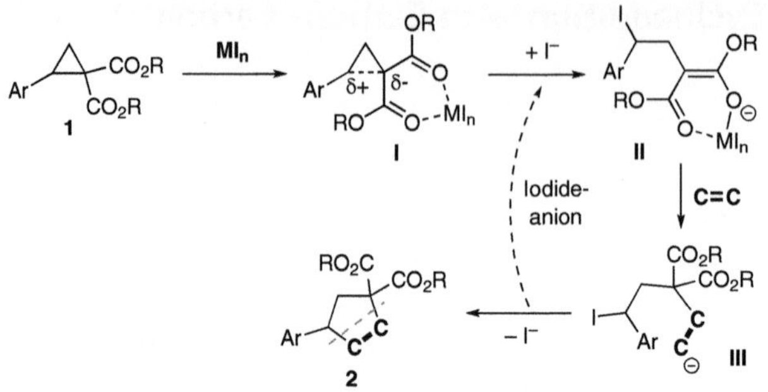

Figure 3.4 Catalysis using metal iodides (MgI_2, CaI_2, ZnI_2).

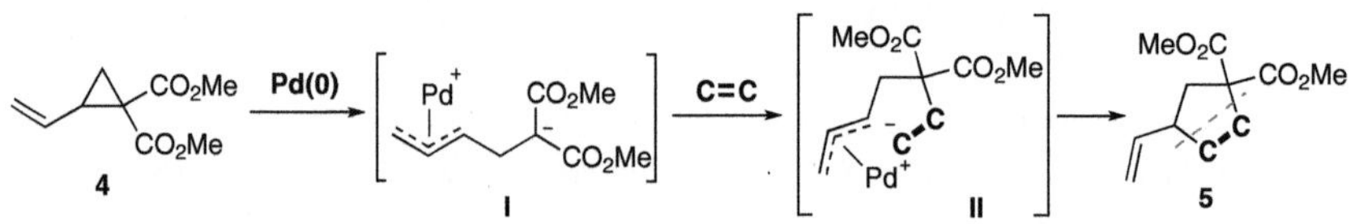

Figure 3.5 Pd(0)-catalyzed formal [3+2]-cycloaddition of the vinyl D–A cyclopropanes **4**.

are also known using base catalysis, the use of organocatalysts, examples of radical processes, as well as the use of photocatalysis methods. Many of these processes are already difficult to attribute to the classical chemistry of D–A cyclopropanes, and they often require the use of specific types of substrates.

3.2.2 Formal [3+2]-Cycloaddition with C=C Double Bond

1,3-Dipolar cycloaddition reactions of activated cyclopropanes to alkenes, with the formation of five-membered carbocycles have been known for a very long time. The first examples appeared almost 60 years ago at the turn of the 1960s–1970s and can be considered the first reactions related to the classical well-known D–A cyclopropane chemistry. The scheme shows some early examples of reactions of activated cyclopropanes, developed in 1966–1991, with electron-rich double bonds in compounds such as enamines, enol ethers, ketene acetals, as well as some electron-poor double bonds and non-activated olefins. The first processes focused on the use of fairly simple and reactive substrates that easily enter into formal cycloaddition processes when the most common 2-substituted 1,1-cyclopropane dicarboxylates of our time were not yet actually used. At the same time, Lewis acids, as well as thermal activation, were also often used to activate cyclopropanes. Also, a noticeable bias was observed toward the development of synthetic methods for obtaining natural compounds or their analogs, in which five-membered carbocycles are very common; for example, the terpene class [9–14]. An illustrative example is also the formal [3+2]-cycloaddition of ethyl 2,2-dimethoxycyclopropanecarboxylate **23** to tetracyanoethylene **24**, with the formation of a polyfunctional cyclopentanecarboxylate **25** with good stereoselectivity, which proceeds even at room temperature even in the absence of an activator, due to the high reactivity of the alkene component and cyclopropane itself [15]. Separately, we can note the very important work of B. Snider in 1986, which set a certain tone in the modern chemistry of D–A cyclopropanes, in which the interaction processes of 1,1-cyclopropane dicarboxylates **21** were first developed and described in detail, including several simple 2-alkyl substituted representatives, with various non-activated alkenes using $EtAlCl_2$ as the Lewis acid activator [16]. Also noteworthy are the first examples of Pd(0)-catalyzed cycloaddition of vinylcyclopropanes **4** with acceptor groups to Michael acceptors, as described by J. Tsuji [17] (Figure 3.6).

Further, after 2000 and up to the present day, the processes of formal [3+2]-cycloaddition of D–A cyclopropanes to double carbon bonds continued to develop rapidly. Many dozens of papers have already been published, and many various examples of such transformations are known, including their various applied aspects and applications in total syntheses. Moreover, a great diversity is found both in the structures of D–A cyclopropanes and alkene substrates, and in the very methods of activation and opening of the three-membered ring.

I. Yokoe in 2001 described the processes of formal [3+2]-cycloaddition of methoxycyclopropane dicarboxylates **27** and various allylsilanes **28** upon activation with a strong Lewis acid, titanium(IV) tetrachloride. The yields of the final cyclopentanes **29** reached 70%, but the diastereoselectivity of the process was very moderate [18] (Figure 3.7).

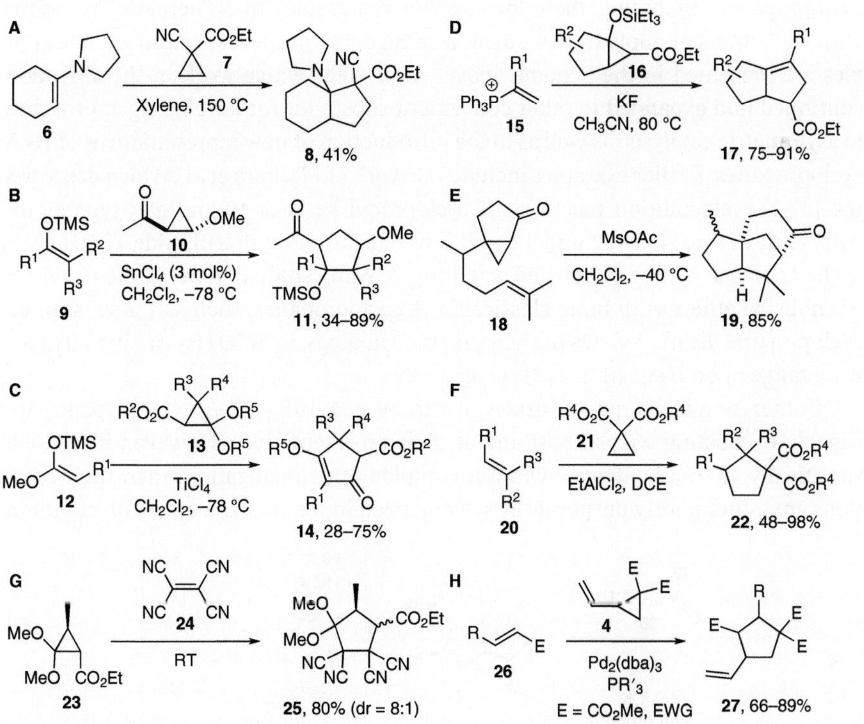

Figure 3.6 First reactions and examples of the [3+2]-cycloaddition of alkenes to activated cyclopropanes.

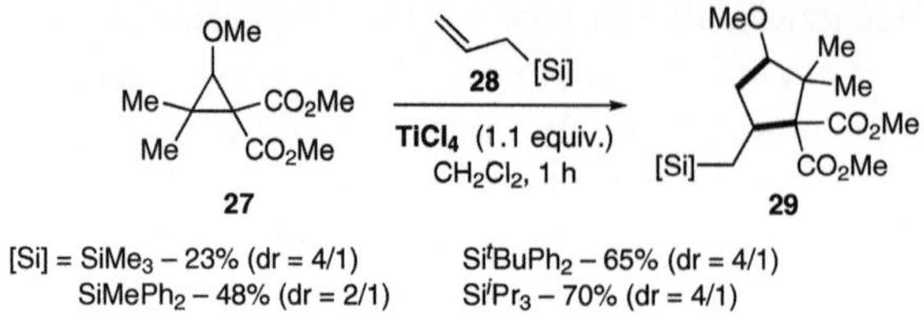

Figure 3.7 [3+2]-Cycloaddition with allylsilanes **28**.

Another type of such transformations is the well-known work of Liu and Montgomery (2006), which describes the processes of [3+2]-cyclodimerization of cyclopropyl ketones **10**, as well as their formal [3+2]-cycloaddition reactions with α, β-unsaturated ketones **30** when initiating the opening of a three-membered ring under the conditions of catalysis with zero-valent nickel-generated in the system of nickel(0) dicyclooctadienyl, N-heterocyclic carbene (NHC), and titanium(IV) tetra-*tert*-butoxide [19]. Subsequently, such metal complex approaches were widely developed separately, but in this chapter, they will be considered in passing, since this is a somewhat separate and distinct field of chemistry from D–A cyclopropanes with its own regularities (Figure 3.8).

Silyl ethers of enols **12, 33** are a very convenient interceptor for the generated 1,3-zwitterions, which are reactive and easily interact with various types of D–A cyclopropanes, including their low-activity representatives. Therefore, reactions with silyl ethers of enols were studied in some detail in the last century, and examples are presented in the Scheme above. After 2000, active work in this direction continued and expanded to other convenient substrates for developing approaches to asymmetric catalysis, as well as to the introduction of new representatives of D–A cyclopropanes. Earlier examples include the work of M. Ihara et al., which describes the [3+2]-cycloaddition reactions of cyclopropyl ketones **10** and alkoxycyclopropane monocarboxylates **32** under conditions of catalysis with triflimide Tf_2NH or its cyclic analog as a strong Brønsted acid [20]. Z. Wang studied in detail the reactions of enol silyl ethers with more classical D–A cyclopropanes, such as 1,2-substituted cyclopropane dicarboxylates or their diketone analogs, in $Sc(OTf)_3$ catalysis using a wide range of both substrates [21] (Figure 3.9).

Of other simple alkene substrates, it can be noted that D–A cyclopropanes are capable of reacting with almost any of their representatives. However, if we look specifically at simple alkenes without complex functionalization, then their reactions are rather poorly purposefully systematized in published works with common

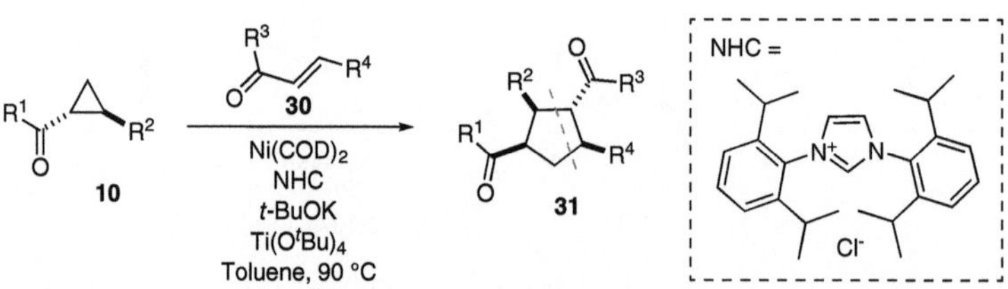

Figure 3.8 Montgomery's Ni(0)-catalyzed [3+2]-cycloaddition of cyclopropyl ketones **10**.

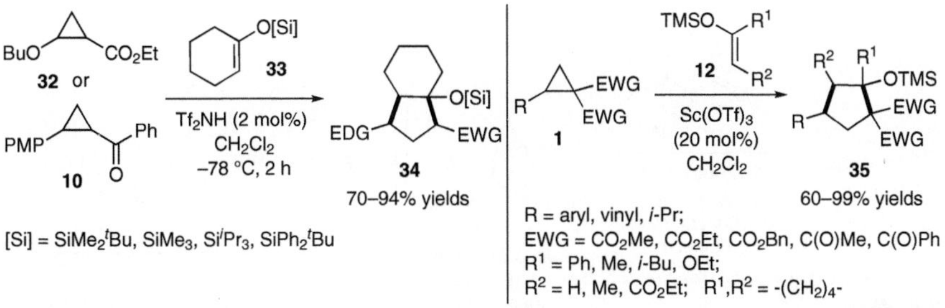

Figure 3.9 Some reactions with silyl esters of enols.

types of D–A cyclopropanes. Since such substrates were studied quite a lot in the last century, and many scattered examples were published, the novelty and interest in such processes in the 2000s, without an applied bias were already significantly reduced. Not all processes have been studied, given the significant evidence of predicting the course of such transformations.

One recent example of such work is the systematic study of the interaction of cyclopropane dicarboxylates **1** with diarylethenes **36** published by the L. Wang group. In addition to cyclopentanes **37**, the formation of acyclic addition products has also been described [22] (Figure 3.10).

One of the most important directions in the development of formal [3+2]-cycloaddition reactions with alkenes in the last decade is the development of general approaches to asymmetric catalysis for the preparation of optically pure cyclopentane derivatives, which significantly expands and increases the interest in this field of chemistry and corresponds to the general trends in its development. The need for just asymmetric catalysis emphasizes the often impossibility of using optically pure cyclopropanes to obtain optically pure products due to the easy occurrence of racemization processes under catalytic conditions of the processes.

Several alternative approaches have been developed for asymmetric catalysis, corresponding to the methods of activation of various types of D–A cyclopropanes. For example, an important work published by the B. Trost group in 2012 presents the development of a method for vinylcyclopropanes **4** in formal cycloaddition reactions with Michael acceptors. Correspondingly, Pd(0) catalysis and the most common chiral phosphine ligands in such cases, phosphine derivatives of diaminocyclohexane, are used for this. However, a particular difficulty for the D–A processes of cyclopropanes is the detailed selection of the reaction conditions and

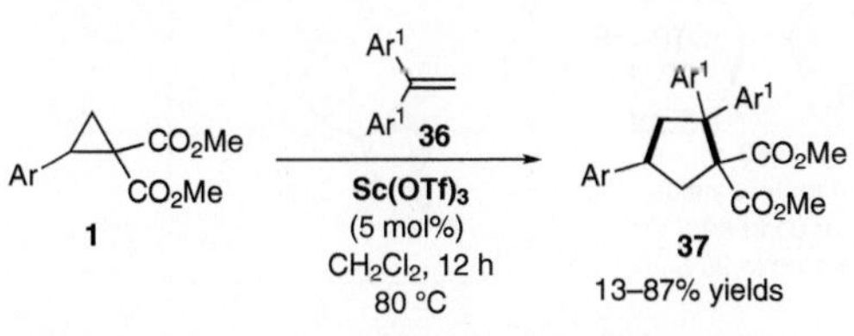

Figure 3.10 Reaction with diarylethenes **36**.

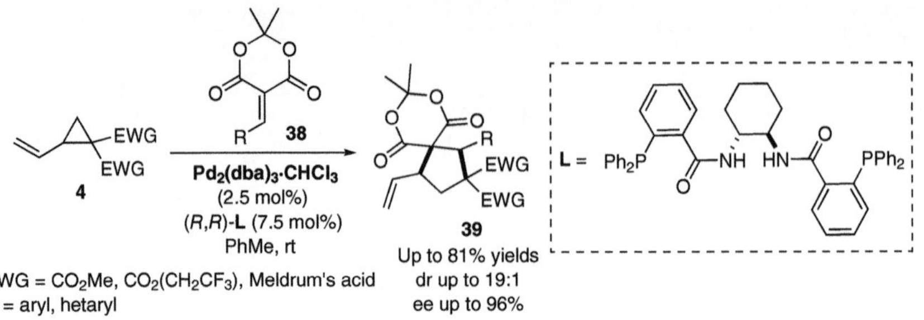

Figure 3.11 B. Trost's asymmetric approach to formal [3+2]-cycloaddition for vinyl cyclopropanes **4**.

substituents in both starting substrates, in order to obtain high enantiomeric excess (*ee*) values, which is often a nontrivial task [23]. (Figure 3.11) Another approach has been developed in parallel by the Y. Tang group for cyclopropane dicarboxylates **1** using Lewis acid catalysis. Reactive cyclic silyl ethers of enols **33** were used as alkene substrates, and an efficient system of copper perchlorate and substituted bisoxazoline ligands was proposed for asymmetric catalysis. Also, sterically loaded 2-adamantyl esters of the starting cyclopropanes were used to achieve the best *ee* values [24] (Figure 3.12).

An extension of this concept was demonstrated by J. Waser for protected aminocyclopropane dicarboxylates **46** using the same chiral catalytic system based on copper perchlorate and bisoxazoline ligands and enol esters, in combination with carefully selected succinimide substituents in D–A cyclopropanes [25]. In general, this is one example from a series of works by the J. Waser group, devoted to the development and introduction to the chemistry of D–A cyclopropanes of new classes based on 2-amino cyclopropane dicarboxylates **41**, **44**, with various protective groups for an amine (usually succinimide and phthalimide), including derivatives nitrogenous bases (thymine and uracil), and allowing to obtain combination products **43**, **45**, **47**, with a protected amino group. The Lewis acid catalysis is also used for these cyclopropanes, and reactions have been developed with various substrates, of which alkenes are among the basic ones [26, 27] (Figure 3.13).

Another direction in the development of formal [3+2]-cycloaddition reactions to alkenes in the last decade has been toward the use of complex polyfunctional

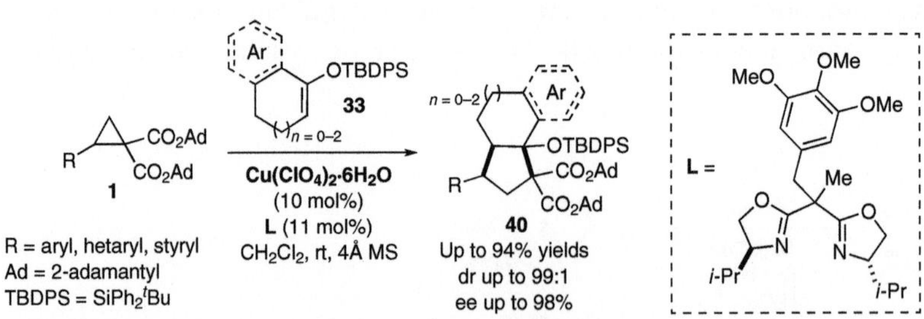

Figure 3.12 Y. Tang's asymmetric approach to formal [3+2]-cycloaddition for cyclopropane diesters **1**.

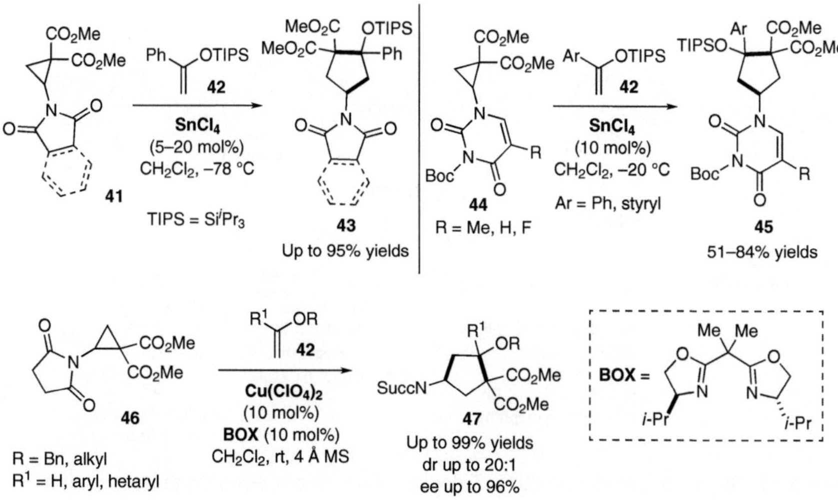

Figure 3.13 J. Waser's [3+2]-cycloaddition of amino-based D–A cyclopropanes and its asymmetric approach.

substrates for the targeted preparation of complex polycyclic or polysubstituted molecules for further applied use in various directions. For example, Trushkov and coworkers describe in detail the interactions of D-A cyclopropanes **1** with various cyclic and acyclic diene systems **48**, many of which proceed as formal [3+2] cycloaddition at the C=C double bond [28]. Hyland et al. described the addition of vinylcyclopropanes **4** to the C=C double bond of nitro-substituted enines **51** under the conditions of palladium catalysis [29] (Figure 3.14).

In a series of works by Prof. Banerjee's group, processes were developed for the addition of D–A cyclopropanes **1** to various heterodiene systems containing nitrogen atoms, such as *N*-sulfoxy-substituted 1,3-azadienes **52**, benzofuran-based azadienes **53**, enamines with carbonyl substituents **54**, and others. In general, a lot depends on the particular heterodiene component used, but many reactions also proceed as formal [3+2]-cycloaddition at C=C bonds, and lead to polyfunctionalized cyclopentane derivatives **55–57**. In reactions with diene and heterodiene

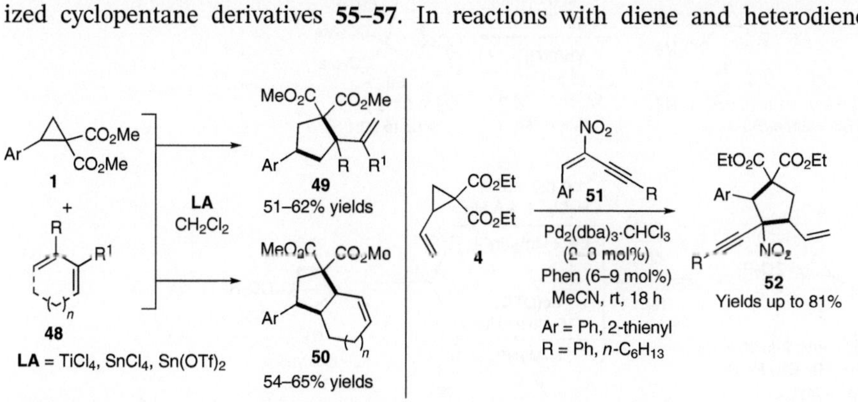

Figure 3.14 [3+2]-Cycloaddition to dienes and nitroenynes.

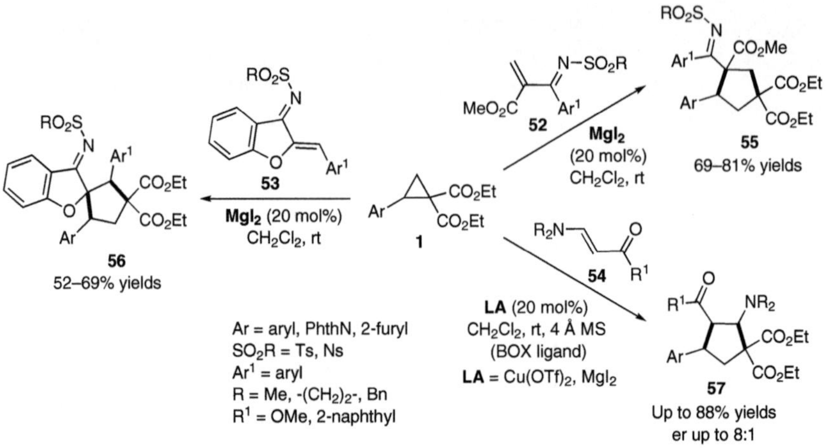

Figure 3.15 Banerjee's reaction with C=C double bond in complex nitrogen-containing heterodiene systems.

systems, cyclopropanes are usually activated using Lewis acids such as MgI_2, $Cu(OTf)_2$, $Sn(OTf)_2$, $SnCl_4$, and $TiCl_4$ [30–32] (Figure 3.15).

M. Doyle with coworkers presented an approach to [3+2]-cycloaddition of silyl ethers of enols containing a diazo group **58** and D–A cyclopropanes **1**, proceeding under mild catalysis with ytterbium(III) triflate with the retention of the diazo group in the final cyclopentane product **59**, which allows using it in final products for their further functionalization [33]. Similar processes for vinyl azides **60**, also proceeding with the retention of the azido group in the final cyclopentane **61**, and its use for further functionalization, are described in the works of the P. Banerjee and S. Chiba groups [34, 35] (Figure 3.16).

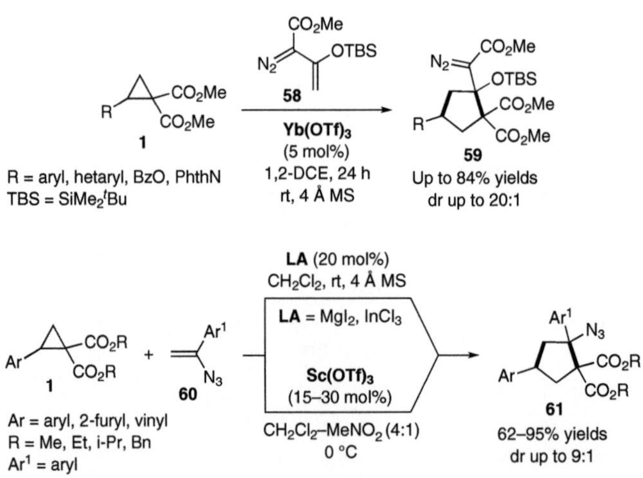

Figure 3.16 Reactions with substituted vinyl azides and vinyl diazo compounds.

Along with Lewis acids, for the activation of D–A cyclopropanes and their introduction into formal [3+2]-cycloaddition reactions with functionalized alkenes, basic catalysis is also quite widely used. Accordingly, this imposes certain requirements on the structures of the initial D–A cyclopropanes, as well as alkene components, which are usually Michael acceptors, such as α, β-unsaturated dicarbonyl compounds or their analogs with cyano groups, as well as nitroolefins, and so on, because basic catalysis assumes that the reaction proceeds through carbanion intermediates. The work of K. Takeda et al. demonstrates an interesting example of [3+2]-cycloaddition of activated cyclopropanes **62** and Knoevenagel condensation products **63**, with the α-*para*-tosyl anion generated in the presence of LiHMDS, acting as the donor moiety of cyclopropane [36]. K. Jorgensen with coworkers developed an asymmetric approach for base-catalyzed formal [3+2]-cycloaddition between substituted dicyanocyclopropanes **65** and nitroalkenes **66** using a chiral organocatalyst based on disubstituted thiourea [37]. A study by C. Wang et al. demonstrated the possibility of interaction of tetrasubstituted activated cyclopropanes **68** and Knoevenagel adducts **69** in the presence of 1,8-Diazabicyclo(5.4.0)undec-7-ene (DBU), which also proceeds as [3+2]-cycloaddition [38] (Figure 3.17).

Modern methods of photocatalysis, including those in combination with Lewis acids, are also used to activate D–A cyclopropanes. Although many similar processes are rather difficult to attribute to the classical chemistry of D–A cyclopropanes, several works can still be considered in this aspect. The study by T. Yoon et al. presents an intramolecular variant of [3+2]-cycloaddition of olefins with a cyclopropyl ketone fragment in its structure **71**, proceeding under the action of lanthanum(III) triflate in the presence of a ruthenium-based photocatalyst under visible light irradiation [39]. Later, Z. Lu with coworkers developed an intermolecular visible-light photocatalyzed [3+2]-cycloaddition of styrene derivatives **74** and nitrocyclopropanecarboxylates **73** in the presence of a similar ruthenium-based photocatalyst [40]. The mechanisms of the processes and their patterns are discussed in detail in the relevant works (Figure 3.18).

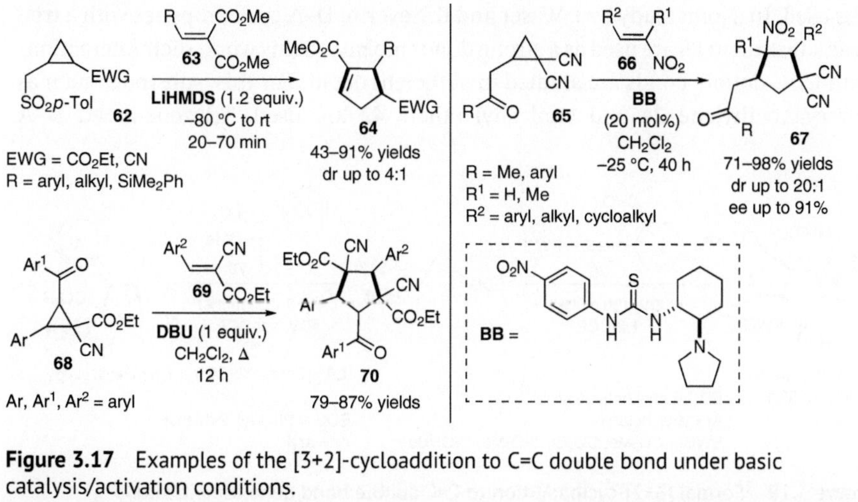

Figure 3.17 Examples of the [3+2]-cycloaddition to C=C double bond under basic catalysis/activation conditions.

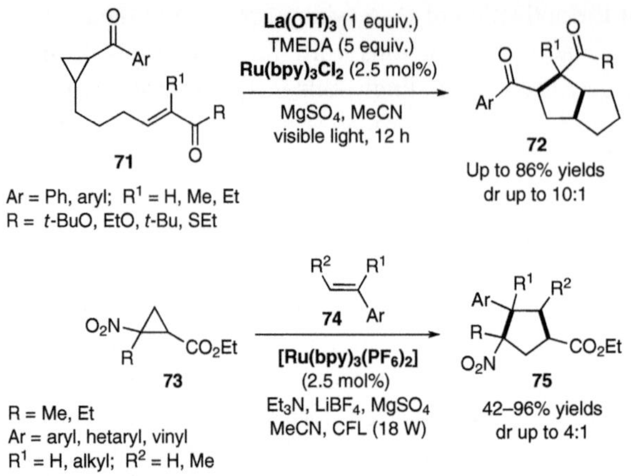

Figure 3.18 Photocatalytic [3+2]-cycloaddition of activated cyclopropanes to C=C double bond.

In the works devoted to the chemistry of D–A cyclopropanes, published during the last two or three years, a large number of formal [3+2] cycloadditions to the C=C double bond are also encountered, touching on various interesting aspects of it.

In a number of studies, complex substrates are used as the alkene component, in which the double bond required for addition is generated in situ during the reaction, which allows expanding the range of substrates and simplifying the synthetic methodology. These reactions include the works of S. Zhu, describing [3+2]-cycloaddition under the action of ZnI_2 to indenones generated from *ortho*-ethynylbenzaldehydes **76**, as well as the work of Trushkov and Voskressensky, devoted to the D–A reaction of cyclopropanes **1** with arylketals **78** under the action of $ZnCl_2$ at microwave irradiation [41, 42] (Figure 3.19).

Another direction is the expansion of the range of used D–A cyclopropanes and methods of their activation, including synthetic improvements in the methodologies used. In a joint study by J. Waser and K. Severin, D–A cyclopropanes with a triazene substituent **80** are used as a strong donor moiety. In this work, their interactions with C=C double bonds are studied in sufficient detail. For this, substrates such as tetracyanoethylene **24** and enol silyl ethers **42** are used. Triazene-based D–A

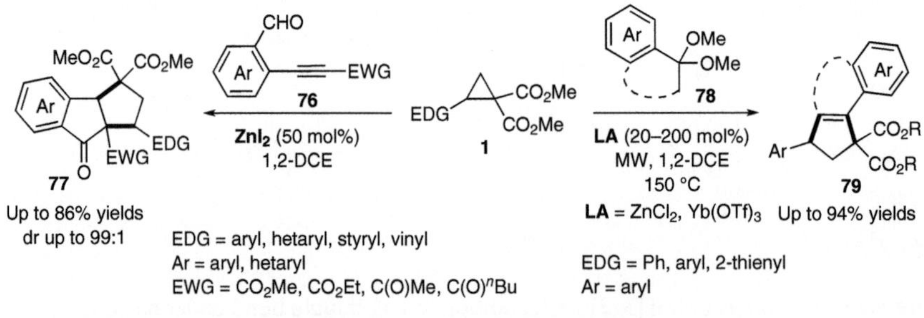

Figure 3.19 Formal [3+2]-cycloaddition to C=C double bond generated in situ.

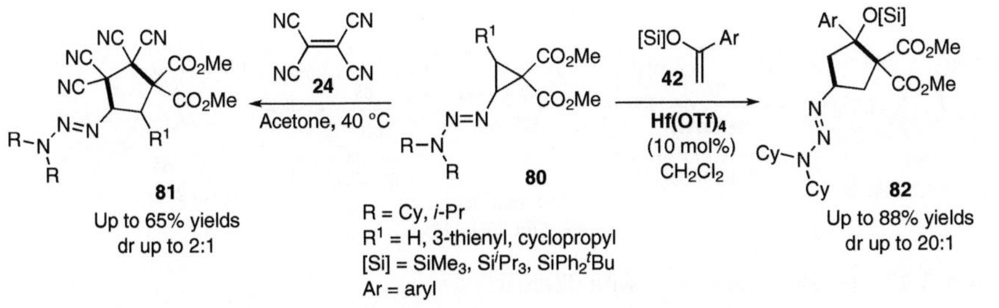

Figure 3.20 Using triazene derivatives **80** as D–A cyclopropanes in reactions with alkenes.

cyclopropanes are reactive and easily react under mild conditions or even without a catalyst [43] (Figure 3.20).

A recent work by the D. Werz group describes the process of formal [3+2]-cycloaddition of D–A cyclopropanes **1** with acceptor olefins based on Michael acceptors with a cyclic fragment **83**, such as derivatives of oxazolone, indandione, barbiturate, and Meldrum's acid. Despite the use of widely used D–A cyclopropanes, in this study, a rather unusual method of activation by calcium iodide CaI_2 is used. This activation occurs due to the addition of the iodide anion to cyclopropane, followed by a nucleophilic attack on the olefin (Michael acceptor), and subsequent cyclization with the elimination of the iodide anion. That is, the actual cycloaddition catalyst is the iodide anion, which makes it possible to introduce acceptor olefins into the reaction, in contrast to the use of donor olefins in classical Lewis acid catalysis through 1,3-zwitterions [44] (Figure 3.21).

Another recent work by D. Werz developed a protocol for [3+2]-cycloaddition between cyclopropane dicarboxylates **1** and cyclic dithioacetals **85**. At the same time, a similar protocol for acyclic dithioacetals **87** was developed in the work of the M. Wang group. Both processes are perfectly catalyzed by $Sc(OTf)_3$; sulfur-containing spiro derivatives **86** or thiocyclopentenes **88** can be obtained as products [45, 46] (Figure 3.22).

Finally, there is a series of works devoted to the use of oxindole derivatives in formal [3+2]-cycloaddition reactions, both as a D–A cyclopropane component **91** and as an alkene substrate **89**. The interest in these transformations is due to the

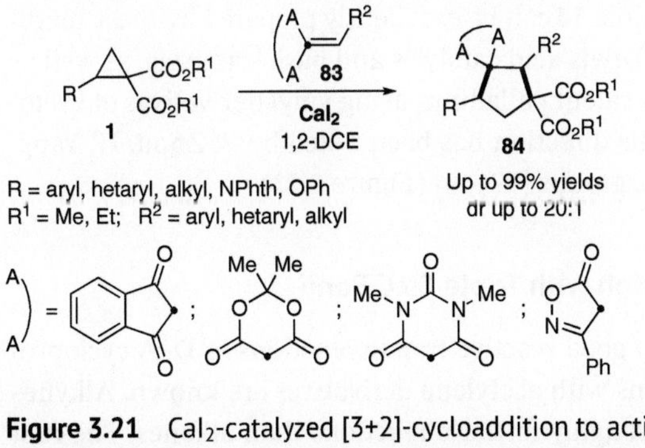

Figure 3.21 CaI_2-catalyzed [3+2]-cycloaddition to activated acceptor olefins.

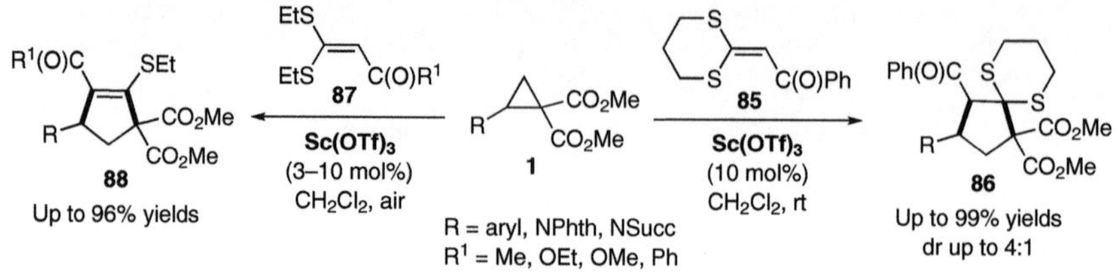

Figure 3.22 [3+2]-Cycloaddition with dithioacetals.

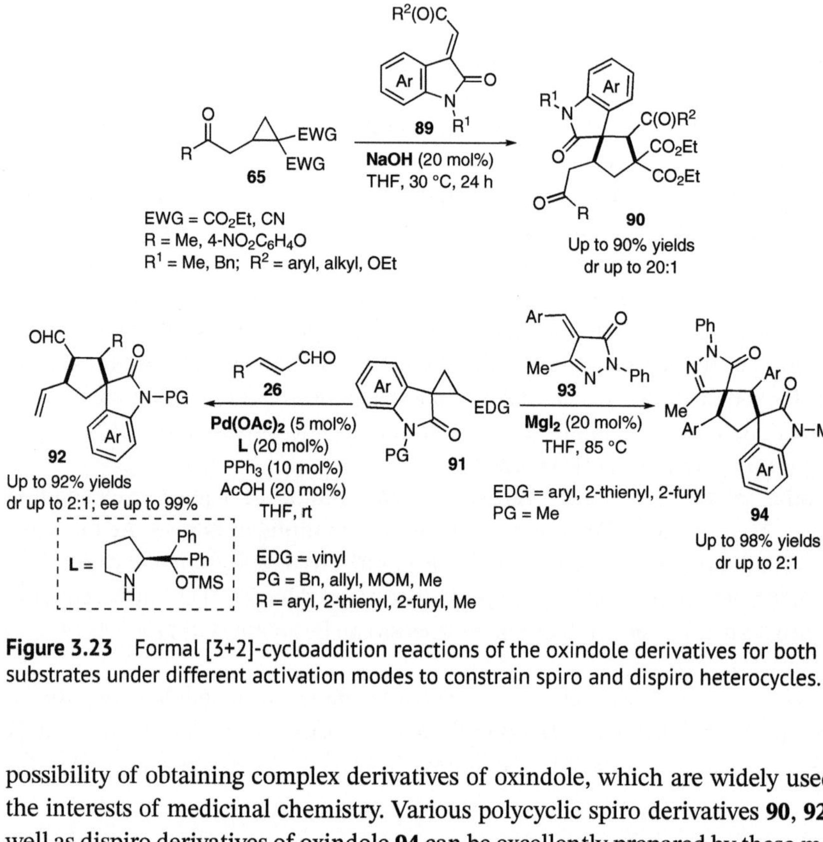

Figure 3.23 Formal [3+2]-cycloaddition reactions of the oxindole derivatives for both substrates under different activation modes to constrain spiro and dispiro heterocycles.

possibility of obtaining complex derivatives of oxindole, which are widely used in the interests of medicinal chemistry. Various polycyclic spiro derivatives **90**, **92**, as well as dispiro derivatives of oxindole **94** can be excellently prepared by these methods. As activation methods, both Lewis acid catalysis and basic catalysis, as well as metal complex catalysis with zero-valent palladium using vinyl derivatives of cyclopropane, are used. Research in this direction has been noted by B. Zhou, H. Yang, J. Saha, and E. Budynina scientific groups [47–49] (Figure 3.23).

3.2.3 Formal [3+2]-Cycloaddition with Triple C≡C Bond

The C≡C triple bond is also a very good reactive trap in reactions of D–A cyclopropanes, although not many reactions with acetylene derivatives are known. Alkynes have been studied much less thoroughly in these reactions than alkenes. The first

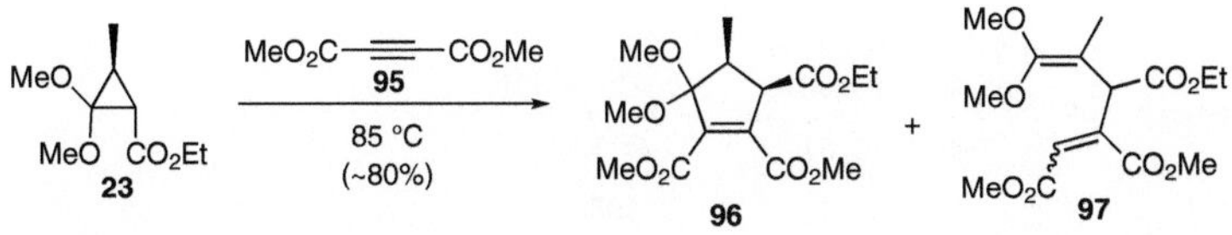

Figure 3.24 First example of the reaction of activated cyclopropanes with C≡C triple bond.

example of such a process is given by the reaction of ethyl 2,2-dimethoxycyclopropanecarboxylate **23** with acetylenedicarboxylate **95** reported by Graziano and Cimminiello in 1992. The reaction occurs at elevated temperatures in the absence of a catalyst and results in two different compounds in approximately equal amounts, including products of formal [3+2]-cycloaddition to the triple bond [50] (Figure 3.24).

In 2004, Yadav and Sriramurthy reported the reaction of terminal acetylenes **99** with silylcyclopropanes **98**. Their work was the first example of a systematic study of the reaction of D–A cyclopropanes and alkynes. The reaction occurs as a formal [3+2]-cycloaddition, to give substituted cyclopentenes **100**. The use of a strong Lewis acid is required; titanium(IV) tetrachloride was used in this case. The yields of products **100** can reach 85%, while the diastereomer ratio is up to 20/1. The synthetic protocol they developed is also successfully applicable in the intramolecular version, which makes it possible to obtain, for example, annulated indanone **102** [51] (Figure 3.25).

Yet another type of process is represented by the [3+2]-cycloaddition of cyclopropyl ketones **103**, which occurs under metal complex catalysis conditions. Processes of this type were developed by Ogoshi et al. In 2011, they extended this approach to the reaction of cyclopropyl ketones **103** with acetylenes **104** in the presence of a zero-valent nickel catalyst. This method provides a very good stereoselectivity [52]. The study by Qi and Ready expanded the idea of [3+2]-cycloaddition reactions of D–A cyclopropanes **106** and alkynes **107**, including silyl ethers of internal alkynes. This method made it possible to obtain substituted cyclopentenones **108** in fairly good yields after direct treatment of the reaction mixture with pyridinium hydrofluoride,

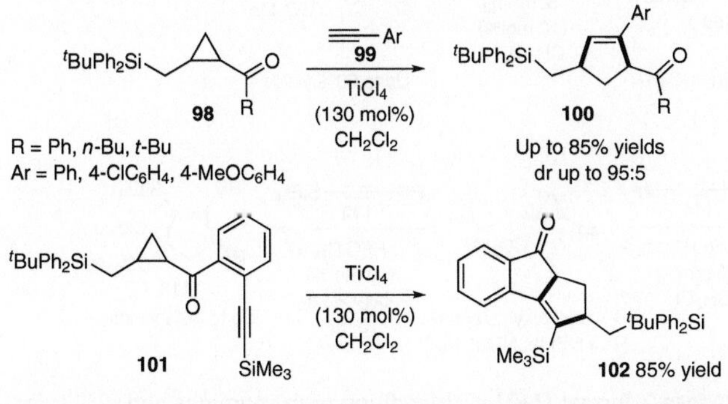

Figure 3.25 One of the first systematic formal [3+2]-cycloaddition process with alkynes.

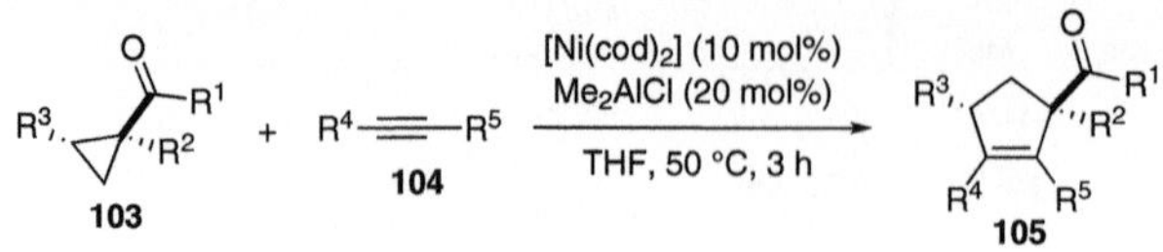

Figure 3.26 Ni(0)-Catalyzed formal [3+2]-cycloaddition for cyclopropylketones **103**.

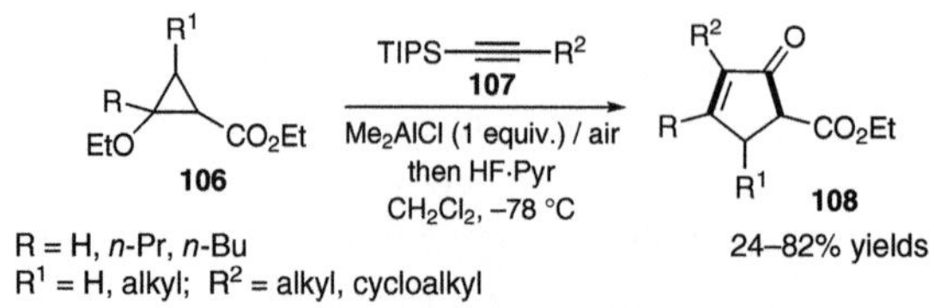

Figure 3.27 Cycloaddition reaction of alkoxy cyclopropanecarboxylates **106**.

which resulted in the removal of the silyl group followed by β-elimination. Cyclopropanes with alkoxy donor groups in the cyclopropane moiety were used as the D–A cyclopropanes [53] (Figures 3.26 and 3.27).

J. Johnson et al. developed a general approach to the [3+2]-annulation of D–A cyclopropanes **109** using ynamides **110**. The process is performed using scandium(III) triflate and provides nearly quantitative yields of the target cyclopentene sulfonamides **111**. This process successfully occurs in the case of D–A cyclopropanes, with donor substituents at the aromatic ring, with a heteroaromatic moiety, or with styryl substituents [54]. A little later, Waser et al. demonstrated an approach to [3+2]-cycloaddition between thioalkynes **112, 114** and D–A cyclopropanes **1** in the presence of Hf(OTf)$_4$, giving thiocyclopentene derivatives **113**. The yields of the target products varied from 36% to 96%. Moreover, the same team developed an approach where the carbonyl group of phthalimide cyclopropanedicarboxylate underwent cyclization to give a tetracyclic bridged compound **115** [55] (Figure 3.28).

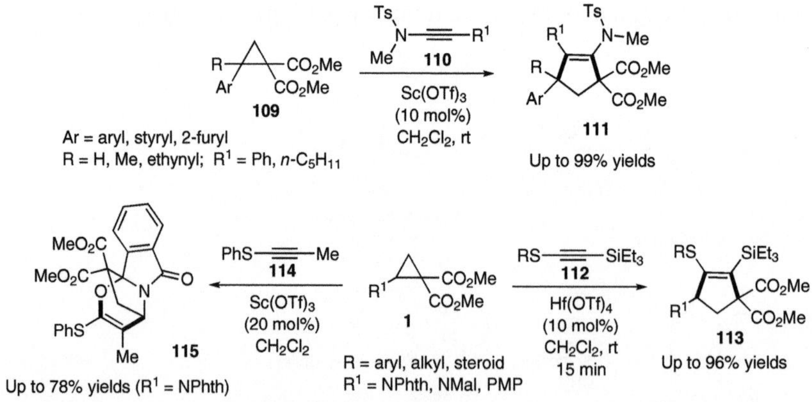

Figure 3.28 Johnson's and Waser's formal [3+2]-cycloaddition with ynamides and thioalkynes.

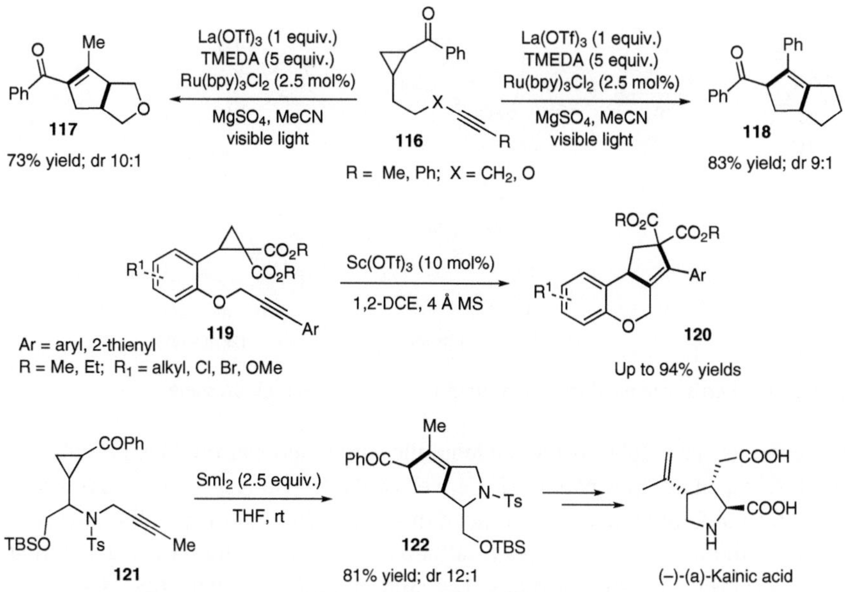

Figure 3.29 Intramolecular formal [3+2]-cycloaddition with alkynes.

Yoon et al. studied an intramolecular version of formal [3+2]-cycloaddition of alkynes comprising a cyclopropylketone moiety **116**. This process occurs in the presence of lanthanum triflate, and a ruthenium-based photocatalyst under visible light irradiation. The yields of cyclic products **117**, **118** were up to 83% with good diastereoselectivity. Moreover, migration of the multiple bonds toward the acceptor group was observed in the reaction with an alkyne with a terminal methyl group [39]. Liang et al. suggested a method for synthesizing the cyclopenta[*c*]chromene skeleton **120** by a related reaction of intramolecular [3+2]-cycloaddition of D–A cyclopropanes **119** with a C≡C triple bond catalyzed by scandium triflate [56]. Li et al. used an intramolecular version of [3+2]-cycloaddition in the complete synthesis of (−)-(α)-kainic acid. Samarium(II) iodide was used as the mediator in the cyclization process. This process occurred with high diastereoselectivity and in high yields [57] (Figure 3.29).

3.2.4 [3+2]-Annulation with Aromatic C=C Bond

The addition reactions of D–A cyclopropanes **1** to aromatic multiple bonds in heterocycles occurring as [3+2]-annulation with dearomatization and the formation of annulated heterocyclic compounds have been studied rather well. Moreover, the reactions with indole have been studied most thoroughly, since polycyclic derivatives based on the indole skeleton are of considerable interest, both as biologically active compounds and in the complete synthesis of natural substances and their analogs. Indole derivatives are widespread in nature. The first examples related to the development of new methods for the synthesis of polycyclic indole heterocycles

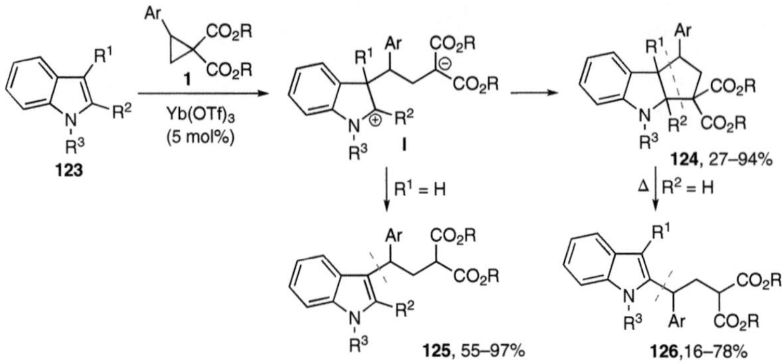

Figure 3.30 Kerr's studies of the addition of indoles to D–A cyclopropanes.

involve the use of indoles as the nucleophilic components in the [3+2]-cycloaddition of activated cyclopropanes. M. Kerr et al. developed a series of processes in 1997–2002 and found that the outcome of the reaction strongly depends on the substituents in the starting indole. As a result, three main directions of the reaction that gave [3+2]-cycloaddition products **124** and acyclic products **125**, **126** were found [58–60] (Figure 3.30).

Subsequently, many scientific teams continued to study and develop processes based on the reactions of various D–A cyclopropanes with indole derivatives, including approaches to asymmetric catalysis and intramolecular transformations. H. Junjappa presented a variant for trapping 1,3-zwitterions, which are generated from cyclopropyl ketones **10** and cyclopropanedicarboxylates **1**, by indole derivatives **127** via [3+2]-annulation in the presence of boron trifluoride etherate to give cyclopentaindolines **129**, **130** in 70% to 93% yields [61]. In turn, Pagenkopf et al. developed this process for alkoxycyclopropanecarboxylates **16** as D–A cyclopropanes. Trimethylsilyl triflate in nitromethane was used as the activator. This reaction occurred with high diastereoselectivity and also gave tricyclic indole derivatives **131** [62]. Y. Tang developed an efficient asymmetric catalytic approach for the [3+2]-annulation with indoles in the presence of copper(II) salts, namely, $Cu(OTf)_2$ or $Cu(SbF_6)_2$, and substituted bisoxazolines as chiral ligands, which allowed a high degree of enantiomeric enrichment of products **132** (up to 98 : 2) to be reached [63]. More recently, Jerome Waser suggested to use D–A cyclopropanes based on aminocyclopropane monoesters **128** in the presence of catalytic amounts of *in situ*-generated Et_3SiNTf_2. Careful optimization of the activating amino group was performed to implement the process efficiently and open the cyclopropane ring. TsMeN was found to be the best option that made it possible to perform the reaction with a wide range of indole and cyclopropane substrates [64] (Figure 3.31).

Y. Tang et al. developed an efficient protocol for the intramolecular [3+2]-cycloaddition of indoles **134** whose structures comprised a D–A cyclopropane moiety. The process occurs efficiently in the presence of $Cu(SbF_6)_2$ and a chelating ligand to give high yields of the final tetracyclic cyclopentane-fused spiroindolines **135** with good diastereoselectivity and enantiospecificity. The use of differently substituted ester groups allowed controlled syntheses of various diastereomers of the resulting

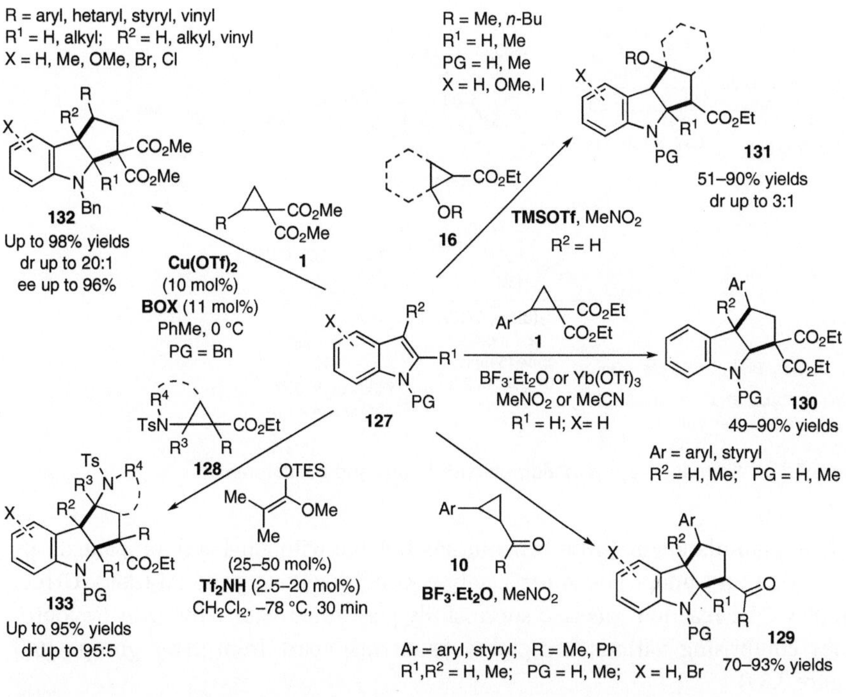

Figure 3.31 [3+2]-Annulation of various D–A cyclopropanes with indoles.

compounds to be performed. The $CO_2(i\text{Pr})$ group (for the *trans* isomer) and the $CO_2(2\text{-Ad})$ group (for the *cis* isomer) showed the best results [65] (Figure 3.32).

Regarding other heterocyclic systems, [3+2]-annulation with derivatives of furans, benzofurans, benzothiazoles, some pyridines, and β-naphthols was studied. The C=C double bond in the furan ring is modified most easily since it has the lowest aromaticity and is most reactive. For example, Trushkov et al. presented a method for the functionalization of 2,5-dimethylfuran **136** with D–A cyclopropanes **1** by single or double formal [3+2]-cycloaddition at C=C bonds to give bi- and tricyclic systems **137**, **138** based on tetrahydrofuran, as well as by more complex cascade processes. Lewis acids such as tin(IV) chloride and ytterbium(III) triflate were used as the catalysts in these reactions [66]. M. Vitale's study presents a related method

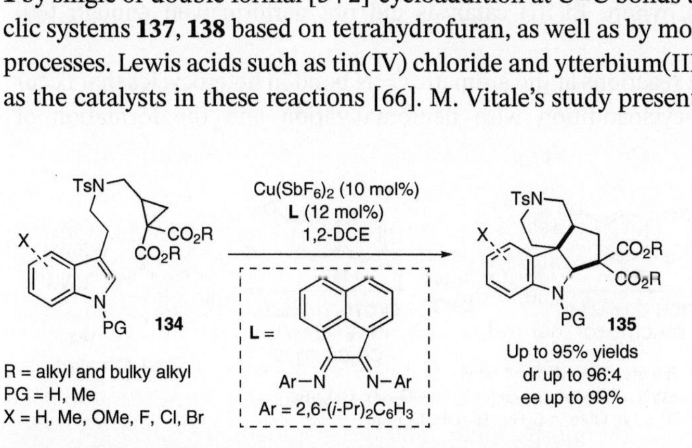

Figure 3.32 Intramolecular [3+2]-annulation with indoles.

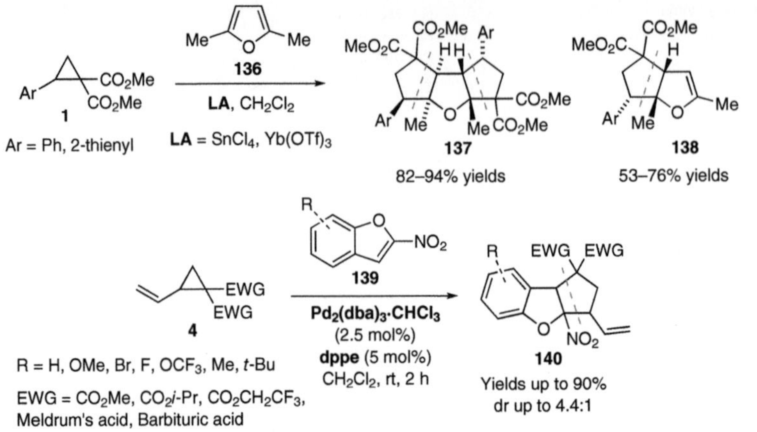

Figure 3.33 Formal [3+2]-cycloaddition with furans and benzofurans.

for [3+2]-annulation of 2-nitrobenzofurans **139**, but with vinyl cyclopropane dicarboxylates **4** and under palladium catalysis conditions using the $Pd_2(dba)_3$·$CHCl_3$ complex. The reaction was also successfully performed with vinyl D–A cyclopropanes comprising various acceptor substituents apart from ester groups [67] (Figure 3.33).

Processes of [3+2]-annulation of 2-naphthols **141** in reactions with D–A cyclopropanes **1** are also known. In this case, elimination of an OH group occurs, which results in the reduction of the naphthalene aromatic system to give derivatives of cyclopentanaphthalenes **142**. Various conditions for this type of process are known. Prof. Biju used a Bi(III)-based catalytic system: $Bi(OTf)_3$–KPF_6, whereas the standard $Sc(OTf)_3$ only led to a simple Friedel–Crafts type reaction without cyclization. At the same time, the reaction was very tolerant to the electronic effects of substituents in the naphthalene core under Bi(III) catalysis conditions [68]. Y. Zhao et al. used a metal-free catalytic system based on a Brønsted acid, TfOH, that showed similar synthetic patterns. However, it is especially successful with vinylcyclopropane dicarboxylates, where Bi(III) catalysis did not perform well enough [69] (Figure 3.34).

A few examples of reactions at the aromatic C=N bond in heterocycles that occur as a formal [3+2]-cycloaddition with dearomatization and the formation of

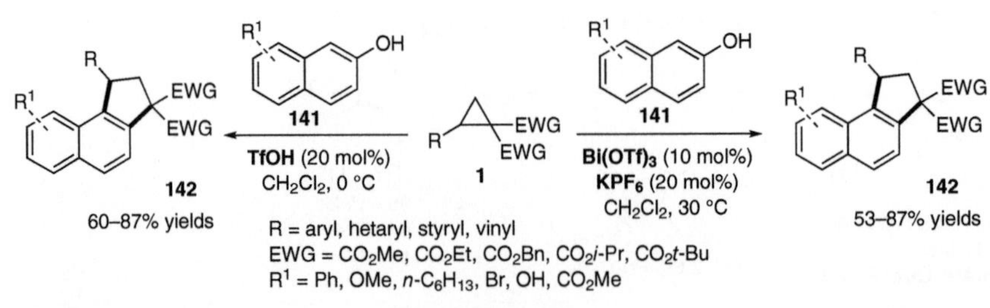

Figure 3.34 [3+2]-Annulation with β-naphthols.

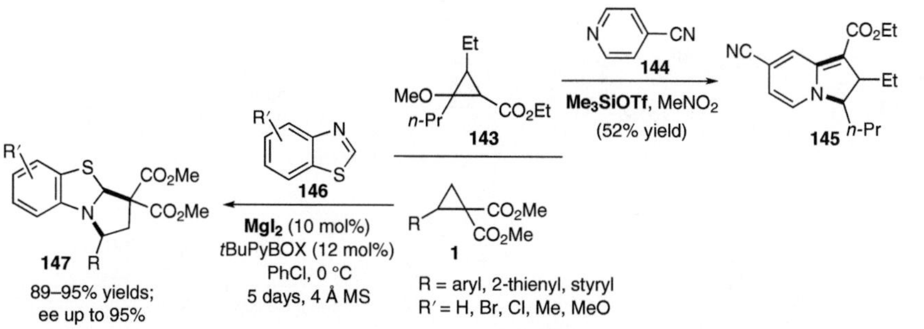

Figure 3.35 [3+2]-Annulation with aromatic C=N double bond in 4-CN-pyridine and benzothiazoles.

heterocyclic annulated systems are known. For example, examples with 4-cyano-pyridine **144** and substituted benzothiazoles **146** were reported [70, 71] (Figure 3.35).

3.2.5 [3+2]-Annulation of D–A Cyclopropanes Involving Aryl/Heteroaryl Donor Substituent

A special place in the wide variety of reactions of formal [3+2]-cycloaddition of D–A cyclopropanes **1** is occupied by the processes of so-called [3+2]-annulation. They proceed with annulation along the aromatic ring, which is used as a donor group for the activation of cyclopropane, while the malonyl center at the acceptor groups remains unused, in contrast to most other processes. There are very few examples of such processes at the moment, and most of them were developed by Prof. Trushkov's group, but their distinctive features make them worthy of a separate consideration. Usually, these processes occur during the interactions of D–A cyclopropanes with double and triple C–C bonds, or during dimerization. These processes generate active carbocationic centers capable of efficient intramolecular electrophilic substitution at the aromatic ring to the *ortho* position according to the S*E*Ar type. Moreover, for the selective direction of the process in the desired direction, the use of additional donor substituents in the benzene ring is usually required, for example, MeO groups are especially suitable. It is also preferable to use heteroaromatic substituents as donor groups in D–A cyclopropane. Older similar processes of [3+3]-annulation and [4+3]-annulation are also known, proceeding in a similar way and described by the same features (see Figure 3.36).

In 2013, the selective [3+2]-annulation of D–A cyclopropanes **1** and polysubstituted alkenes **148** in the presence of Lewis acids was demonstrated. As noted above, this process directly involves the donor group of cyclopropanes, which is attached to the main three-carbon fragment involved in cycloaddition. Derivatives of indanes **149** or their heterocyclic analogs are obtained as reaction products [72]. The known examples of a similar reaction occurring during the dimerization of D–A cyclopropanes **1** in the presence of $Sn(OTf)_2$, in which [3+2]-annulation is observed between β-styrylmalonate formed in situ and the starting cyclopropane. In this case, this dimerization can also be selectively carried out in a cross-variant between two

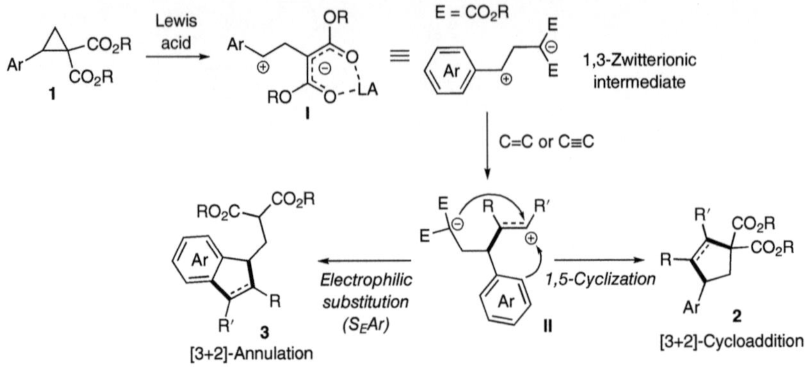

Figure 3.36 [3+2]-Annulation reactions of D–A cyclopropanes with multiple C-C bonds, and their main mechanistic features.

different cyclopropane molecules [73]. In a later study, a similar [3+2]-annulation of arylcyclopropanedicarboxylates **1** and internal alkynes **104** was developed in the presence of Lewis or Brønsted acids. This also requires the use of aryl D–A cyclopropanes with donor substituents in the aromatic fragment. The reaction products are polysubstituted derivatives of indene **151**, which can be used as dyes with high extinction coefficients [74] (Figures 3.37–3.39).

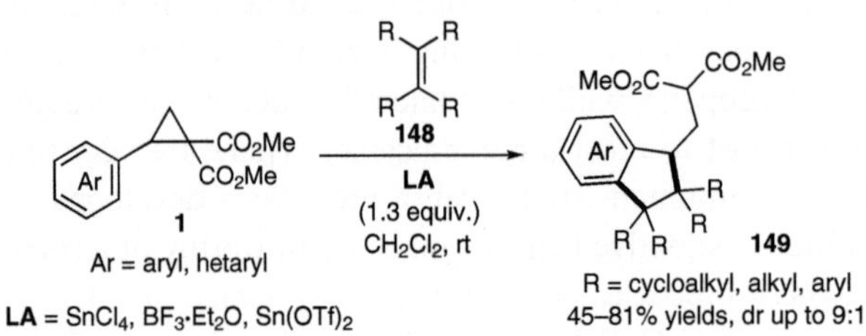

Figure 3.37 [3+2]-Annulation of D–A cyclopropanes **1** with alkenes.

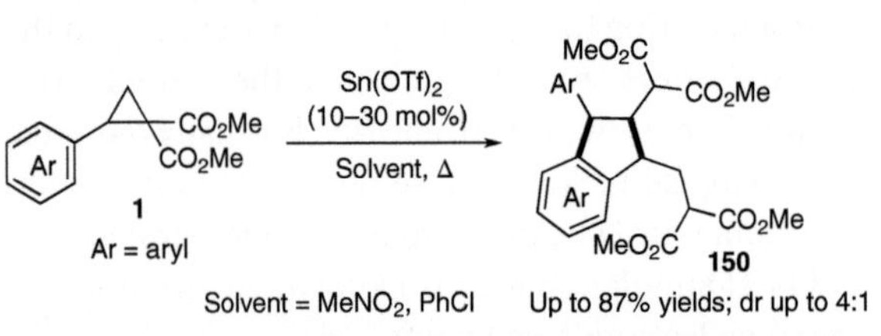

Figure 3.38 [3+2]-Annulation dimerization process of D–A cyclopropanes **1**.

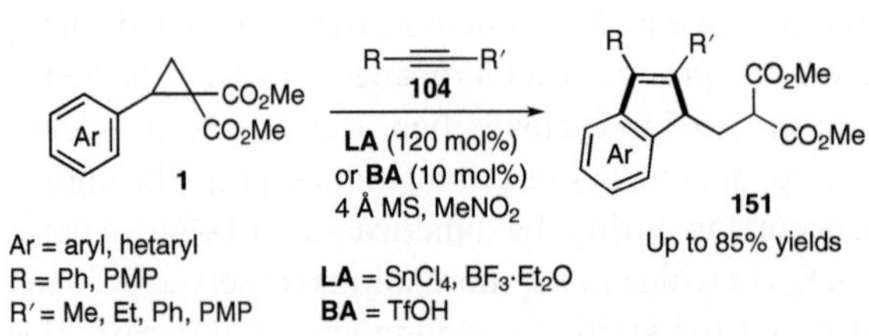

Figure 3.39 [3+2]-Annulation of with acetylenes.

3.3 Formal [3+2]-Cycloaddition with C=O and C=N Double Bond

3.3.1 Formal [3+2]-Cycloaddition with C=O Double Bond

Of all the variety of reactions, those of D–A cyclopropanes with carbonyl compounds have been studied especially well. Moreover, they are among the first reactions that started the development of the chemistry of these compounds. The process is of extremely general nature and nearly always occurs as a formal [3+2]-cycloaddition to the C=O double bond to give a tetrahydrofuran ring; substrates of diverse types undergo this, although other types of processes are also known in some cases. The first examples of [3+2]-cycloaddition of aldehydes to activated cyclopropanes were reported by H. U. Reissig and I. Kuwajima over 40 years ago, back in 1981 and 1977, respectively [75, 76]. After that, these studies slowly continued to develop, and the use of these reactions gradually became more and more popular. The next turning point came in 2005 when J. Johnson published a [3+2]-cycloaddition with 2-substituted cyclopropane-1,1-dicarboxylates that are now the most popular types of D–A cyclopropanes (along with M. Kerr's [3+3]-reaction of the same cyclopropanes with nitrones reported in 2003) [77, 78]. In fact, the publication of these reactions in 2003–2005 began the modern era and the chemistry of D–A cyclopropanes. From that time on, this term was introduced into widespread use [1]. Therefore, the significance of [3+2]-cycloaddition reactions with aldehydes can hardly be overestimated. Sometimes this reaction with cyclopropane-1,1-dicarboxylates is also called Johnson's reaction since it became very popular subsequently. At present, more than a hundred works on the reaction of [3+2]-cycloaddition with aldehydes in various implementations have been published. It is one of the most recognizable and representative processes in the chemistry of D–A cyclopropanes. Moreover, the reaction is now widely used in practice since it produces tetrahydrofuran derivatives that are very useful in syntheses.

The majority of early studies mainly focused on the use of D–A cyclopropanes comprising alkoxy groups **152**, **155**, **157** and cyclopropylketones **159**. It was sometimes rather difficult to reach high diastereoselectivity in those reactions, especially in the case of sterically loaded cyclic systems. Nevertheless, the reactions occurred successfully with various cyclopropanes and often required highly reactive tin and titanium chlorides as the Lewis acids [75, 79–82] (Figure 3.40).

As mentioned above, since 2005, J. Johnson et al. began to intensively study the formal [3+2]-cycloaddition D–A cyclopropanes **1** and aldehydes **160** catalyzed by Lewis acids. Various aromatic, heteroaromatic, as well as alkenyl and alkyl substituents were used as substituents in the initial substrates. A wide range of Lewis acids has been studied using model reactions of 2-phenylcyclopropane-1,1-dicarboxylate with benzaldehyde, as well as some others, leading to the formation of substituted tetrahydrofurans. The product usually forms with high *cis*-diastereoselectivity. It should be noted that this reaction is well-catalyzed by almost any Lewis acid, although the yield of tetrahydrofurans may not be so high. However, triflates of a number of transition and post-transition metals, such as $Sc(OTf)_3$, $Yb(OTf)_3$,

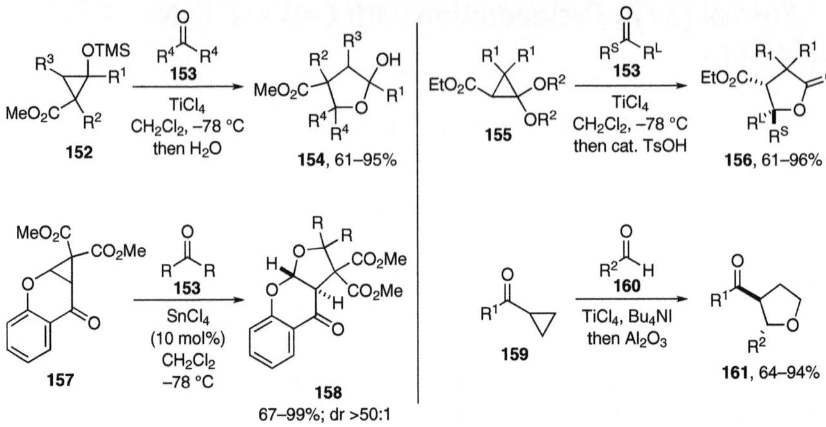

Figure 3.40 Initial reactions of D–A cyclopropanes with aldehydes and ketones.

$Ce(OTf)_3$, $Hf(OTf)_4$, $Cu(OTf)_2$, $Zn(OTf)_2$, $Sn(OTf)_2$, and others, turned out to be the most preferable among them. At the same time, tin(II) triflate was most widely used for catalysis, and for a number of substrates – hafnium(IV) triflate, for less reactive substrates, such as aliphatic aldehydes or cyclopropanes with vinyl or alkyl substituents – $SnCl_4$ and $AlCl_3$ [77, 78, 83, 84].

The process is very easy to experimentally implement. For example, the reaction of 2-phenylcyclopropane-1,1-dicarboxylate with benzaldehyde proceeds very easily using as little as 5% anhydrous $Sn(OTf)_2$ at room temperature and simply mixing the reagents in a suitable solvent (usually methylene chloride), leading to quantitative yields of a tetrahydrofuran product that does not actually require additional purification. At the same time, traces of water, which could potentially interfere with the reaction, are easily removed by adding molecular sieves.

The mechanism of this process is classical in the chemistry of D–A cyclopropanes. It is implemented through the coordination of the Lewis acid at two ester groups, which leads to the activation of the σ-bond of the cyclopropane ring and its subsequent opening, generating 1,3-zwitterionic intermediates, which enter into a stepwise [3+2]-cycloaddition reaction with C=O double bond of aldehydes. In this case, we can consider two main extreme options from the possible ways of the process: with the complete opening of the three-membered ring and the generation of 1,3-zwitterionic intermediates as such, or without complete opening, but only with the polarization of the C–C bond, making it sufficient for nucleophilic attack by an aldehyde molecule. It is believed that polarization of the bond of the cyclopropane ring without opening occurs when milder reaction conditions and less active Lewis acids, such as metal triflates, and full opening occurs when using stronger Lewis acids, such as titanium or tin(IV) chlorides. Although in reality, it turns out to be quite difficult to distinguish one type of mechanism from another. Their significant difference lies, among other things, in the stereochemical result of the reaction, especially when optically active starting cyclopropanes are used since the generation of carbocations leads to significant racemization (Figures 3.41 and 3.42).

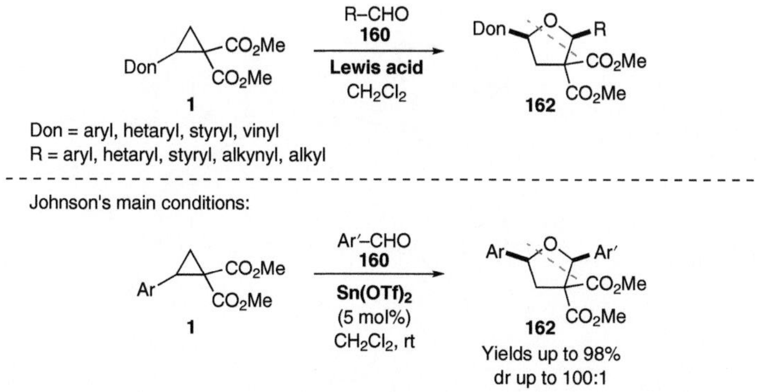

Figure 3.41 Johnson's formal [3+2]-cycloaddition of cyclopropanedicarboxylates with aldehydes.

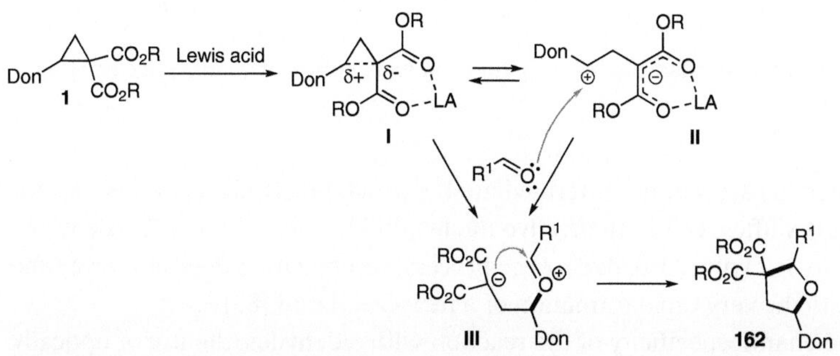

Figure 3.42 General mechanism of the reaction.

In the work of Johnson et al., a wide range of both substrates with different substituents was studied, and a detailed analysis of a number of dipolarophiles was carried out. The following can be noted from the main regularities: aromatic aldehydes **160** containing donor groups react much more easily than aromatic aldehydes with electron-withdrawing groups. However, when $Hf(OTf)_4$ is used as a catalyst, these aldehydes can also be easily introduced into reactions with cyclopropanes. The effect of substituents in the D–A donor group of cyclopropanes **1** was also studied. Electron-donating aromatic substituents also significantly accelerate [3+2]-cycloaddition, stabilizing the carbocationic center formed as a result of the opening of the cyclopropane ring.

The Johnson [3+2]-cycloaddition reaction with aldehydes proceeds with high diastereoselectivity and stereospecificity, which makes it possible to obtain enantiomerically pure tetrahydrofuran derivatives **162** starting from chiral starting cyclopropanes **1**. To explain the abnormally high diastereoselectivity and the absence of racemization, Johnson and colleagues performed a detailed study of the reaction mechanism using deuterium labels (Figure 3.43), under mild catalysis with tin(II) or hafnium(IV) triflates. Activation of the σ-bond of the cyclopropane ring under such conditions occurs without the actual opening of the three-membered ring and

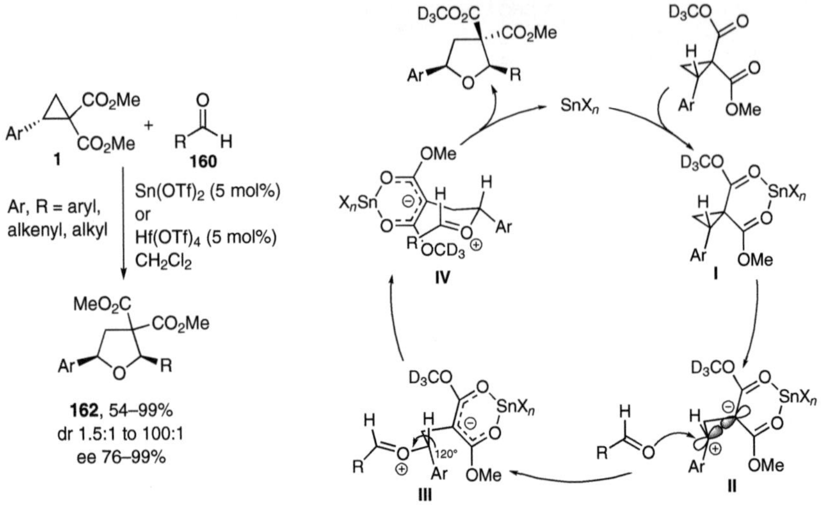

Figure 3.43 Detailed mechanistic investigation and explanation of enantiospecificity by Johnson's group.

the formation of a carbocation (intermediate II), which practically prevents racemization, but it is sufficient for an effective nucleophilic attack by an aldehyde molecule, similar to S_N2 type. Also, during the process, racemization does not have time to occur due to the very rapid formation of a new C—C bond [83].

Despite the enantiospecificity of the reaction with aldehydes, the use of optically pure starting cyclopropanes does not always allow one to transfer chirality and obtain an optically pure product. In cases where less reactive starting D–A cyclopropanes or aldehydes are used, the use of more severe reaction conditions and stronger Lewis acids is required, which leads to significant or complete racemization of the starting cyclopropane during the process.

The next step was the development, by the same authors in 2009, of a catalytic asymmetric variant of the cycloaddition reaction of cyclopropane dicarboxylates **1** to aldehydes **160**, which is one of the first for D–A cyclopropanes (along with the reaction with nitrones and some others). The ability to convert one enantiomer of cyclopropane to another under the action of Lewis acids makes it possible to implement an effective dynamic kinetic asymmetric resolution (DyKAT) protocol when using racemic cyclopropanes, converting all cyclopropanes into an optically pure product **162** with high *ee* values (up to >98–99%). A MgI_2 complex with chiral PyBOX ligands is used as a catalyst [84, 85]. This protocol and the combinations of Lewis acids and chiral ligands used, have proved to be excellent for various substrates and later became the most common and classic for D–A cyclopropanes and their various asymmetric processes (Figure 3.44).

The process of formal [3+2]-cycloaddition of 2-vinyl cyclopropane dicarboxylates **4** to aldehydes **160** can also be considered separately since it proceeds by a different mechanism. This process was also first developed by the J. Johnson group in 2008. It represents another approach to the creation of cyclic systems based on D–A

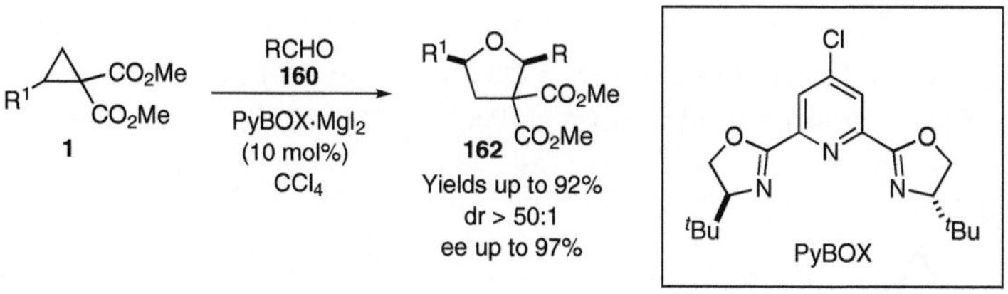

Figure 3.44 Asymmetric [3+2]-cycloaddition with aldehydes (DyKAT).

cyclopropanes. The method is based on the formation of stabilized palladium-allyl zwitterionic intermediates (**I** and **II**), which implies the use of the corresponding vinylcyclopropanes **4** as starting D–A cyclopropanes. This therefore requires the use of Pd(0) metal complex catalysis, which initiates the activation and opening of the three-membered ring upon initial coordination with the vinyl group and not with acceptor ester substituents. Otherwise, the patterns are broadly similar to those for the analogous Lewis acid-catalyzed process for cyclopropane dicarboxylates. The process is also of a general nature and is well implemented for various aromatic as well as aliphatic aldehydes [86] (Figure 3.45).

The formal [3+2]-cycloaddition to the aldehyde's C=O double bond, producing a tetrahydrofuran ring, is a very general process that can involve diverse types of substrates. After the first publications and a series of works published by J. Johnson's team, various scientific groups began to actively develop and study this process. The main focus was on the use of various types of D–A cyclopropanes in this reaction. For example, almost any aldehydes, including aromatic and heteroaromatic aldehydes, aliphatic aldehydes, aldehydes with various functional groups, more complex aldehydes based on natural and biological compounds, building blocks, and so on, as well as ketones enter this reaction under a wide range of conditions. The specifics of the reaction are usually governed by the D–A cyclopropane used. Catalysis by various Lewis acids or sometimes Brønsted acids, often under rather mild conditions, is used in the majority of cases. It is this reaction with aldehydes that is often used as a model process for optimizing the use of new types of D–A cyclopropanes. Studies on the [3+2]-cycloaddition with aldehydes are reported in the works of a number of scientific teams, such as those of V. Yadav, J. Johnson, D. Werz, J. Waser, M. Kerr, B. Pagenkopf, Z. Wang, Y. Tang, P. Banerjee, I. Trushkov, R. Novikov, Y. Tomilov, M. Niggemann, N. Maulide, Y, Hua, J. Chai, H. Shao, Z. Chai, J. Zhou, A. Mikhaylov, D. Rawat, and others, many of whom are very well known in the chemistry of activated cyclopropanes and related substrates [25, 27, 87–101].

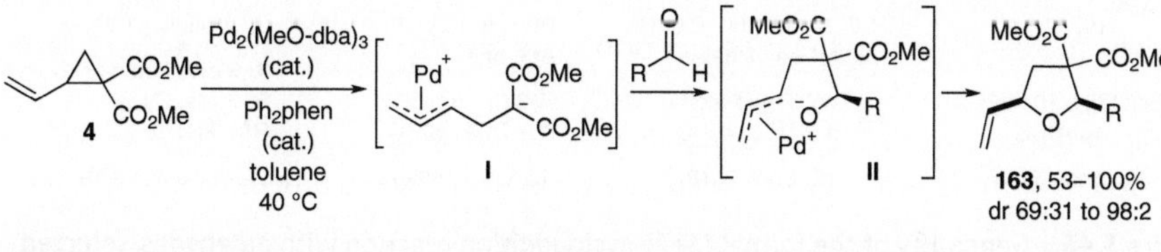

Figure 3.45 Pd(0)-Catalyzed [3+2]-cycloaddition of vinylcyclopropanes.

Also, the [3+2]-cycloaddition process with aldehydes is very convenient for testing the use of new catalysts in the chemistry of D–A cyclopropanes (for example, cobalt porphyrins [CoTPP] or gallium phthalocyanines [PcGa]) [102, 103] (Figure 3.46).

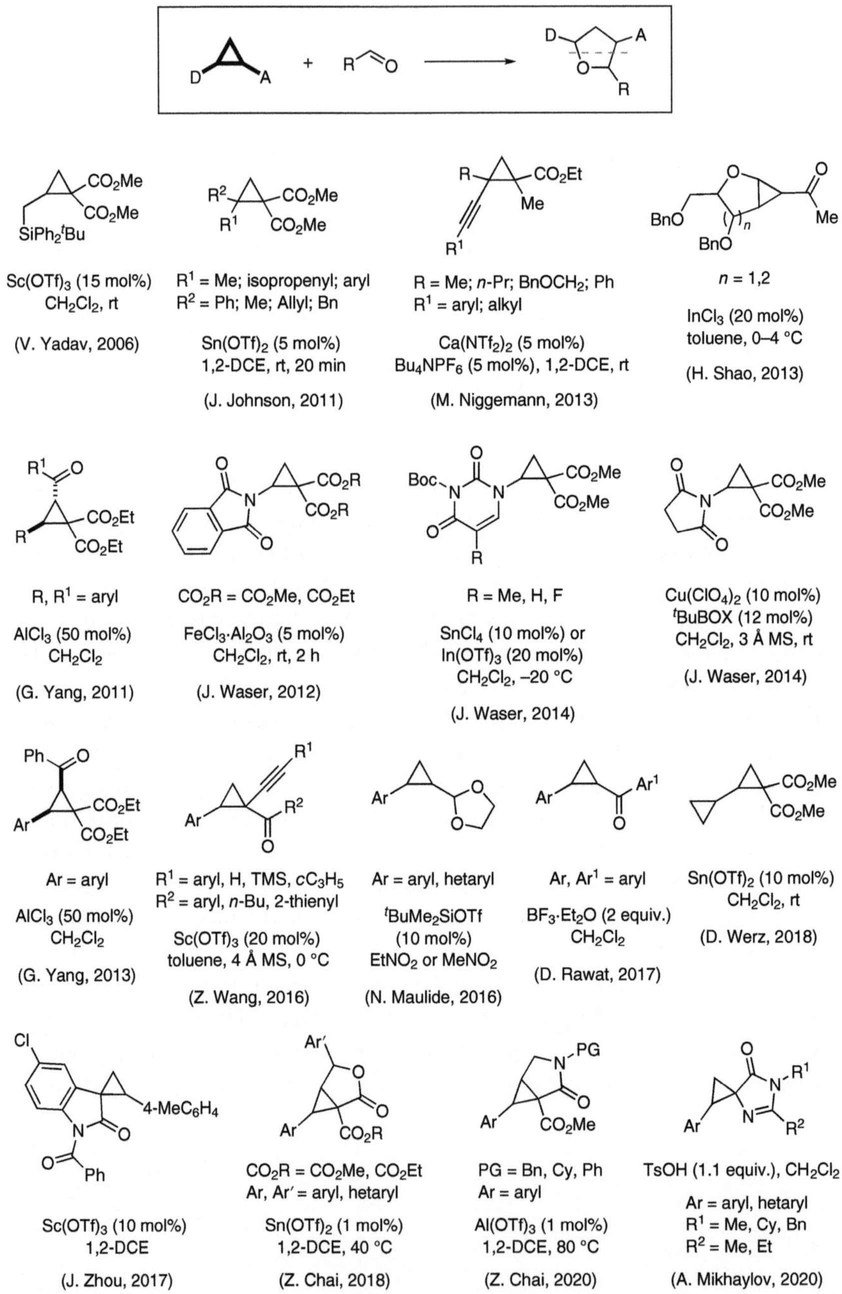

Figure 3.46 Generality of the formal [3+2]-cycloaddition reaction with aldehydes, selected examples of different D–A cyclopropanes.

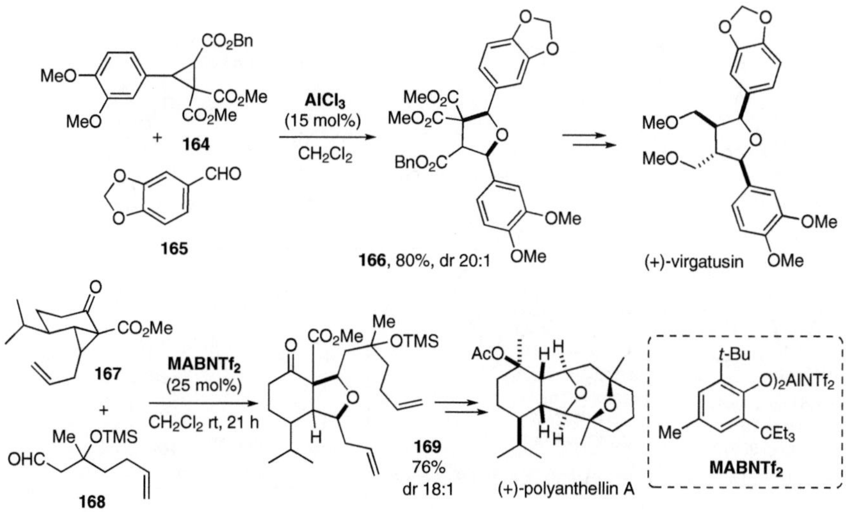

Figure 3.47 Examples of the application of [3+2]-cycloaddition with aldehydes in total synthesis.

The reactions of formal [3+2]-cycloaddition of aldehydes and D–A cyclopropanes have found wide use in the complete synthesis of natural compounds since the tetrahydrofuran ring and its derivatives are widespread in nature. The use of [3+2]-cycloaddition as a key step in the complete syntheses of compounds such as (+)-virgatusin and (+)-polyanthellin-A in the studies by Jeffrey Johnson can be noted as simple and illustrative examples [104, 105] (Figure 3.47).

Ketones exhibit properties similar to aldehydes as dipolarophiles in the formal [3+2]-cycloaddition with D–A cyclopropanes **1**, although their activity may be somewhat lower in some cases. The same Lewis acids are usually employed for ketones as for aldehydes, while the reaction diastereoselectivity remains at the same high level, as it was reported in the works of many scientific teams, such as those of J. Johnson, V. Yadav, J. Waser, and others. Highly substituted tetrahydrofuran derivatives **170** can be obtained in the reaction [27, 83, 87]. Both simple acyclic ketones and more complex cyclic representatives, as well as other substituted ketones **171**, readily react in this process. Various spiro-linked tetrahydrofurans **172**, **174** are obtained as the products in the case of cyclic ketones. For example, even cyclopropenones **173** can be successfully used in the reaction [106]. Special synthetic protocols were also developed for [3+2]-cycloaddition reactions with ketones performed in the enantioselective catalytic version. Y. Tang et al. developed such a protocol for reactions with cyclic ketones under catalysis with copper(II) triflate and bis-oxazoline ligands (SaBOX) [107]. Various functionalized D–A cyclopropanes, for example, those based on γ-butyrolactone, are also successfully used in reactions with ketones [99] (Figure 3.48).

A series of works by Z. Wang's team report a number of interesting intramolecular processes based on formal [3+2]-cycloaddition of the C=O double bond to D–A cyclopropanes to give complex polycyclic systems with a bridging oxygen atom.

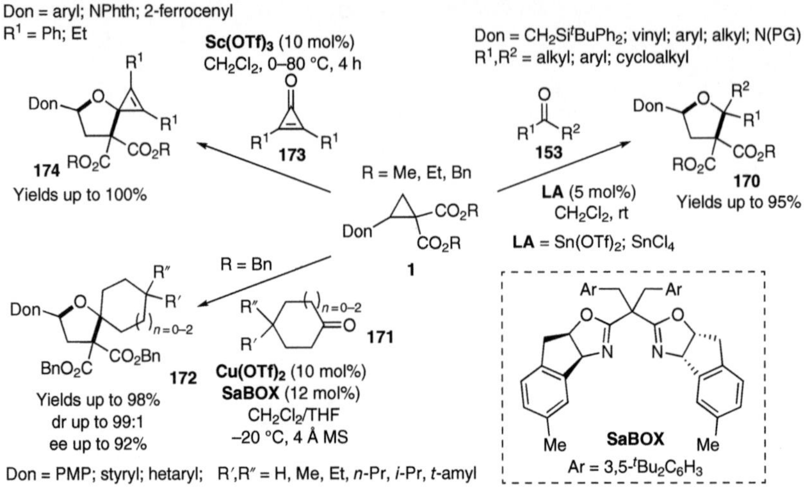

Figure 3.48 Formal [3+2]-cycloaddition with ketones.

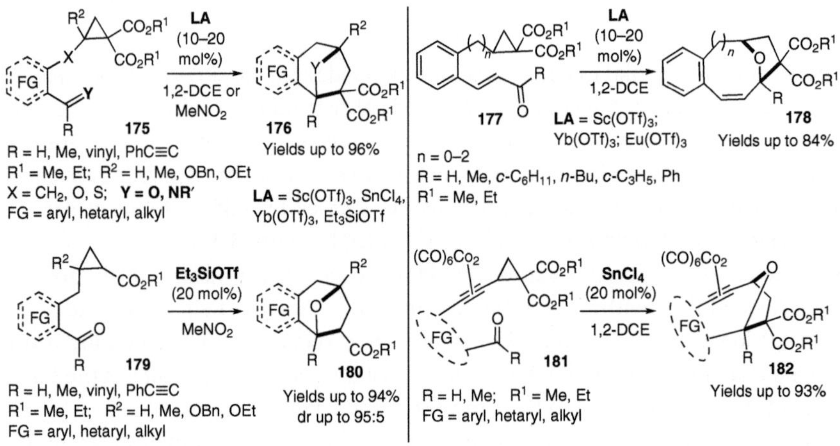

Figure 3.49 Intramolecular formal [3+2]-cycloaddition with C=O double bond.

Aromatic, alkyl, heteronuclear, heteroaromatic, α,β-unsaturated compounds, and moieties with a triple bond serve as bridges between the carbonyl group and the activated cyclopropane moiety. The range of the developed processes is very wide. The reactions mainly involve aldehydes and ketones, but they can also be successfully expanded to include the corresponding imines, where the C=N double bond is involved in the process [108–111] (Figure 3.49).

3.3.2 Formal [3+2]-Cycloaddition with C=N Double Bond

Reactions of formal [3+2]-cycloaddition of activated cyclopropanes with imines result in pyrrolidine heterocycles that are part of many biologically active and natural compounds. For this reason, this class of reactions is of great importance and has been studied very well to date. Studies of these processes began back at the dawn of the chemistry of D–A cyclopropanes, starting from the end of the past century. At the moment, dozens of papers on this subject have been published, and the variety of reactions performed is simply enormous. The process of [3+2]-cycloaddition of cyclopropanes to the double C=N bond is general in nature and readily occurs under mild conditions. Moreover, it is tolerant to the majority of functional groups, so a wide variety of substrate types can be successfully used in it. These factors determine its widespread use in modern organic synthesis.

One of the first synthetically significant processes was developed by Carreira et al. in 1999 for the synthesis of indole spiro derivatives **184** using MgI_2 as the catalyst and their subsequent utilization in complete syntheses [112]. In 2002, Olsson et al. demonstrated a similar approach for reactions of cyclopropyl ketones **185** with imines also promoted by iodide anions [113]. Both processes have a mechanism that somewhat differs from the majority of modern variants and occur not via a formal 1,3-zwitterionic intermediate, but via the initial nucleophilic addition of an iodide anion, which determines the choice of the Lewis acid. This reaction path is also encountered in a number of later reactions of D–A cyclopropanes (Figures 3.50 and 3.51).

At the same time, the cycloaddition of dimethoxycyclopropanecarboxylates to aldehyde hydrazones in the presence of titanium tetrachloride should be noted as one of the earlier examples (1990). In this case, the reaction involves hydrolysis of two alkoxy groups to give isomeric substituted pyrrolidones in high yields [114] (Figure 3.52).

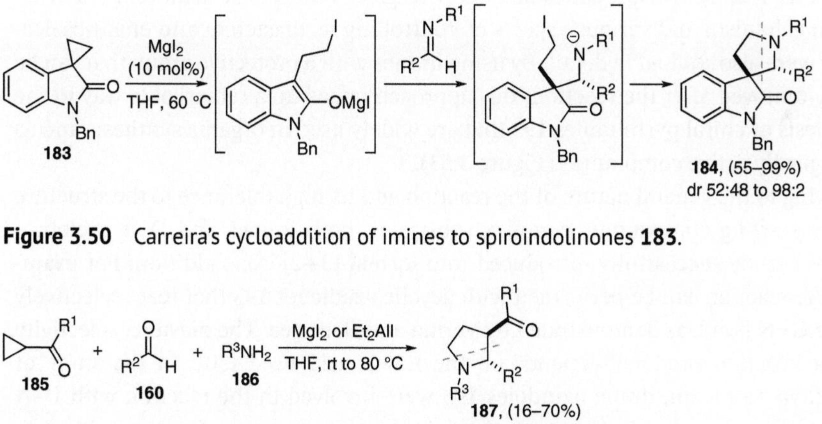

Figure 3.50 Carreira's cycloaddition of imines to spiroindolinones **183**.

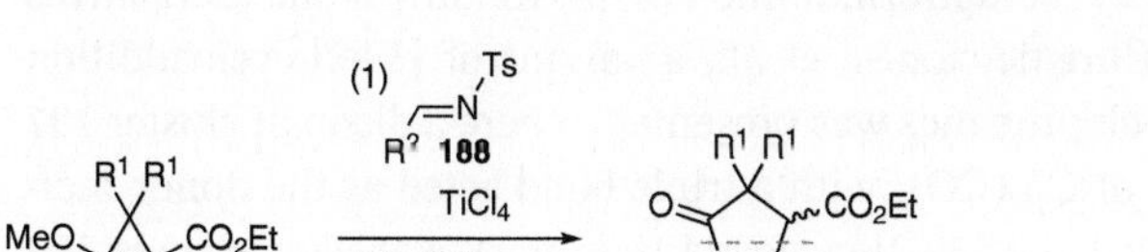

Figure 3.51 Olsson's cycloaddition of imines to cyclopropyl ketones **185**.

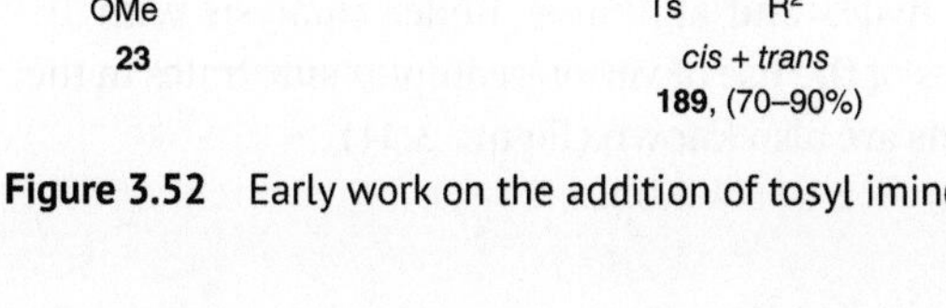

Figure 3.52 Early work on the addition of tosyl imines to activated cyclopropanes.

In subsequent years, the [3+2]-cycloaddition with imines was intensively developed and extended, among other compounds, to cyclopropanedicarboxylates with a donor substituent at position 2 that is most commonly used as classical D–A cyclopropanes **1**. After that, the wide use of this process in organic synthesis began. The widespread advancement of this reaction was facilitated by studies of the teams well-known in this field of chemistry, such as the teams of M. Kerr and J. Johnson, as well as X. L. Sun and Z. Wang, and were later joined by many others [108, 115–118]. The synthetic features and mechanism of the reaction resemble those of the reaction with aldehydes. Moreover, it also has a predominantly high *cis*-selectivity and a very wide scope of starting substrates (hundreds of various combinations are known). Both imines **190** with electron-donor and electron-acceptor substituents readily undergo the reaction. Usually, mild conditions are used with Sc, Yb triflates, magnesium iodide, or similar Lewis acids as the catalysts. Many imines are insufficiently stable, and it is inconvenient to work with them in their individual form. Therefore, the approach in which imines were generated in situ from the corresponding aldehydes and ketones and then immediately reacted with a D–A cyclopropane was found to be very successful [115]. In 2010, J. Johnson et al. developed an efficient asymmetric protocol for this reaction in the most practically useful dynamic kinetic asymmetric resolution (DyKAT) variant, involving racemic cyclopropanes [118]. MgI_2 complexes with chiral PyBOX ligands were used as the catalysts. At the moment, this is one of the most classical protocols for asymmetric reactions of D–A cyclopropanes that allows high *ee* values to be achieved. The reaction mechanism and various aspects of controlling its diastereo- and enantioselectivity were also studied in detail. By using imines with a protective group that can be easily removed after the reaction, this approach opens up a remarkable way to the synthesis of chiral pyrrolidines **192** that are widely used in organic synthesis and as biologically active compounds (Figure 3.53).

Owing to the general nature of the reaction and its high tolerance to the structure of the starting compounds, complex substrates, both imines and D–A cyclopropanes, can be successfully introduced into formal [3+2]-cycloaddition. For example, the reaction can be performed with acyclic azadienes **193** that react selectively at the C=N bond, as demonstrated by Verma and Banerjee. The diastereoselectivity of the reaction products depends on the Lewis acid used [30]. In the study of E. Budynina's team, imine oxindoles **195** were involved in the reaction with D–A cyclopropanes **1** as the imine component. A wide range of cyclopropanes with various acceptor groups have been successfully used in the reactions, which made it possible to obtain a variety of spiro[indoline-3,2′-pyrrolidin]-2-one derivatives **196** [119]. In a study by Christie, Jones, et al., a variant of [3+2]-cycloaddition between imines **160** and cyclopropanes was presented, where a dicobalt cluster **197** formed upon coordination of $Co_2(CO)_8$ with a triple bond acted as the donor moiety [120]. Chai et al. studied the [3+2]-cycloaddition of D–A cyclopropanes **199** based on γ-butyrolactam with aldehydes and aldimines under catalysis with 1% $Al(OTf)_3$ [100]. Many other examples of the use of various complex substrates in the [3+2]-cycloaddition with C=N bonds are also known (Figure 3.54).

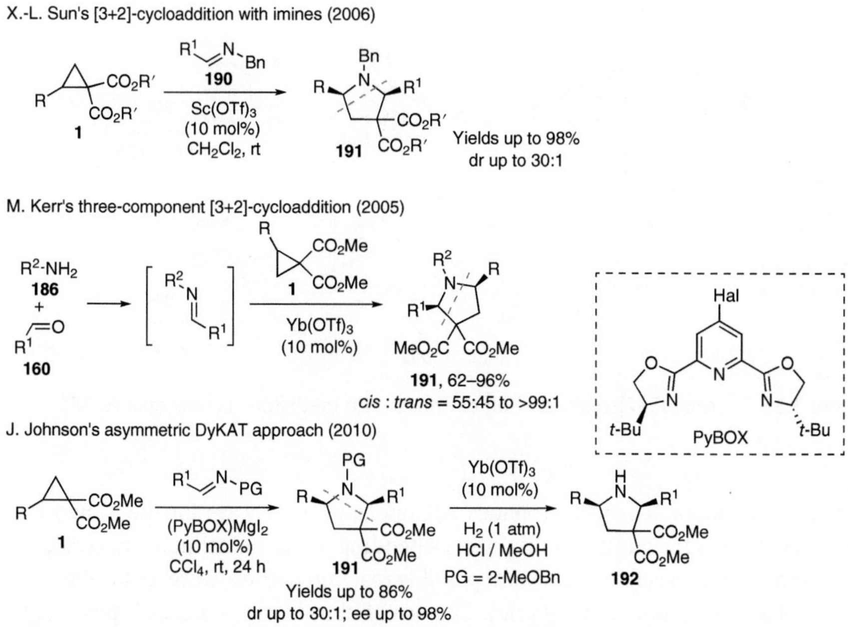

Figure 3.53 Founding works of the [3+2]-cycloaddition of imines with cyclopropane dicarboxylates.

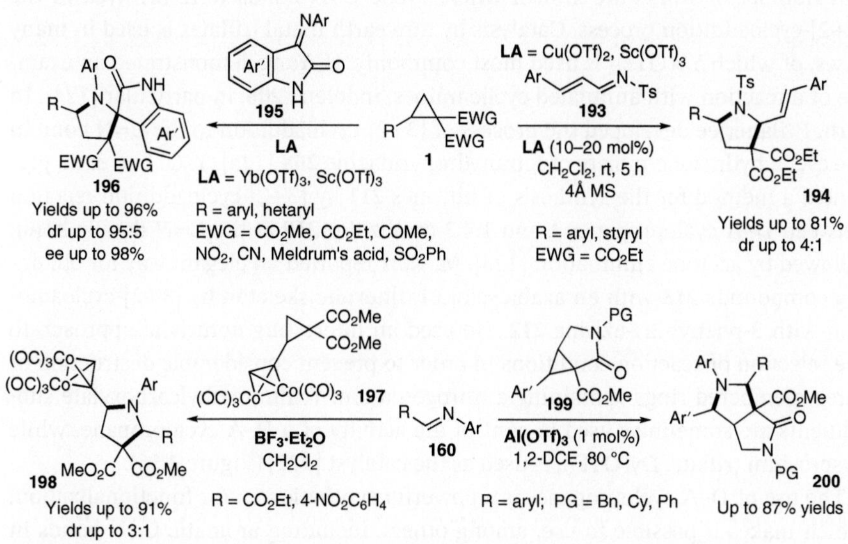

Figure 3.54 Selected examples of the imine formal [3+2]-cycloaddition with complex substrates.

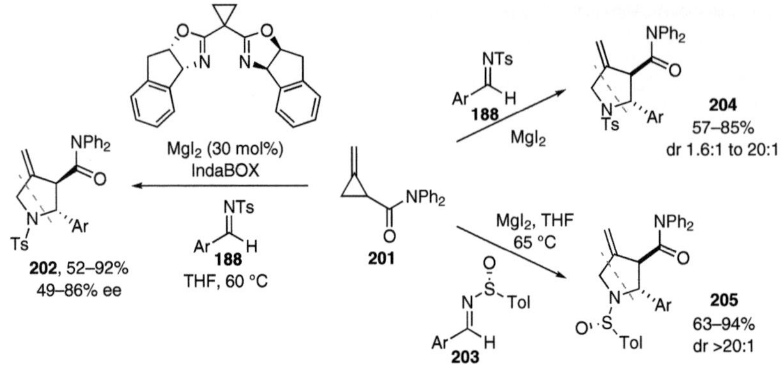

Figure 3.55 Formal [3+2]-cycloaddition of imines with methylene cyclopropanes **201**.

The use of methylenecyclopropanes **201** opens up new ways for their chemical conversion upon the addition of imines. Depending on the substituents in cyclopropane and the Lewis acid used, several possible reaction pathways are realized to give five- and six-membered heterocycles. The use of chiral imines makes it possible to obtain optically active pyrrolidines **202**. Moreover, an asymmetric catalytic variant of this reaction using a chiral IndaBOX ligand was developed. Magnesium iodide is the best catalyst, although aluminum salts can also be used in some cases [121–123] (Figure 3.55).

A number of works are known where cyclic C=N bonds were involved in the [3+2]-cycloaddition process. Catalysis by rare earth metal triflates is used in many cases, of which $Yb(OTf)_3$ is used most commonly. H. Yang demonstrated an example of a reaction with annulated cyclic imines, indolene **206**, in particular [124]. In turn, P. Banerjee developed the process of [3+2]-cycloaddition to the C=N bond in the cyclic hydrazone moiety of tetrahydropyridazine **208** [125]. Y. C. Luo et al. presented a method for the synthesis of nitrones **211** by [3+2]-cycloaddition reaction between D–A cyclopropanes **1** and 1,4,2-dioxazoles **210** at the C=N double bond, followed by acetone elimination [126]. M. Kerr reported an elegant way for obtaining compounds **213** with an azabicyclo[3.1.0]hexane skeleton by [3+2]-cycloaddition with 3-phenyl-2*H*-azirine **212**. He used an interesting nontrivial approach to the selection of reaction conditions in order to prevent considerable destruction of three-membered rings containing a nitrogen atom. Trifluoroethylcarboxylate substituents are sometimes used to control the activity of a D–A cyclopropane, while dysprosium triflate, $Dy(OTf)_3$, is used as the catalyst [127] (Figure 3.56).

The use of D–A cyclopropanes is a powerful synthetic tool for functionalization, which makes it possible to use, among others, including aromatic C=N bonds in heterocyclic systems. Pagenkopf et al. presented an example of [3+2]-annulation of pyridines at the C=N bond to give dihydroindolizine derivatives [70]. S. L. You et al. developed a process for asymmetric annulation of benzothiazoles using chiral complexes of MgI_2 with PyBox ligands [71] (Figure 3.57).

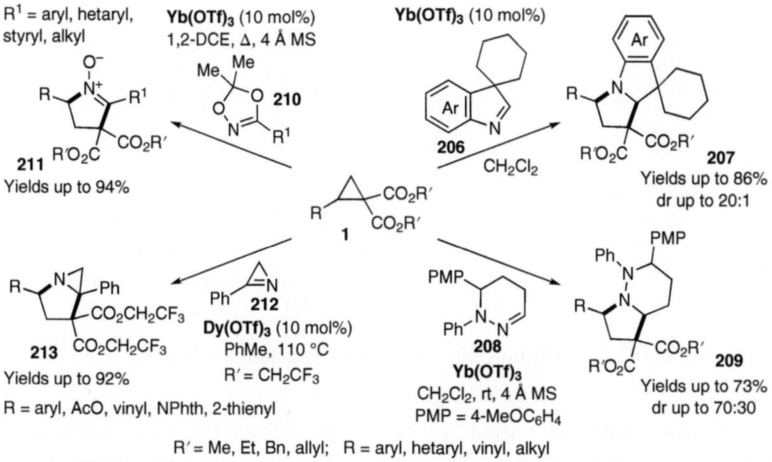

Figure 3.56 Formal [3+2]-cycloaddition with cyclic C=N double bonds.

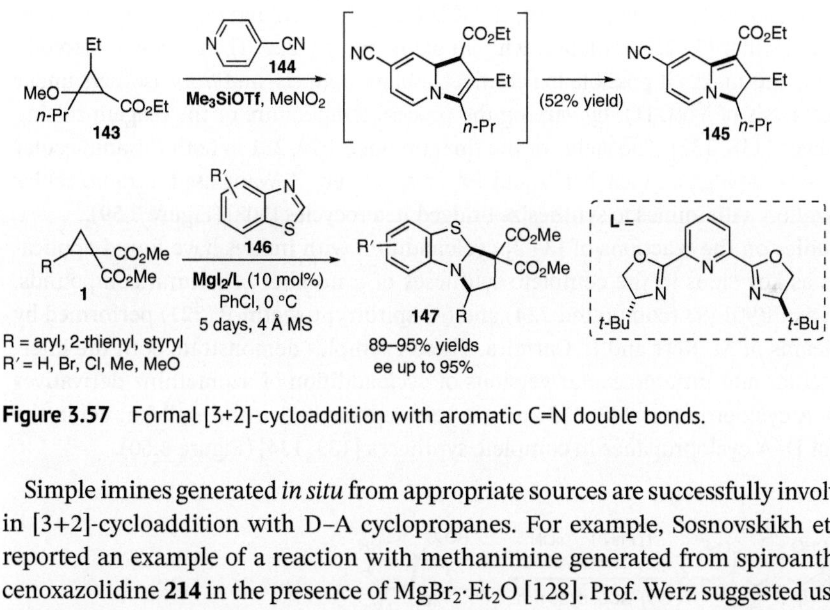

Figure 3.57 Formal [3+2]-cycloaddition with aromatic C=N double bonds.

Simple imines generated *in situ* from appropriate sources are successfully involved in [3+2]-cycloaddition with D–A cyclopropanes. For example, Sosnovskikh et al. reported an example of a reaction with methanimine generated from spiroanthracenoxazolidine **214** in the presence of $MgBr_2{\cdot}Et_2O$ [128]. Prof. Werz suggested using substituted 1,3,5-triazines **216** as sources of imines in the presence of MgI_2 [98, 129]. J. K. Liu et al. also applied 1,3,5-triazines **216** as sources of imines, but under different catalytic conditions, in particular, using $AlCl_3$ as the Lewis acid [130]. It should be noted that in order to make 1,3,5-triazines efficiently act as sources of imines under the reaction conditions, rather a fine selection of a suitable Lewis acid is required since they can also react by other pathways. In general, the [3+2]-cycloaddition processes considered above are tolerant to variations of the initial substrates and the electronic effects of substituents in them, while pyrrolidines **215**, **217** are usually formed in high yields (Figure 3.58).

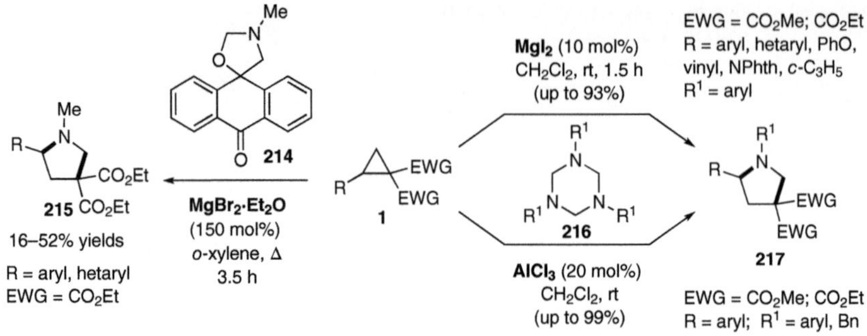

Figure 3.58 Formal [3+2]-cycloaddition with simple imines generated in situ.

Apart from the intermolecular variant of [3+2]-cycloaddition between imines and D–A cyclopropanes considered above, examples of intramolecular transformations also exist. H. Yang presented a method for synthesizing hexahydropyrroloquinoline derivatives **219** by intramolecular cyclization of cyclopropane **218** and an imine moiety in a molecule in the presence of $TiCl_4$ [131]. M. Kerr reported an approach to intramolecular [3+2]-annulation with an oxime moiety **220**. The process is stereodivergent and makes it possible to selectively obtain both *cis*- and *trans*-isomers under the catalysis of $Yb(OTf)_3$ by varying the process temperature or the reagent mixing sequence [116, 132]. The yields of the final products **219**, **221** in both intramolecular processes are high and reach 93% and 99%, respectively. Z. Wang used intramolecular cyclization with imines to synthesize bridged heterocycles [108] (Figure 3.59).

Studies on the reactions of [3+2]-cycloaddition with imines have found applications as key steps in the complete syntheses of a number of natural compounds, such as FR901483 (compound **224**) and (–)-spirotryprostatin-B (**227**) performed by the teams of M. Kerr and E. Carreira. These examples demonstrate both the intermolecular and intramolecular versions of cycloaddition of azomethine derivatives to D–A cyclopropanes **222**, **225** and are very representative and well-known in the use of D–A cyclopropanes in complete syntheses [133, 134] (Figure 3.60).

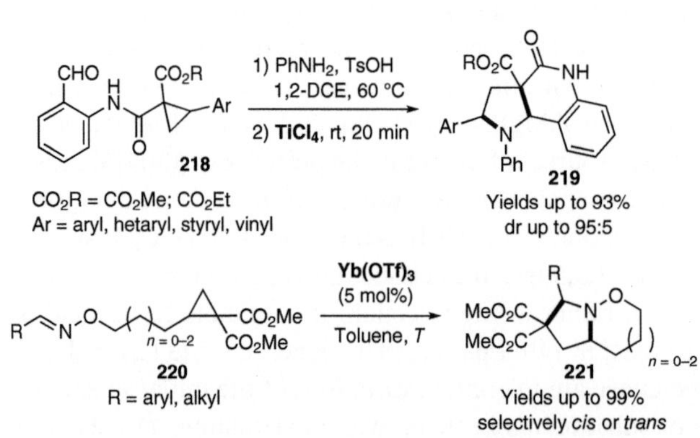

Figure 3.59 Intramolecular formal [3+2]-cycloaddition with C=N double bond.

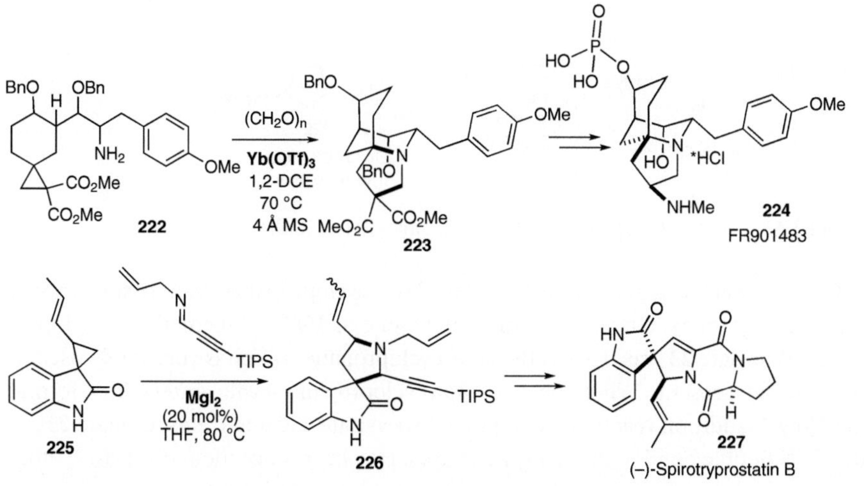

Figure 3.60 Examples of the application of [3+2]-cycloaddition with C=N double bond in total synthesis.

3.4 Formal [3+2]-Cycloaddition with Other Heteroatom X=Y Double Bonds

3.4.1 Formal [3+2]-Cycloaddition with Cumulenes and Heterocumulenes

Despite its apparent relative exoticism, the reactions of D–A cyclopropanes with cumulene and heterocumulene systems have been studied in great detail. (Hetero) cumulenes are very interesting substrates for the construction of carbo- and especially heterocyclic systems. In this case, the overwhelming majority of processes proceed as formal [3+2]-cycloaddition of a 1,3-zwitterionic intermediate generated under the reaction conditions, resulting in the formation of five-membered rings. Reactions with substrates starting with allenes and ending with various heterocumulenes with oxygen, nitrogen, and sulfur atoms have been studied. The first work on reactions with allenes was published in 2004. Studies on the reactions of "classical" D–A cyclopropanes with heterocumulenes started much later, since 2012, and the main work was carried out in 2016–2021. The only example of earlier work is the pioneering work of M. Graziano with 2,2-dimethoxy cyclopropanecarboxylate, carried out back in 1987–1989.

A 2004 paper by V. Yadav and V. Sriramurthy describes a formal [3+2]-cycloaddition between D–A cyclopropanes **98** and allenylsilanes **228** in the presence of Lewis acids such as $TiCl_4$ and Et_2AlCl. The process proceeded with high regio- and diastereoselectivity, leading to the formation of silyl-substituted methylidene cyclopentanes **229** in yields up to 90%. The same work also showed the possibility of formal [3+3]-cycloaddition using this reagent system [135] (Figure 3.61).

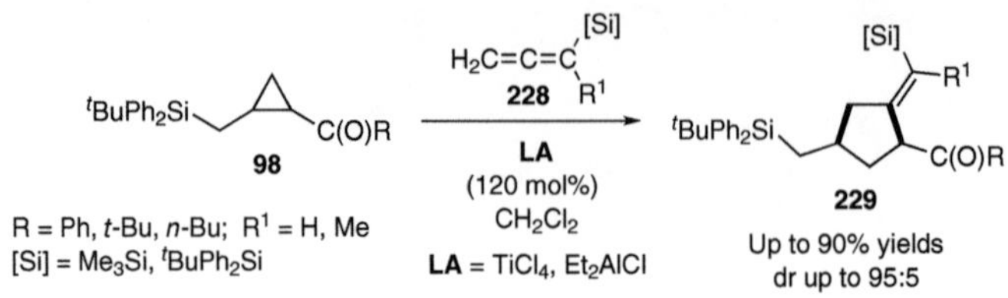

Figure 3.61 Formal [3+2]-cycloaddition with silyl allenes **228**.

The reactions of heterocumulenes with D–A cyclopropanes were first described much earlier, in M. Graziano's pioneering work in 1987. 2,2-Dimethoxy cyclopropanecarboxylate **23** was used as the D–A cyclopropane, which is very rarely used in this role at present. When heated, this cyclopropane easily enters into formal [3+2]-cycloaddition reactions with phenylisocyanate **230** and thioisocyanate **231** at the C=N double bond without any activators. However, since then, until 2012, there have been no mentions of such transformations [136, 137] (Figure 3.62).

Formal [3+2]-cycloadditions with ketenes were recently developed by N. Kerrigan's group. Moreover, the reaction can proceed at both the C=O double bond and the C–C single bond. An efficient regioselective process of interaction between ketenes **234** and standard D–A cyclopropanes **1** at the C=C double bond was developed to form substituted cyclopentanones **235**. The authors used a rather unusual catalytic system of Lewis acids $InBr_3$–$EtAlCl_2$. In addition, optically active starting cyclopropanes with the retention of chirality can be successfully used in this reaction [138]. Also, N. Kerrigan et al. developed a protocol for the interaction of vinyl and styryl D–A cyclopropanes **236** and ketenes **234**, with selectivity for the carbonyl group

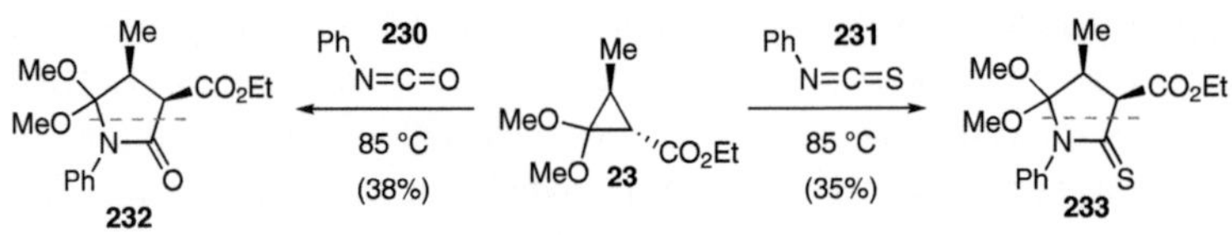

Figure 3.62 First examples of the reactions with phenyl isocyanate and isothiocyanate.

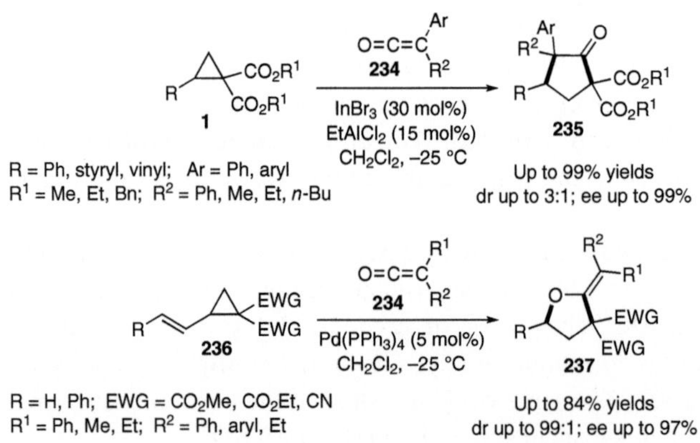

Figure 3.63 Formal [3+2]-cycloaddition with ketenes **234**.

under palladium catalysis conditions $Pd(PPh_3)_4$. The authors also managed to develop an enantioselective version of this reaction [139] (Figure 3.63).

The work of D. Werz's group demonstrates an elegant method for intercepting *in situ*-generated thioketenes **238** using D–A cyclopropanes **1** through formal [3+2]-cycloaddition reactions under the action of scandium(III) triflate. A wide range of D–A cyclopropanes, both with an aromatic fragment and with alkyl and amine substituents, successfully enter into the reaction, proceeding exclusively at the thiocarbonyl group [140]. Subsequent studies by these authors expanded the limits of applicability of the previously developed method and introduced new substituted thioketenes **240** into these transformations [141]. The possibility of formal [3+2]-cycloaddition between phenyl isothiocyanate and a D–A cyclopropane with a cyclopropyl fragment as a donor was also demonstrated [98]. A. Vidal et al., in turn, developed an effective method for the interception of ketenimines **242** using the formal [3+2]-cycloaddition reaction of D–A cyclopropanes **1**, proceeding along the C=N bond [142] (Figure 3.64).

B. Stolz et al. presented examples of formal [3+2]-cycloaddition of D–A cyclopropanes **1** with a number of heterocumulenes – isocyanates **244**, isothiocyanates **245**, as well as carbodimides **246**, which have resulted in a whole range of new, previously undescribed heterocyclic products **247–249**. The developed processes were successfully carried out in the presence of Lewis acids such as $FeCl_3$ and $Sn(OTf)_2$. These processes are very tolerant to the electronic effects of substituents of substrates, proceed with high regioselectivity, and yield end products up to 99% [143]. In a later work by I. Shibata et al., there is an example of a similar interaction of D–A cyclopropanes with vinyl and methylene fragments as donors with isocyanates **244** under catalysis conditions using the $MgBr_2$–Bu_2SnI_2 system [144]. The study of Z. Chai's group added knowledge about the formal [3+2]-cycloaddition of isothiocyanates **245** and carbodimides **246** to D–A cyclopropanes based on γ-butyrolactone under $FeCl_3$ catalysis conditions [145] (Figure 3.65).

In the recent joint work of I. Trushkov's and D. Werz's groups in 2021, it was possible to carry out an interaction between the source of isothiocyanic acid HNCS **250** and a wide range of D–A cyclopropanes **1**. The specificity of this process is that it is

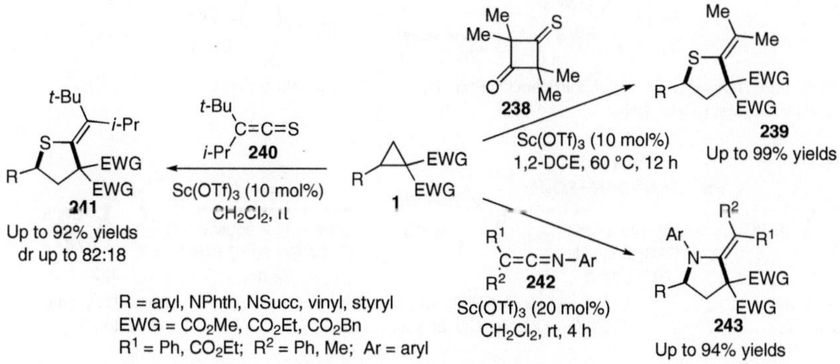

Figure 3.64 Reactions of D–A cyclopropanes with thioketenes and ketenimines.

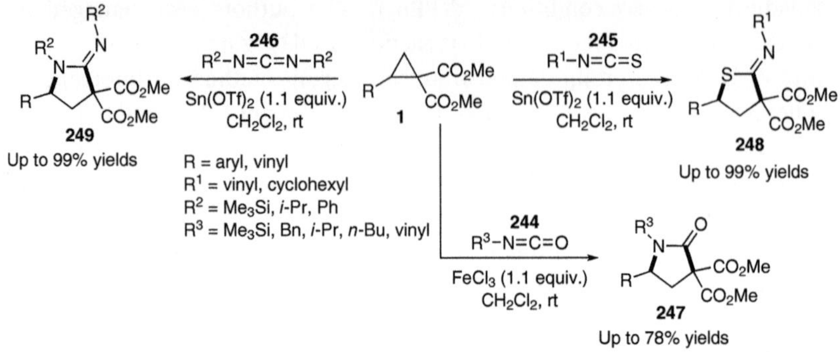

Figure 3.65 Formal [3+2]-cycloaddition with isocyanates, isothiocyanates, and carbodimides.

effectively carried out in an ionic liquid – 1-methylimidazolium thiocyanate, which simultaneously acts both as a source of HNCS and as a catalyst for the process. In this case, the reaction proceeds with exceptional regioselectivity for the C=N bond [146]. Concurrently, in the same 2021, D. Werz's group developed a general approach for the introduction of the S=N double bond into D–A cyclopropanes **1** by the formal [3+2]-cycloaddition reaction, catalyzed by $GaCl_3$ or MgI_2, depending on the substrate. *N*-Sulfinylamines **252** and sulfur diimide derivatives **254** were used as sources of S=N bonds [147] (Figure 3.66).

3.4.2 Formal [3+2]-Cycloaddition with SCN and SeCN

Similar transformations also take place upon the addition of thiocyanates and selenocyanates, which have tautomeric forms with a triple bond C≡N in their structure. This determines the direction of the reaction and differentiates it from the addition of thiocyanic acid, described earlier. P. Singh, A. Goswami et al., proposed organic thio- and selenocyanates **256** as dipolarophiles in the formal [3+2]-cycloaddition reaction with

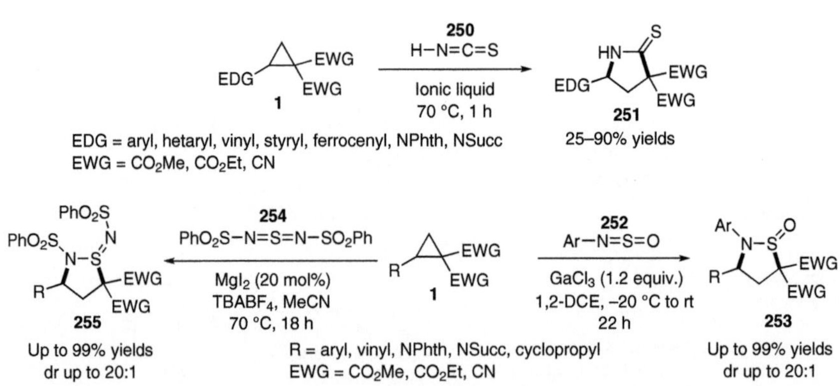

Figure 3.66 Addition of D–A cyclopropanes to thiocyanic acid, sulfinylamines, and sulfur diimides.

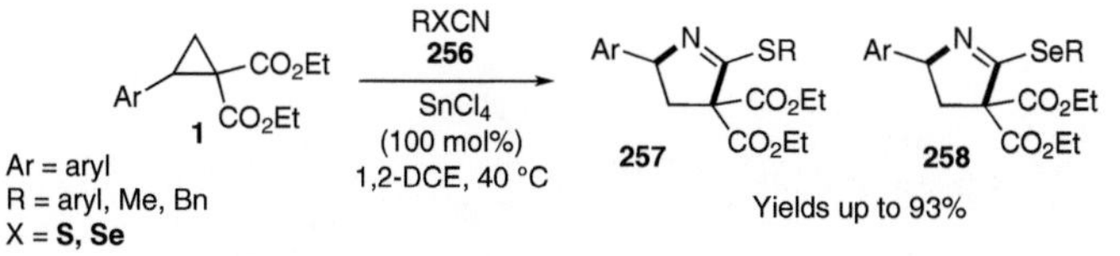

Figure 3.67 Formal [3+2]-cycloaddition with organic thiocyanates **256**.

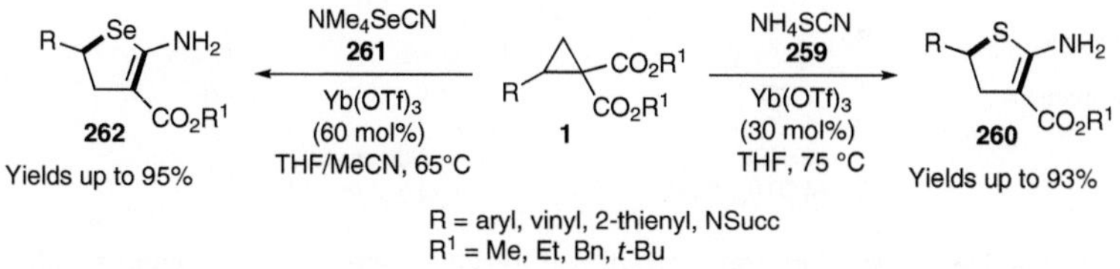

Figure 3.68 Formal [3+2]-cycloaddition with inorganic thiocyanates **259** and selenocyanates **261**.

D–A cyclopropanes **1**, which proceeds through the C≡N triple bond, by analogy with nitriles, in the presence of tin $SnCl_4$ [148]. The processes of addition of inorganic thiocyanates (thiocyanates) **259** and selenocyanates **261** to D–A cyclopropanes **1** were developed in the works of D. Werz's and I. Trushkov's groups. The direction of these reactions differs from organic representatives, and proceeds as nucleophilic addition at the chalcogenide atom, with further cyclization at the C≡N group, as a result of which the chalcogen atom is in the cycle, in contrast to the products of the A. Goswami's group. Also, during the process, one of the ester groups is cleaved off. Inorganic thio- and selenocyanates are quite active substrates and reactions proceed under mild conditions with $Yb(OTf)_3$ catalysis. Reactions with selenocyanates were subsequently used to obtain substituted selenophenes [149, 150] (Figures 3.67 and 3.68).

3.4.3 Formal [3+2]-Cycloaddition with C=S and C=Se Double Bonds

The chemistry of compounds with carbon–chalcogen double bonds in reactions with D–A cyclopropanes is of considerable interest due to heterocyclic products of these transformations that have various practical applications. The first works were carried out quite recently, in 2017, and D. Werz's group can definitely be considered pioneers of research in this field. Thus, the authors studied in detail the reactions with thiocarbonyl and selenocarbonyl compounds **263**, and the corresponding synthetic protocols were developed. These reactions proceed as the classical formal [3+2]-cycloaddition of D–A cyclopropanes **1** to multiple bonds with the transfer of enantioselectivity from the starting cyclopropanes, and aluminum(III) salts are used as catalysts. The reaction is general in nature and tolerant to functional groups, which is proved, for example, by the successful use of thiocarbonyl derivatives of ferrocene **266** [151]. Various cyclopropanes, including nitrogenous, alkyl, and cyclopropyl derivatives, as well as various chalcogen substrates have been successfully used as substrates in this transformation [98, 152]. Reactions of D–A cyclopropanes with tellurium derivatives have not yet been studied (Figure 3.69).

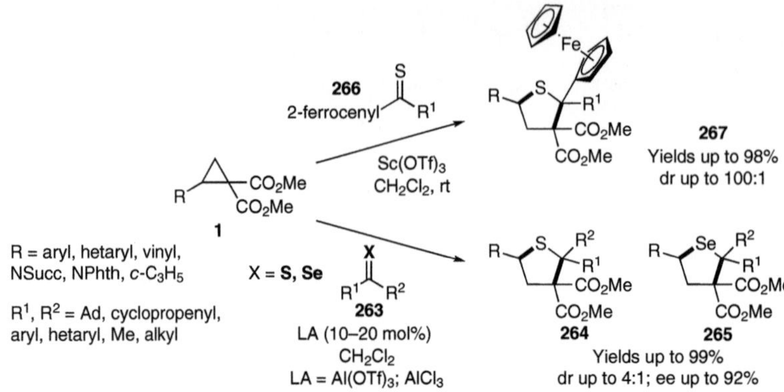

Figure 3.69 Formal [3+2]-cycloaddition with thioketones and selenocarbonyl compounds.

There are also known later works from other research groups in this direction, which have picked up this topic. T. Ohshima et al. succeeded in performing selective [3+2]-cycloaddition at the C=S double bond between various D–A cyclopropanes **1** and thioethers **268**, catalyzed by iron(III) triflate, resulting in the formation of alkoxytetrahydrothiophenes **269** [153]. In another study, H. M. Guo et al. presented an example of [3+2]-cycloaddition of thiourea **270** to D–A cyclopropanes **1**. The $Yb(OTf)_3$–Rb_2CO_3 system acts as a catalyst; the process proceeded with the elimination of one of the acceptor groups, leading to the formation of aminodihydrothiophene derivatives **271** [154] (Figure 3.70).

3.4.4 Formal [3+2]-Cycloaddition with N=O and N=N Double Bonds

Reactions of formal [3+2]-cycloaddition of heteronuclear electrophiles X=Y with the activation of cyclopropanes have been little studied. Only a few works on the involvement of N=O and N=N double bonds in the reaction with D–A cyclopropanes are known. There is also one known work devoted to reactions with an N=S double bond in heterocumulene structures, such as *N*-sulfinylamines and sulfur diimide derivatives (as discussed above in Section 3.4.3).

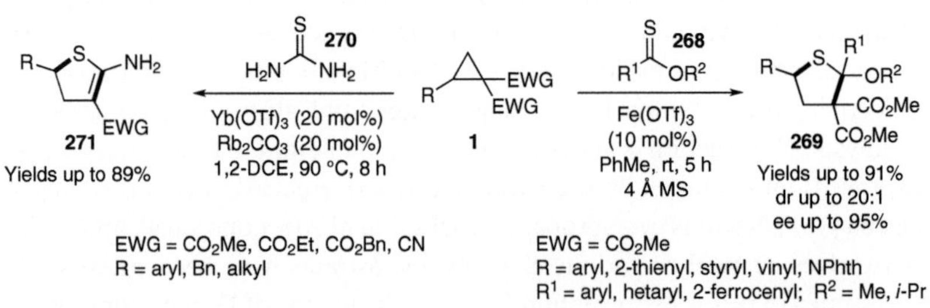

Figure 3.70 Formal [3+2]-cycloaddition with thioesters **268** and thioureas **270**.

Thus, carbon-free dipolarophiles, in particular, nitrosyl chloride, in the reaction of [3+2]-cycloaddition with D–A cyclopropanes **23**, were first proposed by M. Graziano, M. Iesce et al. This process proceeded with good yields of the target isoxazoles **272** in the absence of a catalyst, using highly active cyclopropanes such as 2,2-dimethoxycyclopropanecarboxylates [155]. A study by A. Studer's group presented nitrosoarenes **273** as dipolarophiles under the catalysis of $MgBr_2$. It turned out that many standard Lewis acids, which were used in previous works, give a low yield of the final products **274**, and the electronic effects of substituents in the aryl fragment of the electrophilic substrate have the opposite effect in comparison with the previously described dipolarophiles. Also, nitrosoarenes **273** were able to be involved in more complex cascade transformations [156]. R. Varshnaya and P. Banerjee presented their version of the formal [3+2]-cycloaddition between D–A cyclopropanes **1** and nitrosocarbonyl compounds **275**, generated in situ from hydroxycarbamates in the presence of MnO_2 and MgI_2. Isoxalidines **276** are the products of this reaction [157] (Figure 3.71).

Other X=Y electrophiles in [3+2]-cycloaddition reactions with activated cyclopropanes **23** are also very little studied. There are isolated cases of such reactions for the N=N double bond in diazenes. Thus, dimethoxycyclopropane **23** without any activators easily undergoes cycloaddition reactions with azodicarboxylate **277** and phenyltriazolinedione **278**, resulting in the formation of the corresponding five-membered nitrogen heterocycles **279, 280** with good yields and high stereoselectivity [136, 137]. The study by V. Korotkov and Armin de Meijere, describes azodicarboxylates, azobenzenes, and 4-phenyl-1,2,4-triazoline-3,5-dione (PTAD) **281** in the reaction of [3+2]-cycloaddition with D–A cyclopropanes **1**. It is curious that the authors used gallium trichloride ($GaCl_3$), as a selective and rather specific catalyst, which is not so often used in such transformations. The yields of pyrazolidines **282** are not so high and reach only 63% [158]. Many other types of substrates with N=N double bonds do not undergo formal [3+2]-cycloaddition. For example, PTAD and its analogs usually undergo more complex reactions with fragmentation [159]. Derivatives of 1-pyrazolines are capable of entering into a wide range of different transformations, with the exception of [3+2]-cycloaddition [160] (Figure 3.72).

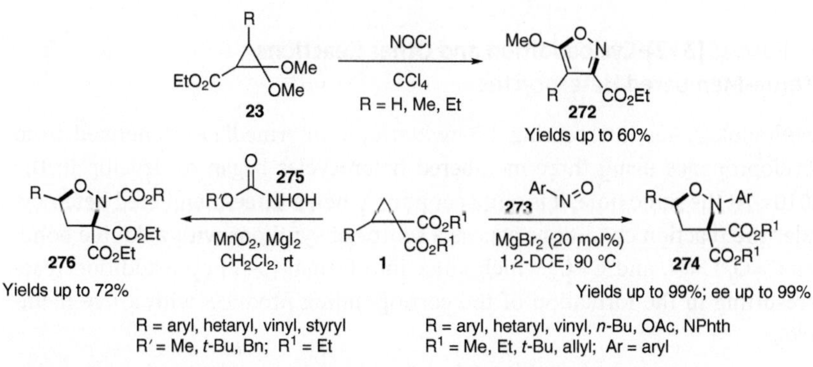

Figure 3.71 Formal [3+2]-cycloaddition with nitroso compounds.

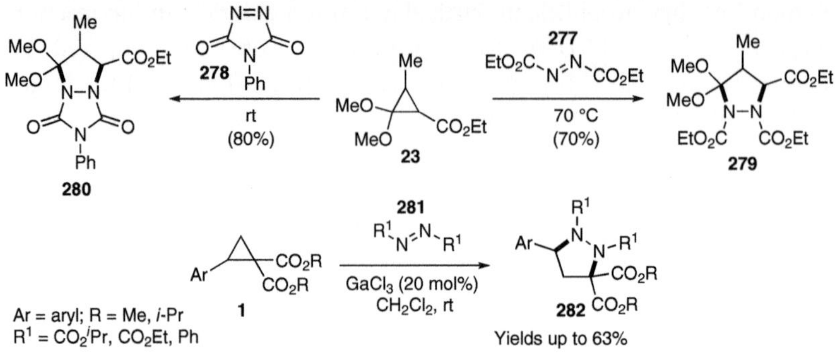

Figure 3.72 Formal [3+2]-cycloaddition with N=N double bond.

3.4.5 Formal [3+2]-Cycloaddition with C≡N Triple Bonds in Nitriles

The B. Pagenkopf's group is a pioneer in the field of using nitriles as dipolarophiles with a C≡N triple bond in formal [3+2]-cycloaddition reactions to D–A cyclopropanes under the catalysis conditions of trimethylsilyl triflate, Me_3SiOTf. In these studies, a wide range of different D–A cyclopropanes **16**, **106**, **128**, and **283** were used, various ways of obtaining heterocyclic systems with nitrogen atoms were discovered, and the application of this approach for the complete synthesis of (±)-goniomycin (compound **293**) was shown [161–166] (Figure 3.73).

The later works of K. Srinivasan's, I. Trushkov's, and Z. Wang's groups supplemented the concept of the reactions of the D–A cyclopropanes **109**, **295** with nitriles, which were initiated by $SnCl_4$ and TfOH. These reactions usually proceed as formal [3+2]-cycloaddition to the C≡N triple bond to form various 3,4-dihydro-2*H*-pyrrole derivatives **294**, **296**. In an attempt to carry out the reaction in an enantioselective variant, a racemate of the final dihydropyrrole was obtained from optically pure cyclopropanes, which is due to the use of strong Lewis acid [167–169]. K. Srinivasan et al. also presented the processes involving the interaction of nitriles **284** with D–A cyclopropanes based on γ-butyrolactone **297** in the presence of $SnCl_4$ as a Lewis acid, to obtain tetrahydro-1*H*-furo[3,4-*c*]pyrrol-1-ones **298** [170, 171] (Figure 3.74).

3.4.6 Formal [3+2]-Cycloaddition and Other Reactions with Three-Membered Heterocycles

The methodology for intercepting 1,3-zwitterionic intermediates generated from D–A cyclopropanes using three-membered heterocycles began to develop in the mid-2010s. At the same time, it is quite common when a three-membered heterocycle under the reaction conditions acts as a source of synthons with a double bond, such as C=O, C=N, and C=C, which enter into formal [3+2]-cycloaddition reactions, resulting in the formation of the corresponding products with a five-membered ring.

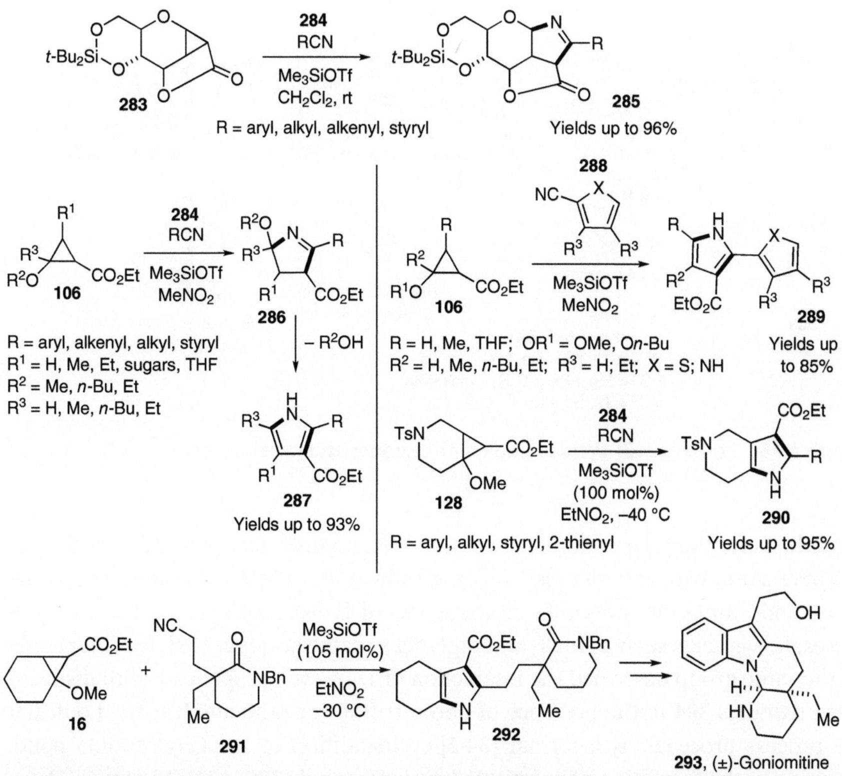

Figure 3.73 Reactions of various D–A cyclopropanes with nitriles developed by B. Pagenkopf group.

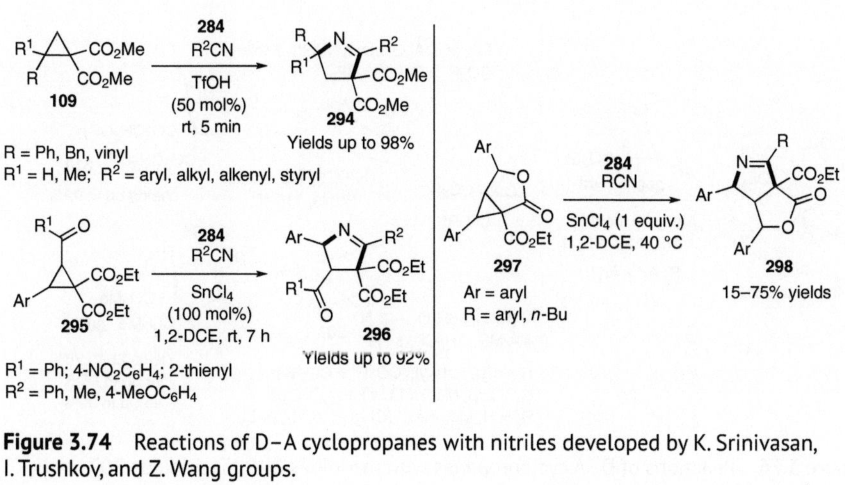

Figure 3.74 Reactions of D–A cyclopropanes with nitriles developed by K. Srinivasan, I. Trushkov, and Z. Wang groups.

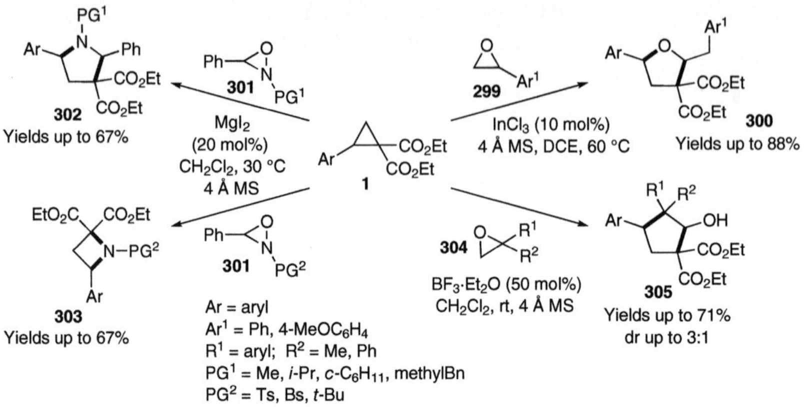

Figure 3.75 Formal [3+2]-cycloaddition with oxiranes and oxaziridines.

Methods for the preparation of substituted tetrahydrofurans **300** and pyrrolidines **302** in reaction with oxiranes **299** and oxaziridines **301**, which act as sources of aldehydes and imines, are presented in the works of P. Banerjee's group. In these processes, Lewis acids such as $InCl_3$ and MgI_2 act as catalysts [172, 173]. Another study by the same group described the interaction of D–A cyclopropanes **1** with disubstituted oxiranes **304** in the presence of boron trifluoride etherate. It turned out that the process proceeds as a formal [3+2]-cycloaddition to the C=C double bond, resulting in the formation of substituted cyclopentanols **305** [174] (Figure 3.75).

More complex cascade processes with aziridines **306**, [175] formal [3+2]-cycloaddition with diaziridines **307**, [176], as well as formal [3+2]-cycloaddition at the C=N double bond with 2*H*-azirines **212** [127] are also known (These transformations are described in more detail in other relevant sections.) (Figure 3.76).

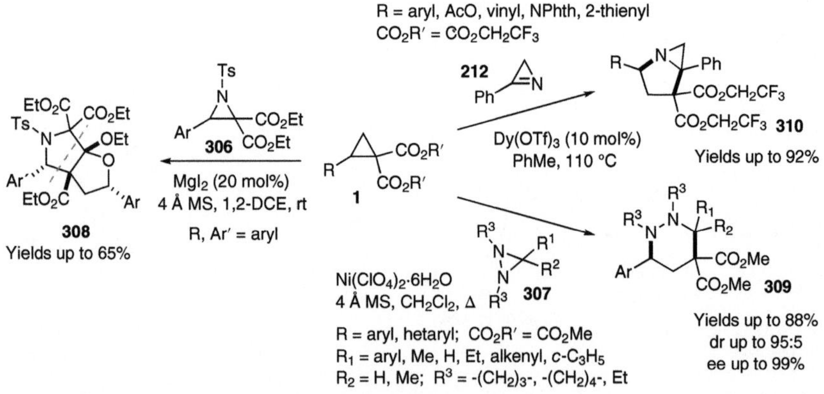

Figure 3.76 Reactions of D–A cyclopropanes with aziridines **306**, diaziridines **307**, and 2*H*-azirines **212**.

3.5 Formal [3+3]-Cycloaddition and Annulation Reactions of D–A Cyclopropanes

3.5.1 General Aspects

Formal [3+2]-cycloaddition reactions of D–A cyclopropanes **1** were known as far back as the last century, beginning in the 1970s–1980s and quite a lot of examples with different types of substrates were known before the 2000s. In contrast to [3+2]-cycloaddition reactions, [3+3]-cycloaddition/annulation reactions of D–A cyclopropanes started to develop much later. Similar reactions involving nitrones and activated cyclopropanes were described by Kerr et al. in a series of papers (links from Section 3.5.5), starting in 2003. And then, these types of transformations quickly gained popularity and turned into a very important synthetic approach for creating six-membered carbo- and heterocyclic systems. At present, [3+3]-annulation reactions of D–A cyclopropanes **1** are one of the main and most representative types of transformations for these, and many dozens of reactions and a huge number of works with various substrates are known. This, however, is not surprising, since the main type of reactivity for D–A cyclopropanes is positioned acting as 1,3-zwitterions. And they perform well in this role in coupling reactions with various other 1,3-dipoles and other compounds acting as 1,3-synthons. These reactions with 1,3-dipoles as nitrones, nitronates, nitrilimines, azomethine imines, and related dipoles, azides, diazo compounds, and so on, as well as various aromatic, heteroaromatic substrates, and substrates with multiple bonds act as three-carbon 1,3-synthons. The mechanisms of such transformations are typical for cycloaddition reactions of D–A cyclopropanes **1** and proceed according to a stepwise ionic mechanism consisting of the following stages (Figure 3.77): Lewis acid coordination/opening of a three-membered ring/nucleophilic addition to the 1,3-dipole/1,6-cyclization, leading to the final ring closure in the product.

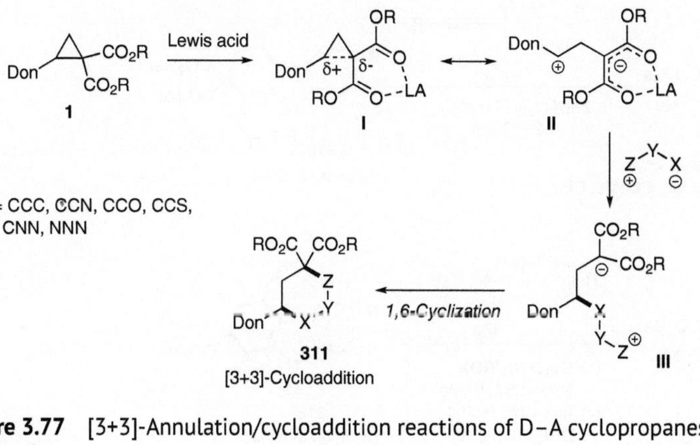

Figure 3.77 [3+3]-Annulation/cycloaddition reactions of D–A cyclopropanes and their mechanism.

3.5.2 [3+3]-Annulation with Aromatic Substrates as 1,3-Synthons

There are many [3+3]-annulation reactions with aromatic substrates that act as 1,3-synthons. In these cases, the annulation occurs at the α-position of the substituent on the aromatic ring and in its *ortho*-position (or a similar position for heteroaromatic substituents). The largest number of reactions is known for indole derivatives, which are especially actively studied because of the high practical attractiveness of the heterocyclic products of such transformations. Thus, in the work of Kerr et al., an example of [3+3]-annulation of cyclopropane dicarboxylates **1** and indoles **312** containing an alkyne fragment in the 2nd position is presented. This process is catalyzed by a rather strong Lewis acid based on Zn^{2+} cations, zinc triflimide $Zn(NTf_2)_2$, in an amount of only 5 mol%, which allows the entire process to be carried out in one stage, although there are other less efficient catalytic systems [177]. Y. Tang describes a similar process for unsubstituted indolylacetylenes **314**, implemented in an enantioselective variant and carried out stepwise in two stages. For this, the following catalytic system is successively used: "$Cu(OTf)_2$/BOX – $InBr_3$/DBU." The authors achieved good *ee* (94%) [178]. M. Ghorai developed a similar method for intercepting the D–A of cyclopropanes **1** using Knoevenagel adducts and nitroolefins with an indole fragment (instead of indolyl acetylenes) **316**. Such transformations proceed as a [3+3]-annulation reaction, are excellently catalyzed by ytterbium(III) triflate, and are well-implemented, including for sterically loaded substrates [179] (Figures 3.78 and 3.79).

In the most recent works of 2021, E. Ruijter described the processes of [3+3]-annulation of indolecarbaldehydes **319**, **322**, and related nitrogenous heterocycles with various types of D–A cyclopropanes **318**, **321** containing a phosphonate acceptor group. The latter extends their synthetic functionality and allows the phosphonate group to be used to create an additional double bond via Horner–Wadsworth–Emmons (HWE) olefination. For this, a double catalytic system "$Sc(OTf)_3$–Cs_2CO_3" is used for aryl-substituted cyclopropanes and a catalytic system based on Pd(0) for

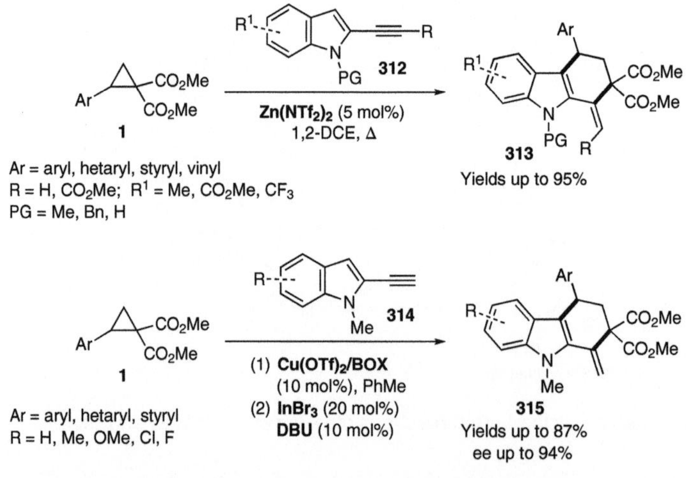

Figure 3.78 [3+3]-Annulation with indolyl acetylenes.

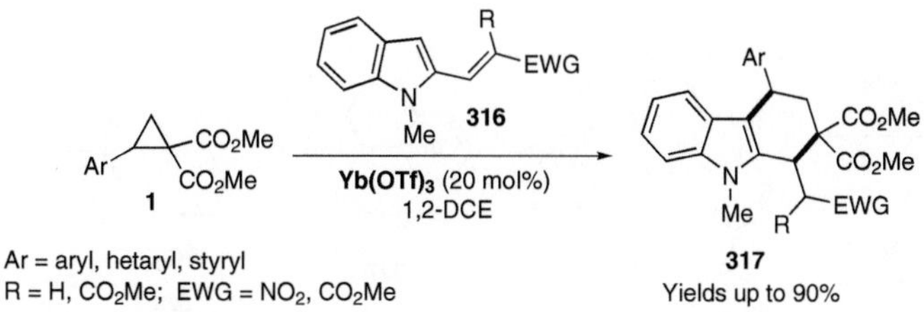

Figure 3.79 [3+3]-Annulation with acceptor indolyl olefins **316**.

vinyl cyclopropanes. The use of various nitrogenous heterocycles in the annulation process makes it possible to implement various types of coupling with cyclopropanes and to carry out annulation both at the NH-group and at the CH-bond of the heterocycle [180, 181]. P. Banerjee, in his study, presented an example of [3+3]-annulation with a rearrangement between D–A cyclopropanes **1** and carbinols with an indole fragment **324**. The process efficiently proceeds with $InCl_3$ catalysis and was accompanied by good yields of final tetrahydrocarbazole products **325**, which can be easily transformed into the corresponding substituted carbazoles [182] (Figures 3.80 and 3.81).

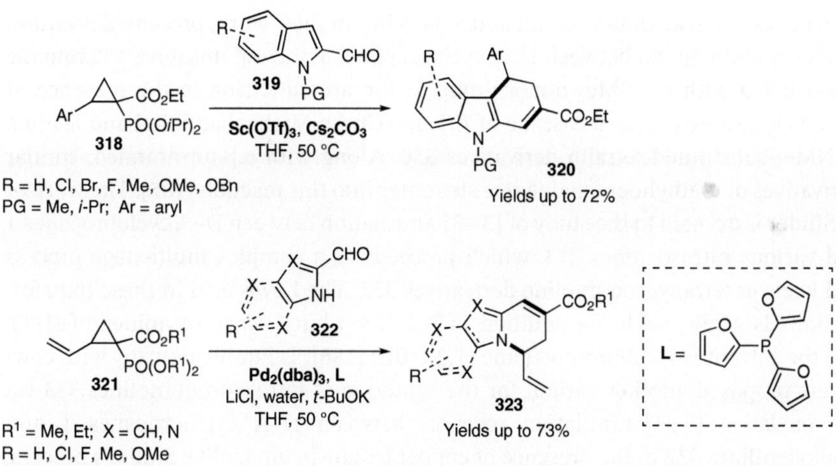

Figure 3.80 [3+3]-Annulation of phosphonate-derived D–A cyclopropanes with indole carbaldehydes.

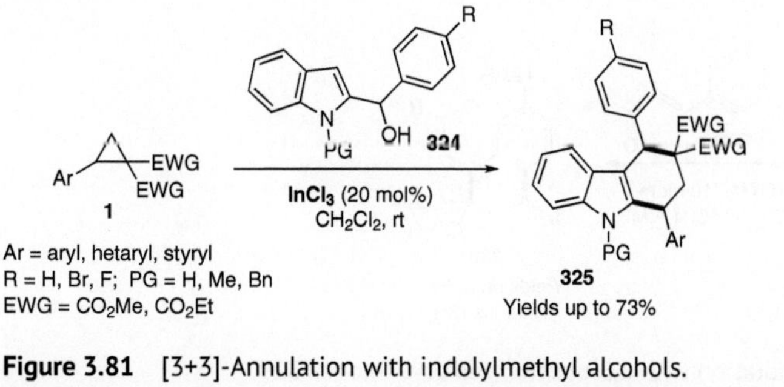

Figure 3.81 [3+3]-Annulation with indolylmethyl alcohols.

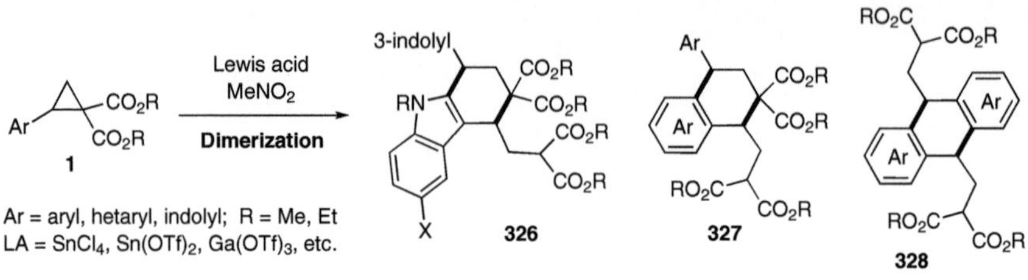

Figure 3.82 Dimerization of D–A cyclopropanes with [3+3]-annulation on aromatic ring.

Many D–A cyclopropanes **1** themselves can also participate in the [3+3]-annulation reaction at the aromatic ring in the absence of substrates, with the involvement of an aryl or heteroaryl substituent in the 2nd position. Such cyclodimerization processes have been actively studied by the group of I. Trushkov and also by the group of Novikov and Tomilov. Moreover, a lot of data on cyclodimerization reactions were obtained on 2-indolyl-substituted cyclopropanes and analogous cyclopropane derivatives of furan and thiophene [183, 184] (Figure 3.82).

Other [3+3]-annulation reactions with aromatic substrates that are not indole derivatives are also known. S. Sin and S. G. Kim, in their work, presented a variant of such an interaction between D–A cyclopropanes **1** and α,β-unsaturated aromatic ketones **329** with an NMe_2 donor group in the aromatic ring in the presence of $Yb(OTf)_3$, proceeding as a cascade of Friedel–Crafts/Michael addition and leading to NMe_2-substituted tetralin derivatives **330**. Along with α,β-unsaturated, similar derivatives of methylidenemalonates also enter into this reaction [185]. The work of A. Studer is devoted to the study of [3+3]-annulation between D–A cyclopropanes **1** and various nitrosoarenes **273**, which proceeds as a complex multi-stage process and leads to tetrahydroquinoline derivatives **331**. The Lewis acid in these transformations is $MgBr_2$ with the addition of 3,5-di-*tert*-butyl-4-hydroxytoluene (BHT), and the solvent is 1,2-dibromoethane (1,2-DBE) [186]. T. Punniyamurthy with coworkers proposed another variant for the synthesis of tetrahydroquinolines **333** via the oxidative [3+3]-annulation reaction between D–A cyclopropanes **1** and *N*-alkylanilines **332** in the presence of copper triflate in air. Unlike Studer's process, the products have a different arrangement of substituents from the original cyclopropane, which is very useful in practical terms [187] (Figures 3.83 and 3.84).

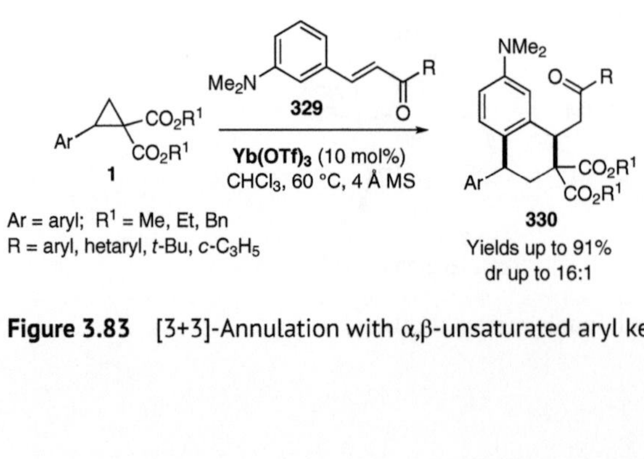

Figure 3.83 [3+3]-Annulation with α,β-unsaturated aryl ketones **329**.

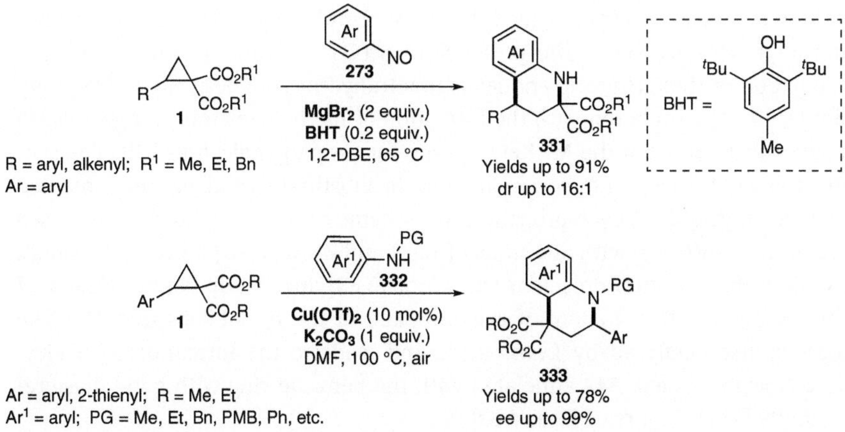

Figure 3.84 [3+3]-Annulation with anilines and nitrosoarenes.

3.5.3 [3+3]-Annulation with Allenes, Allyl, and Propargyl Derivatives

V. Yadav and V. Sriramurthy provided an example of formal [3+3]-cycloaddition of D–A cyclopropanes **98** to allenylsilanes **228** in the presence of Lewis acids such as $TiCl_4$ and Et_2AlCl. The process proceeds with the migration of the silyl fragment of cumulene, resulting in it acting as a three-carbon synthon in the coupling reaction [135]. M. Kerr et al. presented a method for the preparation of substituted methylenecyclohexanes **336** by [3+3]-annulation of D–A cyclopropanes **1** to chloromethyl allylsilanes **335**, which are usually used as 1,3-three-carbon synthons in the synthesis. The process is carried out sequentially in two stages through cyclopropane opening/ring closure. In the first step, $TiCl_4$ is used, and in the second, sodium hydride is used for the final cyclization at the malonyl fragment [188] (Figure 3.85).

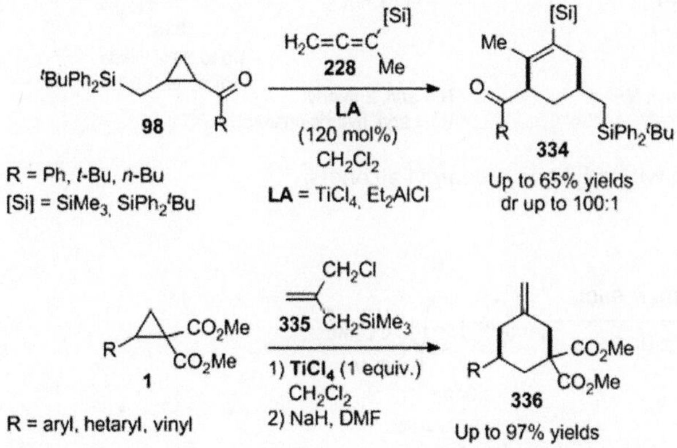

Figure 3.85 Formal [3+3]-cycloaddition with silylallenes **228** and chloromethyl allylsilanes **335**.

K. Mondal and S. Pan investigated the [3+3]-annulations of D–A cyclopropanes **1** and allyl alcohols **337** in the presence of $Sc(OTf)_3$, which proceeds with the participation of the OH group and leads to tetrahydropyran derivatives **338** [189]. A similar [3+3]-annulation with the formation of methylenetetrahydropyrans **340** was described earlier in the M. Kerr's group for propargyl alcohols **339**. The synthetic protocol turned out to be very tolerant to substituents and was implemented for a wide range of D–A cyclopropanes. At the same time, a complex system is used for catalysis – $In(OTf)_3$ with additions of Et_3N and $ZnBr_2$ [190]. Finally, Z. Wang's group described an intramolecular variant of [3+3]-annulation between the allyl fragment and the D–A system of cyclopropane in the initial substrate **341**. The process is also catalyzed by Lewis acids and leads to the formation of a bicyclo[2.2.2]octane system **342** annelated with the benzene ring with good diastereoselectivity [191] (Figures 3.86 and 3.87).

3.5.4 [3+3]-Annulation with Mercaptoacetaldehyde

A series of studies are known on the use of mercaptoacetaldehyde derivatives in reactions with D–A cyclopropanes developed by K. Srinivasan, X. Feng, and J. Zhou. In this case, mercaptoacetaldehyde **343**, introduced into the reaction in the form of a stable dimer (1,4-dithiane-2,5-diol) **344**, acts as a source of 1,3-synthon with a sulfur atom in the processes of formal [3+3]-cycloaddition to D–A cyclopropanes proceeding stepwise, which leads to the formation of derivatives of tetrahydrothiopyran **345, 348, 349**. That is, the main proceeding process in all cases is the same, but the original D–A cyclopropanes differ greatly in nature. The first information on the use of the acetaldehyde **343** as a dipolarophile in these reactions appeared in the work of K. Srinivasan, who used in this role 2-aryl cyclopropanedicarboxylates **295**

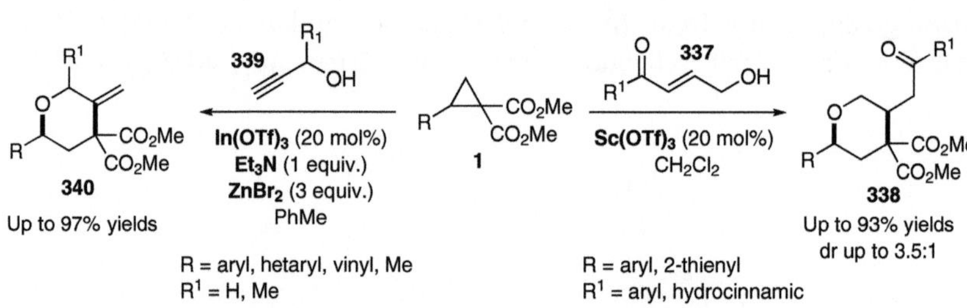

Figure 3.86 [3+3]-Annulation with allyl and propargyl alcohols.

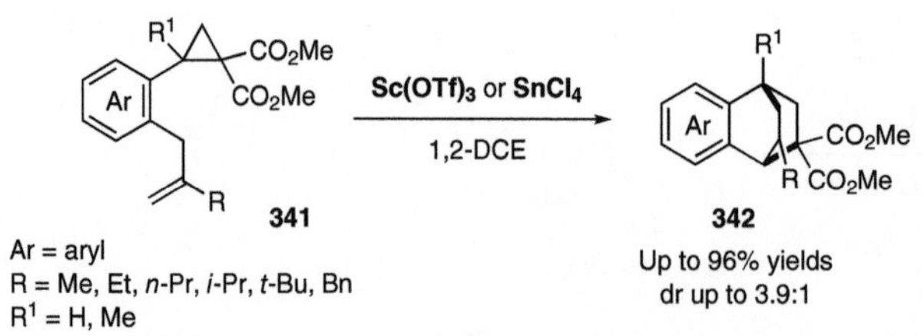

Figure 3.87 Intramolecular [3+3]-annulation with allyl derivatives.

with an additional arylketo group in the 3rd position under the action of $AlCl_3$. The reaction products were used further to obtain substituted thiophenecarbaldehydes **346** [192]. X. Feng managed to carry out the interaction between 2-aryl-cyclopropane-1,1-diketones **347**, and using enantioselective catalysis. The process was catalyzed by scandium(III) triflate and chiral *N,N'*-dioxide as a ligand [193, 194]. J. Zhou's group presented their version of the process using cyclopropane with an oxindole moiety as one acceptor and an *N*-phosphonate group as the other acceptor. This time, indium(III) chloride turned out to be the most effective catalyst [97] (Figure 3.88).

3.5.5 [3+3]-Cycloaddition with Nitrones and Nitronates

Nitrones are particularly good interceptors for the 1,3-zwitterion generated from D–A cyclopropanes, and in many respects, they are classical, since it was from them that the well-known modern chemistry of D–A cyclopropanes began to develop. The first works were made by the Michael Kerr group back in 2003, and this direction quickly gained popularity [195–200]. In general, nitrones are convenient and widespread 1,3-dipoles used in various cycloaddition reactions in organic synthesis. In this regard, D–A cyclopropanes **1** were no exceptions; it turned out that they perfectly entered into [3+3]-cycloaddition processes with various nitrones **350**. The activation of D–A cyclopropanes is usually carried out under mild conditions with Lewis acid catalysis. The most widespread Lewis acids in these reactions include

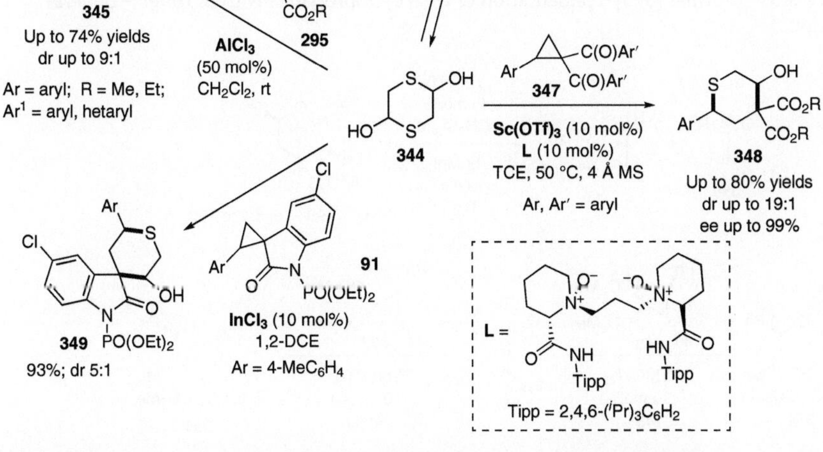

Figure 3.88 [3+3]-Annulation reactions with mercaptoacetaldehyde **343**.

$Yb(OTf)_3$, $Ni(ClO_4)_2$, MgI_2, and others. Usually, the reactions are carried out with acyclic nitrones **350**, which are easily obtained in various ways, for example, from the corresponding aldehydes and hydroxylamines, or by other methods. Tetrahydrooxazine heterocycles **351** are formed as reaction products in high yields and *cis*-selectivity. It is assumed that these transformations proceed according to a stepwise mechanism, like other D–A reactions of cyclopropanes. The use of disubstituted cyclopropane dicarboxylates **353** also exhibits high stereoselectivity. In addition, optically pure starting cyclopropanes can be used in these transformations without any loss of *ee*. The formed tetrahydrooxazine heterocycles **354** are widely used in further syntheses. In addition, in further works, the authors (C. Jasperse group, Y. Tang group) managed to carry out processes in an enantioselective variant using chiral ligands based on oxazolines (Figures 3.89 and 3.90). In 2005, the C. Jasperse group developed the first example of a catalytic asymmetric [3+3]-cycloaddition reaction of D–A cyclopropanes **21** using a Ni catalyst with a chiral Ph-DBFOX ligand. Later, Y. Tang with coworkers developed a slightly different approach to the asymmetric synthesis of tetrahydrooxazines **351** using the same initial substrates, also using nickel catalysis, but with a different tris-oxazoline ligand (TOX) (Figure 3.91) [201] [202]. Johansen and Kerr succeeded in using [3+3]-cycloaddition between D-A cyclopropanes **236** and nitrones **356** to obtain a building block **357** for the synthesis of yuremamine core, which occurs in a number of natural compounds (Figure 3.92). Further work by M. Kerr et al., K. Nolin et al., D. Werz et al., and others expanded the concept of applicability of [3+3]-cycloaddition processes in D–A reactions of cyclopropanes and acyclic nitrones [98, 203–205].

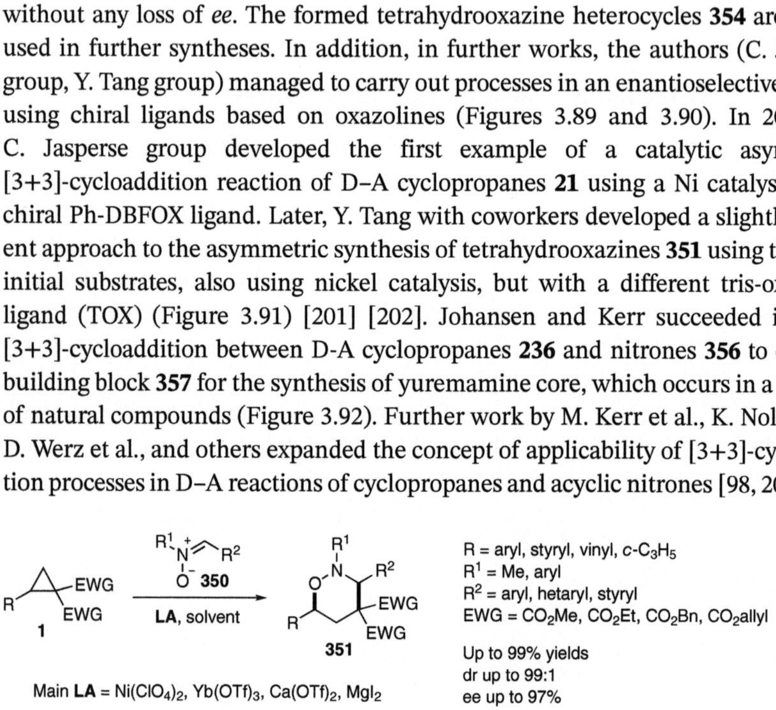

Figure 3.89 Formal [3+3]-cycloaddition of D–A cyclopropanes with nitrones – general reaction.

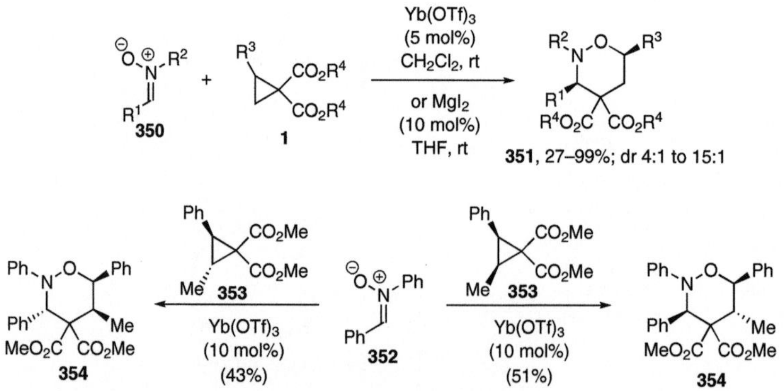

Figure 3.90 Kerr's examples of individual reactions.

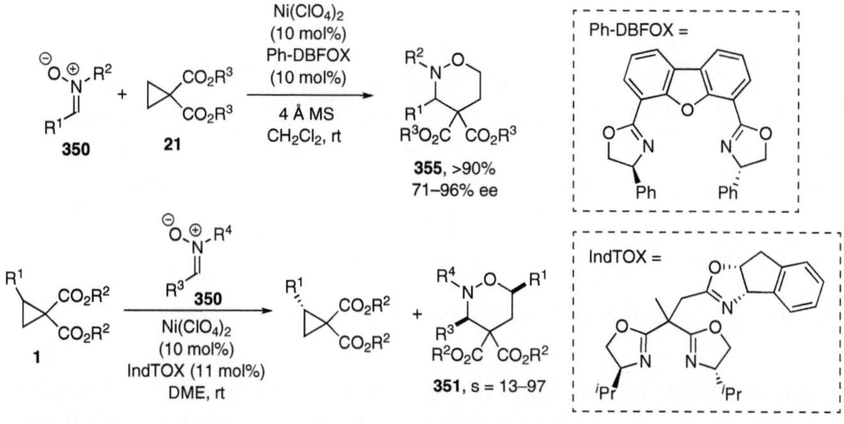

Figure 3.91 Asymmetric approach to formal [3+3]-cycloaddition with nitrones **350**.

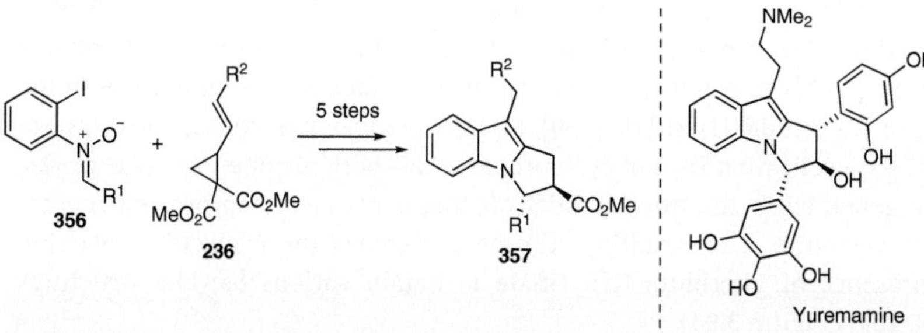

Figure 3.92 Example of application in total synthesis.

As can be seen, the reactions of [3+3]-cycloaddition of nitrones with D–A cyclopropanes have been studied very well since 2003. This is due to the fact that tetrahydrooxazines **351**, the products of these transformations, can be easily transformed with high stereoselectivity into polysubstituted pyrrolidines, which are found as fragments in many natural compounds.

In subsequent years, the methodology of [3+3] cycloaddition of nitrones and D–A cyclopropanes continued to be actively developed and extended to more complex substrates, as well as to applications in applied syntheses. For example, a study by Zhang et al. extended the methodology of [3+3]-cycloaddition of nitrones **350** to polysubstituted D–A cyclopropanes **358** with an alkyne fragment, which is successfully realized in the presence of $Sc(OTf)_3$ and phenanthroline as a ligand [206]. The J. Zhou group used cyclopropanes **91** with an oxindole moiety as one acceptor and an *N*-phosphonate group as another acceptor in the same [3+3]-cycloaddition reaction with nitrones **350**. Nickel(II) triflate proved to be the most effective catalyst, which allows obtaining spiro derivatives of 2-indalinones **360** in high yields [97] (Figure 3.93).

In another study, various methods of *in situ* generation of functionalized nitrones and their intramolecular transformations. Dhote and Ramana's work

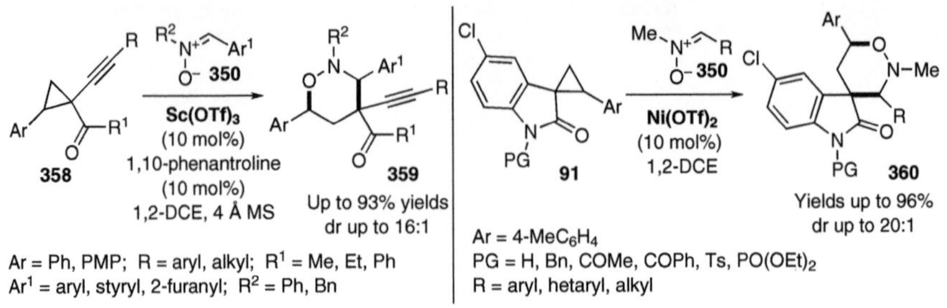

Figure 3.93 Reactions with complex cyclopropane substrates.

demonstrated [3+3] cycloaddition between D–A cyclopropanes **1** and bicyclic nitrones generated via in situ intramolecular cyclization of *ortho*-nitro arylalkines **361** in the presence of $AuCl_3$ [207]. A study by Pagenkopf and the Kerr group demonstrated an elegant method of generating acyclic nitrones through the reaction between D–A cyclopropanes **1** and nitrosoarenes **273** followed by further fragmentation. The generated nitrone interacts with the second cyclopropane molecule by [3+3] cycloaddition reaction. The described process undergoes in the presence of ytterbium(III) triflate [159]. M. Kerr group demonstrated the intramolecular [3+3]-cyclization D–A of cyclopropanes **364** with nitrones. In these works, nitrone is generated *in situ* from an aldehyde fragment in a cyclopropane structure and hydroxylamine. The resulting nitrone undergoes the cyclization reaction in the presence of ytterbium(III) triflate to obtain various bicyclic structures **365** [208, 209] (Figure 3.94).

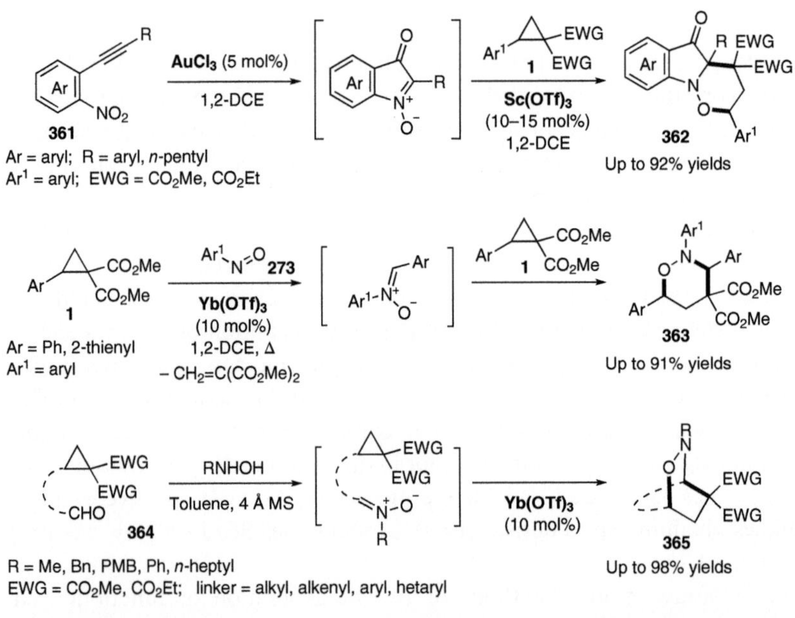

Figure 3.94 Generation of nitrones in situ and intramolecular [3+3]-cycloaddition.

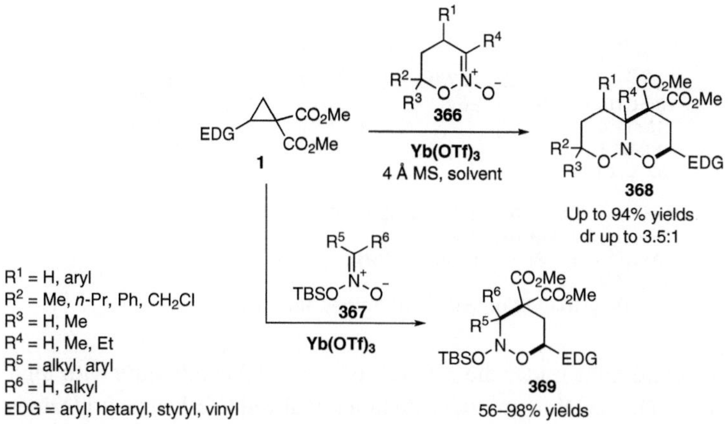

Figure 3.95 Formal [3+3]-cycloaddition with cyclic and acyclic nitronates.

In a series of joint studies by Tabolin, Mikhaylov, Ioffe, Novikov, and Tomilov, general protocols for formal [3+3]-cycloaddition reactions between D–A cyclopropanes **1** and cyclic and acyclic nitronates **366**, **367**, which usually proceed under ytterbium(III) triflate catalysis, were developed. The total coverage of substrates and functional groups used was found to be very broad. The general patterns of reactions are similar to those with nitrones. As a result, various cyclic nitronates and their derivatives are obtained as products, which can be used in further reactions, for example, as 1,3-dipoles, as well as in other transformations, which are very diverse, well-developed, and useful for the subsequent functionalization of nitronate derivatives **368**, **369** [210–212] (Figure 3.95).

3.5.6 [3+3]-Annulation/Cycloaddition with Dinitrogen Substrates

The application of various substrates with two nitrogen atoms as 1,3-synthons in different capacities in formal [3+3]-cycloaddition reactions with D–A cyclopropanes are well studied. The use of hydrazides and their analogs in formal [3+3]-cycloaddition reactions was first proposed by D. Werz et al. under the catalysis with $TiCl_4$ and imidazole for *in situ* generation of nitrile imines as intercepted 1,3-dipoles. Hydrazoyl chlorides **370** of various natures were reacted with cyclopropanedicarboxylates **1** to give tetrahydropyridazine derivatives **371** [213]. In the work of T. Punniyamurthy, another variant of the synthesis of tetrahydropyridazines **373** by means of the formal [3+3]-cycloaddition reaction of D–A cyclopropanes **1** to hydrazones **372** under oxidizing conditions is proposed. This process can be successfully carried out in an enantioselective manner [214] (Figure 3.96).

I. Trushkov's group has developed an elegant process of combining two different types of three-membered rings, proceeding as formal [3+3]-cycloaddition, while generating 1,3-zwitterionic intermediates from both starting substrates. Diaziridines **307**, three-membered rings with two nitrogen atoms, which open in a similar manner, were used as the second component to D–A cyclopropanes **1**. The process takes

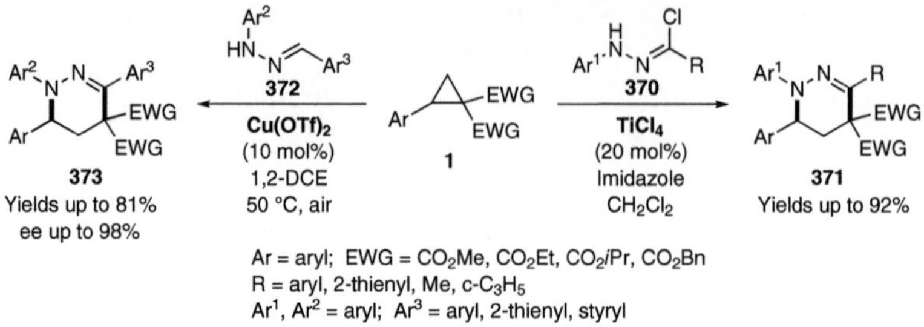

Figure 3.96 Formal [3+3]-cycloaddition with nitrilimines and hydrazones.

place under mild conditions under the catalysis of nickel(II) perchlorate with high diastereoselectivity. The reaction products are functionalized tetrahydropyridazines **309**. It should be noted that diaziridines, in this sense, are the only three-membered heterocycles that enter into formal [3+3]-cycloaddition with cyclopropanes. Other three-membered heterocycles react differently (see Section 3.4.6) [176] (Figure 3.97).

There are also several known cycloadditions with other 1,3-dipoles containing two nitrogen atoms. In the new work of R. Sonawane et al., a variant of the interception of D–A cyclopropane **1** by azomethine imines **372**, proceeding as a formal [3+3]-cycloaddition, is presented. The process requires the use of a strong Lewis acid, $EtAlCl_2$, and the yields of the final products are low, which is associated with the usually dominant alternative routes of interaction [215]. In a study by H. Guo et al., a method of interaction with phthalazinium dicyanomethanides **374** under the action of scandium(III) triflate with the formation of annelated six-membered nitrogen heterocycles **375** is proposed [216]. Other similar heterocycles **377** are obtained by $Ni(ClO_4)_2$-catalyzed formal [3+3]-cycloaddition with aromatic azomethine imines **376**, developed by the A. Charette's group [217] (Figure 3.98).

3.5.7 Formal [3+3]-Cycloaddition with Azides and Diazo Compounds

Azides **378** as 1,3-dipoles for coupling with generated 1,3-zwitterionic intermediates in formal [3+3]-cycloaddition reactions with D–A cyclopropanes **1** were successfully used in the work of P. F. Xu's group. The authors managed to involve a wide range of cyclopropanes in the developed process. To introduce azides into the reaction, stringent conditions are required: $TiCl_4$ is the initiator of this process. The yields of the

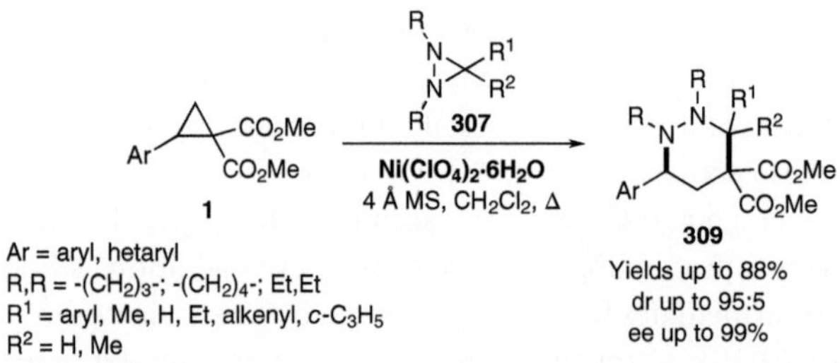

Figure 3.97 Formal [3+3]-cycloaddition with diaziridines **307**.

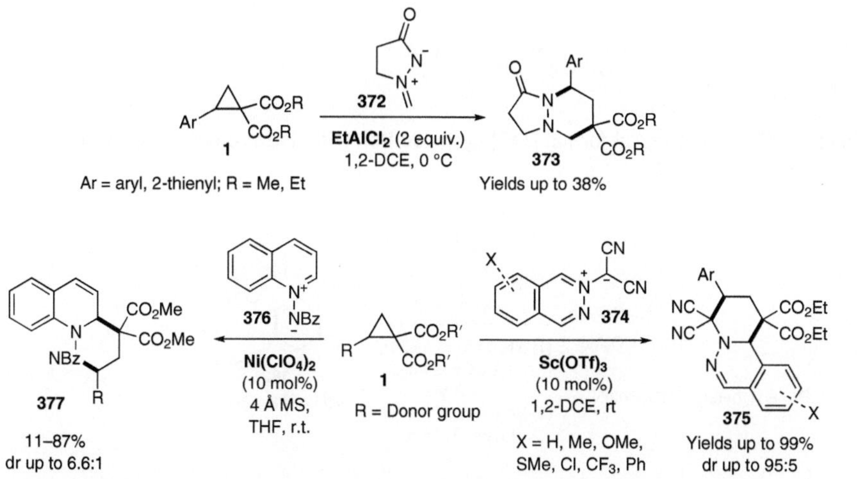

Figure 3.98 [3+3]-Annulation processes of D–A cyclopropanes **1** with azomethine imines **372**, aromatic azomethine imines **376**, and phthalazinium dicyanomethanides **374**.

final tetrahydrotriazines **379** ranged from 49% to 94%. At the same time, during the reaction, the extrusion of the nitrogen molecule does not occur, although this can be achieved if necessary [218] (Figure 3.99).

Diazo compounds are other commonly used 1,3-dipoles in cycloaddition reactions, but they are not nearly as good as azides. Despite their active study in reactions with D–A cyclopropanes **1**, so far it has not been possible to obtain tetrahydropyridazines **381** or their isomeric forms, which are the simplest reaction products of the formal [3+3]-cycloaddition of a diazo compound **380** as a 1,3-dipole. In many respects, the problem lies in the instability of the diazo compounds themselves under the conditions of inherent catalysis by Lewis acids. But as a result, many other directions of interaction with diazo compounds were realized [219, 220]. Thus, M. Doyle, a distinguished specialist in diazo compounds, presented an approach to another formal [3+3]-cycloaddition of silyl ethers of enols **58** containing a diazo group and D–A cyclopropanes **1**. The process proceeds efficiently in the presence of the $Yb(OTf)_3$–$Rh_2(cap)_4$ catalytic system and is sufficiently tolerant to the electronic effects of donor substituents in cyclopropane. At the same time, diazoester formally acts as a source of three-carbon synthon with the elimination of the nitrogen molecule. In reality, the process proceeds in two stages: through the initial addition of cyclopropane to the double C=C bond and further rhodium-catalyzed rearrangement [33]. Another very elegant way of using diazocarbonyl compounds **383**, **384** in

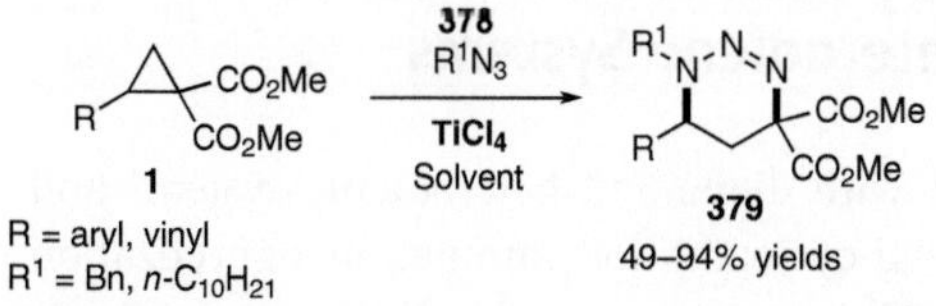

Figure 3.99 Formal [3+3]-cycloaddition with azides.

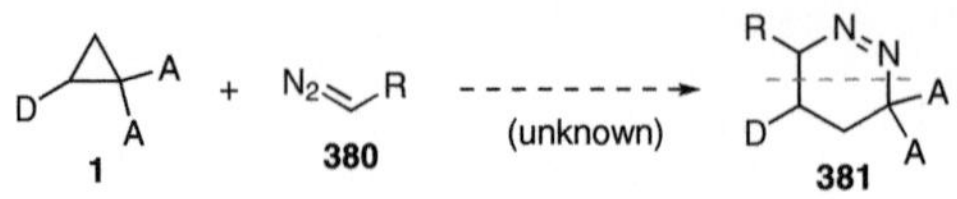

Figure 3.100 Hypothetical formal [3+3]-cycloaddition with diazo compounds, not yet experimentally realized.

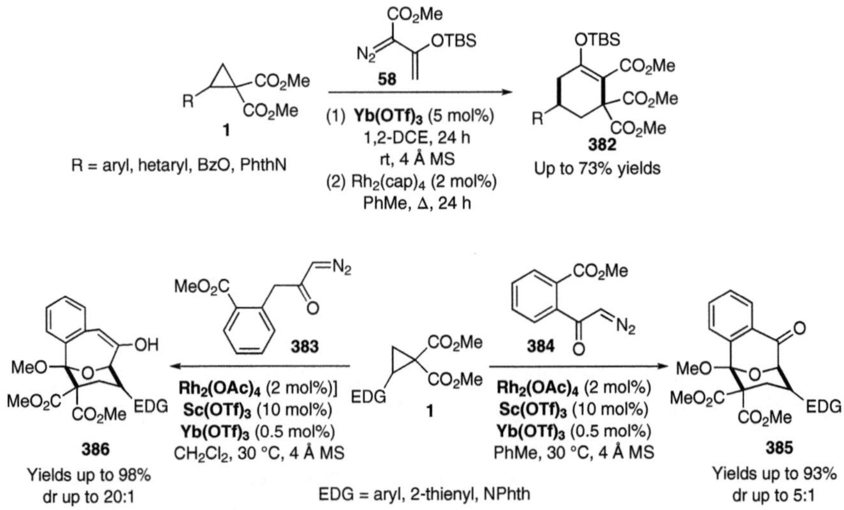

Figure 3.101 [3+3]-Annulation processes with different 1,3-synthones generated from diazo compounds.

cycloaddition reactions with D–A cyclopropanes **1** is presented in the work of D. Werz's group. Diazocarbonyl compounds **383, 384** are used to generate carbonyl ylides – 1,3-zwitterions with an oxygen atom in the middle, which are introduced into further formal [3+3]-cycloaddition with cyclopropanes. To generate carbonyl ylides, an intramolecular process of diazoketone decomposition is used, which requires catalytic amounts of Rh(I), and therefore, for the reaction as a whole, a double catalytic system $Rh_2(OAc)_4$–$Sc(OTf)_3$ is used, similar to what Doyle used. Even better ones are achieved by doping scandium triflate by 5% ytterbium. The implementation of this process leads to the synthesis of polysubstituted tricyclic pyrans **385, 386** [221] (Figures 3.100 and 3.101).

3.6 Reactions of Formal [4+3]-Cycloaddition and Annulation with Diene and Heterodiene Systems

The reactions of D–A cyclopropanes **1** with diene and heterodienic systems and their analogs, which occur as formal [4+3]-cycloaddition, annulation, or cyclization reactions, are of particular interest since they open new approaches to the synthesis of a wide range of various seven-membered cyclic structures that are not so easy to

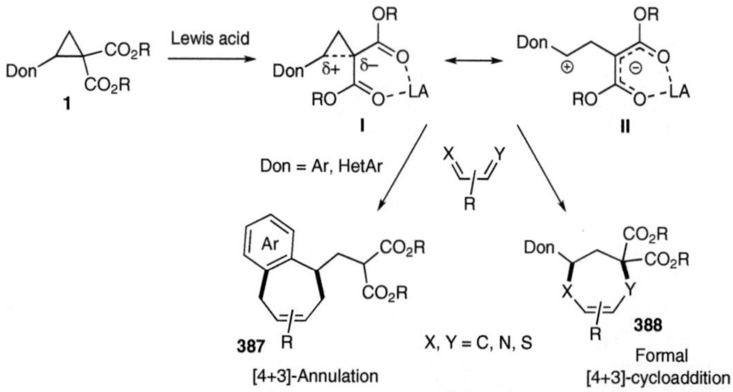

Figure 3.102 Formal [4+3]-cycloaddition and annulation reactions of D–A cyclopropanes with dienes/heterodienes.

obtain even now. At first approximation, the formal [4+3]-cycloaddition is a homo-analog of the Diels–Alder reactions which, however, nearly always occurs stepwise by an ionic mechanism [222]. In this case, the D–A cyclopropane acts as a 1,3-zwitterionic synthon that is coupled with a diene system. In terms of the mechanism, these reactions are typical of D–A cyclopropanes and occur by opening the three-membered ring to give a 1,3-zwitterionic intermediate; hence, they require catalysis with Lewis acids. Conceptually, further, there are two main pathways of [4+3]-coupling, that is, with cyclization on the malonyl moiety or with annulation at the aromatic ring. In the latter case, an annulated system with a seven-membered ring is formed [223]. The [4+3]-cyclization reactions of D–A cyclopropanes often occur in good yields and under mild conditions; therefore, they are very attractive for building seven-membered rings (Figure 3.102).

3.6.1 Dienes as Traps for 1,3-Zwitterions

The first conceptual data on the reactions of dienes and D–A cyclopropanes **1** were obtained by Trushkov et al. in 2008. Isobenzofuran **391**, which is well known in this role and is widely used as a highly reactive diene system in organic synthesis, was used as a cyclic diene in the reactions with D–A cyclopropanes **1**. The process occurs as [4+3]-cycloaddition, ytterbium(III) triflate is used as the process catalyst, and the yields of the final products **392** reach 92%. It should be noted that, as a rule, D–A cyclopropanes **1** with donor groups in the aromatic moiety successfully react in this process [222, 224]. In another work published by the same team, a formal [4+3]-cycloaddition reaction of D–A cyclopropanes **1** and other cyclic diene systems **389**, anthracene and tetracene in particular, in the presence of $TiCl_4$ was developed. The process was carried out with a series of aromatic and heteroaromatic D–A cyclopropanes, resulting in fairly good yields of final products **390**. In the case of D–A cyclopropanes **1** with a thiophene moiety or several methoxy substituents at the aromatic ring, cyclization also involves these moieties in the D–A cyclopropane [225] (Figure 3.103).

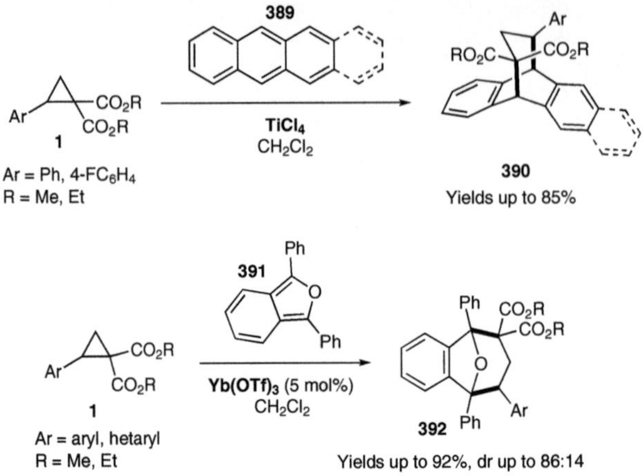

Figure 3.103 Formal [4+3]-cycloaddition of D–A cyclopropanes to anthracene/tetracene and isobenzofuran.

Y. Tang et al. presented a more general version of formal [4+3]-cycloaddition between D–A cyclopropanes and conjugated dienes containing a donor $OSiPh_2{}^tBu$ group (OTBDPS), including cyclic and acyclic dienes **393**, **394**. They succeeded in developing an efficient cyclization process, also in the enantioselective version, using Cu(II) salts as the catalyst and cyclohexyl-TOX or BOX as the chiral ligands. The yields of the target products **395**, **396** reached 96%, and this process is mainly successful with cyclopropanes **1** containing donor substituents at the aromatic ring [226]. It should be noted that many simple cyclic and acyclic dienes without an activating donor group do not enter [4+3]-cyclization with D–A cyclopropanes but react like ordinary alkenes in simpler [3+2]-reactions [28] (Figure 3.104).

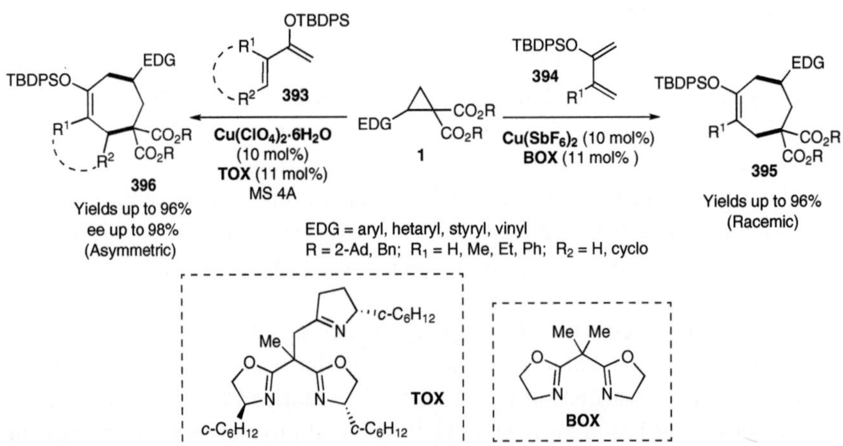

Figure 3.104 Cu(II)-catalyzed formal [4+3]-cycloaddition with dienes and its asymmetric version.

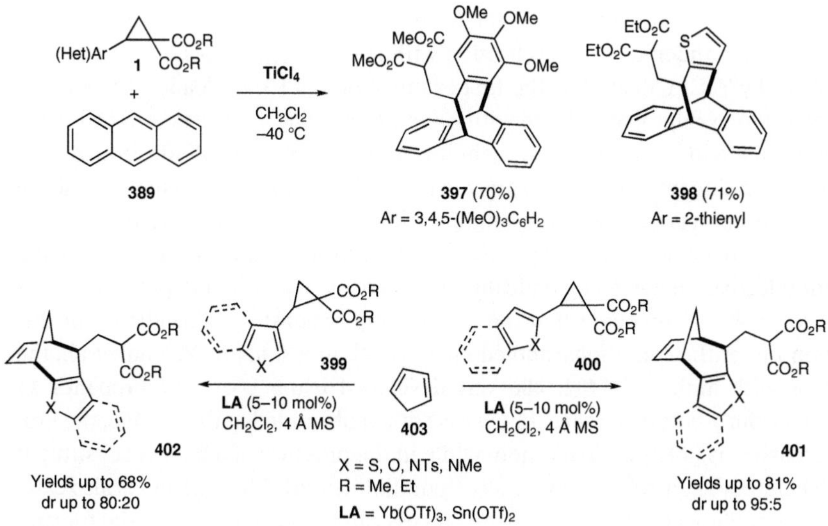

Figure 3.105 [4+3]-Annulation at aromatic rings with anthracene and cyclopentadiene.

Apart from the direct formal [4+3]-cycloaddition that is a homo-analogue of the Diels–Alder reaction, the related less common [4+3]-cyclization of D–A cyclopropanes with annulation at the aromatic ring was also implemented. In particular, the [4+3]-annulation with cyclopentadiene **403** under "mild" catalysis with $Yb(OTf)_3$ or $Sn(OTf)_2$ was studied in detail. As a rule, this reaction involves D–A cyclopropanes **399**, **400** with a heterocyclic moiety as a strong donor. The cyclization direction can be controlled by varying the position of the cyclopropyl moiety in the heterocycle; in this case, more complex cascade reactions giving tri- and tetracyclic annulated systems can be performed. The yields of the final products **401**, **402** are up to 81% [223]. In the case of D–A cyclopropanes **1** with a thiophenyl substituent or with a few methoxy groups at the phenyl ring, a similar [4+3]-cyclization with annulation at the aromatic ring of a D–A cyclopropane can be carried out using anthracene **389** as the diene component (Figure 3.105).

3.6.2 Reactions of [4+3]-Cyclization with Heterodiene Systems and Their Analogs

Since formal [4+3]-cycloaddition reactions with 1,3-dienes were found to occur very well with D–A cyclopropanes, quite a few approaches were developed in the subsequent years (2016–2021) to perform [4+3]-cyclization reactions with the participation of heterodiene systems since this is a very convenient and attractive tool for creating seven-membered heterocycles, many of which are not easy to obtain by alternative methods. Suitable heterodiene systems are structures with nitrogen and sulfur atoms in various positions, such as substrates as cyclic azadienes, anthranils, thiodienes, and bisthioquinones. Related cyclization reactions can also occur with triazines, salicylic, and anthranilic aldehydes. On the other hand, α, β-unsaturated ketones do not undergo [4+3]-cyclization with D–A cyclopropanes.

A study by Y. C. Luo et al. is the first example of a [4+3]-cycloaddition reaction between D–A cyclopropanes **1** and anthranils **404** as cyclic azadienes. A broad range of anthranils and cyclopropanes, including those with an alkyl substituent as a donor, are readily involved in this process to give final products **405** in yields above 90%. The reaction is catalyzed by scandium triflate $Sc(OTf)_3$, which is a typical catalyst for these reactions. The authors also successfully performed this reaction in the enantioselective version using optically active starting D–A cyclopropanes **1** for this purpose. It should be noted that the use of simpler substituted azadienes in this reaction under the conditions studied failed [227]. In a study by Z. Chai et al., the reaction with anthranils **404**, **406** was developed using D–A cyclopropanes **11** obtained from γ-butyrolactones. In the case of unsubstituted anthranil **406**, the process occurred as [4+3]-cycloaddition, while in the presence of a phenyl substituent in the heterocycle, it resulted in tetrahydroquinolines **407**. The reaction is also catalyzed by scandium(III) triflate, and the process can be carried out with the transfer of chirality of the starting D–A cyclopropane [228] (Figure 3.106).

Of particular interest are the reactions with thiodienes developed by D. Werz's team, which open access to seven-membered sulfur-containing heterocycles. For example, thiochalcones **409** were successfully used in [4+3]-cycloaddition to D–A cyclopropanes **1**. The process is catalyzed by $Sc(OTf)_3$. It is tolerant to various functional groups in D–A cyclopropanes and thiochalcones, and the yields of the final tetrahydrothiepins **410** reach 87% [229]. In another work, the researchers succeeded in performing formal [4+3]-cycloaddition reactions with D–A cyclopropanes **1** using benzodithioloneimines **411** as sources of cyclohexadienedithiones **413** that served as dipolarophiles in the reaction. The process is of general scope. The electronic effects of substituents in the starting substrates weakly affect the process,

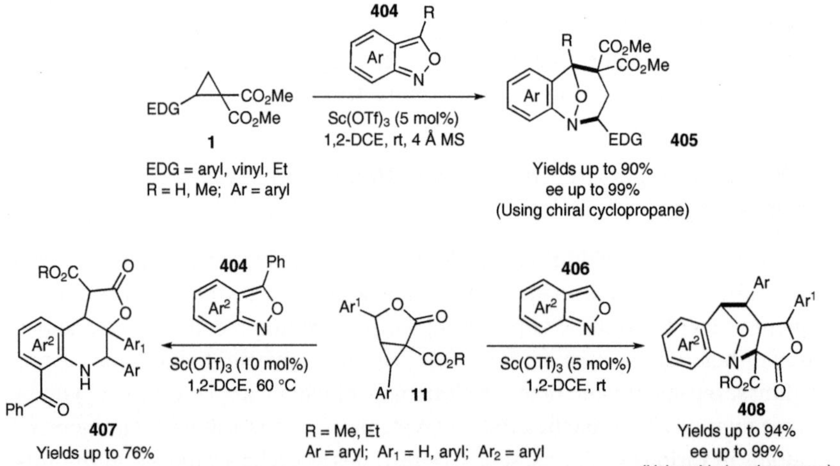

Figure 3.106 Formal [4+3]-cycloadditions with anthraniles.

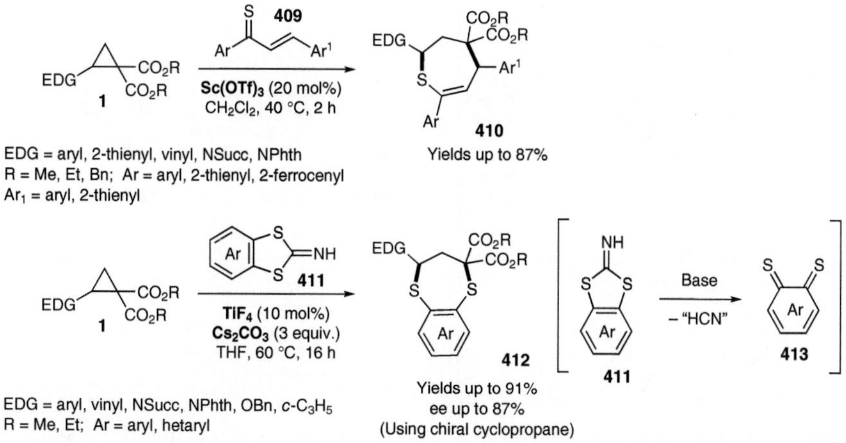

Figure 3.107 Formal [4+3]-cycloadditions with thiodienes and bistioquinones.

which is a great advantage. The reaction results in dihydrobenzodithiepine derivatives **412** in up to 91% yields. A double catalytic system "TiF_4 – Cs_2CO_3", which is unusual in the chemistry of D–A cyclopropanes, is used as the catalyst. It should also be noted that the reaction occurs with the retention of the asymmetric center configuration in the initial cyclopropane [98, 230] (Figure 3.107).

In other studies published by D. Werz et al., they reported an interesting example of cyclization of D–A cyclopropane **1** with triazines **216**, catalyzed by $Sc(OTf)_3$. It should be noted that the process occurs as a formal [4+3]-cycloaddition, with the formal dimer of *N*-arylmethaneimine **415** acting as the 1,4-synthon. A wide range of both D–A cyclopropanes and triazines were involved in the process. In these cases, the yields of the final products were up to 94%. 1,3-Diazepines **414** were subsequently used in the aminal cleavage process in order to obtain the products of 1-amino-3-aminomethylation of the starting D–A cyclopropanes **1** [231] (Figure 3.108).

A study by E. Ruijter et al. (2021) presented a reaction of salicylic aldehydes **416** and vinyl-substituted D–A cyclopropanes **321**, in which the phosphonate group

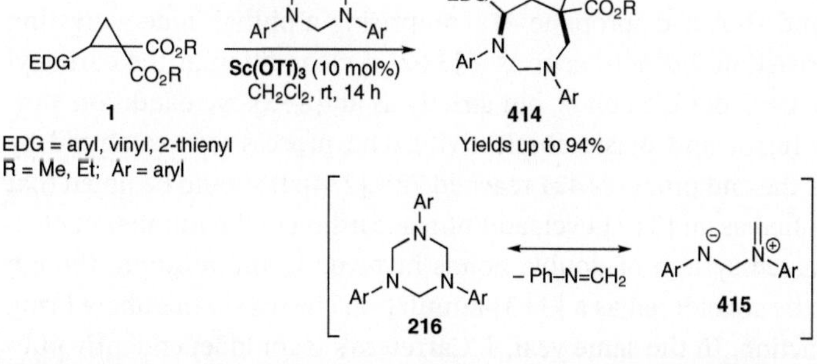

Figure 3.108 Formal [4+3]-cycloaddition with triaryl triazines **216**.

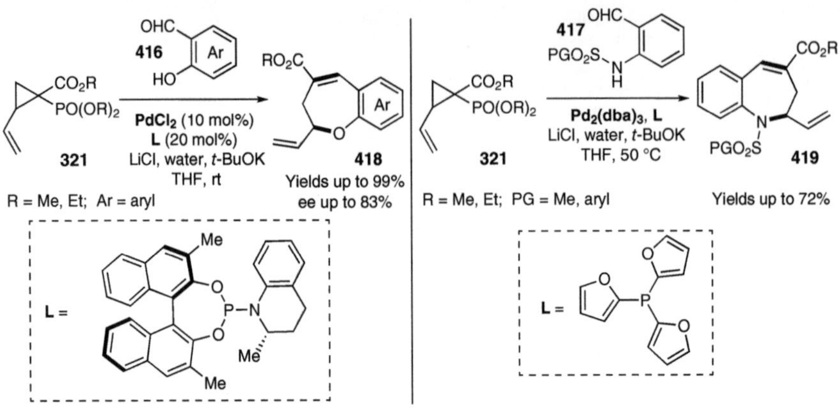

Figure 3.109 [4+3]-Annulation reactions of vinyl D–A cyclopropanes with salicylic and anthranylaldehydes.

acted as one of the acceptor moieties. The [4+3]-annulation occurs with a wide range of initial substrates. The yields of the final product **418** reach 99%. In this process, palladium catalysis was used for cyclopropane opening. Its use was found to be necessary due to the presence of a reactive vinyl group in cyclopropane. This process was successfully implemented in the enantioselective version using chiral phosphonate ligands [232]. Later on, the same team expanded their approach to reactions of anthranyl aldehydes **417** with D–A cyclopropanes **321**, where the same phosphonate group acted as an acceptor moiety. The formal [3+4]-annulation can also be performed with a wide range of initial substrates. The yields of the final product **419** reached 72%. $Pd_2(dba)_3$ with the addition of the trifurylphosphine ligand served as the catalyst for this process [233] (Figure 3.109).

3.7 Other Formal [*n*+*m*]-Cycloaddition and Annulation Processes

3.7.1 Formal [8+3]-Cycloaddition Reactions

M. Sierra et al. presented an example of the reaction of 2-substituted tropones **420** as dipolarophiles and D–A cyclopropane **41**, comprising a phthalimide activating donor group. The reaction did not occur as a [3+2]-cycloaddition at the carbonyl group or one of the C=C double bonds, but strictly as an [8+3]-cycloaddition with exceptionally high regio- and diastereoselectivity. The process was catalyzed by $SnCl_4$. The yields of the end products **421** reached 72% [234]. It should be noted that this process is classified as an [8+3]-cycloaddition according to the number of electrons of the conjugated system of double bonds involved in the reaction, though formally it can also be considered as a [3+3]-annulation since a six-membered ring is formed in the reaction. In the same year, J. Carretero's team independently published a similar process for usual 2-substituted cyclopropanedicarboxylates **1** with

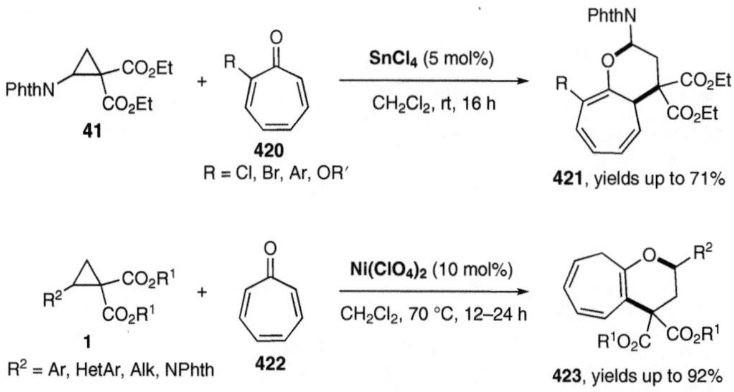

Figure 3.110 Formal [8+3]-cycloadditions of D–A cyclopropanes with tropones **420, 422**.

an unsubstituted tropone **422** catalyzed by nickel perchlorate $Ni(ClO_4)_2$ that also occurred as an [8+3]-cycloaddition followed by migration of double bonds in the cycloheptatriene ring [235] (Figure 3.110).

3.7.2 Other Formal Stepwise "High-Order" Cycloaddition/Annulation Reactions

A few other reactions that do not fit into the classification presented above and occur with the formation of seven- and eight-membered rings are also known for D–A cyclopropanes **1**. For example, processes of stepwise addition of D–A cyclopropanes to vinyl and allyl derivatives of indole with acceptor groups at the double bond were reported. X. T. Xu et al. succeeded in performing the [3+4]-annulation of Knoevenagel adducts comprising an indole moiety **424** with D–A cyclopropane **1**, with possible efficient employment of a wide range of D–A cyclopropanes to give high yields of end products **426** with high diastereoselectivity. The process is catalyzed by ytterbium(III) triflate [236]. A. Nishida et al. developed a similar protocol for a formal [5+3]-cycloaddition of D–A cyclopropanes **1** and substituted indoles **425** with an allylcarboxylate substituent at position 2. The process can be carried out either by a *one-pot* technique or sequentially to give an acyclic product. The yields of end products **427** reached 91% [237] (Figure 3.111).

Recent publications from Novikov and Tomilov's team reported a rare example of formal [6+2]-cycloaddition of 4-phenyl-1,2,4-triazoline-3,5-dione (PTAD) **278** to D–A cyclopropanes **428, 431** based on the bicyclopropane system to give a 1,2-diazacyclooctane moiety [238, 239]. These substrates are highly reactive and occur under mild catalysis conditions using 5–10% Yb triflate. Apart from the formal [6+2]-cycloaddition, a number of other processes occur in this case, including those accompanied by further transformations with the addition of two or three PTAD moieties. Moreover, the same team reported a rare selective example of [4+3]-cyclodimerization of (9-anthracenyl)cyclopropanedicarboxylate **433** in the presence of $GaCl_3$ to give a tetracyclic 1,2,3,4-tetrahydrocyclohepta[*de*]anthracene skeleton **434** [240] (Figures 3.112 and 3.113).

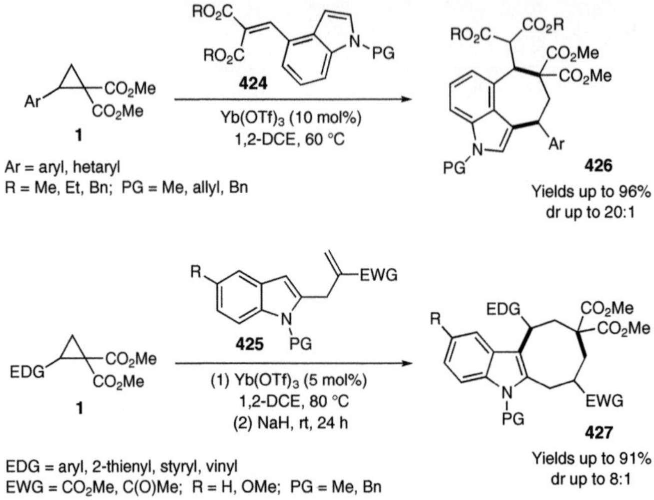

Figure 3.111 Formal stepwise [4+3]- and [5+3]-annulation reactions of D–A cyclopropanes to vinyl and allyl indole derivatives.

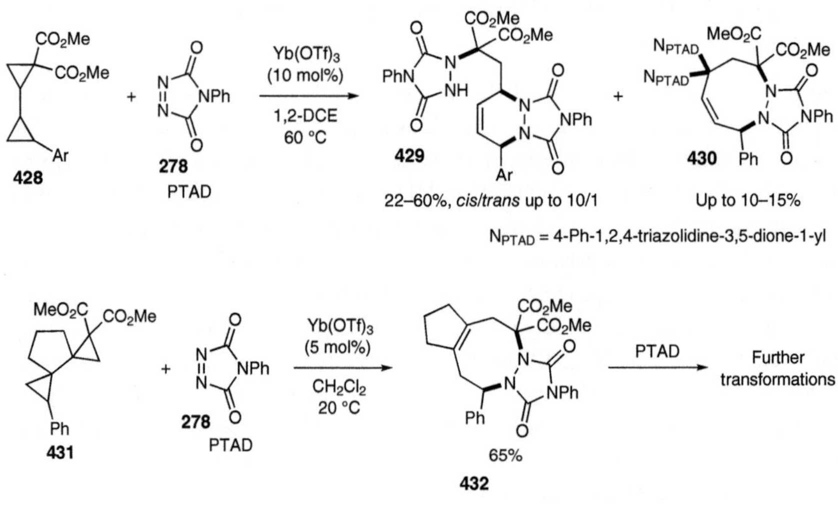

Figure 3.112 Formal [6+2]-cycloadditions of the activated D–A bi(cyclopropanes) to PTAD **278**.

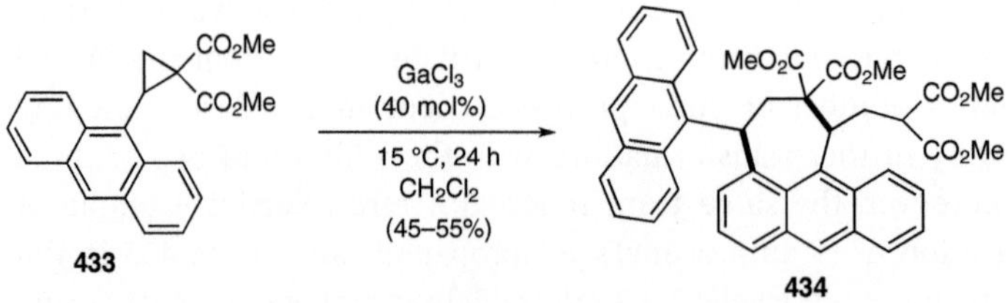

Figure 3.113 [4+3]-Cyclodimerization of anthracene-derived D–A cyclopropane **433**.

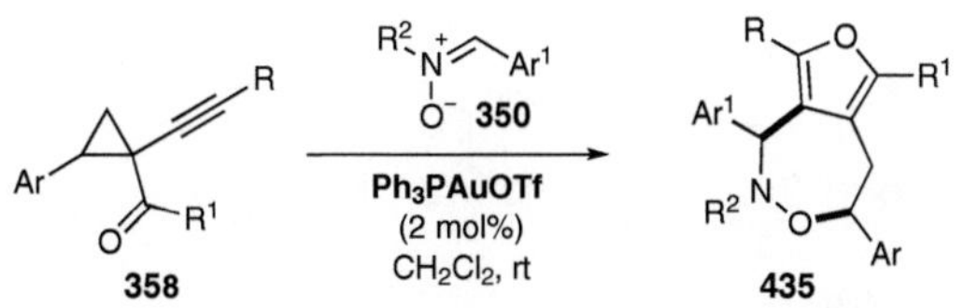

Figure 3.114 Formal [4+3]-cycloaddition with nitrones under Au(I)$^+$ catalysis conditions.

J. Zhang et al. developed a methodology for the formal [4+3]-cycloaddition of nitrones **350** with polysubstituted D–A cyclopropanes **358**, in which an acetylene moiety was present at the cyclopropane ring. The process was successfully implemented under metal complex catalysis conditions using a gold complex, $Ph_3PAuOTf$. The yields of the final product **435** reached 99%, and the process occurred with extremely high diastereoselectivity [206]. In these reactions, rearrangement also occurred in addition to the formal cycloaddition. Moreover, cyclopropane itself acted as a source of a four-carbon synthon in the cyclization (Figure 3.114).

T. Hayashi et al. described Pd-catalyzed [4+3]-annulation between cyclopropanedinitriles **436** and 5-methyliden-3-methoxycarbonyl-3-phenyltetrahydropyran-2-one **437**. The reaction proceeds through the formation of allylpalladium intermediate complex which then undergoes decarboxylation. This process leads to the formation of 1,4-zwitterionic intermediate, which reacts with cyclopropanedinitrile to form substituted cycloheptane **438** [241] (Figure 3.115).

3.7.3 Formal [3+1]- and [3+1+1]-Cycloadditions

Formal [3+1]-cycloaddition processes involving D–A cyclopropanes are very rare, and only a few reactions can be attributed to them at the moment. From the formal point of view, a carbene intermediate should add to a D–A cyclopropane molecule

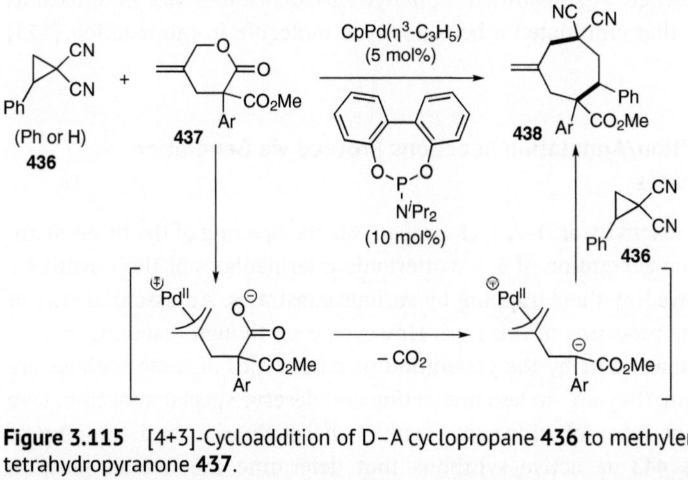

Figure 3.115 [4+3]-Cycloaddition of D–A cyclopropane **436** to methylene tetrahydropyranone **437**.

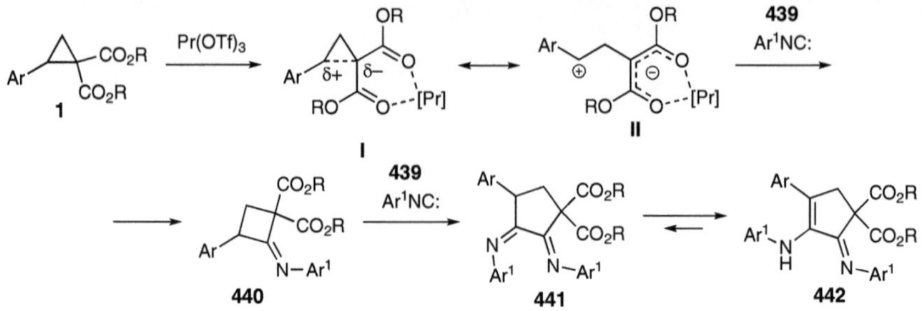

Figure 3.116 Formal [3+1]- and [3+1+1]-cycloadditions of isonitriles to D–A cyclopropanes.

in the course of [3+1]-cycloaddition, but reactions of this kind were found to be unrealizable in real life. The most popular and convenient variant for the generation of carbene intermediates via various diazo compounds combined with D–A cyclopropanes results in a wide range of various reactions but not the formal [3+1]-addition of carbene [33] [219–221] [242]. However, a few reactions of this kind are known. In 2006, Armin de Meijere's team implemented a process of addition of isonitriles **439** to D–A cyclopropanes **1** catalysis by praseodymium triflate $Pr(OTf)_3$ under mild conditions [243]. In this case, isonitrile exhibits the properties of a carbene and adds to a D–A cyclopropane by the [3+1]-cycloaddition type to give an unstable four-membered ring, which immediately adds the second isonitrile molecule in a similar way to give a five-membered ring **441** resulting from formal [3+1+1]-cycloaddition. (Figure 3.116) A couple of formal [3+1]-cycloadditions of nitrene synthons to D–A cyclopropanes, resulting in the formation of azacyclobutane skeleton, are also known. However, in reality, processes of this kind occur either stepwise in two stages or by a more complex multistage mechanism. In 2014, P. F. Xu et al. succeeded in achieving this goal by first entering a D–A cyclopropane **1** into formal [3+3]-cycloaddition with azides **378** and then performing their thermal de-diazotization [218]. Later, P. Banerjee et al. suggested an alternative approach, where *N*-substituted 3-phenyl-1,2-oxaziridines **301** were used as nitrene synthons that eliminated a benzaldehyde molecule in the reaction [173] (Figure 3.117).

3.7.4 Cycloaddition/Annulation Reactions Proceed via Generation of β-Styrylmalonates

The main type of reactivity of D–A cyclopropanes is the opening of the three-membered ring with the generation of 1,3-zwitterionic intermediates or their synthetic equivalents, followed by their trapping by various substrates. An essential part of this chapter covers processes of this type. However, even though reactions of D–A cyclopropanes **1** that occur by the generation of other types of intermediates are much less common, they are no less interesting and deserve special attention. One of these important types of processes involves the generation and trapping of β-styrylmalonates **443** as active synthons that determine the overall reaction

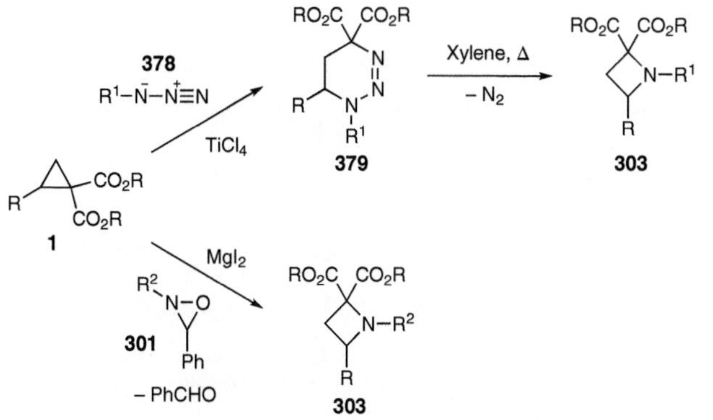

Figure 3.117 Approaches to obtain azetidine derivatives **303** *via* formal addition of nitrene synthon to D–A cyclopropanes **1**.

pathway. β-Styrylmalonates **443** are isomers of D–A cyclopropanes **1** formed through direct isomerization by opening the three-membered ring in a complex with a Lewis acid followed by proton migration. They can be generated directly in situ in the course of the reaction or can be obtained separately. At the moment, there is only one convenient way to obtain them, exactly from D–A cyclopropanes. Depending on the substituents, Me_3SiOTf, $Sn(OTf)_2$, or $GaCl_3$ are used as Lewis acids for isomerization [244] [245]. In terms of the mechanism of the subsequent reactions with substrates, styrylmalonates **443** resemble D–A cyclopropanes **1** and are also activated due to complexation with Lewis acids at two ester groups, and often reacting as malonyl anion. The subsequent pathway of their reactions with substrates usually involves an ionic stepwise mechanism via formal cycloaddition or annulation at the C=C double bond of the β-styrylmalonate, although more complex cascade cyclization processes also occur (Figures 3.118 and 3.119).

Not so many examples of reactions of D–A cyclopropanes that occur through the generation of β-styrylmalonates are known to date. In the study by Prof. Banerjee et al., cyclic chalconimines **444, 445** were used as traps for D–A cyclopropanes **1**. The process was performed in the presence of MgI_2 as the catalyst. It occurred as [4+2]-cycloaddition by the hetero-Diels–Alder reaction of intermediate β-styrylmalonates **443** generated in situ. The yields of end products **446**, **447** were up to 86%; the reaction showed moderate diastereoselectivity. The process

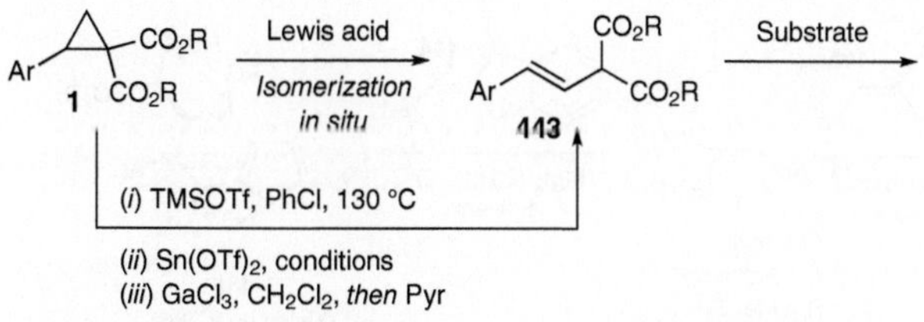

Figure 3.118 The type of reactivity of D–A cyclopropanes **1** realized *via* generation of isomeric styrylmalonates **443** as key intermediates.

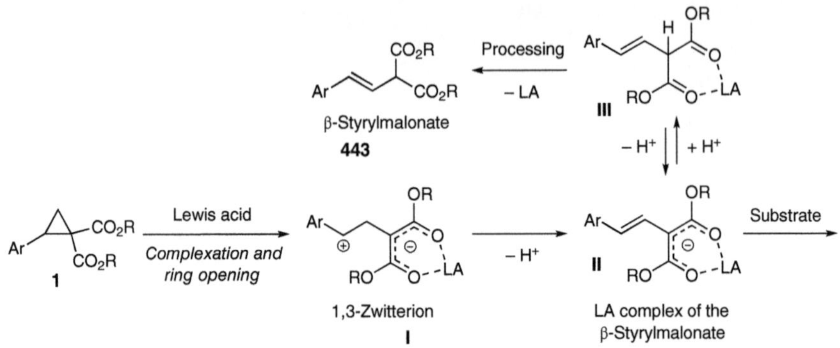

Figure 3.119 General mechanistic features of styrylmalonate-type processes.

was successfully performed with a wide range of substrates [246]. The reaction using pre-synthesized β-styrylmalonates **443** reported by Novikov and Tomilov's team, instead of D–A cyclopropanes **1**, makes it possible to expand the range of substrates even further and use simple α,β-unsaturated imines **448** in this [4+2]-cycloaddition, while in the standard version the latter react with D–A cyclopropanes **1** by the [3+2]-cycloaddition type [247] (Figure 3.120).

In 2013–2021, the teams of R. Sonawane, L. Wang, and I. Trushkov developed reactions of formal [3+2]-annulation and cycloaddition of D–A cyclopropanes **1** that also occurred via generation of β-styrylmalonates **443**, in which D–A cyclopropanes acted as two-carbon synthons in ring formation. *N*-Benzylic sulfonamides **450** [248] and azomethine imines **372** [215] were used as 1,3-substrates to give indane **451** and diazabicyclooctane **452** derivatives, respectively. Moreover, I. Trushkov's team developed a process of homo- and cross-[3+2]-cyclodimerization of D–A cyclopropanes **1** where one of the two molecules reacted as a styrylmalonate synthon, also to give indane derivatives [73]. Yet another type of process involves the reaction of β-styrylmalonates **443** with aromatic aldehydes **160**, which has been

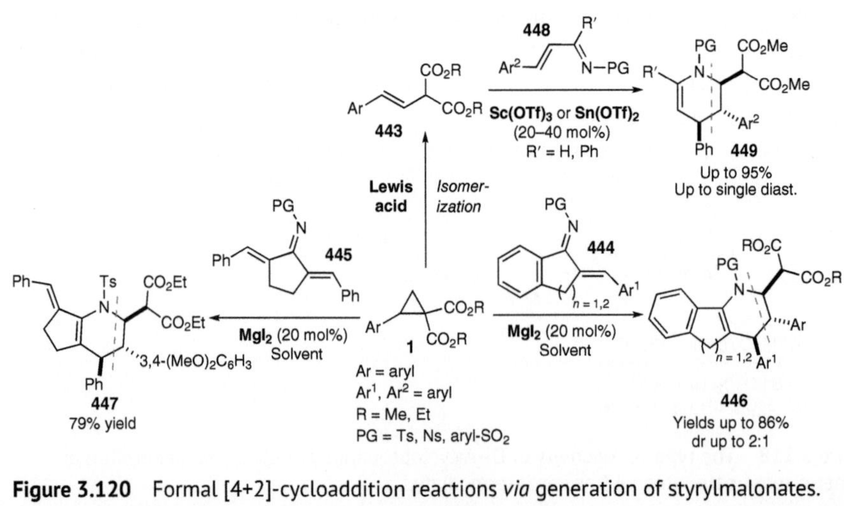

Figure 3.120 Formal [4+2]-cycloaddition reactions *via* generation of styrylmalonates.

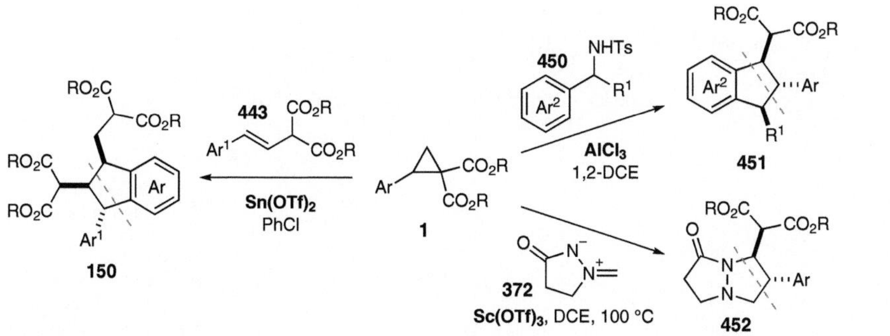

Figure 3.121 Formal [3+2]-cycloaddition/annulation reactions *via* generation of styrylmalonates **443**.

intensely developed by Novikov and Tomilov's team since 2016. Styrylmalonates **443** pre-synthesized from D–A cyclopropanes **1** are usually employed in this case to give noticeably higher product yields in comparison with generation in situ. However, as a rule, both variants do not change the type of the process. Reactions of this class are usually characterized by the utilization of $GaCl_3$ as a Lewis acid and the occurrence of complex cascade reactions that give five- and six-membered carbocycles and lactones **453–455**, as well as polycyclic systems **456** [249–253] (Figure 3.121). The most typical representatives of this class of processes are shown in Figure 3.122.

3.7.5 $GaCl_3$-Mediated Cycloaddition/Annulation Reactions via Generation of 1,2-Zwitterionic Intermediates

The use of anhydrous gallium trichloride ($GaCl_3$) makes it possible to realize a completely separate type of D–A reactivity of cyclopropanes **1**, which is not observed for other Lewis acids. Due to a combination of a number of unique properties, gallium(III) leads to a rapid 1,2-hydride shift in the classical 1,3-zwitterionic

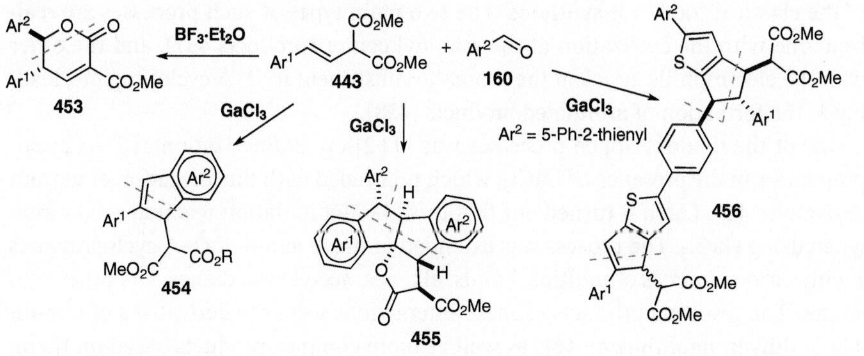

Figure 3.122 Various cascade cyclization reactions of the styrylmalonate **443** with aromatic aldehydes **160**.

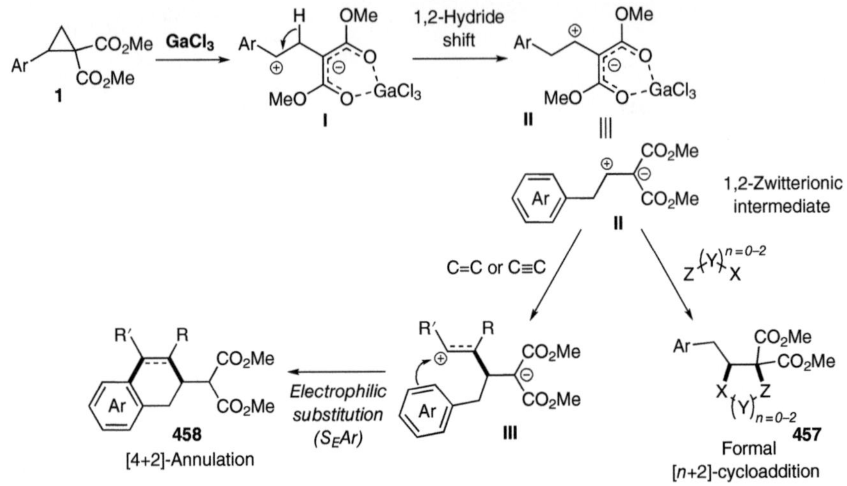

Figure 3.123 1,2-Zwitterionic reactivity type of D–A cyclopropanes in annulation reactions.

intermediates **I**, resulting in the formation of the main reactive 1,2-zwitterionic intermediate **II** as a gallium complex. This intermediate undergoes further reactions with various substrates, especially in the annulation reactions with multiple bonds. The 1,2-zwitterionic intermediate **II** is strongly stabilized by the coordinated gallium cation, in comparison to the previous 1,3-zwitterionic intermediate **I**. This fact explains such a type of reactivity of D–A cyclopropanes **1** in the presence of gallium salts, which is completely different from the classical chemistry of D–A cyclopropanes. This type of reactivity of D–A cyclopropanes is actively developed by the Novikov and Tomilov research group [220] [254–263] (Figure 3.123).

This gallium complex **II** is highly reactive and has a very rich chemistry, including various types of transformations initiated by carbocation attack. One of the key reactions is formal cycloaddition and annulation with various substrates; in particular, reactions with alkenes and acetylenes are well studied. In these reactions, D–A cyclopropanes **1** act as sources of "even" 1,2- and 1,4-zwitterionic synthons instead of the classical "odd" 1,3-synthons. The two main types of such processes are realized, one with the cyclization at the malonyl center (products **457**), and the other with the electrophilic attack at the aromatic substituent in D–A cyclopropane, leading to the formation of annulated products (**458**).

One of the first developed processes was [4+2]-cyclodimerization of D–A cyclopropanes **1** in the presence of $GaCl_3$, which proceeded with the formation of tetralin derivatives **459**. Later, it turned out that this [4+2]-annulation reaction is common when using $GaCl_3$. The process was extended to the reaction of D–A cyclopropanes **1** with various sources of multiple bonds: alkenes, acetylenes, dienes, and other substrates. The resulting products of these interactions were the derivatives of tetralin **460** or dihydronaphthalene **461**, as well as more complex products based on them. Formal [*n*+2]-cycloaddition reactions with the cyclization at the malonyl carbon atom are more characteristic of other types of substrates, since, upon addition of a multiple C–C bond, it should lead to the formation of an unfavorable cyclobutane ring ([2+2]-cycloaddition). So, only one selective example is known in the reaction with bicyclobutylidene. Apparently, various allyl derivatives, azides, and also other sources of 1,3-dipoles react in this direction along the [3+2]-cycloaddition pathway. In addition, dienes also react in this direction – along the path of ionic [4+2]-cycloaddition, as well as diazo compounds – along the path of ionic [2+1]-cycloaddition. It should also be noted that a large number of more complex cascade processes occur based on these transformations [220, 254–263] (Figures 3.124, 3.125, and 3.126).

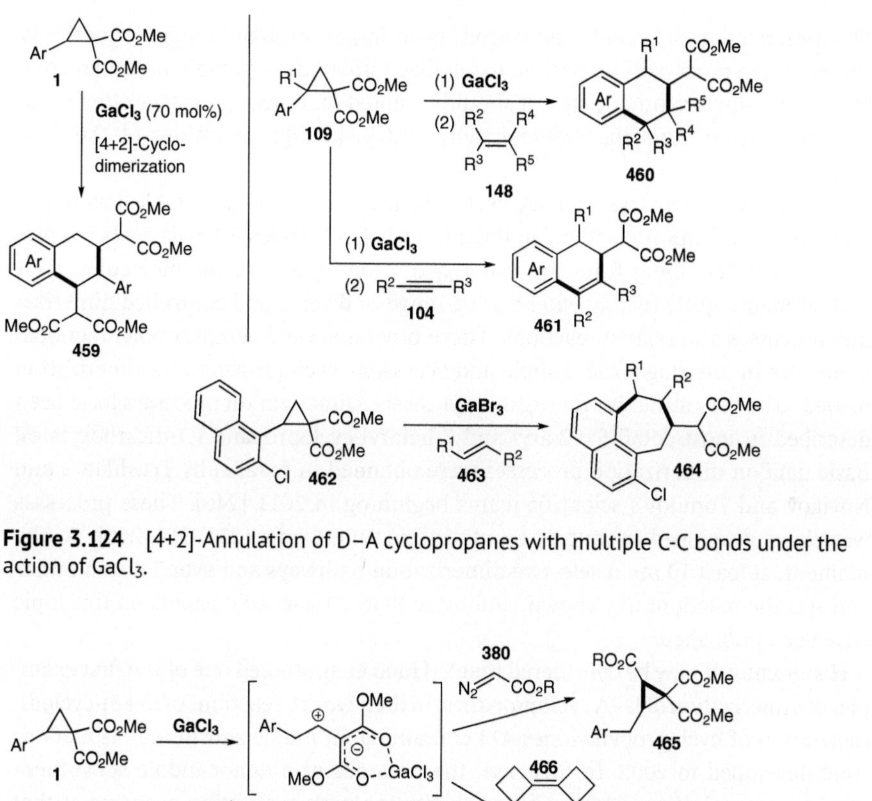

Figure 3.124 [4+2]-Annulation of D–A cyclopropanes with multiple C-C bonds under the action of $GaCl_3$.

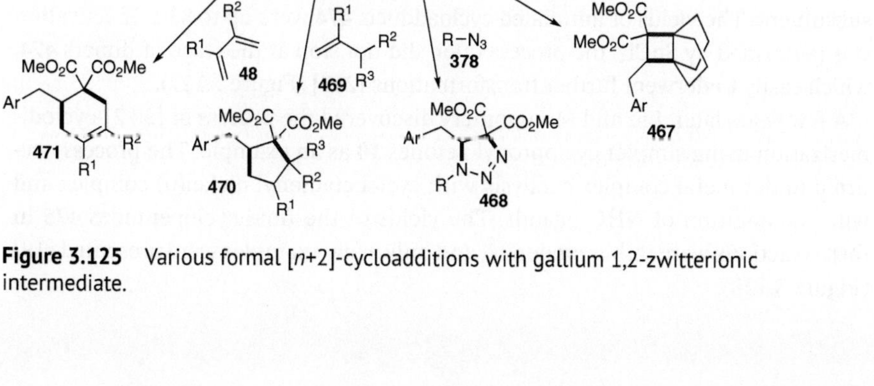

Figure 3.125 Various formal [*n*+2]-cycloadditions with gallium 1,2-zwitterionic intermediate.

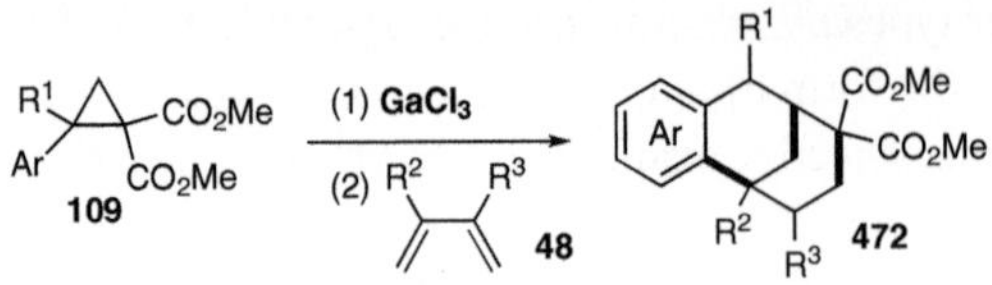

Figure 3.126 $GaCl_3$-Mediated cascade of the double [4+2]-annulation reactions.

3.8 Cyclodimerization Reactions of D–A Cyclopropanes

The dimerization of organic compounds is an important tool in organic synthesis. Dimerization processes can be used to obtain compounds with high molecular complexity in a single synthetic step. It should be noted that these processes often occur with high regio- and diastereoselectivity and yield high quantities of the final products.

One of the trademarks of D–A cyclopropanes is their considerable liability to selective dimerization reactions in the absence of substrates if Lewis acids are present. This differs them from the majority of other cyclopropane molecules. D–A cyclopropanes quite readily enter a wide range of diverse and controlled dimerization processes and related reactions. These processes yield complex polyfunctional molecules in one stage from simple and accessible cyclopropanes, so dimerization proved to be very attractive for organic synthesis. Dimerization processes have been described in most detail for 2-aryl and 2-hetaryl-cyclopropane-1,1-dicarboxylates. Basic data on dimerization processes were obtained in parallel by Trushkov's and Novikov and Tomilov's scientific teams beginning in 2011 [240]. These processes were later developed in studies by other scientific teams [234, 264–266]. At the moment, at least 10 main selective dimerization pathways and over 2 dozen minor and specific reactions are known, and more than 20 scientific papers on this topic have been published.

Historically, it may be considered that V. Huch et al. studied one of the first examples of dimerization of D–A cyclopropanes. In their works, reactions of [3+3]-cyclodimerization of cyclopropylketones **473** containing an indole substituent as a donor were developed in 2002. In this case, the presence of a donor indole substituent played a key role since electrophilic substitution with cyclization occurred at that substituent. The yields of annulated cycloadducts **474** were up to 83%. If activation was performed by $SnCl_4$, the process often did not stop at the stage of dimers **474**, which easily underwent further transformations [267] (Figure 3.127).

A few years later, Liu and Montgomery discovered the process of [3+2]-cyclodimerization using simpler cyclopropyl ketones **10** as an example. The process occurred under metal complex catalysis with cyclooctadienyl nickel(0) complex and with the addition of NHC ligands. The yields of the final cyclopentanes **475** in these reactions were rather moderate, and only a few examples were presented [19] (Figure 3.128).

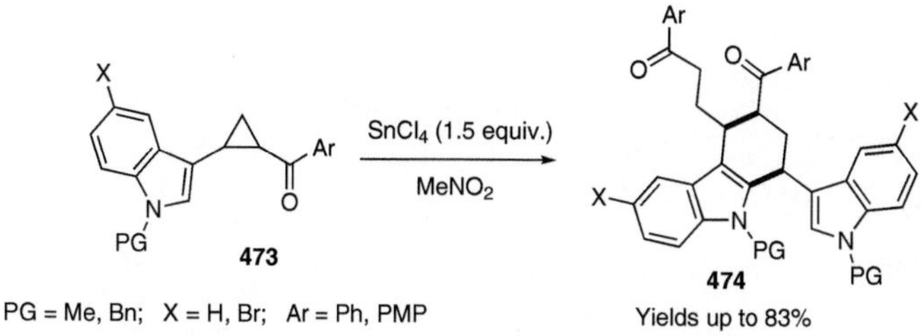

Figure 3.127 [3+3]-Cyclodimerization of (3-indolyl)cyclopropylketones **473**.

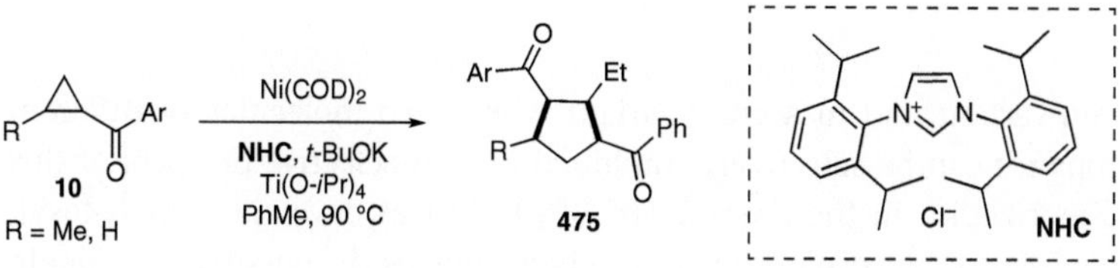

Figure 3.128 Ni(0)-Catalyzed [3+2]-cyclodimerization of some simple cyclopropylketones **10**.

The main studies on the dimerization of D–A cyclopropanes were performed much later, beginning in 2011, independently by I. Trushkov's team and by Novikov and Tomilov's team. Trushkov et al. demonstrated an example of [3+2]-cyclodimerization of D–A cyclopropanes **1** catalyzed by Sn(II) and Sc(III) triflates that generated β-styrylmalonate, which acted as a trap of another cyclopropane molecule. The process occurred in good yields and showed moderate diastereoselectivity in the final cyclopentanes **476** [268]. Concurrently, Novikov, Tomilov et al. reported a different approach for performing the same type of dimerization, namely, in the presence of catalytic amounts of $GaCl_3$ [269] (Figure 3.129).

Later, a method was developed for dimerization of D–A cyclopropanes **1** in the presence of $Sn(OTf)_2$, in which [3+2]-annulation of β-styrylmalonate formed in situ and the original D–A cyclopropane occurs. This reaction occurs with cyclopropanes containing donor groups at the aromatic ring or with heteroaromatic substituents such as thiophene or benzofuran. The yields of the final indanes **150** range from 42% to 87%. Dimerization protocols have been developed both in the

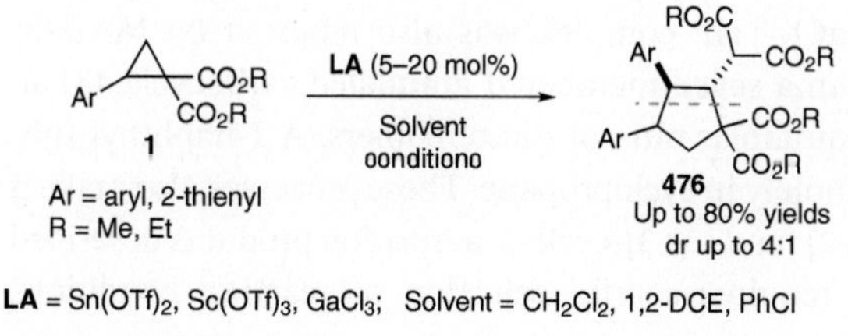

Figure 3.129 [3+2]-Cyclodimerization process of D–A cyclopropanes **1**.

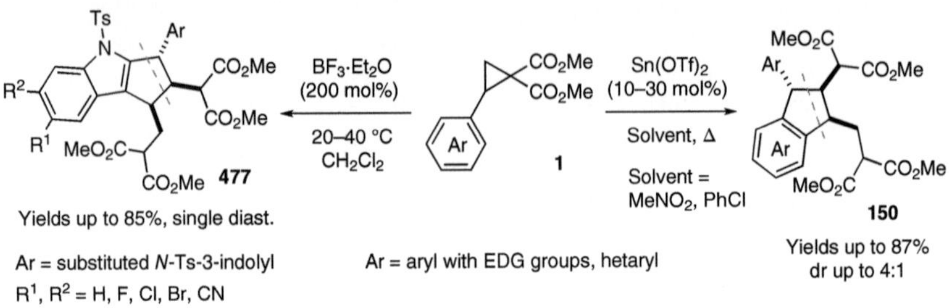

Figure 3.130 [3+2]-Cyclodimerization of D–A cyclopropanes **1** with annulation at the aromatic ring.

standard homo-variant and in a cross-variant where two molecules of different D–A cyclopropanes can be selectively combined [73]. A selective protocol of this [3+2]-cyclodimerization in the presence of $BF_3{\cdot}Et_2O$ was developed for *N*-tosyl-protected 3-indolyl D–A cyclopropanes, which makes it possible to obtain tetrahydrocyclopenta[*b*]indole derivatives **477** [270] (Figure 3.130).

However, the main and most characteristic type of dimerization of D–A cyclopropanes **1** is the [3+3]-cyclization in the presence of various Lewis acids. These processes were thoroughly developed and studied in detail in a series of papers mainly published in 2011–2016. Researchers developed three more main selective pathways for dimerization reactions, namely, direct formal [3+3]-cycloaddition to give cyclohexane derivatives **478**, as well as single or double annulation to the aromatic ring in D–A cyclopropanes. These reactions allowed them to obtain bi- and tricyclic systems **327**, **328**, and moreover, the dimerization direction could be controlled. Numerous diverse examples and reaction conditions were reported, and aspects of their diastereoselectivity were described. It should be noted separately that D–A cyclopropanes with heteroaromatic substituents have a considerable susceptibility to dimerization that opens access to efficient construction of complex annulated heterocyclic systems [183, 184, 240] (Figure 3.131).

Based on the [3+3]-cyclodimerization of 2-phenylcyclopropanedicarboxylate **1**, a protocol for deeper oligomerization realized under the action of the $SnCl_4{\cdot}THF$ complex was developed. In this case, the degree of polymerization and the chain length can be controlled fairly well by the concentrations of the reagents. Moreover, the trimer and tetramer **479** can be obtained quite selectively (Figure 3.132) [269]. An interesting example of selective [4+3]-cyclodimerization of D–A cyclopropane **480** in the presence of the $GaCl_3{\cdot}THF$ complex was also reported by Novikov, Tomilov et al., which resulted in a seven-membered annulated carbocycle **481** in moderate yields and with an equimolar ratio of diastereomers. A 1-naphthyl substituent was used as the donor moiety in cyclopropane. These processes also make it possible to easily obtain the [3+2]- and [3+3]-cyclodimerization products described above; therefore, the process requires careful selection of reaction conditions (Figure 3.133) [271].

The Trushkov's team reported an interesting process of domino-*ipso*-cyclodimerization of D–A cyclopropanes **1** containing an indole or *para*-methoxyphenyl

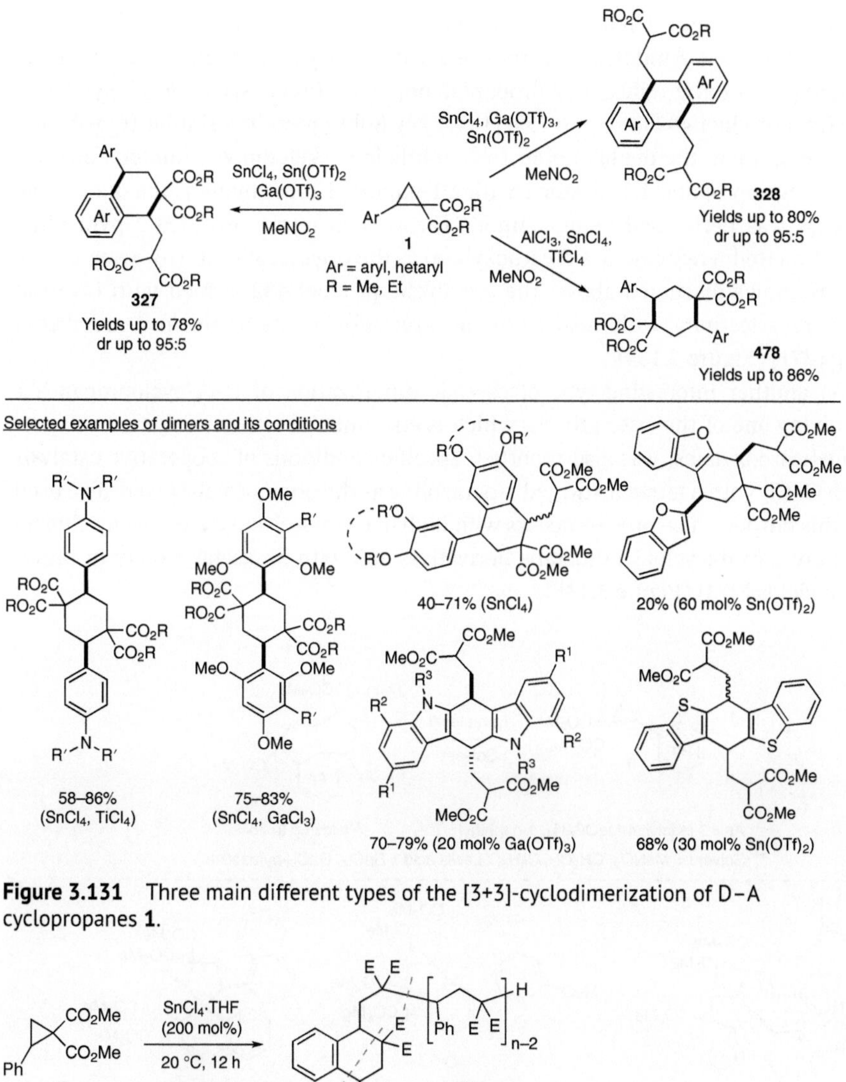

Figure 3.131 Three main different types of the [3+3]-cyclodimerization of D–A cyclopropanes **1**.

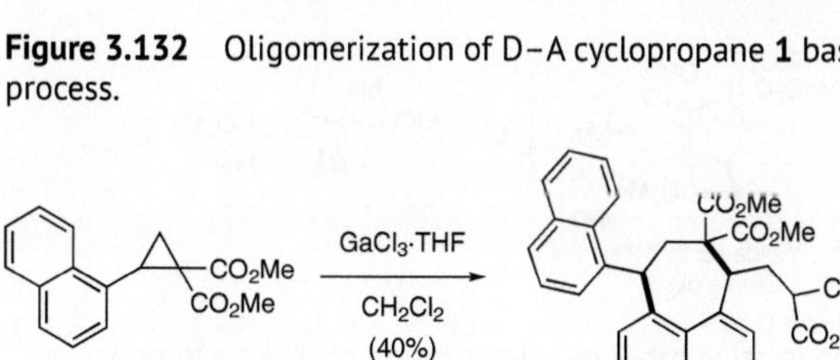

Figure 3.132 Oligomerization of D–A cyclopropane **1** based on [3+3]-cyclodimerization process.

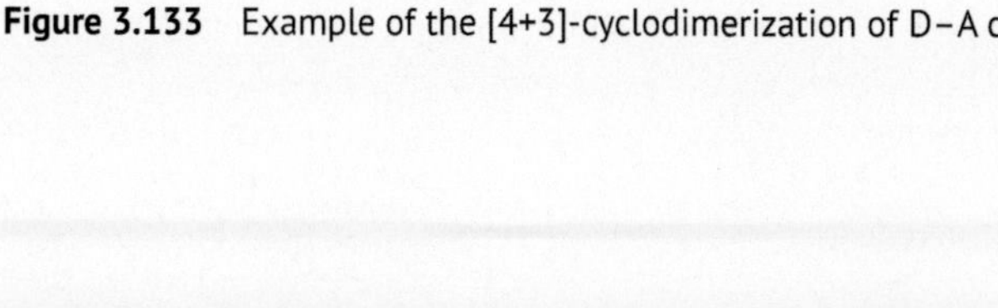

Figure 3.133 Example of the [4+3]-cyclodimerization of D–A cyclopropanes **480**.

moiety as the donor. Tin tetrachloride acts as the catalyst in the process, and moreover, substituted 2-indolylcyclopropane-1,1-dicarboxylates readily enter the reaction to give a tetracyclic octahydropentaleno[1,6a-*b*]indole skeleton. The yields of the final product **482** are up to 75% [183, 272]. Subsequently, a similar transformation resulting in the pentaleno[6*a*,1-*a*]naphthalene skeleton was implemented for a D–A cyclopropane-1,1-diester **1** with a 1-naphthyl substituent as the donor. The process was performed using 20 mol% $GaCl_3$ with tetrasubstituted 1-pyrazoline (3,3-dimethylpyrazoline-3,3-dicarboxylate) as the organocatalyst. However, unlike the examples discussed above, the tetracyclic product **482** formed with inversed diastereoselectivity with respect to the aryl substituent in the five-membered ring [271] (Figure 3.134).

Yet another interesting type of cascade dimerization of D–A cyclopropanes **1** involving one of the ester groups, which is encountered rather rarely for cyclopropanedicarboxylates, was also reported. Specific conditions of cooperative catalysis with $GaCl_3$ with a tetrasubstituted 1-pyrazoline as the organocatalyst were also used for this purpose. The process occurs with high diastereoselectivity and yields dimers that are 2-oxabicyclo[3.3.0]octane derivatives **483** with an acetal moiety in rather good yields [273] (Figure 3.135).

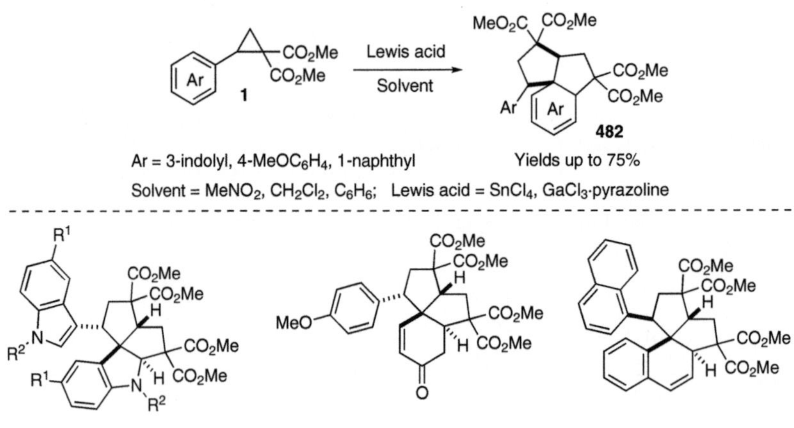

Figure 3.134 Cascade *ipso*-type dimerization of D–A cyclopropanes **1**.

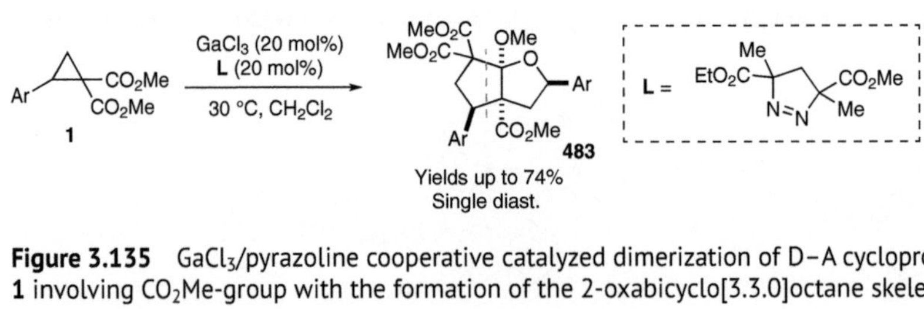

Figure 3.135 $GaCl_3$/pyrazoline cooperative catalyzed dimerization of D–A cyclopropanes **1** involving CO_2Me-group with the formation of the 2-oxabicyclo[3.3.0]octane skeleton **483**.

In terms of the mechanism, the main dimerization processes occur through the formation of zwitterionic intermediates that can be expressed most clearly as synthons **I–III**, which in turn react with each other to give dimers with various structures **150, 327, 328, 476, 478, 482**. Therefore, it is quite convenient to characterize reactions of these types directly by their reaction products. Two pathways of [3+2]-cyclodimerization (with and without annulation), three pathways of [3+3]-cyclodimerization, and the *ipso*-type of cyclodimerization can be distinguished. Each of them may be the main pathway of reactions of the 1,3-zwitterionic intermediate (**I**) initially generated from D–A cyclopropane **1** in the presence of a Lewis acid. The generated 1,3-zwitterionic intermediate **I** dimerizes to intermediate **IV**, which further follows a number of possible transformation pathways, depending on the conditions and substituents used. Yet another pathway of dimerization involves the isomerization of a fraction of D–A cyclopropane **1** into β-styrylmalonate intermediate **II** followed by its reaction with 1,3-zwitterion **I**. It should be noted that the foundations of this classification and analysis of the mechanisms of DAC dimer formation were laid in the works by Trushkov et al [183, 268] (Figure 3.136).

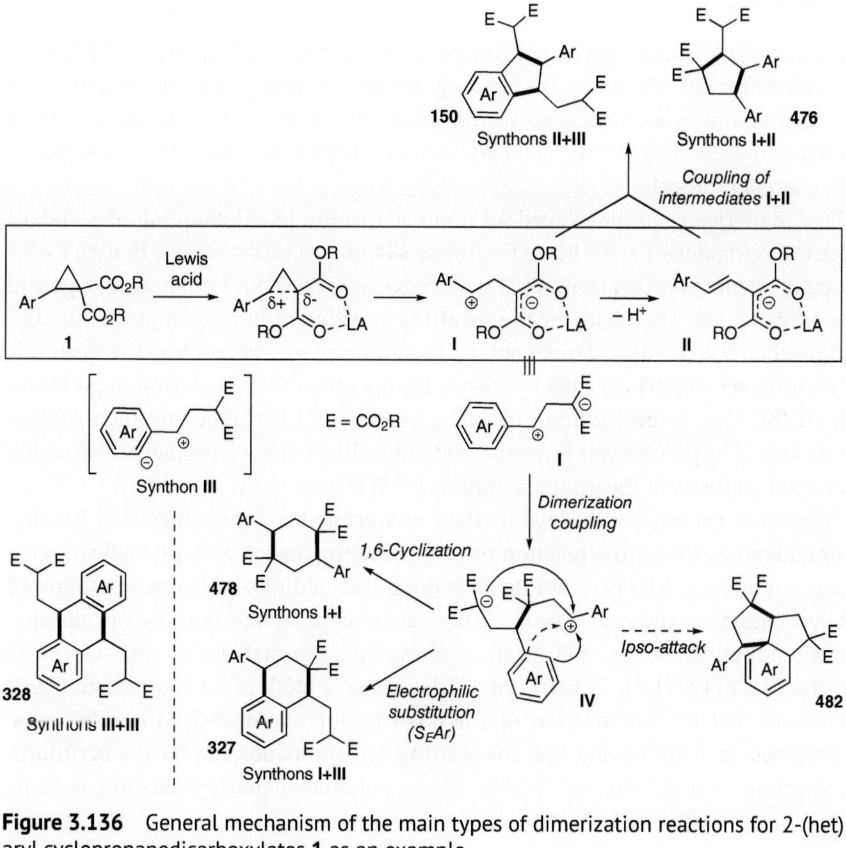

Figure 3.136 General mechanism of the main types of dimerization reactions for 2-(het)aryl cyclopropanedicarboxylates **1** as an example.

3.9 Miscellaneous Reactions, Stepwise Cyclization Reactions, Cyclizations with Involvement of Functional Groups

The chemistry of D–A cyclopropanes is very rich. Since cyclization reactions are one of its main parts, a huge number of types of cyclizations are known, which are rather difficult to reduce to any simple unified classification. The main processes of cycloaddition and annulation are discussed in detail in the previous sections. We will briefly look at the main examples and types of existing reactions with the formation of a new cycle in this section. These transformations include stepwise two-stage cyclization reactions, related processes, more complex cascade cyclizations, cyclizations involving functional groups in cyclopropane, and so on. Some of the transformations are discussed in more detail in other chapters by other authors.

3.9.1 Stepwise Cyclization Using Substrates with Two Nitrogen Atoms

The reactions of coupling with substrates comprising a few nitrogen atoms followed by cyclization are of interest for creating nitrogen heterocycles with two or more nitrogen atoms. A few reactions of this kind are known, with substrates such as substituted ureas, hydrazines, and pyrimidines. They are performed in two stages or by a *one-pot* protocol.

Prof. Banerjee et al. developed an approach to the [3+3]-coupling of standard D–A cyclopropanes **1** with benzyloxyureas **484** in successive stages, that is, cyclopropane opening and cyclization. The process was carried out both in two-stage and *one-pot* versions. The yields of the resulting substituted tetrahydropyrimidinones **486** reached 95% [274]. I. Trushkov's team developed an approach to the synthesis of pyrimidoazepinetriones **489** by successive opening of the D–A ring in cyclopropane **1** followed by cyclization, first using scandium(III) triflate and then sodium methoxide. The process was implemented only with D–A cyclopropanes containing donor substituents in the aromatic moiety [275] (Figure 3.137).

Somewhat earlier, a two-stage method was presented for synthesizing hexahydropyridazines **490** by the reaction of D–A cyclopropanes **1** with phenylhydrazine catalyzed by nickel(II) perchlorate with the initial addition of hydrazine followed by cyclization on the ester group. A similar process for the synthesis of tetrahydroindenopyridazinones **492** from spirocyclopropaninedenones **491** was also reported [276] [277]. H. U. Reissig et al. developed a method for synthesizing pyridazinones **494** by the reaction of activated cyclopropane **493** with substituted hydrazines. It is interesting that the starting cyclopropanes can have a perfluoroalkyl group as a substituent, which allows polyfluorinated pyridazinones to be obtained [278] (Figure 3.138).

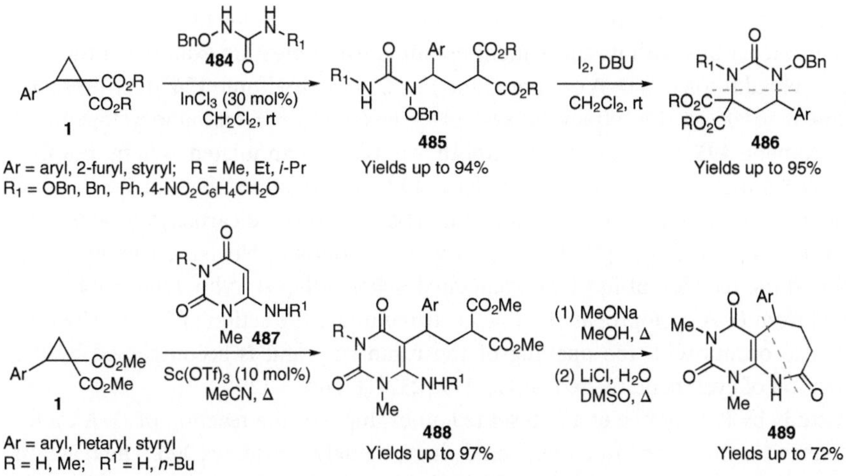

Figure 3.137 Two-step cyclization reactions of BnO-urea **484** and pyrimidines **487**.

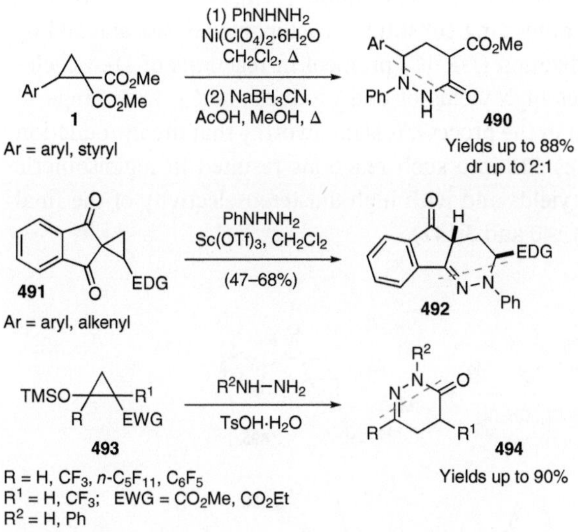

Figure 3.138 Cyclization reactions with hydrazines.

3.9.2 Some Other Cascade and Miscellaneous Formal Cycloaddition Reactions for Cyclopropanedicarboxylates

A series of works by I. Trushkov's team concerning the reactions of D–A cyclopropanes **1** with simple furan derivatives and the use of D–A cyclopropanes with a furyl substituent as the donor moiety appears to be very interesting. The high reactivity of the furan ring makes it possible to perform cascade cyclizations with its participation and reactions with its subsequent opening. An interesting example of the cascade cycloaddition of a D–A cyclopropane **1** to 2,5-dimethylfuran **136** in the presence of $SnCl_4$ to afford the tetracyclic system of hexahydro-3,5-methanocyclopenta[*c*]isochromene **495** was reported. Yet another tetracyclic annulated system, hexahydro-3*H*-5,8-methanocyclopenta[*a*]azulene **497**, was obtained by the reaction of cyclopentadiene **403** with (5-methyl-2-furyl)cyclopropanedicarboxylate **496** with furan ring opening [66] [223]. A recent work by Borisov, Novikov, Tomilov et al. reported the reaction of furyl-2-carbaldehydes **498** with β-styrylmalonates **443** generated from D–A cyclopropanes **1** that occurred under the action of $GaCl_3$. The process also occurs with the opening of the furan ring and is accompanied by the formation of cyclopentene derivatives **499** [252] (Figure 3.139).

A study by P. Banerjee et al. presented an example of the reaction of D–A cyclopropanes **1** with other three-membered rings, namely, aziridines **306**. The reaction does not occur as the expected formal [3+3]-cycloaddition of two generated 1,3-zwitterionic intermediates, but as a more complex cascade process to give furopyrrole derivatives **308**. The process is catalyzed by MgI_2 and involves one of the D–A ester groups in cyclopropane, which is rather a rare example [175]. Ghorai, Tiwari et al. developed an approach for synthesizing substituted cyclopentenes **502** and **503** by the domino-ring-opening-cyclization (DROC) protocol in reactions of D–A cyclopropanes **1**. The reaction gives high yields of end products, and a wide range of cyclopropanes can be involved in the process. It is noteworthy that the introduction of substituted malonodinitriles **501** into such reactions resulted in regioisomeric cyclopentenones **503** in good yields and with high diastereoselectivity of the final products [279–281] (Figures 3.140 and 3.141).

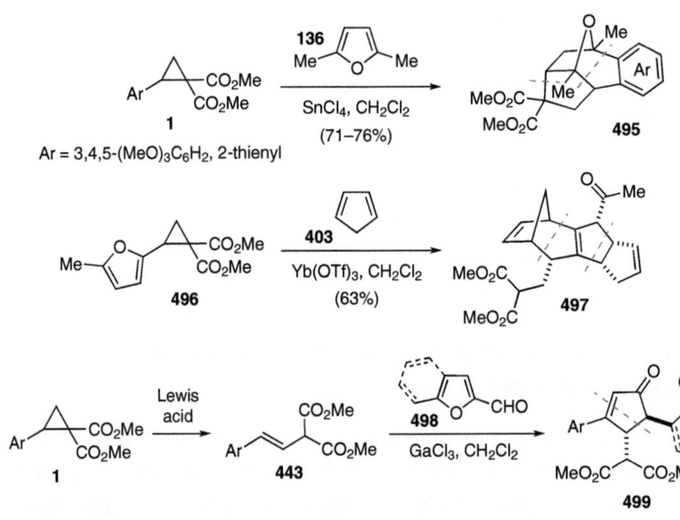

Figure 3.139 Reactions of furane derivatives.

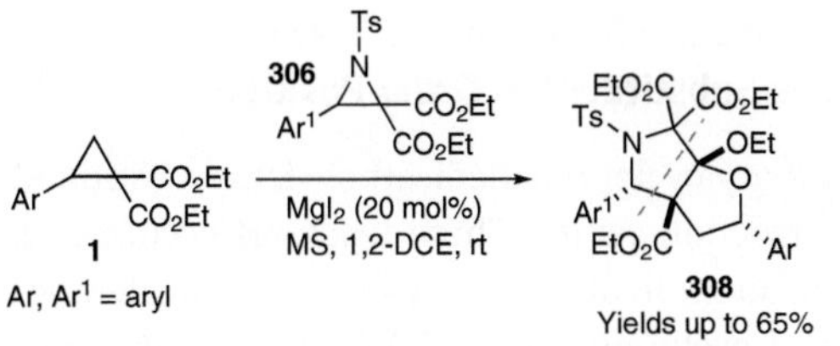

Figure 3.140 Cascade reaction of D–A cyclopropanes **1** with aziridines **306**.

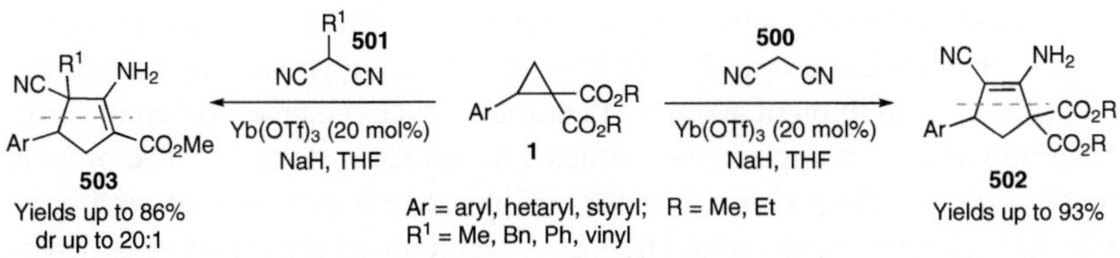

Figure 3.141 Reactions with malonodinitriles **500, 501**.

D. Werz et al. developed a protocol for benzoquinone **504** annulation using D–A cyclopropanes **1** in the presence of $SnCl_2$ followed by oxidation of the product with manganese(IV) oxide. The yields of the final 9-hydroxy-4*H*-cyclopenta[*b*]naphthalen-4-one derivatives **505** that find use as new dyes were up to 88% [282]. Z. He et al. continued a study of the reactions of D–A cyclopropanes **506** with diazenes **507**. It was shown that if an additional acceptor nitro group was incorporated into a D–A cyclopropane **506** and triphenylphosphine was present, a cascade reaction occurred with the conversion of one of the ester groups to give pyrazolines **508**. The process gave high yields of the end products with good diastereoselectivity [283] (Figures 3.142 and 3.143).

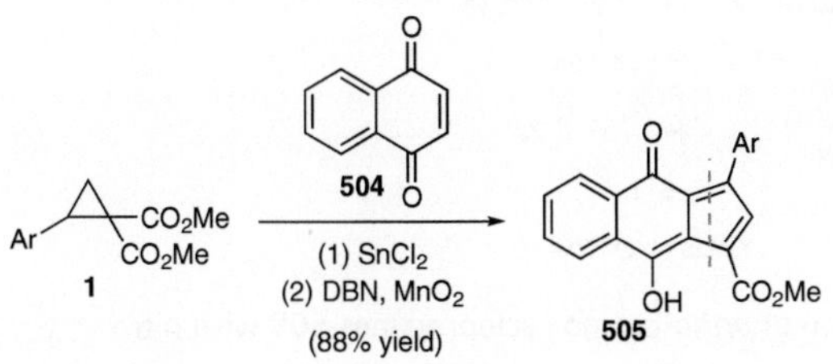

Figure 3.142 Reaction with naphthoquinone.

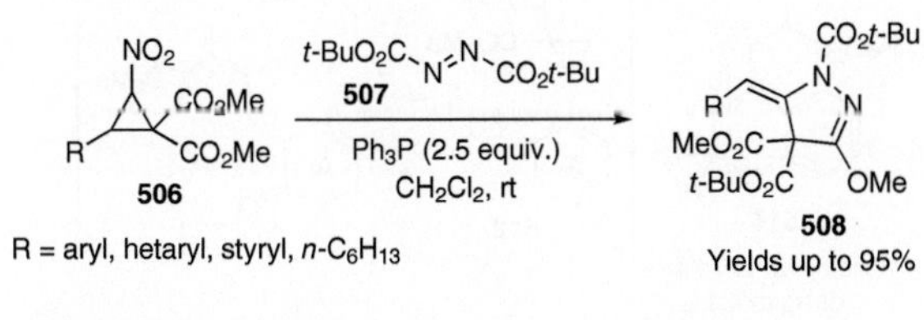

Figure 3.143 Reaction with *tert*-butyl azadicarboxylate.

3.9.3 Formal Cycloaddition and Cyclization Reactions for 2-Aryl D–A Cyclopropanes Containing Active Substituent in *Ortho*-Position

A number of works are known where 2-arylcyclopropanedicarboxylates substituted at the *ortho*-position were used as a reaction center. These reactions occurred as formal cycloaddition/annulation/cyclization. In fact, Mikhaylov, Ioffe et al. demonstrated an approach to the synthesis of cyclic nitronates **511** by C–C coupling of *ortho*-bromo substituted D–A cyclopropanes **509** and primary nitro compounds **510** catalyzed by Pd(0) salts. The reaction products, that is, cyclic annulated nitronates **511** obtained in up to 82% yields, were used in further transformations, including reactions of their cycloaddition (Figure 3.144) [284]. E. Budynina et al. studied the formal [4+2]-cycloaddition of alkenes **513** and D–A cyclopropanes **512** comprising an OH group at the *ortho*-position that acted as the C4-component. The process efficiently occurs in the presence of boron trifluoride etherate, and rather a wide range of alkenes enter the reaction. These D–A cyclopropanes act as sources of synthetic equivalents of *ortho*-quinone methides **I** that are very interesting in reactions of this kind. It appears that these D–A cyclopropanes can also be used in the same role in other reactions, but only this process that results in chromane derivatives is known to date (Figure 3.145) [285].

An acetylene moiety at the *ortho*-position is also used to expand the functional capabilities of 2-arylcyclopropanedicarboxylates. M. Kerr et al. reported a method for synthesizing triazole **516** derivatives by a tandem combination of the opening of D–A cyclopropane **515** and the click-coupling of azides comprising a triple bond [286]. K. Reddy et al. presented a method for synthesizing derivatives of isoquinolines **518** by a similar tandem reaction of the same *ortho*-acetylene containing cyclopropanes **517** with amines. The process is based on the opening of cyclopropane with amines,

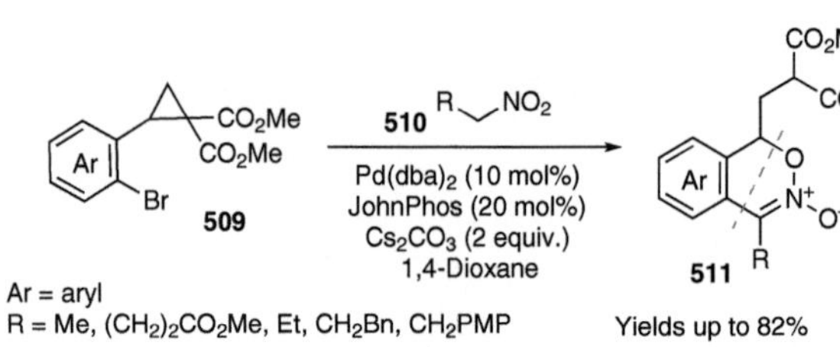

Figure 3.144 Formal [3+3]-annulation of *ortho*-bromo cyclopropanes **509** with nitro compounds **510**.

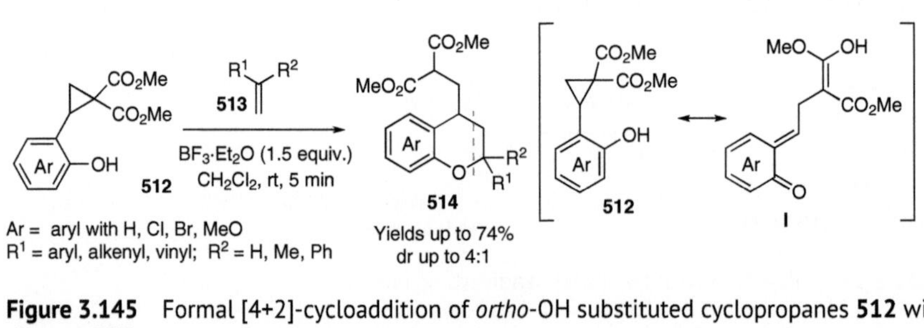

Figure 3.145 Formal [4+2]-cycloaddition of *ortho*-OH substituted cyclopropanes **512** with alkenes **513**.

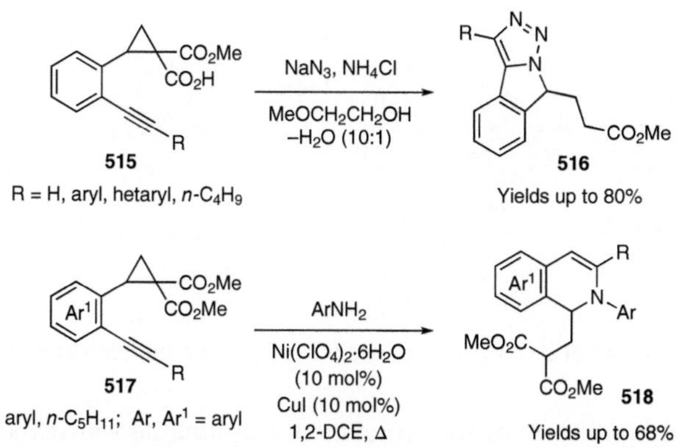

Figure 3.146 Cyclization reactions of 2-aryl D–A cyclopropanes **515, 517** with acetylene fragment in *ortho*-position.

followed by cyclization on the triple bond, and requires double catalysis with Ni(II) and Cu(I)/Ag(I) salts. The reaction successfully occurs with donor substituents. Acceptor substituents in the amine and additional substituents in the D–A cyclopropane decrease the yield of the final products [287] (Figure 3.146).

3.9.4 Cyclization Reactions of D–A Cyclopropanes with Additional CHO Group in Donor Part

The presence of a reactive group, in particular, an aldehyde group, in the donor part of D–A cyclopropane promotes various cyclization reactions. A study by J. Vicario presented a method for the interaction of DAC **519** with α,β- and β,γ-unsaturated compounds **89, 520** in the presence of NHC. The process proceeded as a formal [4+2]-cycloaddition with almost quantitative yields of the final products **521, 522** with high enantio- and diastereoselectivity [288] (Figure 3.147).

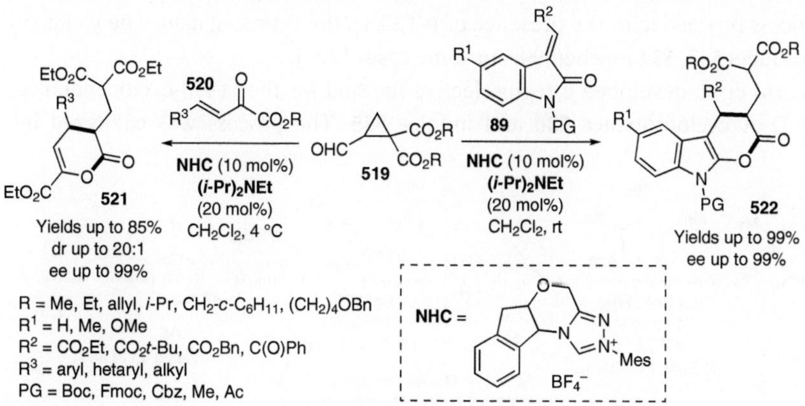

Figure 3.147 Cyclization reactions of D–A cyclopropanes **519** with Michael acceptors **89, 520**.

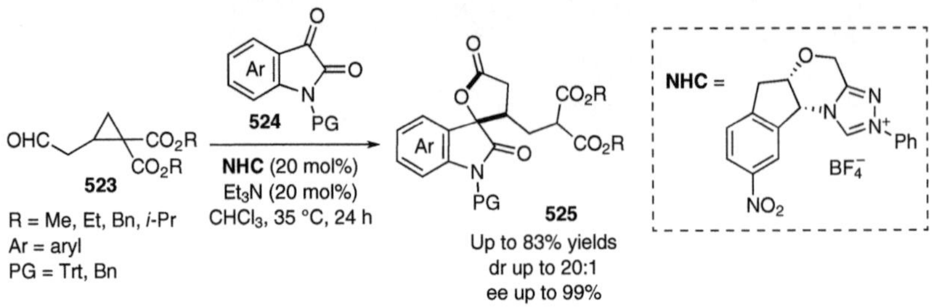

Figure 3.148 Cyclization reaction of D–A cyclopropanes **523** with isatin derivatives **524**.

Y. Chi et al. developed a protocol for the annulation of isatine derivatives **524** with D–A cyclopropane **523**, which have an aldehyde group in the donor part of the cyclopropane. The process successfully proceeded in the presence of NHC. The yields of the final product **525** were up to 83% [289] (Figure 3.148).

3.9.5 Miscellaneous Cyclizations with Phenols and Nitrogen-Containing Heterocycles

In a number of works described below, unusual transformations of D–A cyclopropanes in reactions with phenols and nitrogen-containing heterocycles are presented. It should be noted that these transformations occur as cascades of ring-opening/cyclization reactions. P. F. Xu et al. demonstrated an interaction of D–A cyclopropanes **526** with phenols **527**, **528** containing donor groups. The process proceeds as a formal [4+2] annulation, with the cyclopropane acting as the C4 component. The process proceeded in the presence of $Sc(OTf)_3$. The yields of the final product **529** reached 98%. The authors also successfully developed a protocol for the preparation of spiro-fused quinones **530** with 90% yields of the final products (Figure 3.149) [290].

The Liu work complements the information on [4+2]-annulation reactions in which cyclopropane **526** acts as a C4 component. They developed a protocol for the interaction of D–A cyclopropanes with nitrogen-containing heterocycles **531**, **532**. The process proceeded in the presence of *p*-TSA as the Bronsted acid. The yields of final products **533**, **534** reached 99% in some cases [291].

J. Vicario et al. developed a regioselective method for the [4+2]-cyclocondensation of D–A cyclopropanes **526** and indoles **535**. The process was catalyzed by

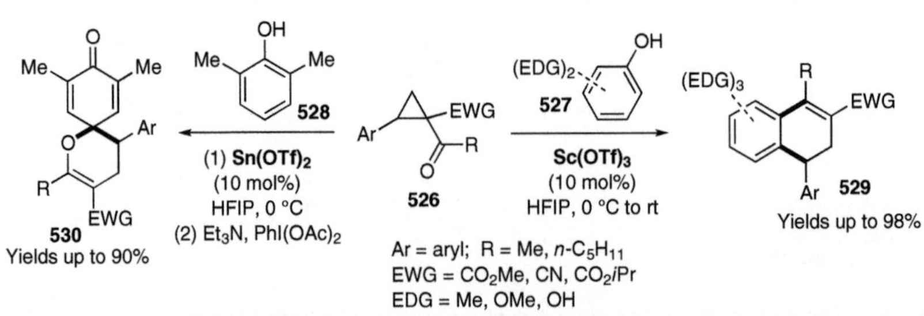

Figure 3.149 Cyclization reactions of D–A cyclopropanes **526** with phenols.

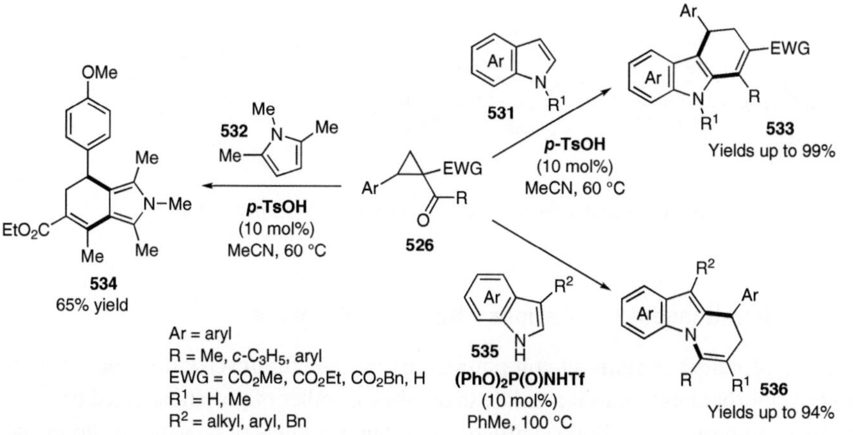

Figure 3.150 Examples of cyclization reactions of D–A cyclopropanes **526** with nitrogen-containing heterocycles.

$(PhO)_2P(O)NHTf$ as the Brønsted acid. The yields of the final product **536** reached 94% (Figure 3.150) [292].

3.9.6 Some Cyclization Reactions of 1,1-Dicyano Cyclopropanes

The presence of reactive cyano groups in the acceptor part of D–A cyclopropanes makes it possible to carry out cascade transformations with various substrates. A study by Q. Fu and C. G. Yan et al demonstrated a method for synthesizing polysubstituted pyrroles **538** by the reaction of cyclopropane **537** and aromatic imines **160** in the presence of $Co(ClO_4)_2$. The yields of the final product **538** were up to 82% (Figure 3.151) [293].

An unusual example of cyclodimerization of D–A cyclopropanes **539** with nitrile groups was presented in a study by C. G. Yan. The process made it possible to obtain polysubstituted indolizine derivatives **540** in a single synthetic step, with good yields of the final products [264].

C. Wang et al. demonstrated a protocol for the annulation of D–A cyclopropanes **539** with thiourea **270** to form substituted pyridines **541**. The process was catalyzed by potassium carbonate, and a wide range of starting substrates with aromatic substituents were involved in the reaction. The yields of the final products **541** reached 95% (Figure 3.152) [294].

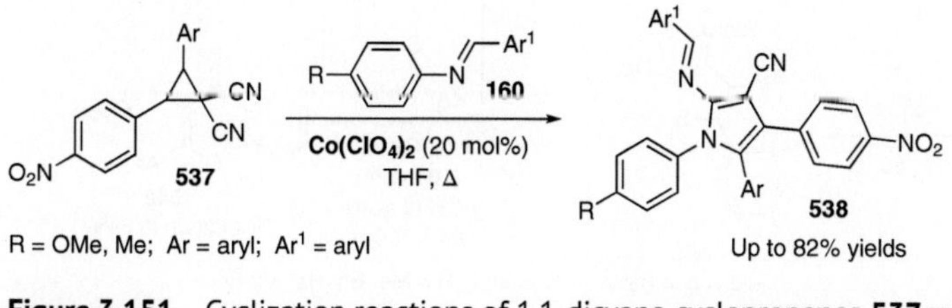

Figure 3.151 Cyclization reactions of 1,1-dicyano cyclopropanes **537** with aromatic imines **160**.

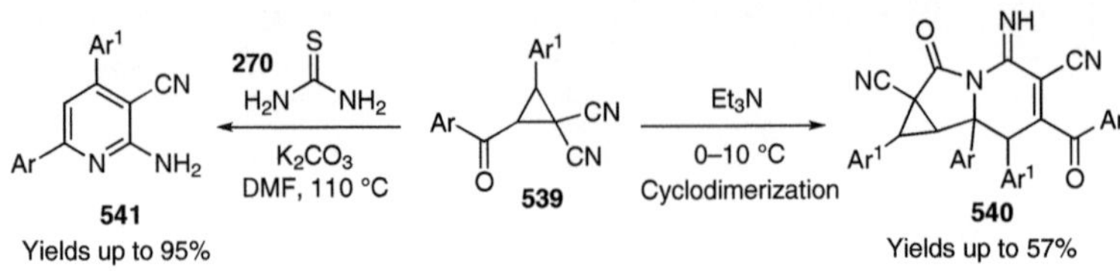

Figure 3.152 Cascade transformations of 1,1-dicyano cyclopropanes **539**.

3.9.7 Miscellaneous Cyclizations with Sulfur Reagents

The use of sulfur-containing compounds in ring-opening/cyclization reactions of D–A cyclopropanes makes it possible to obtain a number of polysubstituted organosulfur compounds in a single synthetic step. The presence of a cyano group in the acceptor moiety of cyclopropane is necessary for providing these transformations. C. Wang et al. presented an example of [4+1]-annulation, in which cyclopropane **542** acts as a C4 component. The process proceeds in the presence of DBU as a base. The yields of the final polysubstituted thiophenes **543** were up to 94% [295]. C. G. Yan et al. carried out the Gewald reaction with cyclopropane **544**, which made it possible to obtain a series of substituted aminothiophenes **545**. The yields of the final thiophenes were up to 83% [296]. Sundaravelu and Sekar proposed a method for the synthesis of thioflavones **547** and thioflavotions **548** from D–A cyclopropanes **546** and xanthan in the presence of copper diacetate. It was possible to involve a wide range of D–A cyclopropanes in the process, both with donor and acceptor groups in the aromatic ring (Figure 3.153) [297].

The work of D. Gladow and H. U. Reissig demonstrated a method for the synthesis of substituted pyrroles **550** and thiophenes **551** from cyclopropanes **549** containing a perfluoroalkyl moiety. This process proceeded with high yields of products and could also be carried out as a *one-pot* reaction (Figure 3.154) [298].

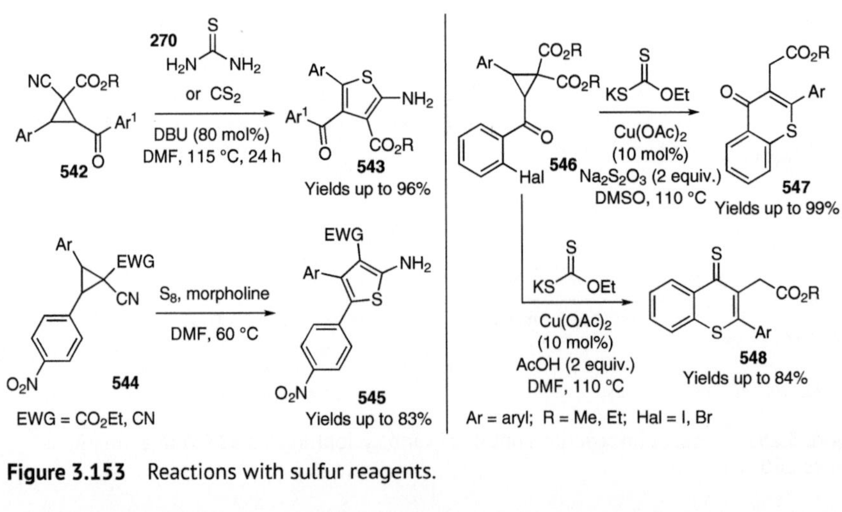

Figure 3.153 Reactions with sulfur reagents.

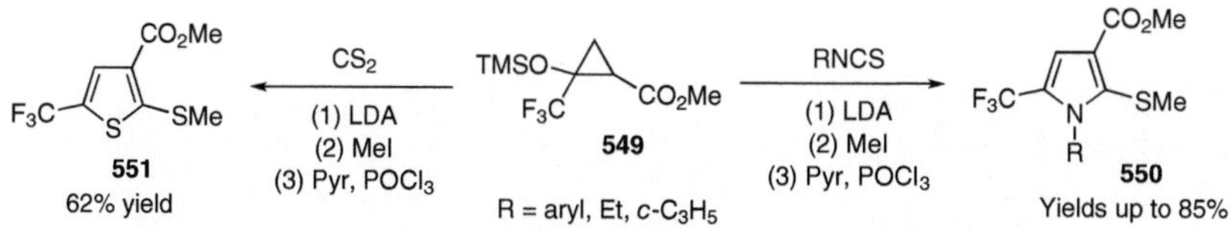

Figure 3.154 Reaction of perfluoroalkyl-substituted cyclopropanes with sulfur reagents.

3.9.8 Cyclizations of Cyclopropanes Containing Carbonyl Group as an Acceptor with Amine Reagents

The presence of a carbonyl group in the acceptor fragment of D–A cyclopropane makes it possible to carry out cascade ring-opening/cyclization reactions in a single synthetic step. S. France et al. developed an efficient protocol for the opening/cyclization of cyclopropane **552** in the presence of amines. This process utilizes a wide range of cyclopropanes and amines The yields of the final pyrrolines **553** were up to 96% [299]. The work of T. Yakura extended this approach to the ring-opening/cyclization of D–A cyclopropanes **554** in the presence of amines. Thus, cyclopropane with a cyclohexane-1,3-dione fragment as an acceptor was used. It should be noted that the process proceeded in the absence of a catalyst with excellent yields of pyrrolines **555** (Figure 3.155) [300].

D. Werz et al. presented a work devoted to the synthesis of pyrroles **557** and **558** from cyclopropanes **556**. A wide range of amines were used in the reaction. Amines containing an acceptor fragment led to the formation of a mono-product **557**, while the remaining amines yielded bis-polysubstituted pyrroles **558** [301] (Figure 3.156).

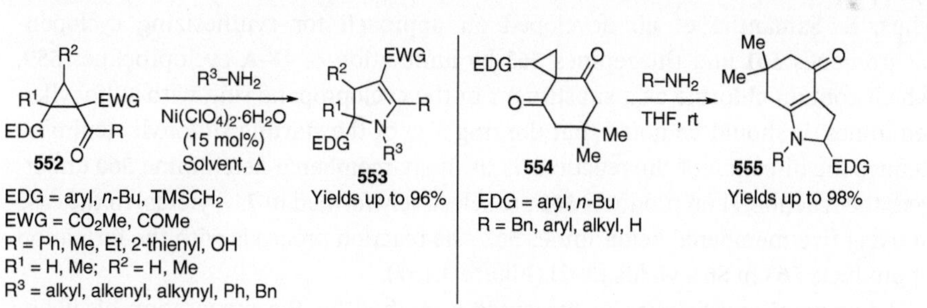

Figure 3.155 Cascade transformations of carbonyl-containing cyclopropanes with amines.

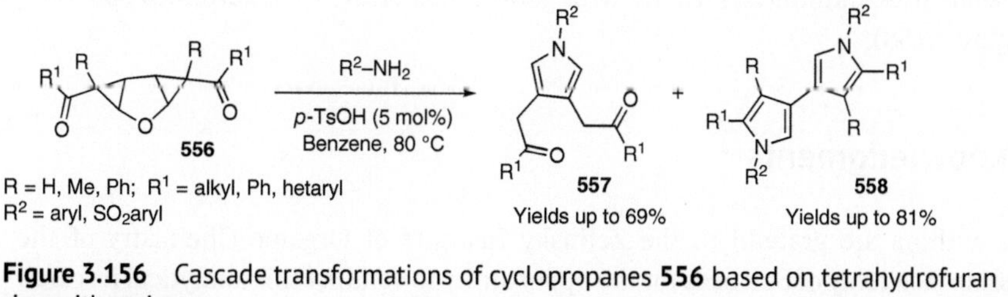

Figure 3.156 Cascade transformations of cyclopropanes **556** based on tetrahydrofuran ring with amines.

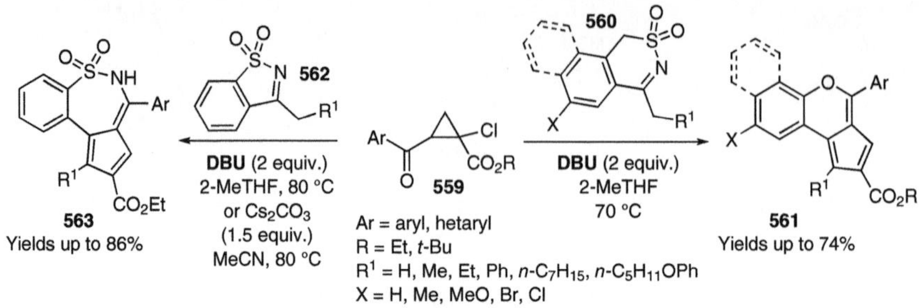

Figure 3.157 Reaction of cyclopropane **559** with sulfonylketenimines.

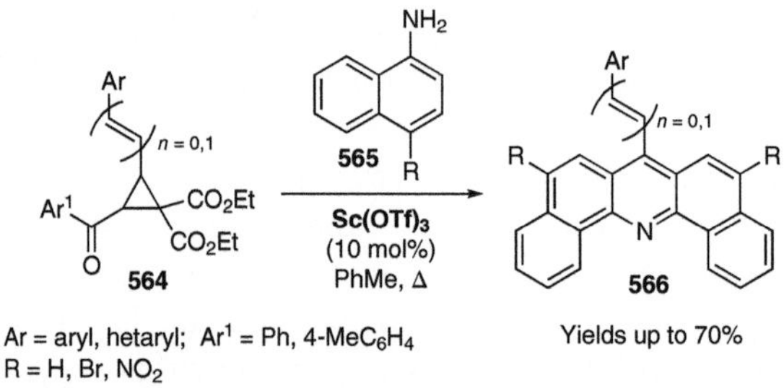

Figure 3.158 Reaction of cyclopropanes **564** with 1-naphthylamines **565**.

3.9.9 Miscellaneous Reactions

There are a number of data in the literature on unusual transformations of D–A cyclopropanes, which are difficult to attribute to specific classes of transformations. Thus, S. Samantha et al. developed an approach for synthesizing cyclopentachromenes **561** and thiazepines **563** by annulation of D–A cyclopropanes **559**, which contain chlorine as a substituent in the cyclopropane ring with sulfonylketenimines. It should be noted that the ring size of the starting sulfonylketenimine defines the direction of the reaction. Thus, if six-membered ketenimine **560** undergoes the reaction, then products **561** are selectively formed in 74% yields. In the case of using five-membered ketenimines **562**, the reaction proceeds with the formation of products **563** in 86% yields [302] (Figure 3.157).

Thangamani and Srinivasan presented a method for the preparation of dibenzacridins **566** by the reaction of substituted 1-naphthylamines **565** and cyclopropanes **564**, which act as a source of the C1 fragment. The process proceeded in the presence of scandium(III) triflate with good yields of the final acridines **566** [303] (Figure 3.158).

Acknowledgments

The authors are grateful to the Zelinsky Institute of Organic Chemistry of the Russian Academy of Sciences for the opportunity to conduct this analytical research and for the support of the Russian Science Foundation (grant No. 22-13-00418).

References

1 Reissig, H.U. and Zimmer, R. (2003). *Chem. Rev.* 103: 1151–1196.
2 Schneider, T.F., Kaschel, J., and Werz, D.B. (2014). *Angew. Chem. Int. Ed.* 53: 5504–5523.
3 Tomilov, Y.V., Menchikov, L.G., Novikov, R.A. et al. (2018). *Russ. Chem. Rev.* 87: 201–250.
4 Pirenne, V., Muriel, B., and Waser, J. (2021). *Chem. Rev.* 121: 227–263.
5 Augustin, A.U. and Werz, D.B. (2021). *Acc. Chem. Res.* 54: 1528–1541.
6 O'Connor, N.R., Wood, J.L., and Stoltz, B.M. (2016). *Isr. J. Chem.* 56: 431–444.
7 Ivanova, O.A. and Trushkov, I.V. (2019). *Chem. Rec.* 19: 2189–2208.
8 Ghosh, K. and Das, S. (2021). *Org. Biomol. Chem.* 19: 965–982.
9 Corey, E.J. and Balanson, R.D. (1973). *Tetrahedron Lett.* 14: 3153–3156.
10 Dolfini, J.E., Menich, K., and Corliss, P. (1966). *Tetrahedron Lett.* 7: 4421–4426.
11 Berkowitz, W.F. and Grenetz, S.C. (1976). *J. Org. Chem.* 41: 10–13.
12 Komatsu, M., Suehiro, I., Horiguchi, Y., and Kuwajima, I. (1991). *Synlett* 11: 771–773.
13 Saigo, K., Shimada, S., Shibasaki, T., and Hasegawa, M. (1990). *Chem. Lett.* 19: 1093–1096.
14 Marino, J.P. and Laborde, E. (1985). *J. Am. Chem. Soc.* 107: 734–735.
15 Graziano, M.L. and Chiosi, S. (1989). *J. Chem. Res. (S)* 2: 44–45.
16 Beal, R.B., Dombroski, M.A., and Snider, B.B. (1986). *J. Org. Chem.* 51: 4391–4399.
17 Shimizu, I., Ohashi, Y., and Tsuji, J. (1985). *Tetrahedron Lett.* 26: 3825–3828.
18 Sugita, Y., Yamadoi, S., Hosoya, H., and Yokoe, I. (2001). *Chem. Pharm. Bull.* 49: 657–658.
19 Liu, L. and Montgomery, J. (2006). *J. Am. Chem. Soc.* 128: 5348–5349.
20 Takasu, K., Nagao, S., and Ihara, M. (2006). *Adv. Synth. Cat.* 348: 2376–2380.
21 Fang, J., Ren, J., and Wang, Z. (2008). *Tetrahedron Lett.* 49: 6659–6662.
22 He, Y., Zhu, X., Hu, C. et al. (2019). *ChemistrySelect* 4: 1437–1440.
23 Trost, B.M., Morris, P.J., and Sprague, S.J. (2012). *J. Am. Chem. Soc.* 134: 17823–17831.
24 Xu, H., Qu, J.-P., Liao, S. et al. (2013). *Angew. Chem. Int. Ed.* 52: 4004–4007.
25 de Nanteuil, F., Serrano, E., Perrotta, D., and Waser, J. (2014). *J. Am. Chem. Soc.* 136: 6239–6242.
26 Serrano, E., de Nanteuil, F., and Waser, J. (2014). *Synlett* 25: 2285–2288.
27 Racine, S., de Nanteuil, F., Serrano, E., and Waser, J. (2014). *Angew. Chem. Int. Ed.* *53*: 8484–8487.
28 Budynina, E.M., Ivanova, O.A., Chagarovskiy, A.O. et al. (2015). *J. Org. Chem.* 80: 12212–12223.
29 Drew, M.A., Tague, A.J., Richardson, C. et al. (2021). *Org. Lett.* 23: 4635–4639.
30 Verma, K. and Banerjee, P. (2017). *Adv. Synth. Catal.* 359: 3848–3854.
31 Verma, K., Taily, I.M., and Banerjee, P. (2019). *Org. Biomol. Chem.* 17: 8149–8152.
32 Verma, K. and Banerjee, P. (2016). *Adv. Synth. Catal.* 358: 2053–2058.
33 Cheng, Q.-Q., Qian, Y., Zavalij, P.Y., and Doyle, M.P. (2015). *Org. Lett.* 17: 3568–3571.

34 Dey, R. and Banerjee, P. (2017). *Org. Lett.* 19: 304–307.

35 Kaga, A., Gandamana, D.A., Tamura, S. et al. (2017). *Synlett* 28: 1091–1095.

36 Sasaki, M., Kondo, Y., Nishio, T., and Takeda, K. (2016). *Org. Lett.* 18: 3858–3861.

37 Blom, J., Vidal-Albalat, A., Jorgensen, J. et al. (2017). *Angew. Chem. Int. Ed.* 56: 11831–11835.

38 Dai, C., Li, M., Chen, M. et al. (2019). *J. Chem. Res.* 43: 43–49.

39 Lu, Z., Shen, M., and Yoon, T.P. (2011). *J. Am. Chem. Soc.* 133: 1162–1164.

40 Wang, C., Ren, X., Xie, H., and Lu, Z. (2015). *Chem. – Eur. J.* 21: 9676–9680.

41 Zhang, J., Jiang, H., and Zhu, S. (2017). *Adv. Synth. Catal.* 359: 2924–2930.

42 Nguyen, H.M., Chand, H.R., Golantsov, N.E. et al. (2020). *Synlett* 31: 295–299.

43 Suleymanov, A.A., Le Du, E., Dong, Z. et al. (2020). *Org. Lett.* 22: 4517–4522.

44 Ahlburg, N.L., Jones, P.G., and Werz, D.B. (2020). *Org. Lett.* 22: 6404–6408.

45 Luecht, A., Kreft, A., Jones, P.G., and Werz, D.B. (2020). *Eur. J. Org. Chem.* 17: 2560–2564.

46 Xu, C., Wei, N., Zhu, D., and Wang, M. (2020). *ChemistrySelect* 5: 11399–11402.

47 Zhu, X., Pan, D., Mou, C. et al. (2020). *Front. Chem.* 8: 542.

48 Xiao, J.A., Li, J.L., Cheng, X.L. et al. (2021). *Chem. Comm.* 57: 4456–4459.

49 Singh, K., Pramanik, S., Hamlin, T.A. et al. (2019). *Chem. Commun.* 55: 7069–7072.

50 Graziano, M.L., Iesce, M.R., Cermola, F., and Cimminiello, G. (1992). *J. Chem. Res. (S)* 1: 4–5.

51 Yadav, V.K. and Sriramurthy, V. (2004). *Angew. Chem. Int. Ed.* 43: 2669–2671.

52 Tamaki, T., Ohashi, M., and Ogoshi, S. (2011). *Angew. Chem. Int. Ed.* 50: 12067–12070.

53 Qi, X. and Ready, J.M. (2008). *Angew. Chem. Int. Ed.* 47: 7068–7070.

54 Mackay, W.D., Fistikci, M., Carris, R.M., and Johnson, J.S. (2014). *Org. Lett.* 16: 1626–1629.

55 Racine, S., Hegedues, B., Scopelliti, R., and Waser, J. (2016). *Chem. – Eur. J.* 22: 11997–12001.

56 Xia, X.-F., Song, X.-R., Liu, X.-Y., and Liang, Y.-M. (2012). *Chem. Asian J.* 7: 1538–1541.

57 Luo, Z., Zhou, B., and Li, Y. (2012). *Org. Lett.* 14: 2540–2543.

58 Harrington, P. and Kerr, M.A. (1997). *Tetrahedron Lett.* 38: 5949–5952.

59 Kerr, M.A. and Keddy, R.G. (1999). *Tetrahedron Lett.* 40: 5671–5675.

60 England, D.B., Kuss, T.D.O., Keddy, R.G., and Kerr, M.A. (2001). *J. Org. Chem.* 66: 4704–4709.

61 Venkatesh, C., Singh, P.P., Ila, H., and Junjappa, H. (2006). *Eur. J. Org. Chem.* 23: 5378–5386.

62 Bajtos, B., Yu, M., Zhao, H., and Pagenkopf, B.L. (2007). *J. Am. Chem. Soc.* 129: 9631–9634.

63 Xiong, H., Xu, H., Liao, S. et al. (2013). *J. Am. Chem. Soc.* 135: 7851–7854.

64 Pirenne, V., Robert, E.G.L., and Waser, J. (2021). *Chem. Sci.* 12: 8706–8712.

65 Zhu, J., Liang, Y., Wang, L. et al. (2014). *J. Am. Chem. Soc.* 136: 6900–6903.

66 Chagarovskiy, A.O., Budynina, E.M., Ivanova, O.A. et al. (2009). *Tetrahedron* 65: 5385–5392.

67 Ling, J., Laugeois, M., Michelet, V. et al. (2018). *Synlett* 29: 928–932.

68 Kaicharla, T., Roy, T., Thangaraj, M. et al. (2016). *Angew. Chem. Int. Ed.* 55: 10061–10064.

69 Zhao, H., Shen, P., Sun, D. et al. (2021). *Front. Chem.* 9: 711257.

70 Morra, N.A., Morales, C.L., Bajtos, B. et al. (2006). *Adv. Synth. Cat.* 348: 2385–2390.

71 Wang, D.-C., Xie, M.-S., Guo, H.-M. et al. (2016). *Angew. Chem. Int. Ed.* 55: 14111–14115.

72 Volkova, Y.A., Budynina, E.M., Kaplun, A.E. et al. (2013). *Chem. – Eur. J.* 19: 6586–6590.

73 Ivanova, O.A., Budynina, E.M., Skvortsov, D.A. et al. (2013). *Chem. Commun.* 49: 11482–11484.

74 Rakhmankulov, E.R., Ivanov, K.L., Budynina, E.M. et al. (2015). *Org. Lett.* 17: 770–773.

75 Reissig, H.U. (1981). *Tetrahedron Lett.* 22: 2981–2984.

76 Nakamura, E. and Kuwajima, I. (1977). *J. Am. Chem. Soc.* 22: 7360–7362.

77 Pohlhaus, P.D. and Johnson, J.S. (2005). *J. Am. Chem. Soc.* 127: 16014–16015.

78 Pohlhaus, P.D. and Johnson, J.S. (2005). *J. Org. Chem.* 70: 1057–1059.

79 Shimada, S., Hashimoto, Y., Sudo, A. et al. (1992). *J. Org. Chem.* 57: 7126–7133.

80 Shimada, S., Hashimoto, Y., and Saigo, K. (1993). *J. Org. Chem.* 58: 5226–5234.

81 Sugita, Y., Kawai, K., and Yokoe, I. (2000). *Heterocycles* 53: 657–664.

82 Han, Z., Uehira, S., Tsuritani, T. et al. (2001). *Tetrahedron* 57: 987–995.

83 Pohlhaus, P.D., Sanders, S.D., Parsons, A.T. et al. (2008). *J. Am. Chem. Soc.* 130: 8642–8650.

84 Campbell, M.J., Johnson, J.S., Parsons, A.T. et al. (2010). *J. Org. Chem.* 75: 6317–6325.

85 Parsons, A.T. and Johnson, J.S. (2009). *J. Am. Chem. Soc.* 131: 3122–3123.

86 Parsons, A.T., Campbell, M.J., and Johnson, J.S. (2008). *Org. Lett.* 10: 2541–2544.

87 Gupta, A. and Yadav, V.K. (2006). *Tetrahedron Lett.* 47: 8043–8047.

88 Smith, A.G., Slade, M.C., and Johnson, J.S. (2011). *Org. Lett.* 13: 1996–1999.

89 Haubenreisser, S., Hensenne, P., Schröder, S., and Niggemann, M. (2013). *Org. Lett.* 15: 2262–2265.

90 Yang, G., Shen, Y., Li, K. et al. (2011). *J. Org. Chem.* 76: 229–233.

91 Yang, G., Sun, Y., Shen, Y. et al. (2013). *J. Org. Chem.* 78: 5393–5400.

92 Ma, X., Tang, Q., Ke, J. et al. (2013). *Org. Lett.* 15: 5170–5173.

93 Benfatti, F., de Nanteuil, F., and Waser, J. (2012). *Org. Lett.* 14: 386–389.

94 Zhang, C., Xu, M., Ren, J., and Wang, Z. (2016). *Eur. J. Org. Chem.* 14: 2467–2478.

95 Sabbatani, J. and Maulide, N. (2016). *Angew. Chem. Int. Ed.* 55: 6780–6783.

96 Gupta, A., Kholiya, R., and Rawat, D.S. (2017). *Asian J. Org. Chem.* 6: 993–997.

97 Xu, P.-W., Liu, J.-K., Shen, L. et al. (2017). *Nat. Commun* 8: 1–8.

98 Kreft, A., Jones, P.G., and Werz, D.B. (2018). *Org. Lett.* 20: 2059–2062.

99 Yang, P., Shen, Y., Feng, M. et al. (2018). *Eur. J. Org. Chem.* 30: 4103–4112.

100 Su, P., Li, H., Chen, W. et al. (2020). *Eur. J. Org. Chem.* 33: 5380–5387.

101 Mikhaylov, A.A., Kuleshov, A.V., Solyev, P.N. et al. (2020). *Org. Lett.* 22: 2740–2745.

102 Shiba, T., Kuroda, D., Kurahashi, T., and Matsubara, S. (2014). *Synlett* 25: 2005–2008.
103 Novikov, R.A., Levina, A.A., Borisov, D.D. et al. (2020). *Organometallics* 39: 2580–2593.
104 Sanders, S.D., Ruiz-Olalla, A., and Johnson, J.S. (2009). *Chem. Commun.* 34: 5135–5137.
105 Campbell, M.J. and Johnson, J.S. (2010). *Synthesis* 16: 2841–2852.
106 Rivero, A.R., Fernández, I., de Arellano, C.R., and Sierra, M.A. (2015). *J. Org. Chem.* 80: 1207–1213.
107 Zheng, Z.-B., Cheng, W.-F., Wang, L. et al. (2020). *Chin. J. Chem.* 38: 1629–1634.
108 Xing, S., Pan, W., Liu, C. et al. (2010). *Angew. Chem. Int. Ed.* 49: 3215–3218.
109 Xing, S., Li, Y., Li, Z. et al. (2011). *Angew. Chem.* 123: 12813–12817.
110 Wang, Z., Chen, S., Ren, J., and Wang, Z. (2015). *Org. Lett.* 17: 4184–4187.
111 Zhang, J., Xing, S., Ren, J. et al. (2015). *Org. Lett.* 17: 218–221.
112 Alper, P.B., Meyers, C., Lerchner, A. et al. (1999). *Angew. Chem. Int. Ed.* 38: 3186–3189.
113 Bertozzi, F., Gustafsson, M., and Olsson, R. (2002). *Org. Lett.* 4: 3147–3150.
114 Saigo, K., Shimada, S., and Hasegawa, M. (1990). *Chem. Lett.* 19: 905–908.
115 Carson, C.A. and Kerr, M.A. (2005). *J. Org. Chem.* 70: 8242–8244.
116 Jackson, S.K., Karadeolian, A., Driega, A.B., and Kerr, M.A. (2008). *J. Am. Chem. Soc.* 130: 4196–4201.
117 Kang, Y.-B., Tang, Y., and Sun, X.-L. (2006). *Org. Biomol. Chem.* 4: 299–301.
118 Parsons, A.T., Smith, A.G., Neel, A.J., and Johnson, J.S. (2010). *J. Am. Chem. Soc.* 132: 9688–9692.
119 Akaev, A.A., Bezzubov, S.I., Desyatkin, V.G. et al. (2019). *J. Org. Chem.* 84: 3340–3356.
120 Christie, S.D.R., Davoile, R.J., and Jones, R.C.F. (2006). *Org. Biomol. Chem.* 4: 2683–2684.
121 Lautens, M. and Han, W.S. (2002). *J. Am. Chem. Soc.* 124: 6312–6316.
122 Scott, M.E., Han, W., and Lautens, M. (2004). *Org. Lett.* 6: 3309–3312.
123 Taillier, C. and Lautens, M. (2007). *Org. Lett.* 9: 591–593.
124 Li, J., Xiao, J.-A., Zhao, S.-J. et al. (2017). *Synthesis* 49: 4292–4298.
125 Dey, R., Kumar, P., and Banerjee, P. (2018). *J. Org. Chem.* 83: 5438–5449.
126 Wang, Z.-H., Zhang, H.-H., Xu, P.-F., and Luo, Y.-C. (2018). *Chem. Commun.* 54: 10128–10131.
127 Tejeda, J.E.C., Irwin, L.C., and Kerr, M.A. (2016). *Org. Lett.* 18: 4738–4741.
128 Buev, E.M., Moshkin, V.S., and Sosnovskikh, V.Y. (2016). *Tetrahedron Lett.* 57: 3731–3734.
129 Garve, L.K.B., Kreft, A., Jones, P.G., and Werz, D.B. (2017). *J. Org. Chem.* 82: 9235–9242.
130 Chu, Z.-Y., Li, N., Liang, D. et al. (2018). *Tetrahedron Lett.* 59: 715–718.
131 Xiao, J.-A., Li, J., Xia, P.-J. et al. (2016). *J. Org. Chem.* 81: 11185–11194.
132 Irwin, L.C., Allen, M.A., Vriesen, M.R., and Kerr, M.A. (2020). *Chem. Eur. J.* 26: 171–175.
133 Carson, C.A. and Kerr, M.A. (2009). *Org. Lett.* 11: 777–779.

134 Marti, C. and Carreira, E.M. (2005). *J. Am. Chem. Soc.* 127: 11505–11515.

135 Yadav, V.K. and Sriramurthy, V. (2004). *Org. Lett.* 6: 4495–4498.

136 Graziano, M.L. and Iesce, M.R. (1987). *J. Chem. Res.* 362–363.

137 Graziano, M.L. and Cimminiello, G. (1989). *J. Chem. Res.* 42–43.

138 Mondal, M., Panda, M., Davis, N.W. et al. (2019). *Chem. Commun.* 55: 13558–13561.

139 Mondal, M., Panda, M., McKee, V., and Kerrigan, N.J. (2019). *J. Org. Chem.* 84: 11983–11991.

140 Augustin, A.U., Busse, M., Jones, P.G., and Werz, D.B. (2018). *Org. Lett.* 20: 820–823.

141 Mloston, G., Kowalczyk, M., Augustin, A.U. et al. (2021). *Eur. J. Org. Chem. Open Access* 2021: 6250–6253.

142 Alajarin, M., Egea, A., Orenes, R.-A., and Vidal, A. (2016). *Org. Biomol. Chem.* 14: 10275–10284.

143 Goldberg, A.F.G., O'Connor, N.R., Craig, R.A., and Stoltz, B.M. (2012). *Org. Lett.* 14: 5314–5317.

144 Tsunoi, S., Maruoka, Y., Suzuki, I., and Shibata, I. (2015). *Org. Lett.* 17: 4010–4013.

145 Feng, M., Yang, P., Yang, G. et al. (2018). *J. Org. Chem.* 83: 174–184.

146 Andreev, I.A., Ratmanova, N.K., Augustin, A.U. et al. (2021). *Angew. Chem. Int. Ed.* 60: 7927–7934.

147 Werz, D.B., Oliver, G.A., Loch, M.N. et al. (2021). *Angew. Chem. Int. Ed. Open Access* 60: 25825–25831.

148 Singh, P.R., Kalaramna, P., Ali, S., and Goswami, A. (2021). *Eur. J. Org. Chem.* 33: 4683–4689.

149 Jacob, A., Barkawitz, P., Andreev, I.A. et al. (2021). *Synlett* 32: 901–904.

150 Jacob, A., Jones, P.G., and Werz, D.B. (2020). *Org. Lett.* 22: 8720–8724.

151 Mloston, G., Kowalczyk, M., Augustin, A.U. et al. (2020). *Beilstein J. Org. Chem.* 16: 1288–1295.

152 Augustin, A.U., Sensse, M., Jones, P.G., and Werz, D.B. (2017). *Angew. Chem. Int. Ed.* 56: 14293–14296.

153 Matsumoto, Y., Nakatake, D., Yazaki, R., and Ohshima, T. (2018). *Chem. – Eur. J.* 24: 6062–6066.

154 Xie, M.-S., Zhao, G.-F., Qin, T. et al. (2019). *Chem. Commun.* 55: 1580–1583.

155 Cermola, F., di Gioia, L., Graziano, M.L., and Iesce, M.R. (2005). *J. Chem. Res.* 10: 677–681.

156 Chakrabarty, S., Chatterjee, I., Wibbeling, B. et al. (2014). *Angew. Chem. Int. Ed.* 53: 5964–5968.

157 Varshnaya, R.K. and Banerjee, P. (2016). *Eur. J. Org. Chem.* 23: 4059–4066.

158 Korotkov, V.S., Larionov, O.V., Hofmeister, A. et al. (2007). *J. Org. Chem.* 72: 7504–7510.

159 Chidley, T., Vemula, N., Carson, C.A. et al. (2016). *Org. Lett.* 18: 2922–2925.

160 Tomilov, Y.V., Novikov, R.A., and Nefedov, O.M. (2010). *Tetrahedron* 66: 9151–9158.

161 Yu, M. and Pagenkopf, B.L. (2003). *Org. Lett.* 5: 5099–5101.

162 Yu, M. and Pagenkopf, B.L. (2003). *J. Am. Chem. Soc.* 125: 8122–8123.

163 Yu, M., Pantos, G.D., Sessler, J.L., and Pagenkopf, B.L. (2004). *Org. Lett.* 6: 1057–1059.

164 Moustafa, M.M.A.R. and Pagenkopf, B.L. (2010). *Org. Lett.* 12: 3168–3171.

165 Morales, C.L. and Pagenkopf, B.L. (2008). *Org. Lett.* 10: 157–159.

166 Bajtos, B. and Pagenkopf, B.L. (2009). *Eur. J. Org. Chem.* 7: 1072–1077.

167 Sathishkannan, G. and Srinivasan, K. (2011). *Org. Lett.* 13: 6002–6005.

168 Chagarovskiy, A.O., Ivanov, K.L., Budynina, E.M. et al. (2012). *Chem. Het. Comp.* 48: 825–827.

169 Cui, B., Ren, J., and Wang, Z. (2014). *J. Org. Chem.* 79: 790–796.

170 Tamilarasan, V.J. and Srinivasan, K. (2019). *J. Org. Chem.* 84: 8782–8787.

171 Tamilarasan, V.J. and Srinivasan, K. (2020). *Eur. J. Org. Chem.* 5: 593–598.

172 Kumar Pandey, A., Ghosh, A., and Banerjee, P. (2015). *Eur. J. Org. Chem.* 11: 2517–2523.

173 Ghosh, A., Mandal, S., Chattaraj, P.K., and Banerjee, P. (2016). *Org. Lett.* 18: 4940–4943.

174 Pandey, A.K., Varshnaya, R.K., and Banerjee, P. (2017). *Eur. J. Org. Chem.* 12: 1647–1656.

175 Ghosh, A., Pandey, A.K., and Banerjee, P. (2015). *J. Org. Chem.* 80: 7235–7242.

176 Chagarovskiy, A.O., Vasin, V.S., Kuznetsov, V.V. et al. (2018). *Angew. Chem. Int. Ed.* 57: 10338–10342.

177 Grover, H.K., Lebold, T.P., and Kerr, M.A. (2011). *Org. Lett.* 13: 220–223.

178 Liu, Q.-J., Yan, W.-G., Wang, L. et al. (2015). *Org. Lett.* 17: 4014–4017.

179 Talukdar, R., Tiwari, D.P., Saha, A., and Ghorai, M.K. (2014). *Org. Lett.* 16: 3954–3957.

180 Faltracco, M., Damian, M., and Ruijter, E. (2021). *Org. Lett.* 23: 7592–7596.

181 Faltracco, M., Strähler, S., Snabilié, D., and Ruijter, E. (2022). *Adv. Synth. Catal.* 364: 53–57.

182 Varshnaya, R.K. and Banerjee, P. (2019). *J. Org. Chem.* 84: 1614–1623.

183 Ivanova, O.A., Budynina, E.M., Chagarovskiy, A.O. et al. (2011). *J. Org. Chem.* 76: 8852–8868.

184 Ivanova, O.A., Budynina, E.M., Khrustalev, V.N. et al. (2016). *Chem. – Eur.J.* 22: 1223–1227.

185 Sin, S. and Kim, S.-G. (2016). *Adv. Synth. Catal.* 358: 2701–2706.

186 Das, S., Chakrabarty, S., Daniliuc, C.G., and Studer, A. (2016). *Org. Lett.* 18: 2784–2787.

187 Das, B.K., Pradhan, S., and Punniyamurthy, T. (2019). *Chem. Commun.* 55: 8083–8086.

188 Sapeta, K. and Kerr, M.A. (2009). *Org. Lett.* 11: 2081–2084.

189 Mondal, K. and Pan, S.C. (2017). *Eur. J. Org. Chem.* 3: 534–537.

190 Leduc, A.B., Lebold, T.P., and Kerr, M.A. (2009). *J. Org. Chem.* 74: 8414–8416.

191 Ma, W., Fang, J., Ren, J., and Wang, Z. (2015). *Org. Lett.* 17: 4180–4183.

192 Sathishkannan, G. and Srinivasan, K. (2014). *Chem. Commun.* 50: 4062–4064.

193 Fu, X., Lin, L., Xia, Y. et al. (2016). *Org. Biomol. Chem.* 14: 5914–5917.

194 Wang, H.P., Zhang, H.H., Hu, X.Q. et al. (2015). *Eur. J. Org. Chem.* 2015: 3486–3494.

195 Young, I.S. and Kerr, M.A. (2003). *Angew. Chem. Int. Ed.* 42: 3023–3026.
196 Young, I.S. and Kerr, M.A. (2004). *Org. Lett.* 6: 139–141.
197 Ganton, M.D. and Kerr, M.A. (2004). *J. Org. Chem.* 69: 8554–8557.
198 Lebold, T.P., Carson, C.A., and Kerr, M.A. (2006). *Synlett* 3: 364–368.
199 Sapeta, K. and Kerr, M.A. (2007). *J. Org. Chem.* 72: 8597–8599.
200 Karadeolian, A. and Kerr, M.A. (2007). *J. Org. Chem.* 72: 10251–10253.
201 Sibi, M.P., Ma, Z.H., and Jasperse, C.P. (2005). *J. Am. Chem. Soc.* 127: 5764–5765.
202 Kang, Y.-B., Sun, X.-L., and Tang, Y. (2007). *Angew. Chem. Int. Ed.* 46: 3918–3921.
203 Johansen, M.B. and Kerr, M.A. (2008). *Org. Lett.* 10: 3497–3500.
204 Humenny, W.J., Kyriacou, P., Sapeta, K. et al. (2012). *Angew. Chem. Int. Ed.* 51: 11088–11091.
205 Braun, C.M., Congdon, E.A., and Nolin, K.A. (2015). *J. Org. Chem.* 80: 1979–1984.
206 Zhang, Y., Liu, F., and Zhang, J. (2010). *Chem. – Eur.J.* 16: 6146–6150.
207 Dhote, P.S. and Ramana, C.V. (2019). *Org. Lett.* 21: 6221–6224.
208 Dias, D.A. and Kerr, M.A. (2009). *Org. Lett.* 11: 3694–3697.
209 Lebold, T.P. and Kerr, M.A. (2010). *Pure Appl. Chem.* 82: 1797–1812.
210 Gorbacheva, E.O., Tabolin, A.A., Novikov, R.A. et al. (2013). *Org. Lett.* 15: 350–353.
211 Tabolin, A.A., Novikov, R.A., Khomutova, Y.A. et al. (2015). *Tetrahedron Lett.* 56: 2102–2105.
212 Mikhaylov, A.A., Novikov, R.A., Khomutova, Y.A. et al. (2014). *Synlett* 25: 2275–2280.
213 Garve, L.K.B., Petzold, M., Jones, P.G., and Werz, D.B. (2016). *Org. Lett.* 18: 564–567.
214 Mishra, M., De, P.B., Pradhan, S., and Punniyamurthy, T. (2019). *J. Org. Chem.* 84: 10901–10910.
215 Mandadapu, R., Dehade, A.S., Shete, S.A. et al. (2021). *Synthesis* 53: 2865–2873.
216 Liu, H., Yuan, C., Wu, Y. et al. (2015). *Org. Lett.* 17: 4220–4223.
217 Perreault, C., Goudreau, S.R., Zimmer, L.E., and Charette, A.B. (2008). *Org. Lett.* 10: 689–692.
218 Zhang, H.-H., Luo, Y.-C., Wang, H.-P. et al. (2014). *Org. Lett.* 16: 4896–4899.
219 Novikov, R.A., Borisov, D.D., and Tomilov, Y.V. (2018). *Russ. Chem. Bull.* 67: 265–273.
220 Borisova, I.A., Ratova, D.-V., Potapov, K.V. et al. (2021). *J. Org. Chem.* 86: 4567–4579.
221 Petzold, M., Jones, P.G., and Werz, D.B. (2019). *Angew. Chem. Int. Ed.* 58: 6225–6229.
222 Ivanova, O.A., Budynina, E.M., Grishin, Y.K. et al. (2008). *Angew. Chem. Int. Ed.* 47: 1107–1110.
223 Ivanova, O.A., Budynina, E.M., Chagarovskiy, A.O. et al. (2011). *Adv. Synth. Cat.* 353: 1125–1134.
224 Chagarovskiy, A.O., Ivanova, O.A., Budynina, E.M. et al. (2013). *Russ. Chem. Bull.* 62: 2407–2423.
225 Ivanova, O.A., Budynina, E.M., Grishin, Y.K. et al. (2008). *Eur. J. Org. Chem.* 31: 5329–5335.
226 Xu, H., Hu, J.-L., Wang, L. et al. (2015). *J. Am. Chem. Soc.* 137: 8006–8009.

227 Wang, Z.-H., Zhang, H.-H., Wang, D.-M. et al. (2017). *Chem. Commun.* 53: 8521–8524.
228 Zhang, X., Feng, M., Yang, G., and Chai, Z. (2020). *J. Org. Chem.* 85: 430–440.
229 Augustin, A.U., Merz, J.L., Jones, P.G. et al. (2019). *Org. Lett. 21*: 9405–9409.
230 Garve, L.K.B., Pawliczek, M., Wallbaum, J. et al. (2016). *Chem. Eur. J.* 22: 521–525.
231 Garve, L.K.B., Jones, P.G., and Werz, D.B. (2017). *Angew. Chem. Int. Ed.* 56: 9226–9230.
232 Faltracco, M., van de Vrande, K.N.A., Dijkstra, M. et al. (2021). *Angew. Chem. Int. Ed.* 60: 14410–14414.
233 Faltracco, M., Strähler, S., Snabilié, D., and Ruijter, E. (2021). *Adv. Synth. Catal. Open Access 364*: 53–57.
234 Rivero, A.R., Fernández, I., and Sierra, M.Á. (2013). *Org. Lett.* 15: 4928–4931.
235 Tejero, R., Ponce, A., Adrio, J., and Carretero, J.C. (2013). *Chem. Commun.* 49: 10406–10408.
236 Li, B.Q., Qiu, Z.-W., Ma, A.-J. et al. (2020). *Org. Lett.* 22: 1903–1907.
237 Okabe, A., Harada, S., Takeda, T., and Nishida, A. (2019). *Eur. J. Org. Chem.* 24: 3916–3920.
238 Potapov, K.V., Denisov, D.A., Glushkova, V.V. et al. (2020). *J. Org. Chem.* 85: 15562–15576.
239 Denisov, D.A., Novikov, R.A., and Tomilov, Y.V. (2021). *Russ. Chem. Bull.* 70: 1568–1574.
240 Novikov, R.A. and Tomilov, Y.V. (2015). *Mendeleev Commun.* 25: 1–10.
241 Shintani, R., Murakami, M., Tsuji, T. et al. (2009). *Org. Lett.* 11: 5642–5645.
242 Mei, L.-Y., Tang, X.-Y., and Shi, M. (2014). *Chem. Eur. J.* 20: 13136–13142.
243 Korotkov, V.S., Larionov, O.V., and de Meijere, A. (2006). *Synthesis* 21: 3542–3546.
244 Chagarovskiy, A.O., Ivanova, O.A., Rakhmankulov, E.R. et al. (2010). *Adv. Synth. Cat.* 352: 3179–3184.
245 Borisov, D.D., Chermashentsev, G.R., Novikov, R.A., and Tomilov, Y.V. (2021). *Synthesis* 53: 2253–2259.
246 Verma, K. and Banerjee, P. (2018). *Adv. Synth. Catal.* 360: 3687–3692.
247 Sergeev, P.G., Novikov, R.A., and Tomilov, Y.V. (2021). *Adv. Synth. Catal.* 363: 5292–5299.
248 Zhu, M., Liu, J., Yu, J. et al. (2014). *Org. Lett.* 16: 1856–1859.
249 Borisov, D.D., Novikov, R.A., Eltysheva, A.S. et al. (2017). *Org. Lett.* 19: 3731–3734.
250 Borisov, D.D., Novikov, R.A., and Tomilov, Y.V. (2016). *Angew. Chem. Int. Ed.* 55: 12233–12237.
251 Borisov, D.D., Chermashentsev, G.R., Novikov, R.A., and Tomilov, Y.V. (2019). *Tetrahedron Lett.* 60: 746–750.
252 Borisov, D.D., Chermashentsev, G.R., Novikov, R.A., and Tomilov, Y.V. (2021). *J. Org. Chem.* 86: 8489–8499.
253 Borisov, D.D., Novikov, R.A., and Tomilov, Y.V. (2021). *J. Org. Chem.* 86: 4457–4471.
254 Novikov, R.A., Balakirev, D.O., Timofeev, V.P., and Tomilov, Y.V. (2012). *Organometallics* 31: 8627–8638.

255 Novikov, R.A., Tarasova, A.V., Korolev, V.A. et al. (2014). *Angew. Chem. Int. Ed.* 53: 3187–3191.

256 Novikov, R.A., Tarasova, A.V., Korolev, V.A. et al. (2015). *J. Org. Chem.* 80: 8225–8235.

257 Novikov, R.A., Tarasova, A.V., Denisov, D.A. et al. (2017). *J. Org. Chem.* 82: 2724–2738.

258 Novikov, R.A., Borisov, D.D., Tarasova, A.V. et al. (2018). *Angew. Chem. Int. Ed.* 57: 10293–10298.

259 Zotova, M.A., Novikov, R.A., Volodin, A.D. et al. (2019). *Eur. J. Org. Chem.* 2019: 4207–4214.

260 Borisova, I.A., Tarasova, A.V., Potapov, K.V. et al. (2019). *Russ. Chem. Bull.* 68: 1504–1509.

261 Zotova, M.A., Novikov, R.A., Shulishov, E.V., and Tomilov, Y.V. (2018). *J. Org. Chem.* 83: 8193–8207.

262 Belaya, M.A., Knyazev, D.A., Novikov, R.A., and Tomilov, Y.V. (2020). *Tetrahedron Lett.* 61: 151990.

263 Belaya, M.A., Knyazev, D.A., Borisov, D.D. et al. (2021). *J. Org. Chem.* 86: 8089–8100.

264 Han, Y., Tang, W.-Q., Guo, L.-Y., and Yan, C.-G. (2014). *Tetrahedron* 70: 6663–6668.

265 Ye, W., Zhou, L., Xue, S. et al. (2015). *Synlett* 26: 1769–1773.

266 Dousset, M., Parrain, J.-L., and Chouraqui, G. (2017). *Eur. J. Org. Chem.* 2017: 5238–5245.

267 Venkatesh, C., Ila, H., Junjappa, H. et al. (2002). *J. Org. Chem.* 67: 9477–9480.

268 Chagarovskiy, A.O., Ivanova, O.A., Budynina, E.M. et al. (2011). *Tetrahedron Lett.* 52: 4421–4425.

269 Novikov, R.A., Korolev, V.A., Timofeev, V.P., and Tomilov, Y.V. (2011). *Tetrahedron Lett.* 52: 4996–4999.

270 Ivanova, O.A., Budynina, E.M., Skvortsov, D.A. et al. (2014). *Synlett* 25: 2289–2292.

271 Novikov, R.A. and Tomilov, Y.V. (2013). *Helv. Chim. Acta* 96: 2068–2080.

272 Ivanova, O.A., Budynina, E.M., Chagarovskiy, A.O. et al. (2011). *Chem. – Eur. J.* 17: 11738–11742.

273 Novikov, R.A., Timofeev, V.P., and Tomilov, Y.V. (2012). *J. Org. Chem.* 77: 5993–6006.

274 Taily, I.M., Saha, D., and Banerjee, P. (2019). *Eur. J. Org. Chem.* 48: 7804–7813.

275 Vartanova, A.E., Levina, I.I., Rybakov, V.B. et al. (2021). *J. Org. Chem.* 86: 12300–12308.

276 Chagarovskiy, A.O., Ivanova, O.A., Shumsky, A.N., and Trushkov, I.V. (2017). *Chem. Het. Comp.* 53: 1220–1227.

277 Chagarovskiy, A.O., Strel'tsova, E.D., Rybakov, V.B. et al. (2019). *Chem. Het. Comp.* 55: 240–245.

278 Gladow, D. and Reissig, H.-U. (2012). *Helv. Chim. Acta* 95: 1818–1830.

279 Ghorai, M.K., Talukdar, R., and Tiwari, D.P. (2013). *Chem. Commun.* 49: 8205–8207.

280 Ghorai, M.K., Talukdar, R., and Tiwari, D.P. (2014). *Org. Lett.* 16: 2204–2207.

281 Saha, A., Bhattacharyya, A., Talukdar, R., and Ghorai, M.K. (2018). *J. Org. Chem.* 83: 2131–2144.
282 Luecht, A., Sobottka, S., Patalag, L.J. et al. (2019). *Chem. – Eur.J.* 25: 10359–10365.
283 Yang, C., Liu, W., He, Z., and He, Z. (2016). *Org. Lett.* 18: 4936–4939.
284 Mikhaylov, A.A., Dilman, A.D., Novikov, R.A. et al. (2016). *Tetrahedron Lett.* 57: 11–14.
285 Ivanov, K.L., Bezzubov, S.I., Melnikov, M.Y., and Budynina, E.M. (2018). *Org. Biomol. Chem.* 16: 3897–3909.
286 Flisar, M.E., Emmett, M.R., and Kerr, M. (2014). *Synlett* 25: 2297–2300.
287 Unnava, R., Chahal, K., and Reddy, K.R. (2021). *Org. Biomol. Chem.* 19: 6025–6029.
288 Prieto, L., Sanchez-Diez, E., Uria, U. et al. (2017). *Adv. Synth. Catal.* 359: 1678–1683.
289 Nie, G., Huang, X., Wang, Z. et al. (2021). *Org. Chem. Front.* 8: 5105–5111.
290 Luo, Y.-C., Ma, H., Hu, X.-Q., and Xu, P.-F. (2017). *J. Org. Chem.* 82: 1013–1023.
291 Liu, C., Zhou, L., Huang, W. et al. (2016). *Tetrahedron* 72: 563–570.
292 Ortega, A., Uria, U., Tejero, T. et al. (2021). *Org. Lett.* 23: 2326–2331.
293 Fu, Q. and Yan, C.-G. (2011). *Tetrahedron Lett.* 52: 4497–4500.
294 Wang, S., Xie, Z., Li, M., and Wang, C. (2020). *ChemistrySelect* 5: 6011–6015.
295 Su, Z., Qian, S., Xue, S., and Wang, C. (2017). *Org. Biomol. Chem.* 15: 7878–7886.
296 Han, Y., Tang, W.-Q., and Yan, C.-G. (2014). *Tetrahedron Lett.* 55: 1441–1443.
297 Sundaravelu, N. and Sekar, G. (2019). *Org. Lett.* 21: 6648–6652.
298 Gladow, D. and Reissig, H.-U. (2014). *J. Org. Chem.* 79: 4492–4502.
299 Martin, M.C., Patil, D.V., and France, S. (2014). *J. Org. Chem.* 79: 3030–3039.
300 Nambu, H., Fukumoto, M., Hirota, W., and Yakura, T. (2014). *Org. Lett.* 16: 4012–4015.
301 Kaschel, J., Schneider, T.F., Kratzert, D. et al. (2013). *Org. Biomol. Chem.* 11: 3494–3509.
302 Prakash, M., Lodhi, R., and Samanta, S. (2021). *J. Org. Chem.* 86: 6721–6733.
303 Thangamani, M. and Srinivasan, K. (2021). *J. Org. Chem.* 86: 1172–1177.

4

Activation of Donor–Acceptor Cyclopropanes under Covalent Organocatalysis: Enamine, Iminium, NHC, Phosphine and Tertiary Amine Catalysis

Efraim Reyes, Liher Prieto, Luisa Carrillo, Uxue Uria, and Jose L. Vicario

University of the Basque Country (UPV/EHU), Department of Organic and Inorganic Chemistry, B° Sarriena s/n. 48940 Leioa, P.O. Box 644, 48080, Bilbao, Spain

CHAPTER MENU

4.1 Introduction

The advent of organocatalysis has released new opportunities in organic synthesis with enormous potential to unveil new reactivity patterns or to implement known reactions under cheaper or safer conditions [1]. The ability of small organic molecules that do not contain a metal center in their active site to activate a given starting material toward a subsequent transformation has become a well-established area in asymmetric catalysis, which often provides a complementary approach to synthetic problems not easily faced through enzymatic and metal catalysis. In particular, those reactions in which the substrate and the catalyst interact through reversible covalent bonding (also referred to as covalent organocatalysis) constitute a particular class among organocatalytic reactions in which the strong nature of the substrate–catalyst interaction greatly enhances the probability to tune the electronic and steric properties of the activated starting material. This is of evident relevance when trying to implement enantioselective variants of a given reaction, given the enhanced ability of the catalyst to transfer stereochemical information through

Donor–Acceptor Cyclopropanes in Organic Synthesis, First Edition.
Edited by Prabal Banerjee and Akkattu T. Biju.

such a strongly bonded intermediate. For this reason, it is not surprising to find that the use of organocatalysts that interact through covalent bonding with the substrate has gained a great deal of popularity when developing catalytic and enantioselective versions of important reactions, including those involving donor–acceptor cyclopropanes (DAC) or acceptor cyclopropanes as substrates [2]. The extraordinary synthetic versatility of this simple concept has been recognized very recently, with the Nobel Prize in Chemistry to B. List and D. W. C. MacMillan for their pioneering contributions to this area [3].

The different possibilities for the activation of a DAC through covalent organocatalysis are illustrated in Scheme 4.1. At first hand, primary and secondary amines, together with *N*-heterocyclic carbenes (NHCs), belong to the privileged class of covalent organocatalysts due to their ability to condense in a reversible manner with carbonyl compounds. The first ones generate an active azomethine intermediate after interacting with an aldehyde or ketone substrate **I** or **II** to form either an electron-donating enamine moiety or an electron-withdrawing iminium ion species, respectively. When combined with the appropriate substitution pattern at the starting cyclopropane derivative, a catalytically generated DAC is generated with the potential to undergo a subsequent chemical transformation in which the aminocatalyst has been involved in the activation of the starting material and also provides the opportunity for asymmetric induction if a chiral aminocatalyst is employed [4]. On the other hand, *N*-heterocyclic carbenes can form a dienol intermediate (also known as Breslow intermediate) or an acyl azolium enolate species upon condensation with substrates **I** and **II**, respectively, also generating either an electron-donor or an electron-withdrawing moiety that, in combination with the

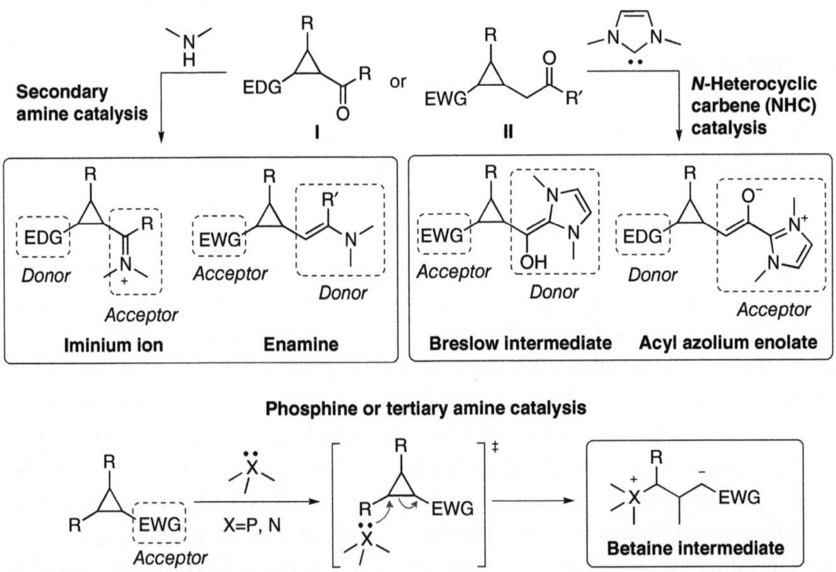

Scheme 4.1 The different possibilities for the activation of cyclopropanes through covalent organocatalysis.

correct substitution pattern at the cyclopropane, again leads to an activated intermediate able to undergo a variety of transformations [5]. In a different context, the ability of phosphines and tertiary amines to participate as nucleophilic catalysts that are able to interact with an electrophilic cyclopropane and therefore form a betaine intermediate is an appealing possibility for the activation of this type of cyclopropanes through covalent interaction with an organocatalyst [6].

In the following sections, the most relevant examples reported in the literature that imply any of these activation manifolds will be discussed following the classification shown in Scheme 4.1.

4.2 Secondary Amine Catalysis: Enamine Activation

The generation of enamine species in cyclopropylacetaldehyde-containing substrates through the condensation of a primary or secondary amine activates the substrate for a cyclopropane ring-opening event to take place. This was pioneered by Jørgensen and coworkers, who initially studied the system with the help of computational tools. Indeed, calculations showed that methods for the formation of the enamine intermediate also implied an increase in C–C bond lengths in several cyclopropane substrates through the formation of the proposed cyclopropylenamine intermediate in comparison with the starting cyclopropylacetaldehyde or its corresponding iminium ion (Figure 4.1) [7]. In particular, the presence of such a catalytically generated electron-donating substituent (the enamine) in combination with electron-withdrawing substituents (R^1 and $R^2 = CO_2Me$) led to the formation of a DAC in which an increase in the bond length of the C–C sigma bond within the cyclopropane moiety was maximized with respect to activated cyclopropylenamine intermediates with other substituents of different nature. Moreover, calculations also showed the presence of a predominant conformation in which the π-orbital of the enamine and the σ^*_{C-C} orbital of the cyclopropyl moiety lied in a parallel arrangement, thus favoring a hypothetical C–C bond cleavage event.

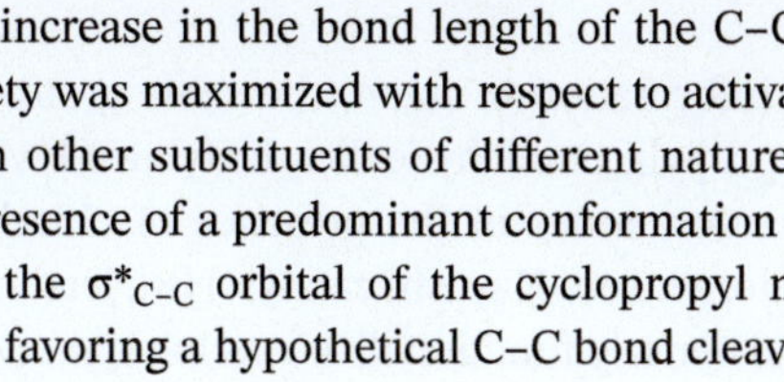

	C^1–C^2 bond length (Å)		
$R^1 = R^2 = H$	1.506	1.500	1.509
$R^1 = H$; $R^2 = CO_2Me$	1.512	1.508	1.530
$R^1 = R^2 = CO_2Me$	1.517	1.522	1.559

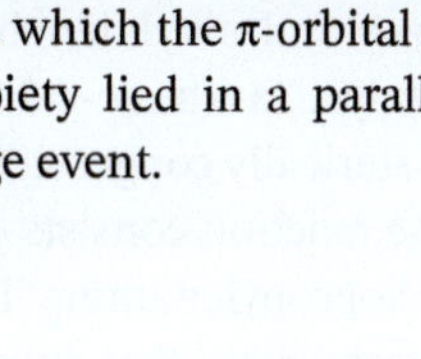

Figure 4.1 Selected bond distances in several cyclopropyl-containing species and the preferred conformation of enamine containing cyclopropanedicarboxylate.

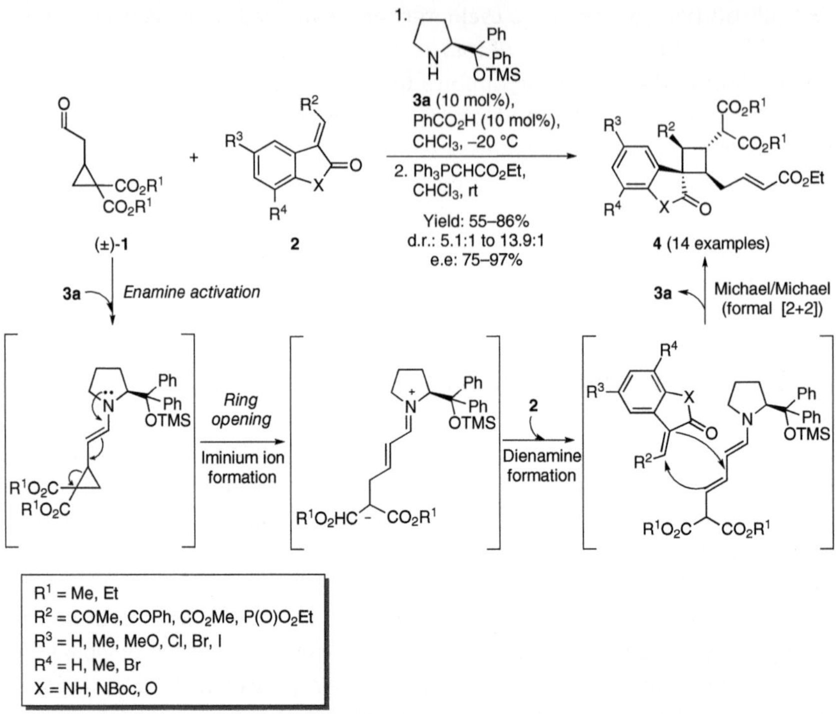

Scheme 4.2 Asymmetric formal [2+2] cycloaddition between cyclopropylacetaldehyde dicarboxylates and alkylideneoxindoles or alkylidenebenzofuranones.

With this preliminary information in hand, the same group demonstrated the concept by reacting cyclopropylacetaldehyde dicarboxylates **1** with alkylideneoxindoles and alkylidenebenzofuranones in the presence of chiral secondary amine catalyst **3a** (the so-called Jorgensen–Hayashi catalyst) [8], leading to the formation of sterically congested and densely functionalized cyclobutanes **4** (Scheme 4.2) [7]. The reaction consists of the initially projected ring-opening event of the activated cyclopropylenamine intermediate, leading to an α,β-unsaturated iminium ion intermediate that subsequently equilibrates with a tautomeric nucleophilic dienamine species. This participates in a formal [2+2] cycloaddition step with the highly electrophilic alkylideneoxindole or benzofuranone reagent in a sequence that is mechanistically explained as a Michael/intramolecular Michael cascade process that generates the cyclobutane scaffold [9]. The final hydrolysis of the cyclobutylenamine intermediate regenerates the catalyst and releases the final product, which was isolated in high yield. In addition, the ability of the chiral catalyst **3a** to block one of the diastereotopic faces of the dienamine intermediate by steric shielding as well as to achieve efficient control over the conformational preferences of the dienamine scaffold turns into an overall process that takes place with high diastereo- and enantioselectivity.

The same type of cyclopropylacetaldehydes **1** has been employed in the direct synthesis of dihydroquinolines **6** through reaction with o-aminobenzaldehydes **5** (Scheme 4.3) [10]. In this particular case, the authors made use of the electrophilic

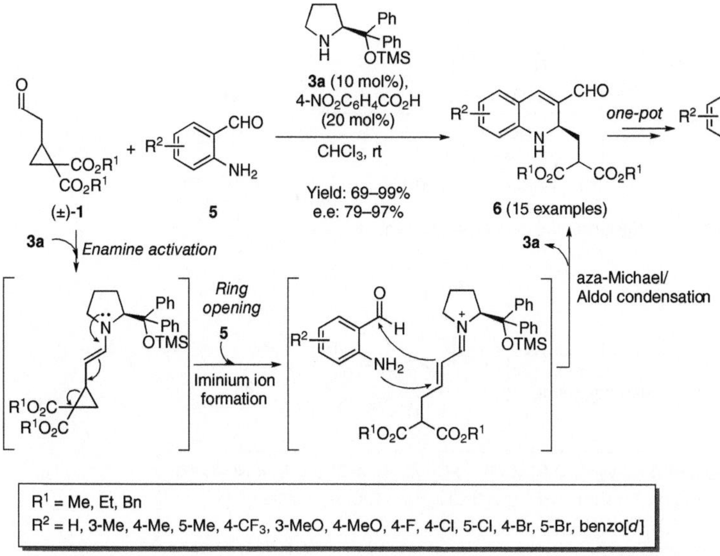

Scheme 4.3 Asymmetric synthesis of dihydroisoquibnolines through cascade ring-opening/aza-Michael reaction/intramolecular aldol condensation.

nature of the unsaturated iminium ion formed after the ring-opening event to initiate a cascade process that involved an initial aza-Michael-type reaction of the aniline moiety followed by intramolecular aldol condensation [11]. As in the previous example, the catalyst was highly effective in providing high stereocontrol to the overall process. Moreover, the dihydroisoquinoline compounds obtained could be easily transformed into synthetically useful pyrroloisoquinolines **7** through a three-step set of reactions that took place in a one-pot manner.

The group of Christmann has also combined the principle of vinilogy [12] with this specific reactivity profile and has developed a particular variant of the Cope rearrangement using cyclopropanes **7** as substrates (Scheme 4.4) [13]. Condensation of the amine catalyst **3a** with the enolizable conjugated enal substituent of the cyclopropane substrate led to the formation of a dienamine intermediate that spontaneously underwent the [3.3] sigmatropic rearrangement with the participation of the vinyl substituent at the C-2 position, allowing the preparation of differently substituted cycloheptadienes **8** in good yields and excellent *cis*-diastereoselectivity. Interestingly, only cis-configured cyclopropane substrates **8** were found to be reactive in this transformation, which is consistent with the proposed sigmatropic rearrangement process. This finding, together with the fact that the reaction was found to be completely enantiospecific, ruled out the possibility of a ring-opening/ring closure alternative mechanism to participate. In fact, the cycloheptadiene products **9** were isolated in all cases as racemic materials, even though the reaction proceeded with the participation of a chiral catalyst such as **3a**. Moreover, when an enantioenriched starting material was employed, its enantiomeric purity was preserved in the final product, demonstrating the enantiospecific nature of the process and further supporting the mechanistic proposal.

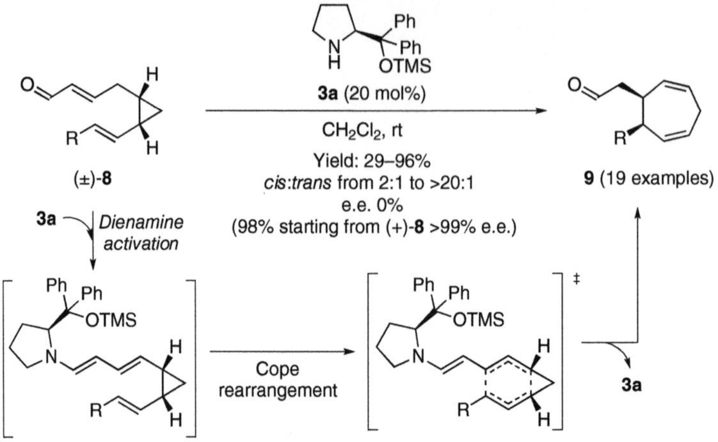

R = H, Ph, 4-MeC_6H_4, 4-(*t*-Bu)C_6H_4, 2-$CF_3C_6H_4$, 3-$CF_3C_6H_4$, 4-$CF_3C_6H_4$, 4-$NO_2C_6H_4$, 4-$MeOC_6H_4$, 2-FC_6H_4, 3-FC_6H_4, 4-FC_6H_4, 3-ClC_6H_4, 4-ClC_6H_4, tiophen-2-yl, CO_2Et, CN, $Ph(CH_2)_2$, $Ph(CH_2)_2CH{=}CH$

Scheme 4.4 Diastereoselective divinylcyclopropane-cycloheptadiene rearrangement using under dienamine activation.

4.3 Secondary Amine Catalysis: Iminium Ion Activation

The first evidence of the possibility of performing a cyclopropane ring-opening reaction through iminium activation was demonstrated by the Wang group in 2007 [14]. During their research directed toward the organocatalytic preparation of differently substituted cyclopropanes such as **12** from enals **10** and dimethyl bromomalonate, they observed the formation of byproduct **13**, which resulted to be formed after a ring-opening event from the cyclopropanecarbaldehyde products **12**. This reaction was further studied in detail and optimized for the selective formation of the ring-opening product, which could be obtained quantitatively and as a single diastereoisomer by careful selection of the base employed. The proposed mechanism consisted of the activation of the formylcyclopropane **12** through iminium ion formation by its condensation with the aminocatalyst **3a**, which subsequently had to undergo α-deprotonation by the Brønsted base additive incorporated into the reaction. This would generate a cyclopropylidenamine intermediate with the ability to undergo the ring-opening reaction through a retro-Michael-type process. This proposal was supported by a set of experiments that, for example, showed that the yield of this ring-opening product **13** was drastically decreased when either the catalyst **3a** or the Bronsted base (NaOAc) were not present in the media and also when *p*-nitrocinnamaldehyde (which can compete with **12** to condense with catalyst **3a**) was added to the reaction mixture (Scheme 4.5).

After this previous example that advanced the possibility for formylcyclopropanes to undergo a ring-opening event through iminium ion activation, several other

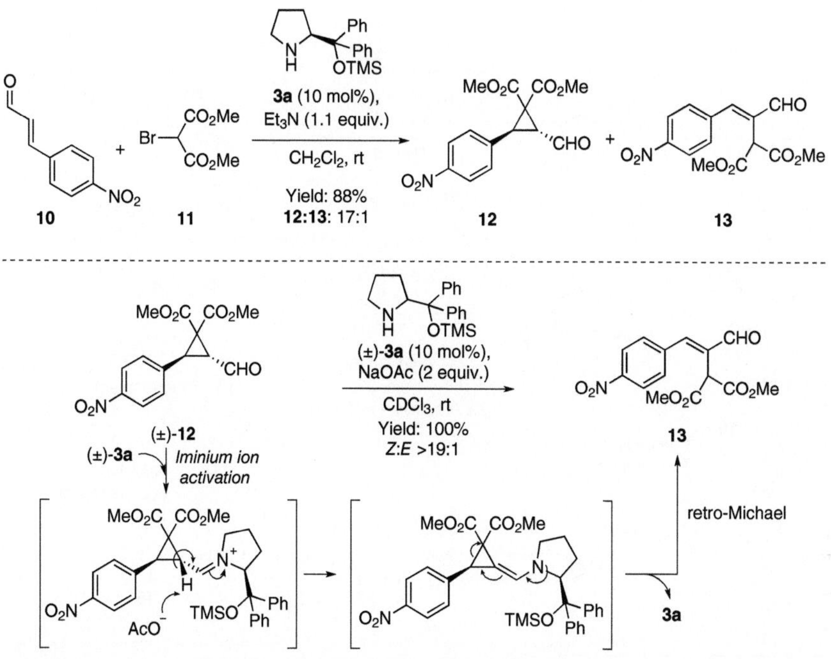

Scheme 4.5 The ring-opening reaction of cyclopropanecarbaldehydes under iminium ion activation.

examples appeared in which the increase of the polarization of the cyclopropane C–C bond upon formation of a cyclopropyliminium ion intermediate was explored to develop a variety of transformations. In fact, the formation of the iminium ion increases the ability of the formylcyclopropane to participate as an electrophile in ring-opening reactions triggered by the attack of an external nucleophile on the polarized cyclopropane C–C bond. The first example of this type of reactivity was described by the Wang group in 2009, who studied the ring-opening reaction of 2-substituted formylcyclopropanes **14** with benzenothiols **15** using (*S*)-proline **16** as a catalyst [15]. This reaction was carried out under very mild reaction conditions and led to the formation of the homo-Michael-type adduct **17** in low to moderate yields. Despite the use of an enantiopure catalyst such as proline, in all cases, racemic materials were obtained. Importantly, functionalized *o*-mercaptosalicylaldehydes **18** could also be employed through this reaction scheme, playing the role of bifunctional nucleophilic/electrophilic reagents able to be involved in a cascade process toward the synthesis of benzo[*b*]thiepines **19**. In this case, the enamine generated after the ring-opening step undergoes intramolecular aldol condensation to provide the final products in moderate to good yields. This reaction was also found to be enantiospecific, as observed when an enantioenriched cyclopropane was used as the starting material. As a consequence, the ring-opening event should have to take place through an S_N2-type reaction on the electrophilic activated carbon atom of the polarized cyclopropyliminium ion intermediate (Scheme 4.6).

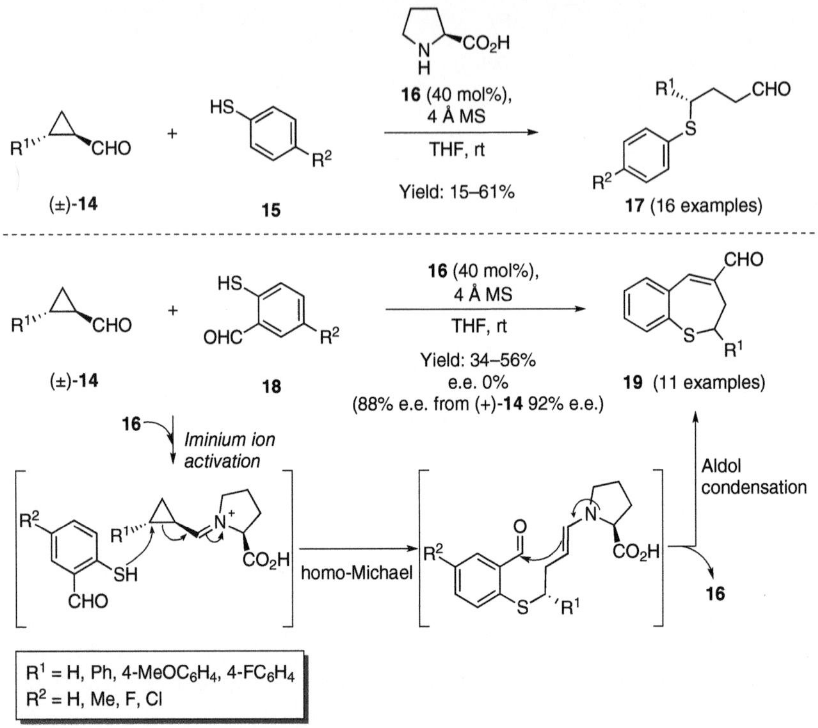

Scheme 4.6 Ring-opening reaction of formylcyclopropanes promoted by benzenethiol derivatives under iminium ion activation.

The first enantioselective version of this type of cyclopropane ring-opening reaction that converted an achiral formylcyclopropane starting material into an enantioenriched addition product was described two years later by the group of Gilmour (Scheme 4.7, top) [16]. In this case, they took advantage of the use of meso-2,3-disubstituted formylcyclopropanes **20** for performing the homo-Michael-type reaction using chloride as the nucleophile that initiates the ring-opening process. Because of the initial addition, the overall process takes place with concomitant desymmetrization of the starting material, in which the chiral imidazolidinone catalyst **23a** is involved in the stereodifferentiation of the two prostereogenic electrophilic carbon atoms of the formylcyclopropyliminium ion intermediate. In addition, an electrophilic chlorine source such as perchlorinated quinone **22** was also incorporated into the reaction scheme for the electrophilic trapping of the intermediate enamine that is formed after the ring-opening event. The overall chloride anion-initiated ring-opening reaction/α-clorination cascade process delivered a series of β-branched α,γ-dichlorinated aliphatic aldehydes **24** with high yields and diastereo- and enantioselectivities. An extension of this methodology was presented by the group of Werz, who reported the direct chlorosulfenylation of the same type of meso-formylcyclopropanes **20**, this time using a variety of arylsulfenyl chlorides **25** as external electrophiles engaged in the reaction with the enamine intermediate generated after the chloride-initiated nucleophilic ring-opening step (Scheme 4.7, bottom) [17]. In this case,

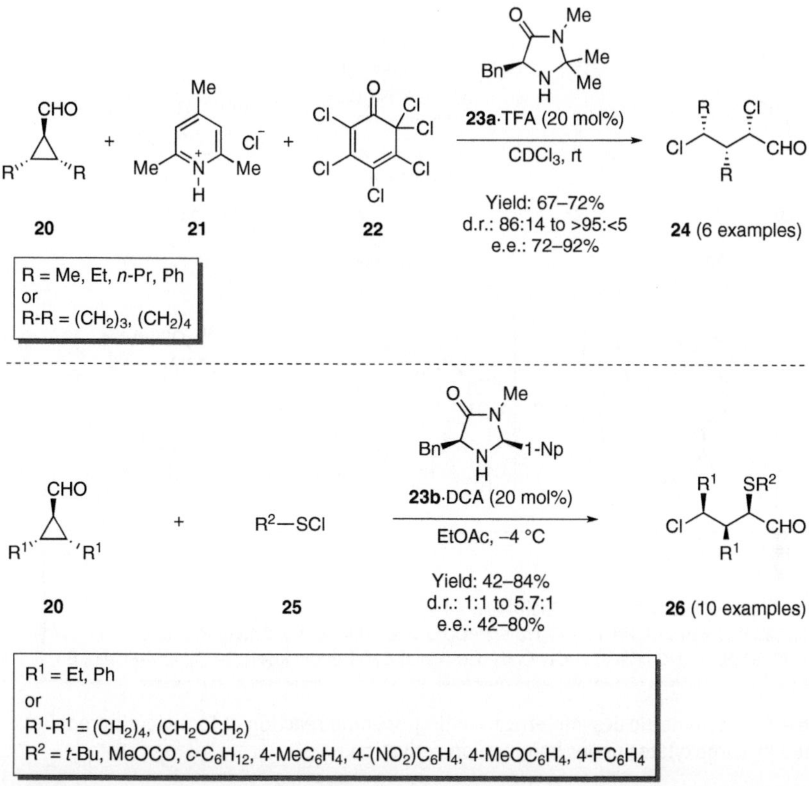

Scheme 4.7 Chloride-initiated asymmetric cyclopropane ring-opening reaction of meso-formylcyclopropanes under iminium ion activation.

imidalodidinone catalyst **23b** was employed as the best performing one, observing that the obtained adducts **26** had the opposite absolute configuration as those obtained in the previous dichlorination reaction catalyzed by **23a**, although in both cases all compounds obtained presented identical relative configuration at the newly generated three contiguous stereogenic centers.

Our group has also demonstrated the possibility of using carboxylates as active nucleophiles able to initiate the ring-opening process under this activation manifold and without the need for an additional electrophilic reagent for the external trapping of the enamine intermediate, which in this case underwent direct hydrolysis to release the catalyst and the final adducts (Scheme 4.8) [18]. This protocol enables the formation of γ-acyloxy-substituted aldehydes **29** with good to excellent yields and high stereoselectivities in most of the cases, using modified diarylprolinol **3b** as the most effective catalyst. The mechanism of the reaction was studied using computational methods that showed that the first step consisted of the formation of the iminium ion as an encounter pair in which the carboxylate anion is stabilized by one molecule of water and close to the iminium ion intermediate. The nucleophilic addition of the carboxylate to this iminium ion was identified as the rate-limiting step, consisting of an inverting-S_N1-type process but without the occurrence of any

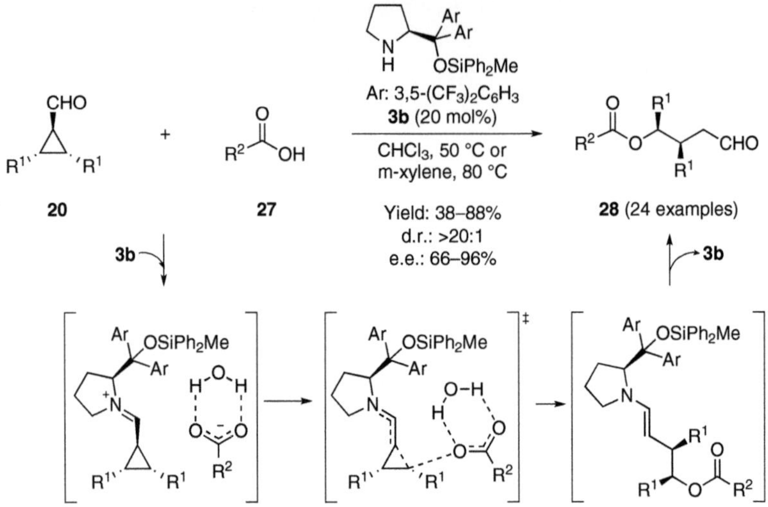

R^1 = Et, Ph
or
R^1-R^1 = $(CH_2)_3$, $(CH_2)_4$, $(CH_2)_5$, benzo[*c*]$(CH_2)_4$, $-CH_2CH{=}CHCH_2-$
R^2 = Me, $ClCH_2$, furan-2-yl, Ph, 2-MeC_6H_4, 4-MeC_6H_4, 2,4,6-$(Me)_3C_6H_2$, 2,4,6-(*i*-Pr)$_3C_6H_2$, 2-$(NO_2)C_6H_4$, 4-$(NO_2)C_6H_4$, 2-$(OH)C_6H_4$, 2,6-$(MeO)_2C_6H_3$, 3-$MeOC_6H_4$, 2-FC_6H_4, 4-FC_6H_4, Bn

Scheme 4.8 Asymmetric desymmetrizative ring-opening reaction of formylcyclopropanes promoted by carboxylates under iminium ion activation.

intermediate (kinetically a single step), that forms the subsequent enamine species with complete control in both the enantioselectivity and diastereoselectivity under catalyst control that is able to differentiate between the two prostereogenic C2 and C3 atoms of the cyclopropane scaffold.

The applicability of this methodology as a tool in synthesis has been demonstrated with the total synthesis of speciosin H (Scheme 4.9), which is a metabolite isolated from the basidiomycete fungus *Hexagonia speciosa* that grows in subtropical areas of China [19]. The initial enantioselective ring-opening step took place with excellent yield and enantiocontrol and at a synthetically useful gram-scale. This ring-opening adduct was converted into the target natural product in an additional 6 steps using standard and reliable reactions.

4.4 NHC Catalysis: Activation Through Breslow Intermediates

The generation of the Breslow intermediate upon condensation of an *N*-heterocyclic carbene with a formylcyclopropane converts the electron-withdrawing formyl group into an electron-donating enaminol moiety. In this sense, a cyclopropane containing both an electron-withdrawing group and a formyl substituent (an acceptor/

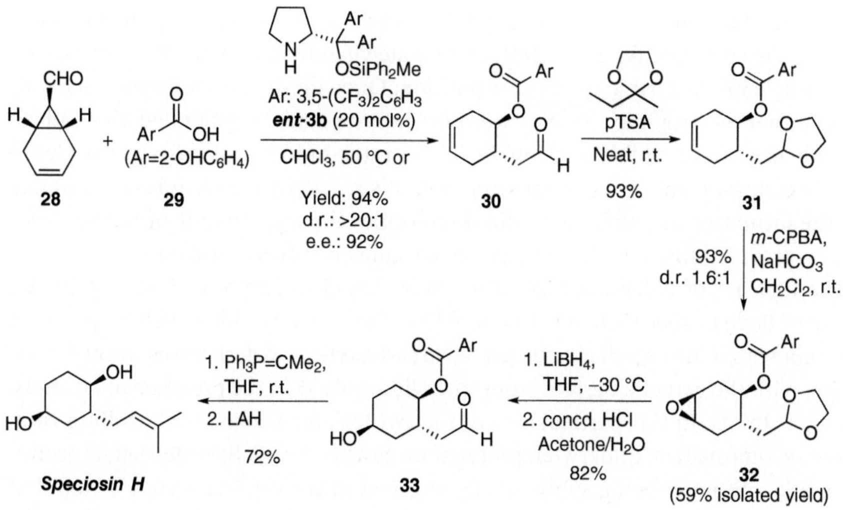

Scheme 4.9 Application of the carboxylate-initiated enantioselective ring-opening of formylcyclopropanes to the total synthesis of (-)-speciosin H.

acceptor cyclopropane) leads to the formation of an activated DAC in which the polarization of the C–C bond has been significantly increased, therefore favoring a ring-opening event that generates an acyclic zwitterionic intermediate (Scheme 4.10).

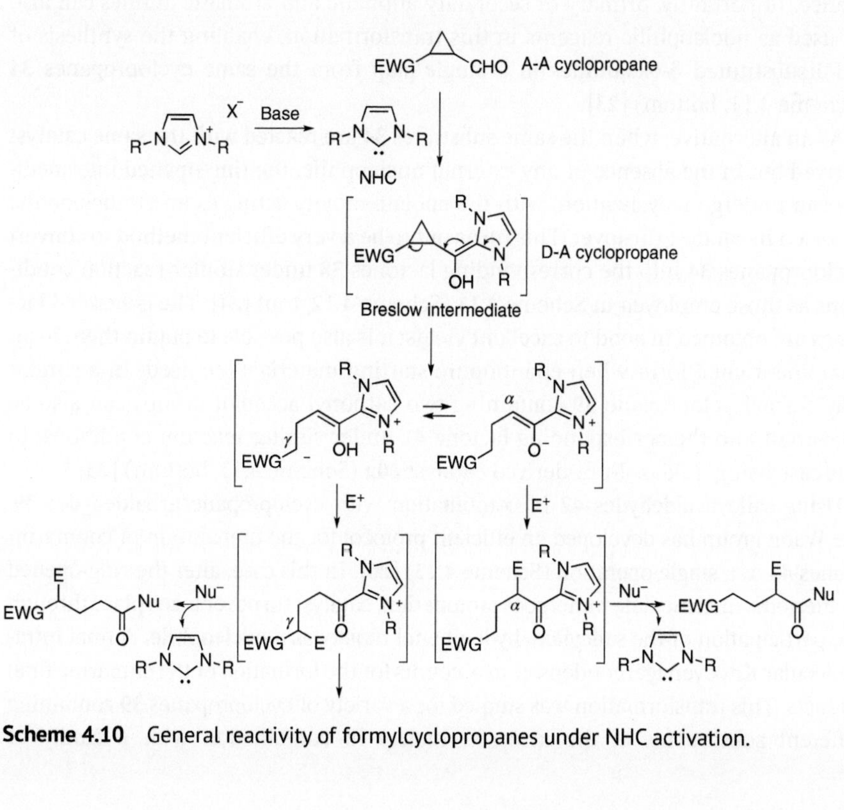

Scheme 4.10 General reactivity of formylcyclopropanes under NHC activation.

This zwitterion can now react through the γ-carbon with an electrophile or equilibrate to the corresponding acyl azolium enolate via proton transfer, the latter showing a nucleophilic character at the α-position. Once these intermediates react with an external electrophile, an acyl azolium intermediate is generated that subsequently has to react with a nucleophile (either inter- or intramolecularly) in order to enable catalyst turnover. As a consequence, this reactivity platform is very suitable for the formation of polifunctionalized molecules through cascade or domino processes, provided that all selectivity issues are conveniently controlled.

The first use of *N*-heterocyclic carbenes in this chemistry was described by the group of Bode in 2006 (Scheme 4.11, top) [20]. In this study, chiral cyclopropanes **34** were converted into α,β-disubstituted δ-oxo carboxylic acid derivatives through activation with the carbene derived from triazolium salt **35** in the presence of alcohols, water, or thiols. In this particular case, the zwitterionic enolate/acyl azolium ring-opening intermediate underwent protonation assisted by the Brønsted acidic pronucleophile, the latter being subsequently engaged in the catalyst turnover step that generates the carboxylic acid moiety (ester, thioester, or carboxylate). The reaction was also found to be enantioconservative, enabling the formation of products **36** when enantioenriched cyclopropanes were used as starting materials with no loss of optical purity. The scope of this reaction was further extended to substrates that contain different types of electron-withdrawing groups, such as 2-nitrocyclopropyl carbaldehyde [20, 21] or 2,4-dinitrophenylcyclopropyl carbaldehyde [22], which demonstrated to be compatible with this methodology, observing similar performance. Importantly, primary or secondary aliphatic and aromatic amines can also be used as nucleophilic reagents in this transformation, enabling the synthesis of α,β-disubstituted δ-oxoamides in a single step from the same cyclopropanes **34** (Scheme 4.11, bottom) [23].

As an alternative, when the same substrates **34** are reacted with the same catalyst derived but in the absence of any external nucleophile, the ring-opened intermediate can undergo recyclization, with the enolate moiety acting as an *O*-nucleophile involved in catalyst turnover. This turns out to be a very efficient method to convert cyclopropanes **34** into the corresponding lactones **38** under similar reaction conditions as those employed in Scheme 4.11 (Scheme 4.12, top) [24]. The generated lactones are obtained in good to excellent yields; it is also possible to obtain them in an enantioenriched form when enantiopure starting materials are used. In a similar way, formylcyclopropane **39** containing two carbonyl acceptor groups can also be converted into the corresponding lactone **41** under similar reaction conditions, in this case using imidazolium-derived catalyst **40a** (Scheme 4.12, bottom) [25].

Using salicylicaldehydes **42** in combination with cyclopropanecarbaldehydes **39**, the Wang group has developed an efficient protocol for the preparation of coumarin-2-ones **43** in a single operation (Scheme 4.13) [26]. In this case, after the ring-opened zwitterionic intermediate undergoes protonation, catalyst turnover takes place through the participation of the salicylaldehyde phenol moiety as a nucleophile. A final intramolecular Knoevenagel condensation accounts for the formation of the cumarine final adducts. This transformation was studied for a variety of cyclopropanes **39** containing different substituents at the ketoester moiety in combination with a family of

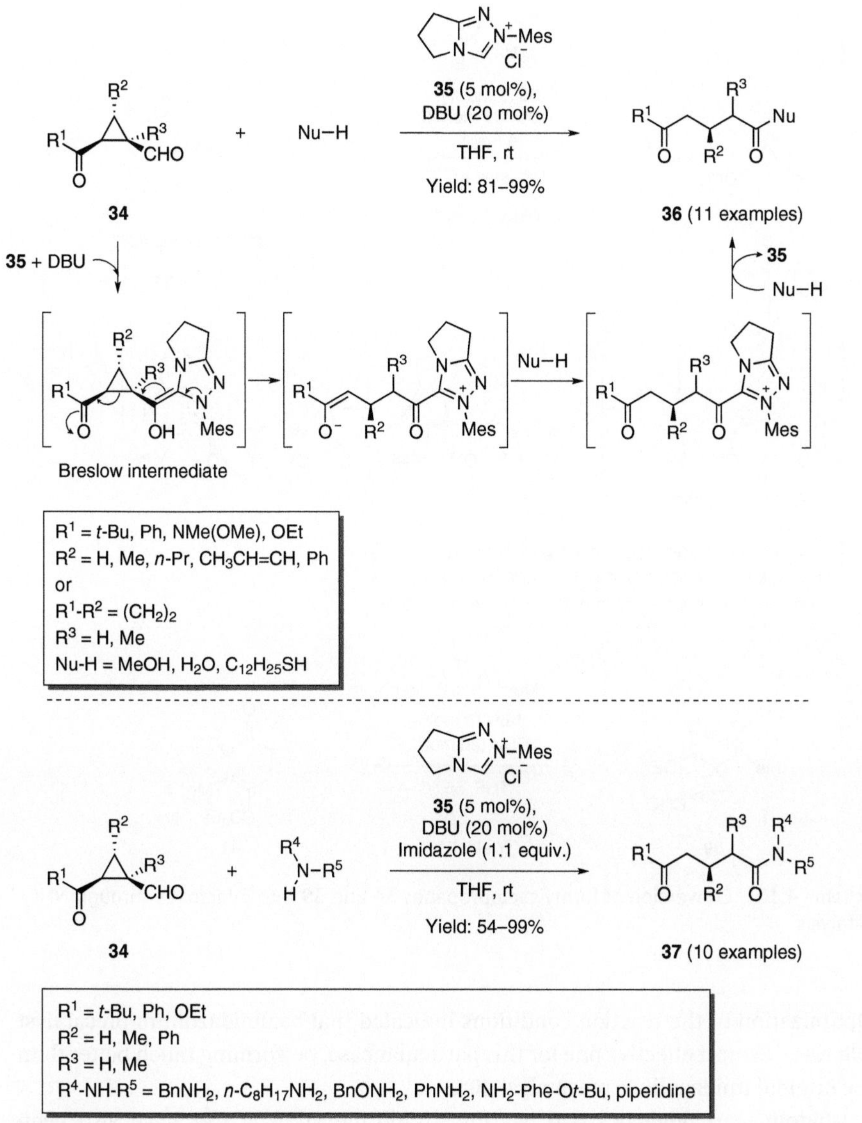

Scheme 4.11 Cyclopropane ring-opening reaction promoted by an NHC and an external nucleophile.

substituted salicylaldehydes **42**, obtaining good yields in almost all cases, which demonstrates the wide scope of the reaction.

An extension of this methodology was presented by the same group using, in this case, 2-indolinecarbaldehydes **44a** as the functionalized pronucleophile engaged in the catalyst turnover step (Scheme 4.14) [26]. This reaction, when applied to differently substituted indoles and cyclopropanecarbaldehydes **39**, allowed the preparation of a wide family of pyrrolo[1,2-*a*]indol-3-ones **45** in good to moderate yields.

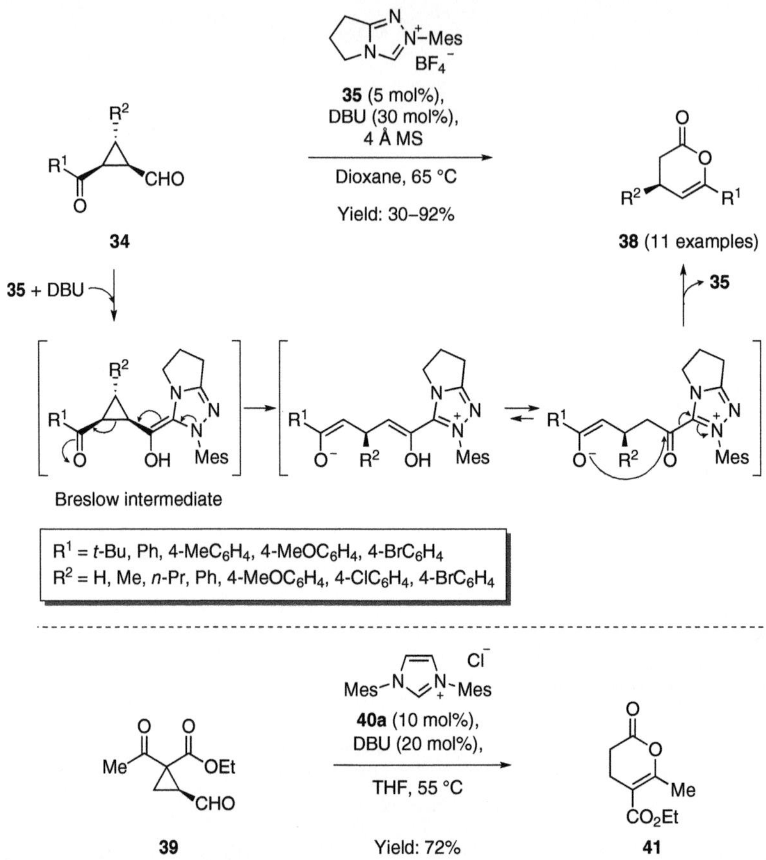

Scheme 4.12 Conversion of formylcyclopropanes **34** and **39** into δ-lactones through NHC catalysis.

Optimization of the reaction conditions indicated that bezimidazolium precatalyst **40b** was the most effective one for this particular case, performing much better than the original imidazolium precatalyst **40a**.

3-Indolinecarbaldehydes **44b** bearing a chlorine atom at C-2 have also been employed as bifunctional reagents able to engage in a similar cascade process when reacting with formylcyclopropanes **39** under similar reaction conditions (Scheme 4.15) [27]. In this particular case, the acyl azolium intermediate formed after the ring-opening event reacted with the NH group of the indolinecarbaldehyde reagent, forming the corresponding amide and releasing the NHC catalyst. Next, intramolecular formal nucleophilic aromatic substitution between the malonate moiety and the 2-chloroindole site (being proposed to occur through conjugate addition/elimination) took place to form the final adduct **46** in moderate to good yields. This reaction involved the use of modified 3-phenyltriazolium salt **40c** as a precatalyst that performed much better than the benzimidazolium salt **40b** used in the previous reaction with simple indolinecarbaldehydes.

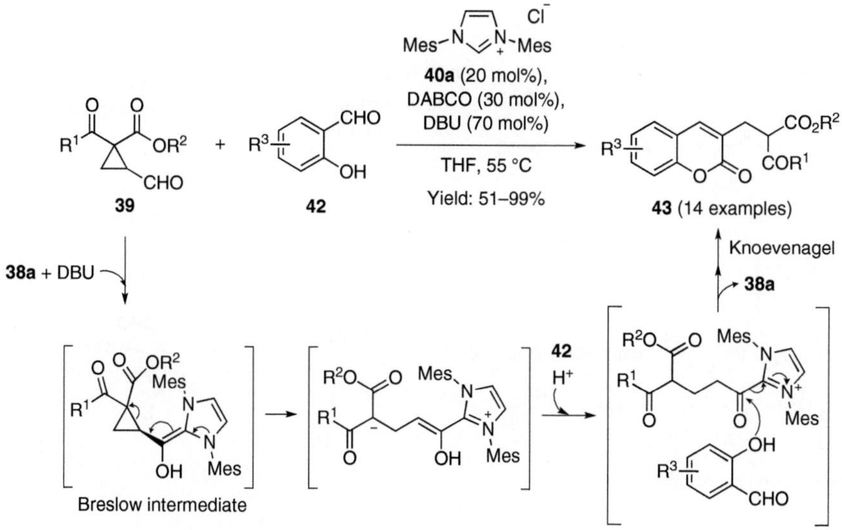

R¹ = OMe, OEt, O*i*-Pr, Me
R² = Me, Et, *i*-Pr
R³ = H, 5-NO_2, 3-Me, 3-MeO, 4-MeO, 5-MeO, 3-OCF_3, 3-F, 5-F, 5-Cl, 5-Br

Scheme 4.13 Reaction of cyclopropanecarbaldehydes and salicilaldehydes under NHC catalysis.

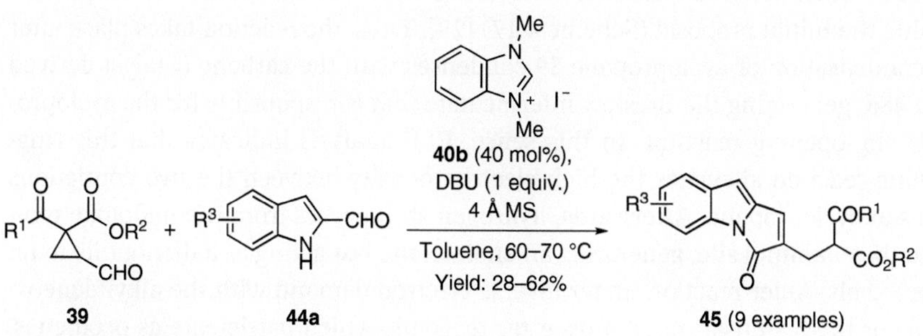

R¹ = OMe, OEt, O*i*-Pr
R² = Me, Et, *i*-Pr
R³ = H, 5-Me, 5-Et, 5-*i*-Pr, 5-MeO, 5-Cl

Scheme 4.14 Reaction of cyclopropanecarbaldehydes and indolinecarbaldehydes under NHC catalysis.

Recently, our group has disclosed the potential use of chiral NHC in this chemistry that could enable the use of racemic formylcyclopropanes **39** as starting materials, leading to enantioenriched final products. In particular, chiral triazolium salt **48a** or its enantiomer was employed as a precatalyst in the reaction between **39** and alkylideneoxindoles or β,γ-unsaturated α-oxoesters for the enantioselective synthesis of pyranoindolones **49** and pyranones **51**, respectively (Scheme 4.16) [28].

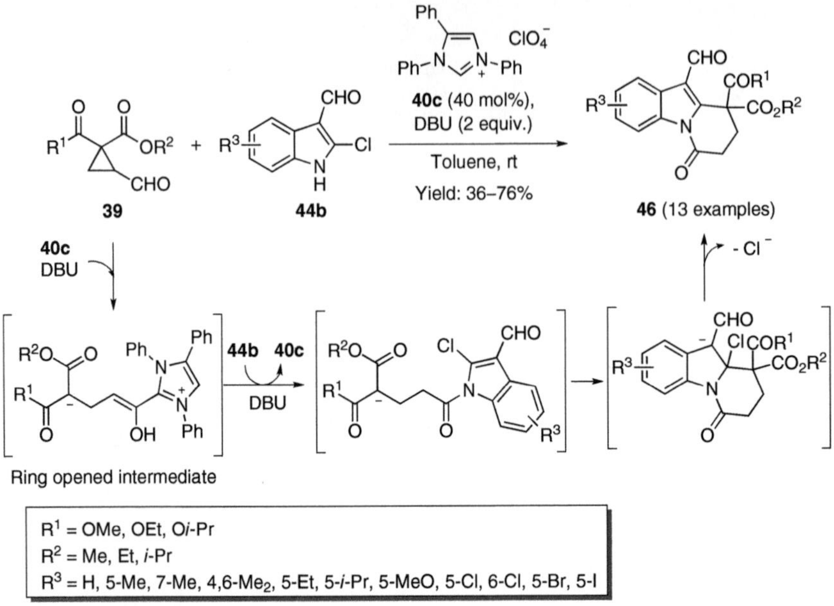

Scheme 4.15 Reaction of formylcyclopropanes and 2-chloroindolinecarbaldehydes under NHC catalysis.

A subsequent report studied the mechanism of this transformation in detail, confirming the initial proposal (Scheme 4.17) [29]. Thus, the reaction takes place after the condensation of cyclopropane **39** condenses with the carbene catalyst derived from **48a**, generating the Breslow intermediate that is responsible for the cyclopropane ring-opening reaction. In this sense, ELF analysis indicates that this ring-opening reaction alleviates the high electron density between the two contiguous atoms at cyclopropane. Afterwards, hydrogen shift occurs from the malonate moiety to the enaminol site, generating an enolate, the last acting as a dienophile in an hetero Diels–Alder reaction under inverse electron demand with the alkylideneoxindole or the β,γ-unsaturated α-oxoester reagents, which participate as oxodienes. A final α-elimination has to take place in order to generate the free carbene catalyst and the final adducts.

In a different approach, 2-acetylcyclopropanecarbaldehydes **52** have been used in enantioselective cascade reactions under chiral NHC activation, in this case being involved as a C4 component in a [4+2] cyclocondensation reaction with highly electrophilic sulfonimines such as **53** (Scheme 4.18) [30]. In this transformation, after the cyclopropane ring-opening event, the formed enolate undergoes an intermolecular Mannich reaction with imine **53**, followed by cyclization through the reaction of the imide moiety with the acyl azolium site, which enables catalyst turnover. The reaction provides a family of benzoisothiazolo[2,3-*a*]pyridin-7-one 5,5-dioxides **54** in good yields and high disastereo- and enantioselectivities. The authors also demonstrate that the reaction involves a DYKAT-type process in which the diastereomeric purity of the starting material **52** does not affect isomer composition of **54**.

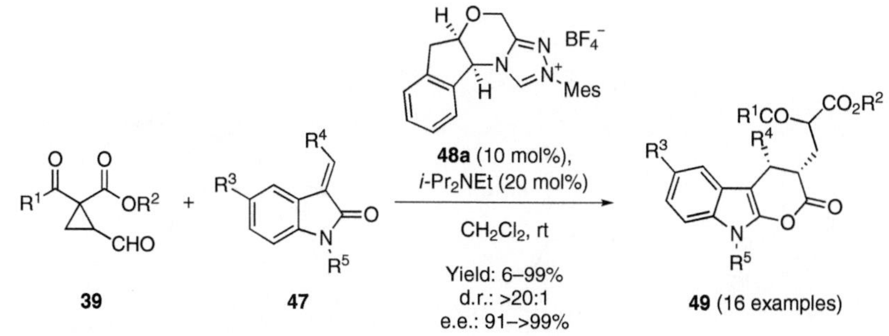

R^1 = OMe, OEt, O*i*-Pr, OAllyl, $OCH_2c\text{-}C_6H_{11}$, $O(CH_2)_4OBn$
R^2 = Me, Et, *i*-Pr, Allyl, $CH_2c\text{-}C_6H_{11}$, $(CH_2)_4OBn$
R^3 = H, 5-Me, 5-MeO
R^4 = CO_2Et, $CO_2t\text{-}Bu$, CO_2Bn, COPh
R^5 = Boc, Fmoc, Cbz, Me, Ac

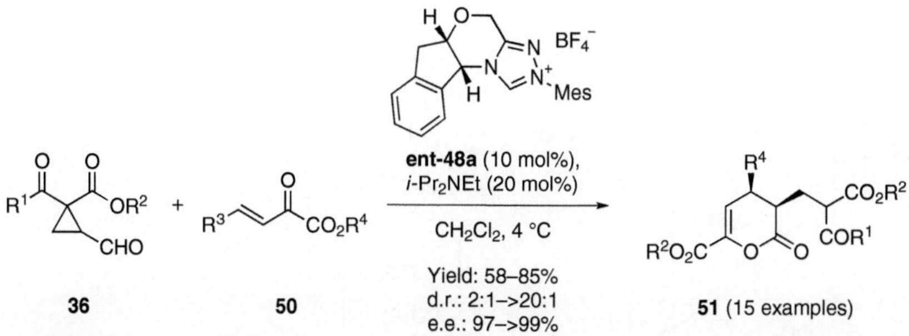

R^1 = OEt
R^2 = Et
R^3 = Me, *i*-Pr, $PhCH_2CH_2$, $BnOCH_2$, Ph, 2-MeC_6H_4, 3-MeC_6H_4, 4-MeC_6H_4, 4-$MeOC_6H_4$, 3,4-$OCH_2O\text{-}C_6H_3$, 4-FC_6H_4, 2-furyl, 2-thienyl
R^4 = Me, Et

Scheme 4.16 Reaction of formylcyclopropanes with alkylideneoxindoles or β,γ-unsaturated α-oxoesters under chiral NHC catalysis for the enantioselective synthesis of δ-lactones.

Additional remarkable features are the fact that the chiral catalyst is able to exert an efficient facial and simple selectivity control despite the fact that it is located somewhat far away from the reaction site. In addition, it is also able to control the face selectivity in the protonation of the enol moiety, which also generates an additional stereogenic center if this position of the starting material is also substituted.

Cyclopropanes **55** have been employed in a photo/NHC co-catalyzed reaction directed to the synthesis of γ-substituted-α,β-unsaturated esters by the group of Ye (Scheme 4.19) [31]. This reaction combines the formation of an activated acyclic dienol-type Breslow intermediate after the ring-opening of the cyclobutane substrate with the generation of a malonyl radical species from α-substituted dialkylbromomalonates **56** upon activation of the latter by $Ru(bpy)_3^{2+}$ by light irradiation in

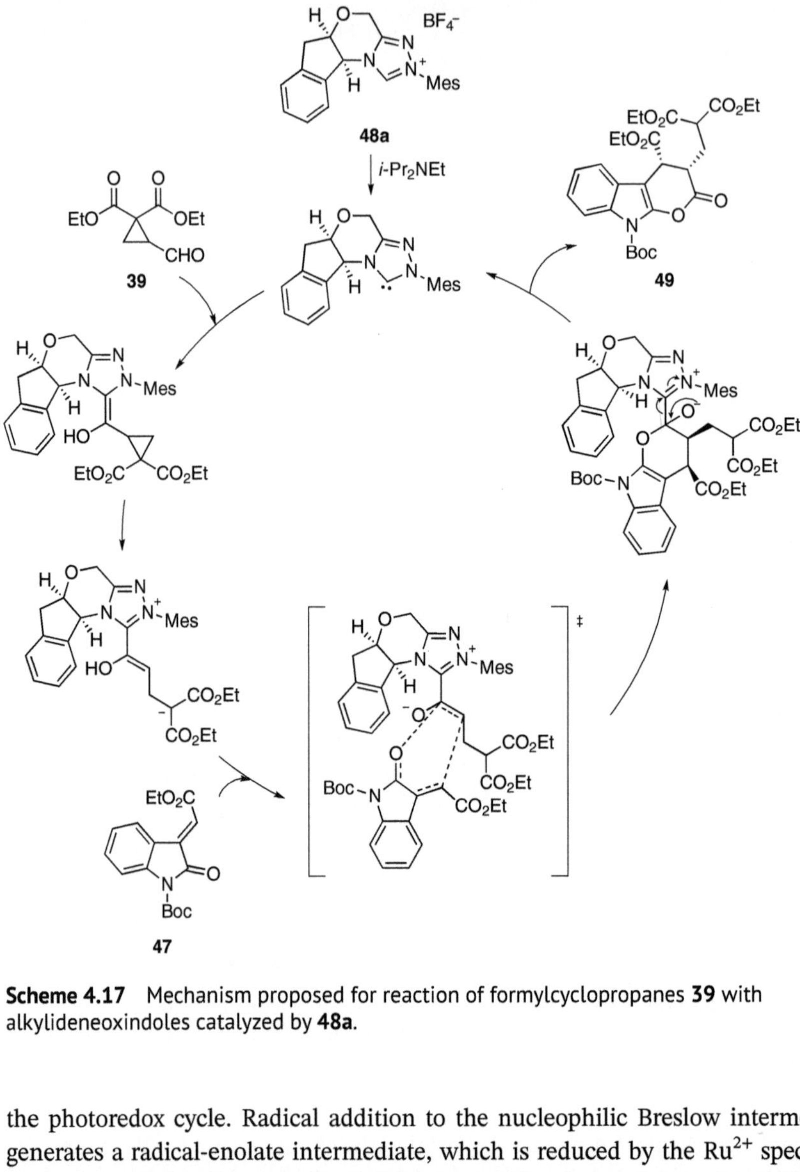

Scheme 4.17 Mechanism proposed for reaction of formylcyclopropanes **39** with alkylideneoxindoles catalyzed by **48a**.

the photoredox cycle. Radical addition to the nucleophilic Breslow intermediate generates a radical-enolate intermediate, which is reduced by the Ru^{2+} species to generate an acyl azolium species and regenerates the photoredox catalyst. The reaction of the former with an alcohol incorporated into the reaction scheme enables the turnover of the carbene catalyst and releases the final product **57**. The reaction is remarkably wide in scope, tolerating a wide variety of substitution patterns at both the bromomalonate reagent and the alcohol additive, showing good to excellent yields in most of the cases. The enantioselective version has also been studied using different chiral NHC catalysts, but observing low enantioselectivities.

Cyclopropylacetaldehydes bearing two electron-withdrawing groups, such as **58**, are another type of cyclopropyl substrates that have been evaluated as starting materials for promoting a cascade reaction initiated by ring-opening under NHC

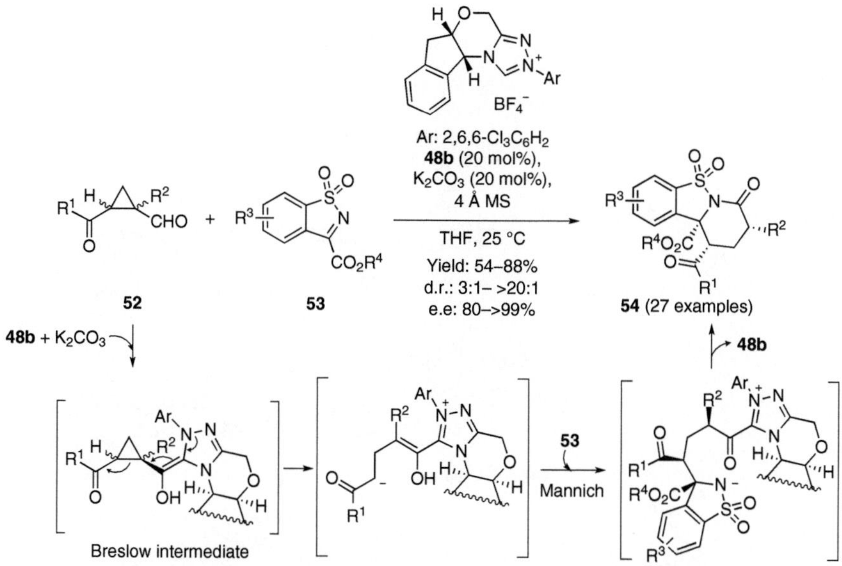

R^1 = thienyl, Ph, 4-(CF_3)C_6H_4, 4-(NO_2)C_6H_4, 4-MeC_6H_4, 2-MeOC_6H_4, 3-MeOC_6H_4, 4-MeOC_6H_4, 2-FC_6H_4, 4-FC_6H_4, 2-ClC_6H_4, 3-ClC_6H_4, 4-ClC_6H_4, 3-BrC_6H_4, 4-BrC_6H_4, 4-IC_6H_4
R^2 = H, Me, Et, 4-(NO_2)$C_6H_4CH_2$
R^3 = H, 5-Me, 5-*i*-Pr, 5-*t*-Bu, 7-CF_3, 5-MeO, 5-F, 7-F, 5-Cl, 7-Cl, bezo[*e*]
R^4 = Me, Et

Scheme 4.18 Reaction of 2-acylcyclopropanecarbaldehydes with sulfonimines under chiral NHC catalysis.

catalysis (Scheme 4.20) [32]. In this particular case, a ring-opening event takes place from a cyclopropylmethylenaminol Breslow intermediate that undergoes intramolecular nucleophilic addition to the electrophilic cyclopropane moiety, generating a cyclopropanol intermediate facilitated by the formation of a stable malonate-type enolate. Proton shift generates an alcoxide anion that undergoes a second spontaneous ring-opening, generating a homoenolate-type enol species, which is the one reacting with the isatin electrophile **59** through 1,2-addition followed by intramolecular reaction to release the catalyst and to generate the spirocyclic lactone moiety. The reaction proceeded generally with moderate to excellent yields and good stereoselectivity for a wide variety of isatin reagents employed using modified chiral triazolium salt **48c** as the most effective precatalyst.

4.5 Phosphine or Tertiary Amine Catalysis

As mentioned earlier, a Lewis base can act as a nucleophilic catalyst that triggers the ring-opening event on a cyclopropane substrate with an electrophilic character. In particular, cyclopropanes with two electron-withdrawing substituents are highly reactive substrates for activation with a tertiary amine or a phosphine. A good

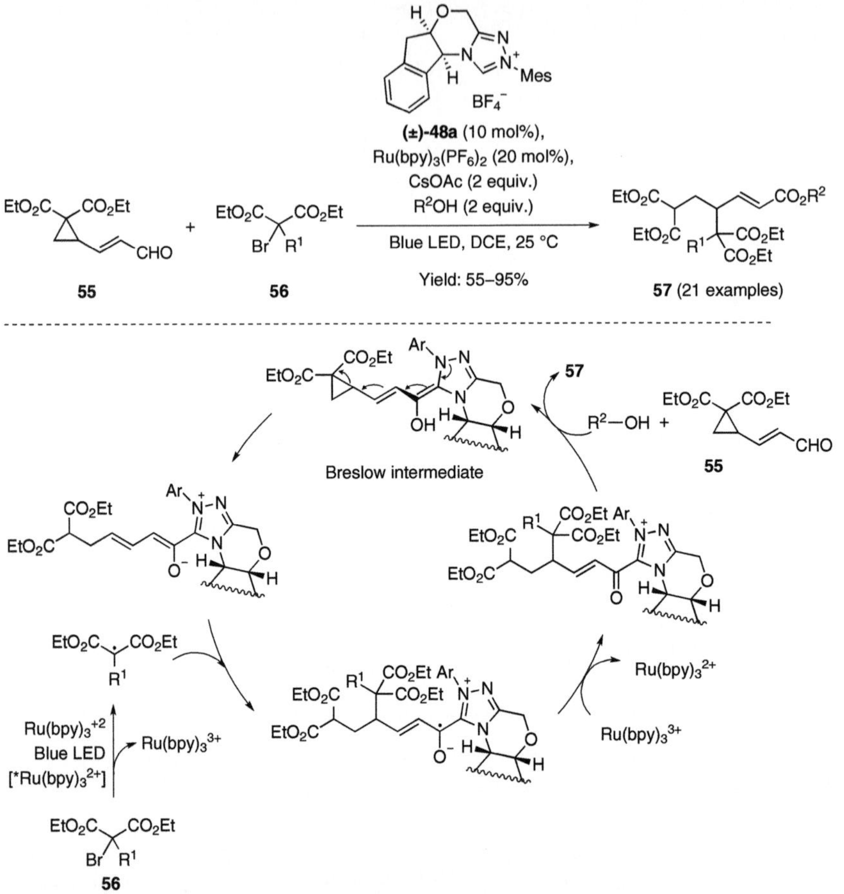

R^1 = Me, n-Pr, n-Bu, i-Bu, Allyl, $EtO_2CCH{=}CHCH_2$, Bn, 4-(Bpin)$C_6H_4CH_2$, 2-Me$C_6H_4CH_2$, 4-Me$C_6H_4CH_2$, 4-(NO_2)$C_6H_4CH_2$, 3-Cl$C_6H_4CH_2$, 4-Br$C_6H_4CH_2$, 2-NpCH_2
R^2 = Me, Et, n-Pr, CF_3CH_2, c-C_3H_5, c-C_4H_7, Cl$(CH_2)_2CH_2$, $CH_2{=}CHCH_2CH_2$

Scheme 4.19 Ring-opening reaction of cyclopropyl acrylaldehydes and subsequent reaction with malonyl radicals under NHC/photoredox catalysis.

example of this behavior is shown in Scheme 4.21, in which electrophilic cyclopropanes **61** have been used as starting materials for performing a Cloke–Wilson rearrangement catalyzed by DABCO (Scheme 4.21) [33]. Experimental data together with additional computational information [34] have demonstrated that the mechanism of the reaction involves the addition of DABCO to the electrophilic C-2 site of the cyclopropane, generating a zwitterionic enolate (that could be isolated) that subsequently underwent cyclization to provide the final dihydrofurane products **62**. Surprisingly, the reaction was not found to be enantiospecific since enantioenriched cyclopropanes delivered racemic products, which was interpreted in terms of possible racemization processes occurring due to the need for high temperatures and a remarkable high amount of DABCO for the reaction to take place.

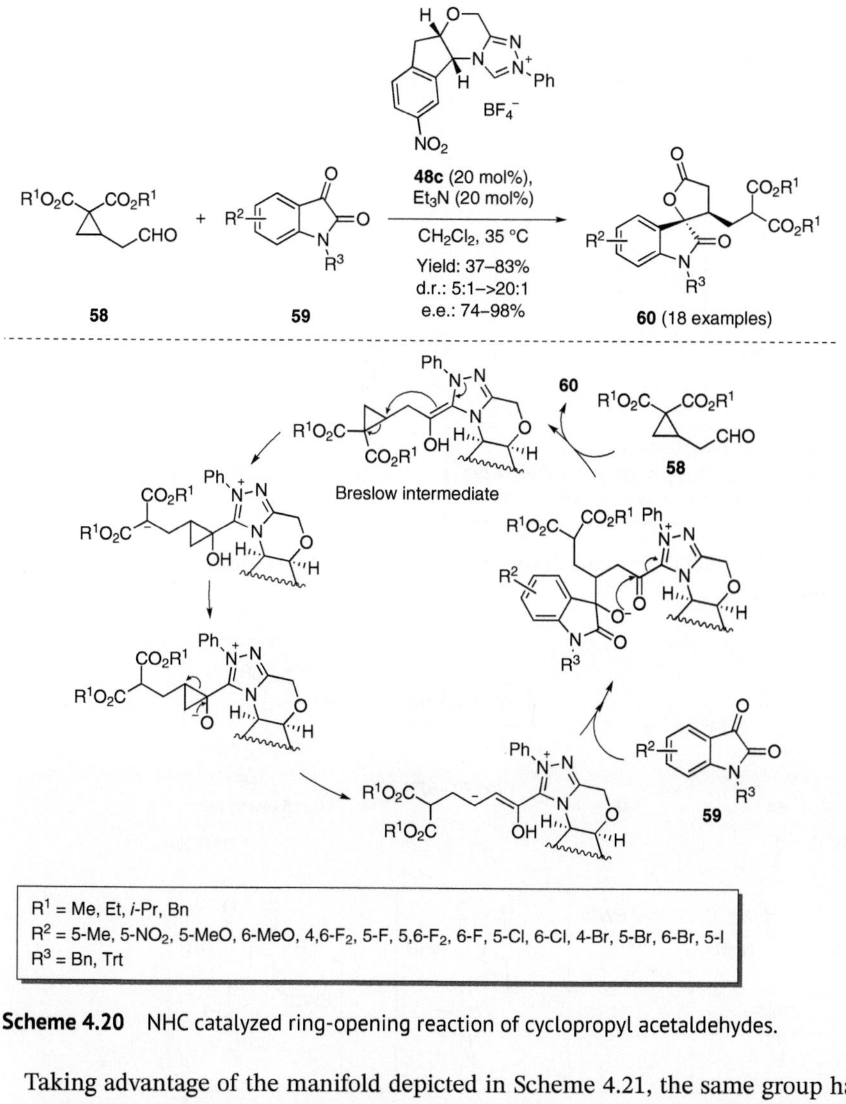

R^1 = Me, Et, *i*-Pr, Bn
R^2 = 5-Me, 5-NO_2, 5-MeO, 6-MeO, 4,6-F_2, 5-F, 5,6-F_2, 6-F, 5-Cl, 6-Cl, 4-Br, 5-Br, 6-Br, 5-I
R^3 = Bn, Trt

Scheme 4.20 NHC catalyzed ring-opening reaction of cyclopropyl acetaldehydes.

Taking advantage of the manifold depicted in Scheme 4.21, the same group has extended this reactivity to the incorporation of an external electrophilic alkene **65** able to react with the zwitterionic enolate formed after the ring-opening event (Scheme 4.22) [35]. In this case, catalyst release took place with the participation of an amide incorporated at one of the electron-withdrawing substituents of the electrophilic cyclopropane reagent **63**, leading to the formation of γ-lactam **66** as the final product of the reaction.

As an alternative possibility, a Michael acceptor was incorporated as one of the electron-withdrawing substituents that activate the electrophilic cyclopropane reagent **67** (Scheme 4.23) [36]. In this type of substrate, DABCO first had to undergo conjugate addition to the Michael acceptor site, and it was the enolate moiety that triggered the ring-opening event in an intramolecular fashion with the participation of the oxygen as the nucleophile. Subsequent ring closure, as in the standard

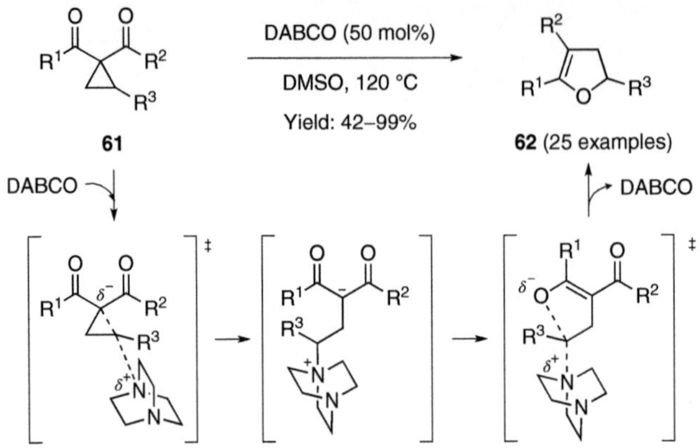

R^1 = Me, Ph, 4-MeC_6H_4, 4-$MeOC_6H_4$
R^2 = Me, Ph, 4-MeC_6H_4, 4-$MeOC_6H_4$, OEt, NHPh
R^3 = H, Me, CH_2=CH, Ph, 4-$MeOC_6H_4$, 4-ClC_6H_4

Scheme 4.21 DABCO catalyzed Cloke–Wislon rearrangement of electrophilic cyclopropanes.

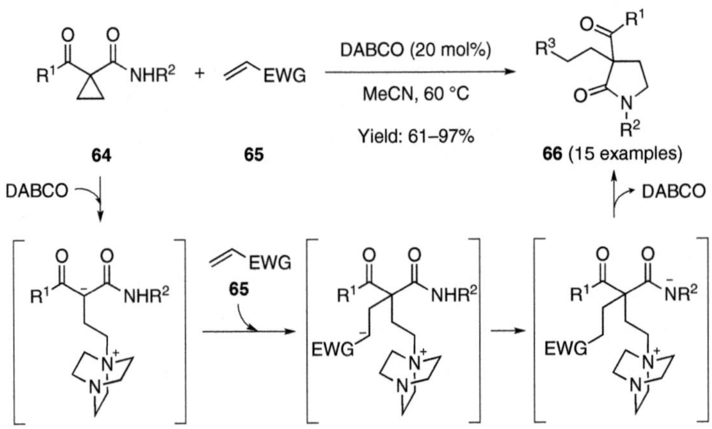

R^1 = Me, Ph
R^2 = Ph, 4-MeC_6H_4, 2,4-$Me_2C_6H_3$, 2-$(NO_2)C_6H_4$, 4-$MeOC_6H_4$, 2-ClC_6H_4, 4-ClC_6H_4, 2-Cl-4-$MeOC_6H_3$, 2-pyridyl, 1-naphthyl
R^3 = CN, CO_2Et, CO_2n-Bu, CO_2t-Bu, SO_2Ph, $CONMe_2$

Scheme 4.22 DABCO-catalyzed cyclopropane ring-opening/Michael/cyclization cascade reaction.

Cloke–Wilson-type rearrangement already shown in Scheme 4.21, delivered the final dihydrofurane products **68**.

Finally, the use of electrophilic vinylcyclpropanes **69** in combination with phosphines as catalysts produces cyclohepenones **70** as rearrangement products in high yields (Scheme 4.24) [37]. This transformation started with the nucleophilic

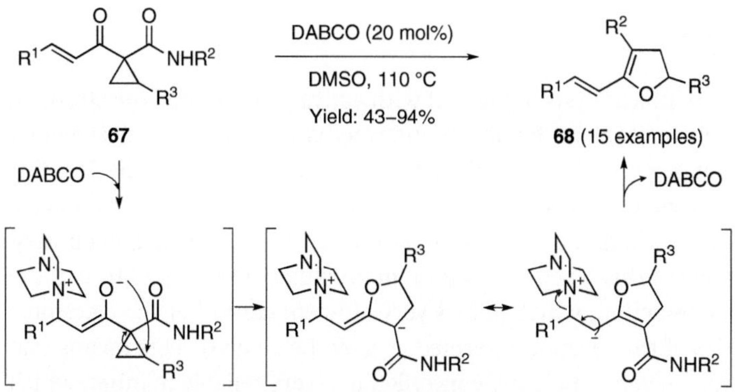

R^1 = 2-thienyl, 2-furyl, Ph, 4-MeC_6H_4, 3-$(NO_2)C_6H_4$, 4-$MeOC_6H_4$, 3,4-$(OCH_2O)C_6H_3$, 2-ClC_6H_4, 4-ClC_6H_4, 4-pyridyl
R^2 = Bn, 4-$MeC_6H_4CH_2$, 4-$MeOC_6H_4CH_2$, 2-$ClC_6H_4CH_2$, 4-$ClC_6H_4CH_2$
R^3 = H, Me

Scheme 4.23 DABCO catalyzed ring-opening reaction of enoylcyclopropanes.

ring-opening reaction of the cyclopropane **69**, triggered by the vinylogous addition of the phosphine to the terminal position of the olefin, and generated a zwitterionic enolate. This subsequently had to equilibrate to a second, less stable enolate, the second undergoing intramolecular S_N2' type displacement, which is preferred to a disfavored 5-*endo*-trig cyclization from the initially generated and more stable enolate. In fact, both enolates coexist in the reaction, as observed by the authors by ^{31}P-NMR. DFT studies have also supported this mechanism [38].

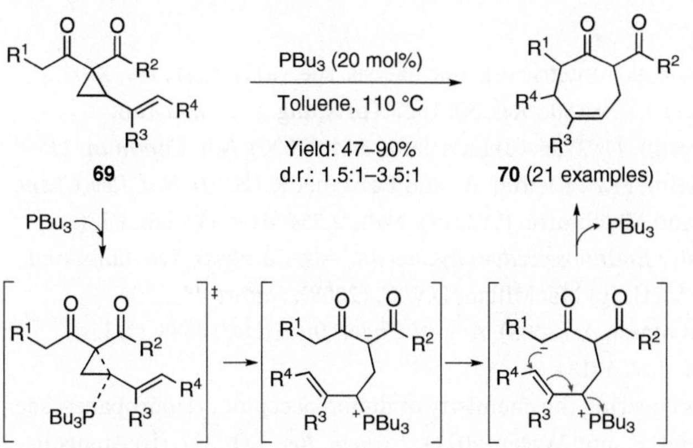

R^1 = H, Me, Et, n-Pr, *i*-Pr, *n*-Bu, *n*-C_6H_{13}, Bn, allyl, Ph, MeO
R^2 = Me, Ph, NHPh, NH(4-ClC_6H_4), NH(4-$MeOC_6H_4$), OMe, OEt, O*i*-Pr
R^3 = H, Me
R^4 = H, *n*-Bu, *n*-C_6H_{13}, *n*-$C_{10}H_{21}$, Ph

Scheme 4.24 Phosphine catalyzed rearrangement of electrophilic vinylcyclopropanes.

4.6 Conclusion

The possibility for organocatalysts to interact with activated cyclopropanes through the formation of new covalent bonds in intermediates that can subsequently undergo ring-opening events has experienced a true renaissance in the last few years and has opened new venues to find untrodden reactivity patterns that enable carrying out transformations that cannot be performed using standard chemistry. An evident advantage of this type of activation manifold is the strong nature of the catalyst–substrate interaction, which allows very efficient control of the electronic and steric properties of the reaction intermediates by the catalyst. This means that selectivity issues can be very efficiently controlled in a very reliable manner, which is particularly evident with respect to the possibility of performing these reactions in an enantioselective fashion. Nevertheless, and despite the impressive advances in the field, the portfolio of reactions that have been discovered or optimized up to date is still somewhat narrow, and therefore, venues for new exciting findings in the near future are fully open.

Acknowledgments

Financial suupport by the Spanish Ministerio de Ciencia, Innovación y Universidades (MCIU) through projects FEDER-PID2020-118422-GB-I00 and FEDER-CTQ2017-83633-P and by the Basque Government (Grupos IT908-16) is gratefully acknowledged.

References

1 For general reviews on asymmetric organocatalysis, see: (a) Han, B., He, X.-H., Liu, Y.-Q. et al. (2021). *Chem. Soc. Rev.* 50: 1522. (b) Xiang, S.-H. and Tan, B. (2020). *Nat. Commun. 11*: 3786. (c) Lassaletta, J.M. (2020). *Nat. Commun. 11*: 3787. (d) van Der Helm, M.P., Klemm, B., and Eelkema, R. (2019). *Nat. Rev. Chem. 3*: 491. (e) Silvi, M. and Melchiorre, P. (2018). *Nature 554*: 41. (f) Dalko, P.I. (ed.) (2013). *Comprehensive Enantioselective Organocatalysis: Catalysts, Reactions, and Applications.* Wiley-VCH. (g) MacMillan, D.W.C. (2008). *Nature 455*: 304. (h) Dondoni, A. and Massi, A. (2008). *Angew. Chem. Int. Ed. 47*: 4638. (i) List, B. (2007). *Chem. Rev. 107*: 5413.

2 For selected reviews covering the chemistry of donor–acceptor cyclopropanes, see: (a) Pirenne, V., Muriel, B., and Waser (2021). *J. Chem. Rev.* 121: 227. (b) Augustin, A.U. and Werz, D.B. (2021). *Acc. Chem. Res. 54*: 1528. (c) Ghosh, K. and Das, S. (2021). *Org. Biomol. Chem. 19*: 965. (d) Xia, Y., Liu, X., and Feng, X. (2020). *Angew. Chem. Int. Ed. 60*: 9192. (e) Werz, D.B. and Biju, A.T. (2019). *Angew. Chem. Int. Ed. 59*: 3385. (f) Gharpure, S.J. and Nanda, L.N. (2017). *Tetrahedron Lett. 58*: 711. (g) Grover, H.K., Emmett, M.R., and Kerr, M.A. (2015). *Org. Biomol. Chem.*

13: 655. (h) Novikov, R.A. and Tomilov, Y.V. (2015). *Mendeleev Commun. 25*: 1. (i) Cavitt, M.A., Phun, L.H., and France, S. (2014). *Chem. Soc. Rev. 43*: 804. (j) Schneider, T.F., Kaschel, J., and Werz, D.B. (2014). *Angew. Chem. Int. Ed. 53*: 5504. (k) Gharpure, S.J., Nanda, L.N., de Nanteuil, F. et al. (2014). *Chem. Commun. 50*: 10912. (l) Tang, P. and Qin, Y. (2012). *Synthesis 44*: 2969. (m) YaMelQnikov, M., Budynina, E.M., Ivanova, O.A., and Trushkov, I.V. (2011). *Mendeleev Commun. 21*: 293. (n) Lebold, T.P. and Kerr, M.A. (2010). *Pure Appl. Chem.* 82: 1797. (o) Carson, C.A. and Kerr, M.A. (2009). *Chem. Soc. Rev. 38*: 3051. (p) Yu, M. and Pagenkopf, B.L. (2005). *Tetrahedron 61*: 321. (q) Reissig, H.-U. and Zimmer, R. (2003). *Chem. Rev. 103*: 1151.For selected reviews covering the chemistry of acceptor cyclopropanes, see: (r) Apel, C. and Chrismann, M. (2021). *Tetrahedron 82*: 131760. (s) Babu, K.R., He, X., and Xu, S. (2020). *Synlett* 31: 117.

3 (a) List, B., Lerner, R.A., and Barbas, C.F. III (2000). *J. Am. Chem. Soc. 122*: 2395. (b) Ahrendt, K.A., Borths, C.J., and MacMillan, D.W.C. (2000). *J. Am. Chem. Soc. 122*: 4243.

4 For selected reviews on aminocatalysis, see: (a) Albrecht, Ł., Jiang, H., and Jørgensen, K.A. (2014). *Chem. Eur. J.* 20: 358. (b) Nielsen, M., Worgull, D., Zweifel, T. et al. (2011). *Chem. Commun. 47*: 632. (c) Bertelsen, S. and Jørgensen, K.A. (2009). *Chem. Soc. Rev. 38*: 2178. (d) Melchiorre, P., Marigo, M., Carlone, A., and Bartoli, G. (2008). *Angew. Chem. Int. Ed. 47*: 6138. (e) Mukherjee, S., Yang, J.W., Hoffmann, S., and List, B. (2007). *Chem. Rev. 107*: 5471. (f) List, B. (2006). *Chem. Commun.* 2006: 819. (g) List, B. (2001). *Synlett* 2001: 1675.

5 For selected reviews covering the use of NHC as organocatalyst, see: (a) Bellotti, P., Koy, M., Hopkinson, M.N., and Glorius, F. (2021). *Nat. Chem. Rev.* 5: 711. (b) Barik, S. and Biju, A.T. (2020). *Chem. Commun. 56*: 15484. (c) Chen, X.-Y., Gao, Z.-H., and Ye, S. (2020). *Acc. Chem. Res. 53*: 690. (d) Das, T.K. and Biju, A.T. (2020). *Chem. Commun. 56*: 8537. (e) Murauski, K.J.R., Jaworski, A.A., and Scheidt, K.A. (2018). *Chem. Soc. Rev. 47*: 1773. (f) Chen, X.-Y., Liu, Q., Chauhan, P., and Enders, D. (2018). *Angew. Chem. Int. Ed. 57*: 3862. (g) Mukherjee, S. and Biju, A.T. (2018). *Chem. Asian J. 13*: 2333. (h) Biju, A.T. (ed.) (2018). *N-Heterocyclic Carbenes in Organocatalysis*. Weinheim: Wiley-VCH. (i) Reyes, E., Uria, U., Carrillo, L., and Vicario, J.L. (2017). *Synthesis 49*: 451. (j) Wang, M.H. and Scheidt, K.A. (2016). *Angew. Chem. Int. Ed. 55*: 14912. (k) Walden, D.M., Ogba, O.M., Johnston, R.C., and Cheong, P.H. (2016). *Acc. Chem. Res. 49*: 1279. (l) Flanigan, D.M., Romanov-Michailidis, F., White, N.A., and Rovis, T. (2015). *Chem. Rev. 115*: 9307. (m) Hopkinson, M.N., Richter, C., Schedler, M., and Glorius, F. (2014). *Nature 510*: 485. (n) Mahatthananchai, J. and Bode, J.W. (2014). *Acc. Chem. Res. 47*: 696. (o) Ryan, S.J., Candish, L., and Lupton, D.W. (2013). *Chem. Soc. Rev. 42*: 4906. (p) Chen, X.Y. and Ye, S. (2013). *Org. Biomol. Chem. 11*: 7991. (q) Bugaut, X. and Glorius, F. (2012). *Chem. Soc. Rev. 41*: 3511. (r) Grossmann, A. and Enders, D. (2012). *Angew. Chem. Int. Ed. 51*: 314. (s) Izquierdo, J., Hutson, G.E., Cohen, D.T., and Scheidt, K.A. (2012). *Angew. Chem. Int. Ed. 51*: 11686. (t) Biju, A.K., Kuhl, N., and Glorius, F. (2011). *Acc. Chem. Res. 44*: 1182. (u) Phillips, E.M., Chan, A., and Scheidt, K.A. (2009). *Aldrichimica Acta 42*: 55. (v) Marion, N.,

Diez-Gonzalez, S., and Nolan, S.P. (2007). *Angew. Chem. Int. Ed. 46*: 2988. (w) Enders, D., Niemeier, O., and Henseler, A. (2007). *Chem. Rev. 107*: 5606.

6 For selected reviews covering the use of tertiary amines and/or phosphines as organocatalysts, see: (a) Ni, H., Chan, W.-L., and Lu, Y. (2018). *Chem. Rev. 118*: 9344. (b) Wang, T., Han, X., Zhong, F. et al. (2016). *Acc. Chem. Res. 49*: 1369. (c) Xiao, Y., Guo, H., and Kwon, O. (2016). *Aldrichimica Acta 49*: 3. (d) Voituriez, A., Marinetti, A., and Gicquel, M. (2015). *Synlett 26*: 142. (e) Wang, Z., Xu, X., and Kwon, O. (2014). *Chem. Soc. Rev. 43*: 2927. (f) Wei, Y. and Shi, M. (2014). *Chem. Asian J.* 9: 2720. (g) Wang, Z. and Kwon, O. (2013). Phosphine organocatalysis as a platform for diversity-oriented synthesis. In: *Diversity-Oriented Synthesis: Basics and Applications in Organic Synthesis, Drug Discovery, and Chemical Biology* (ed. A. Trabocchi) , 97–133. Hoboken, NJ: John Wiley & Sons, Inc. (h) Fan, Y.C. and Kwon, O. (2012). Phosphine catalysis. In: *Science of Synthesis*, Asymmetric Organocatalysis, vol. vol. 1 (ed. B. List) , 723–782. Stuttggart: Georg Thieme. (i) Tian, S.-K., Chen, Y., Hang, J. et al. (2004). *Acc. Chem. Res. 37*: 621. (j) Wei, Y. and Shi, M. (2010). *Acc. Chem. Res. 43*: 1005. (k) Marinetti, A. and Voituriez, A. (2010). *Synlett* 2: 174. (l) Ye, L.-W., Zhou, J., and Tang, Y. (2008). *Chem. Soc. Rev.* 37: 1140. (m) Methot, J.L. and Roush, W.R. (2004). *Adv. Synth. Catal. 346*: 1035. (n) France, S., Guerin, D.J., Miller, S.J., and Lectka, T. (2003). *Chem. Rev. 103*: 2985.

7 Halskov, K.S., Kniep, F., Lauridsen, V.H. et al. (2015). *J. Am. Chem. Soc. 137*: 1685–1691.

8 For selected reviews covering the use of Jørgensen-Hayashi catalyst, see: (a) Reyes-Rodriguez, G.J., Rezayee, N.M., Vidal-Albalat, A., and Jørgensen, K.A. (2019). *Chem. Rev.* 6: 4221. (b) Klier, L., Tur, F., Poulsen, P.H., and Jørgensen, K.A. (2017). *Chem. Soc. Rev. 46*: 1080. (c) Chauhan, P., Mahajan, S., and Enders, D. (2017). *Acc. Chem. Res. 50*: 2809. (d) Donslund, B.S., Johansen, T.K., Poulsen, P.H. et al. (2015). *Angew. Chem. Int. Ed. 54*: 13860. (e) Jiang, H., Albrecht, L., Dickmeiss, G. et al. (2013). TMS-prolinol catalyst in organocatalysis. In: *Comprehensive Enantioselective Organocatalysis* (ed. P.I. Dalko) , 33–50. Weinheim: Wiley-VCH. (f) Gotoh, H. and Hayashi, Y. (2013). Diarylprolinol silyl ethers: development and application as organocatalysts. In: *Sustainable Catalysis* (ed. P.J. Dunn) , 287–316. Hoboken: Wiley. (g) Jensen, K.L., Dickmeiss, G., Jiang, H. et al. (2012). *Acc. Chem. Res. 45*: 248. (h) Marques-Lopez, E. and Herrera, R.P. (2011). *Curr. Org. Chem. 15*: 2311. (i) Lattanzi, A. (2009). *Chem. Commun.* 1452. (j) Mielgo, A. and Palomo, C. (2008). *Chem. Asian J. 6*: 922.

9 For previous [2+2] cycloadditon proceeding under dienamine activation, see: (a) Qi, L.-W., Yang, Y., Gui, Y.-Y. et al. (2014). *Org. Lett. 16*: 6436. (b) Albrecht, Ł., Dickmeiss, G., Acosta, F.C. et al. (2012). *J. Am. Chem. Soc. 134*: 2543. (c) Talavera, G., Reyes, E., Vicario, J.L., and Carrillo, L. (2012). *Angew. Chem. Int. Ed. 51*: 4104.

10 Sanchez-Diez, E., Vesga, D.L., Reyes, E. et al. (2016). *Org. Lett. 18*: 1270.

11 For some examples of aza Michael/aldol condensation over unsaturated aldehydes under iminium ion activation see: (a) Enders, D., Joie, C., and Deckers, K. (2013). *Chem. Eur. J. 19*: 10818. (b) Lee, H.-J. and Cho, C.-W. (2013). *J. Org. Chem. 78*: 3306. (c) Enders, D., Greb, A., Deckers, K. et al. (2012). *Chem. Eur. J. 18*: 10226. (d) Zhang, X., Song, X., Li, H. et al. (2012). *Angew. Chem. Int. Ed. 51*: 7282. (e)

Fernandez, M., Vicario, J.L., Reyes, E. et al. (2012). *Chem. Commun. 48*: 2092. (f) Lee, S.-J., Youn, S.-H., and Cho, C.-W. (2011). *Org. Biomol. Chem. 9*: 7734. (g) Bae, J.-Y., Lee, H.-J., Youn, S.-H. et al. (2010). *Org. Lett. 12*: 4352. (h) Hong, L., Sun, W., Liu, C. et al. (2010). *Chem. Eur. J. 16*: 440. (i) Enders, D., Wang, C., and Raabe, G. (2009). *Synthesis* 41: 4119. (j) Yoshitomi, Y., Arai, H., Makino, K., and Hamada, Y. (2008). *Tetrahedron 64*: 11568. (k) Li, H., Wang, J., Xie, H. et al. (2007). *Org. Lett. 9*: 965. (l) Sundén, H., Rios, R., Ibrahem, I. et al. (2007). *Adv. Synth. Catal. 349*: 827.

12 Fuson, R.C. (1934). *Chem. Rev.* 16: 1.

13 Apel, C., Hartmann, S.S., Lentz, D., and Christmann, M. (2019). *Angew. Chem. Int. Ed. 58*: 5075.

14 Xie, H., Zu, L., Li, H. et al. (2007). *J. Am. Chem. Soc. 129*: 10886.

15 Li, L., Li, Z., and Wang, Q. (2009). *Synlett* 11: 1830.

16 Sparr, C. and Gilmour, R. (2011). *Angew. Chem. Int. Ed. 50*: 8391.

17 Wallbaum, J., Garve, L.K.B., Jones, P.G., and Werz, D.B. (2016). *Chem. Eur. J. 22*: 18756.

18 Diaz, E., Reyes, E., Uria, U. et al. (2018). *Chem. Eur. J. 24*: 8764.

19 Jiang, M.-Y., Zhang, L., Liu, R. et al. (2009). *J. Nat. Prod. 72*: 1405.

20 Sohn, S.S. and Bode, J.W. (2006). *Angew. Chem. Int. Ed. 45*: 6021.

21 Vesely, J., Zhao, G.-L., Bartoszewicz, A., and Cordova, A. (2008). *Tetrahedron Lett. 49*: 4209.

22 Meazza, M., Ashe, M., Shin, H.Y. et al. (2016). *J. Org. Chem. 81*: 3488.

23 Bode, J.W. and Sohn, S.S. (2007). *J. Am. Chem. Soc. 129*: 13798.

24 Li, G.-Q., Dai, L.-X., and You, S.-L. (2009). *Org. Lett. 11*: 1623.

25 Du, D. and Wang, Z. (2008). *Eur. J. Org. Chem.* 2008: 4949.

26 Li, L., Du, D., Ren, J., and Wang, Z. (2011). *Eur. J. Org. Chem.* 2011: 614.

27 Du, D., Li, L., and Wang, Z. (2009). *J. Org. Chem. 74*: 4379.

28 Prieto, L., Sanchez-Diez, E., Uria, U. et al. (2017). *Adv. Synth. Catal. 359*: 1678.

29 Wang, Y., Qiao, Y., Lan, Y., and Wei, D. (2021). *Cat. Sci. Technol. 11*: 332.

30 Lv, J., Xu, J., Pan, X. et al. (2021). *Sci. China Chem. 64*: 985.

31 Day, L. and Ye, S. (2020). *Org. Lett. 22*: 986.

32 Nie, G., Huang, X., Wang, Z. et al. (2021). *Org. Chem. Front. 8*: 5105.

33 Zhang, J., Tang, Y., Wei, W. et al. (2017). *Org. Lett. 19*: 3043.

34 Shi, Q., Wang, Y., and Wei, D. (2018). *Comput. Theor. Chem. 1123*: 20.

35 (a) Lin, S., Li, L., Liang, F., and Liu, Q. (2014). *Chem. Commun.* 50: 10491.for a related example, see (b) Du, D. and Wang, Z. (2008). *Tetrahedron Lett. 49*: 956.

36 Li, M., Lin, S., Dong, Z. et al. (2013). *Org. Lett. 15*: 3978.

37 Wu, J., Tang, Y., Wu, Y. et al. (2018) . *Angew. Chem. Int. Ed. 57*: 6248.

38 Wu, Y., Li, M., Jin, L., and Zhao, X. (2020) . *ACS Omega 5*: 2957.

5

Ring-Opening 1,3-Bisfunctionalization of Donor–Acceptor Cyclopropanes

Avishek Guin and Akkattu T. Biju

Indian Institute of Science, Department of Organic Chemistry, Bangalore, 560012, India

CHAPTER MENU

Donor–Acceptor Cyclopropanes in Organic Synthesis, First Edition.
Edited by Prabal Banerjee and Akkattu T. Biju.

5.1 Introduction

Donor–acceptor (D–A) cyclopropanes are strained molecules, which are highly electrophilic in nature. In 1980, Reissig originally coined the term D–A cyclopropanes [1]. The strain energy for cyclopropane is around 27.5 kcal/mol. A high degree of polarization of the C—C single bond is induced within the D–A cyclopropane system because of the push-pull effect of the nearby electron-donating and electron-accepting units. Polarization can be further enhanced with the existence of Lewis acid [2–8]. Thus, suitable nucleophiles can add to the D–A cyclopropane, which leads to the opening of the strained three-membered ring. Consequently, a stable carbanion is generated at the acceptor terminus of D–A cyclopropane. If the carbanion is quenched with the proton source, it will lead to mono functionalization of D–A cyclopropane, and this chemistry is very well explored in D–A cyclopropane field. But if the *in situ* generated carbanion is slaked with an appropriate electrophilic source, 1,3-bisfunctionalization of D–A cyclopropane is achieved. 1,3-Bisfunctionalization of D–A cyclopropane is highly important because it can introduce two functional groups at the same time. However, it is also highly challenging because of the appropriate nucleophile-electrophile combination used for the process. Sometimes compatibility of nucleophiles and electrophiles in the same reaction flask is an issue. Consequently, very few reactions for ring-opening 1,3-bisfunctionalization of D–A cyclopropane are reported to date. The noteworthy reactions involving ring-opening 1,3-bisfunctionalization of D–A cyclopropane will be discussed in the following sections.

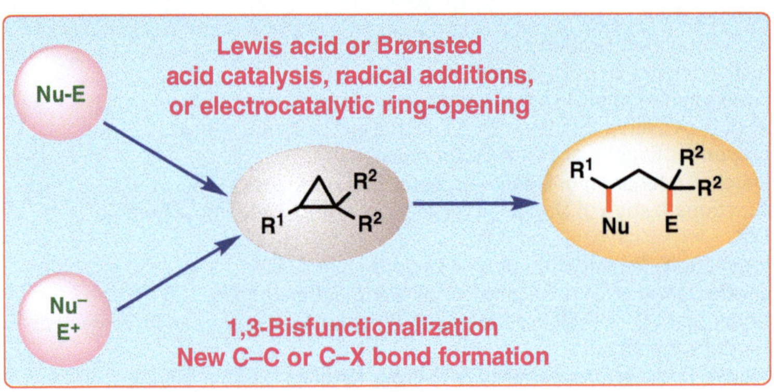

5.2 Enantioselective 1,3-Dichlorination of Formyl Group-Containing Cyclopropanes

Synthetic organic chemists used D–A cyclopropanes as versatile intermediates for more than 40 years. However, selective bisfunctionalization of D–A cyclopropanes was rarely obtained in the early days. In 2011, Sparr and Gilmour smartly designed an enantioselective 1,3-dichlorination of cyclopropanes containing a formyl group [9]. The reaction was initiated with the formation of cyclopropyl iminium intermediate **7**. Nucleophilic chloride ion **2** adds to the γ position of intermediate **7** and generates the effective secondary enamine intermediate **3**. This can be considered as formal umpolung through the use of a secondary amine, to utilize nucleophilic chloride for γ-functionalization, which furnishes intermediate **3**. The enamine intermediate further adds through α-attack to electrophilic chlorine **4** to furnish the intermediate **5**. Hydrolysis of intermediate **5** yields the expected enantioselective 1,3-dichlorination of cyclopropanes **6** (Scheme 5.1). Since enamine formation is the main criterion for this 1,3-dichlorination of cyclopropanes, this reaction is limited to reactive formyl groups as substitutions.

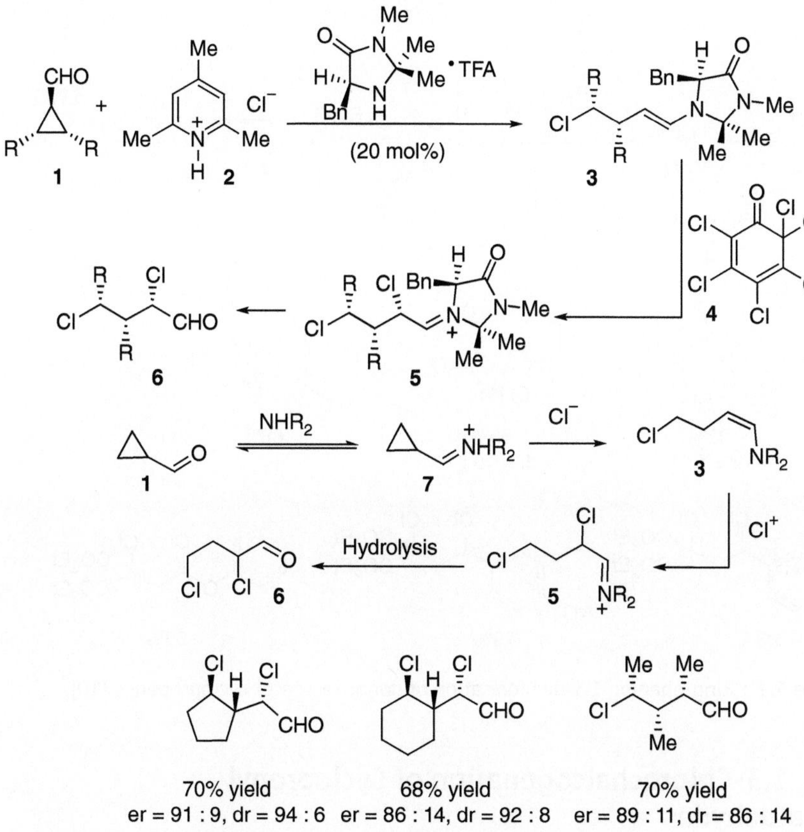

Scheme 5.1 Enantioselective 1,3-dichlorination of cyclopropanes [9].

5.3 Ring-Opening 1,3-Dichlorination of D–A Cyclopropanes

Inspired by the above enantioselective 1,3-dichlorination of cyclopropanes, later Werz and coworkers developed a general method for the 1,3-dichlorination of D–A cyclopropanes using iodobenzene dichloride **9** as a chloride source [10]. Several D–A cyclopropanes were transformed with the readily obtainable iodobenzene dichloride into the corresponding ring-opened 1,3-dichlorinated compounds. Most of the transformations yielded moderate-to-high yields of the desired 1,3-dichlorinated compounds. Dichloromethane was found to be the optimum solvent for this transformation. The reaction proceeds with the homolytic cleavage of the I—Cl bond of iodobenzene dichloride **9** to release the Cl atom. This very reactive radical species adds to the weakest bond between D–A cyclopropanes to generate another delocalized radical intermediate **13**. Intermediate **13** can either abstract Cl from $PhICl_2$ or merge with a Cl atom **11** or with a PhICl radical **12**, releasing PhI (Scheme 5.2).

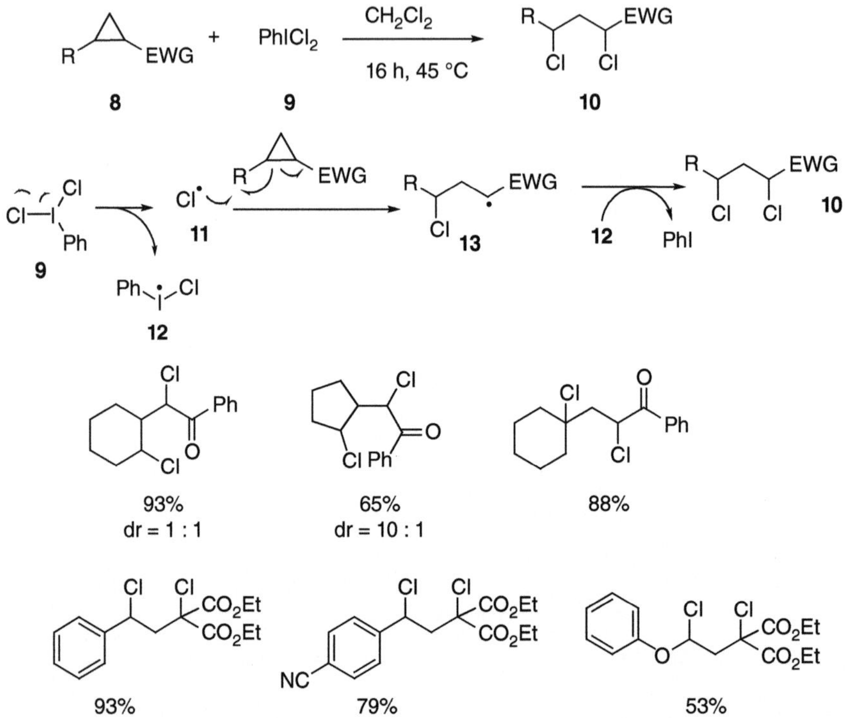

Scheme 5.2 Ring-opening 1,3-dichlorination of donor–acceptor cyclopropanes [10].

5.4 1,3-Chlorochalcogenation of Cyclopropyl Carbaldehydes

Interestingly, Werz and coworkers further developed regio-, diastereo-, and enantioselective 1,3-chlorochalcogenation of cyclopropyl carbaldehydes in 2016 [11]. In the presence of an organocatalyst, meso-cyclopropyl carbaldehydes reacted with sulfenyl or selenyl chlorides to furnish 1,3-chlorochalcogenated products. This conversion was successfully attained by merging iminium–enamine catalysis with aldehyde-substituted cyclopropanes. The target products were delivered through enantioselective desymmetrization reactions, leading to three adjacent stereocenters in complete regioselectivity and moderate-to-high diastereo- and enantioselectivities. Mechanistically, the reaction is similar to 1,3-dichlorination of cyclopropanes as demonstrated by Sparr and Gilmour [9]. The reaction proceeds with the formation of iminium intermediate **17**. Sulfenyl chloride attacks at the 3-position by the chloride ion to the *in situ* generated intermediate **17**, to furnish the enamine intermediate **18**. The further addition of the enamine to positively charged sulfenylium ion, followed by hydrolysis, yields the expected 1,3-chlorochalcogenation product of cyclopropyl carbaldehydes **16** (Scheme 5.3).

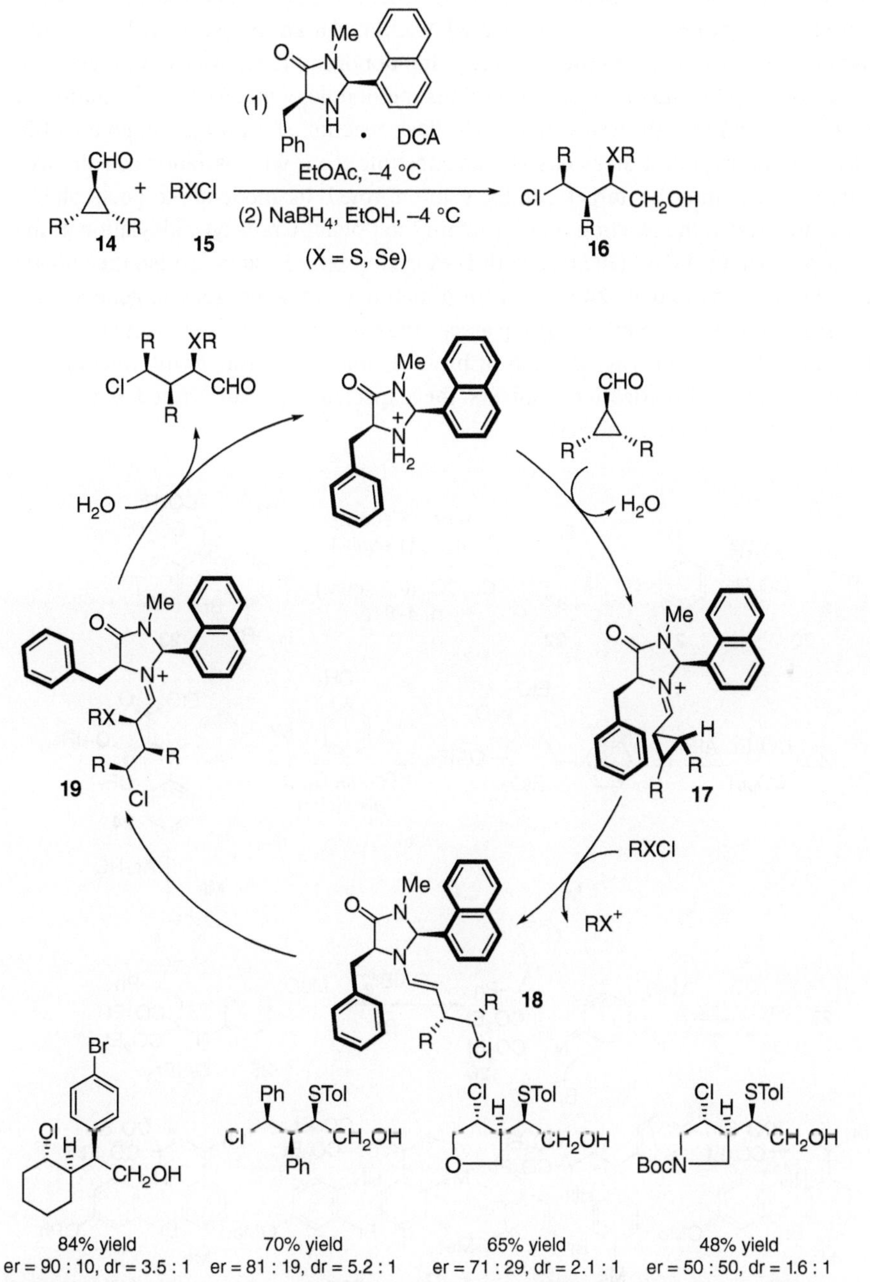

Scheme 5.3 1,3-Chlorochalcogenation of cyclopropyl carbaldehydes [11].

5.5 1,3-Bisfunctionalization of D–A Cyclopropanes with Arenes and Nitrosoarenes

Multicomponent 1,3-bisfunctionalization of D–A cyclopropanes with arenes and nitrosoarenes as the coupling partners was uncovered by Studer and coworkers in 2016 for the synthesis of γ,γ-disubstituted *N*-arylated α-amino esters [12]. $AlBr_3$ acts as both the Lewis acid and the bromide anion donor in the regioselective arene bromination step for this transformation. This methodology involves the formation of C—C, C—N, and C—Br bonds in a single flask reaction. This multicomponent 1,3-bisfunctionalization strategy has a broad substrate scope with variations on all three components, and the target products are formed in moderate-to-good yields. Mechanistically, the reaction proceeds with the Friedel-Crafts type alkylation at the *para*-position of the toluene ring with D–A cyclopropanes **20** to furnish the enolate **24**. The resulting enolate **24** reacts with *p*-methoxy nitrosobenzene to generate the intermediate **25**. The methoxy group assists the cleavage of the N—O bond to form a new intermediate **26**. The addition of bromide ion from $AlBr_3$ to intermediate **26** followed by tautomerization, delivers the expected γ,γ-disubstituted *N*-arylated α-amino esters **23** (Scheme 5.4).

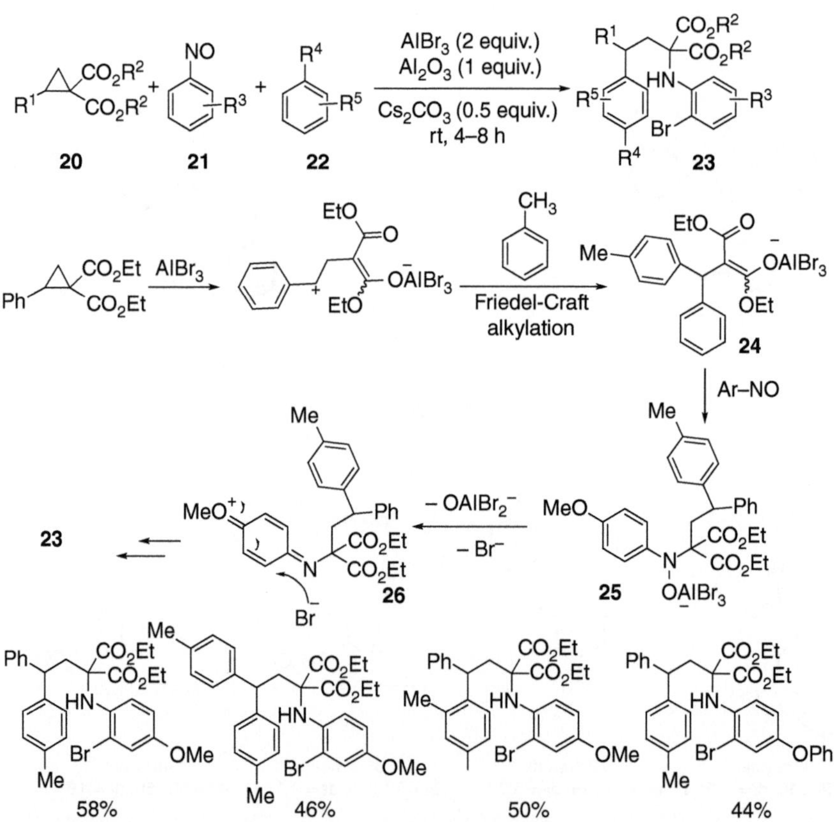

Scheme 5.4 1,3-Bisfunctionalization of D–A cyclopropanes with arenes and nitrosoarenes [12].

5.6 1-Amino-3-Aminomethylation of D–A Cyclopropanes

In 2017, Werz and coworkers demonstrated a facial ring-opening reaction of D–A cyclopropanes to afford 1,4-diamines [13]. The primary reaction was the (4+3) cycloaddition of D–A cyclopropanes **20** with triazinanes **27** in the presence of scandium triflate as a Lewis acid catalyst for the synthesis of 1,3-diazepanes. Acid treatment of 1,3-diazepanes led to the efficient cleavage of the aminal moiety, producing the desired 1,4-diamines **28**. The one-pot reaction was also performed, which includes an amine and an aminomethyl group to the former cyclopropane, with ring-opening at the first and third positions, respectively (Scheme 5.5). The reaction advances under very mild conditions and tolerates many functional groups. A library of 1,4-diamine compounds can be synthesized using this efficient protocol. Mechanistic investigation indicates that the reaction is stereospecific, with an initial S_N2 attack of triazinane opens the activated cyclopropane to yield the expected diamine product **28** (Scheme 5.5).

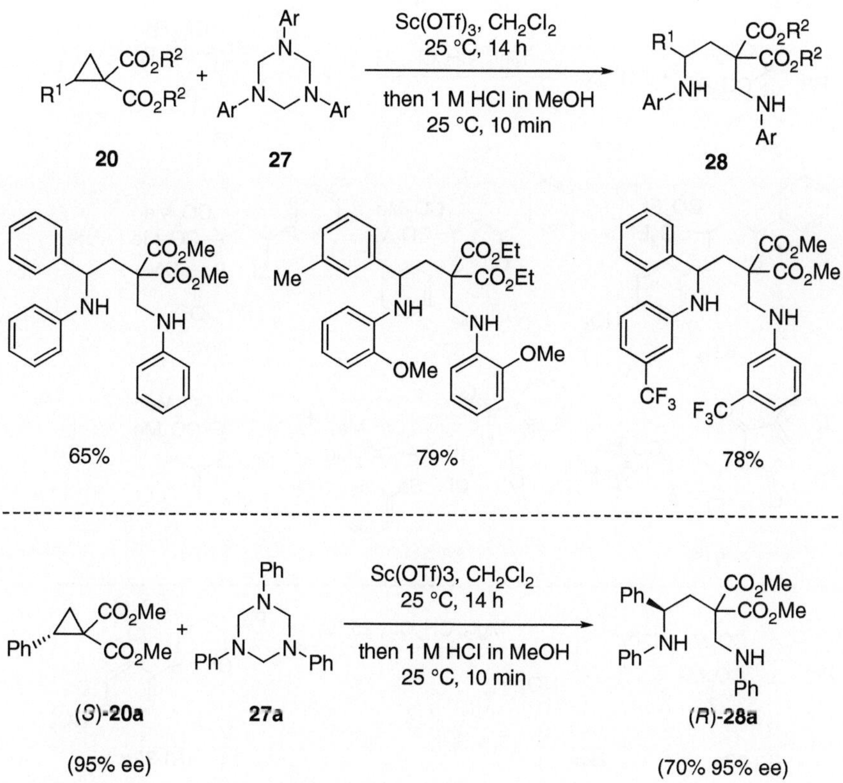

Scheme 5.5 1-Amino-3-aminomethylation of D–A cyclopropanes [13].

5.7 1,3-Halochalcogenation of D–A Cyclopropanes

Inspired by the results of 1,3-chlorochalcogenation of cyclopropyl carbaldehydes, the Werz group developed a more general method for the 1,3-halochalcogenation of D–A cyclopropanes for the synthesis of γ-halogenated α-thiolated malonic diester derivatives **30** [14]. Although in this case no iminium catalysis was involved, MgI_2 as a Lewis acid catalyst was used for the desired transformation. D–A cyclopropanes **20** bearing two geminal carboxylic esters are reacted with chalcogenyl chlorides or bromides **29** to furnish the ring-opened products **30**, having the halogen atoms adjacent to the donor and the chalcogenyl residue next to the two acceptor groups. Several D–A cyclopropanes were transformed with either readily available sulfenyl chlorides, selenyl chlorides, or sulfenyl bromides. The use of several other donors such as oxygen, nitrogen, and even aromatic systems are all well tolerated. In the presence of the enantiopure D–A cyclopropane **20a**, the 1,3-halochalcogenated product **30a** was obtained with inversion of configuration, which indicates that the transformation is stereospecific and proceeds in S_N2-like fashion (Scheme 5.6).

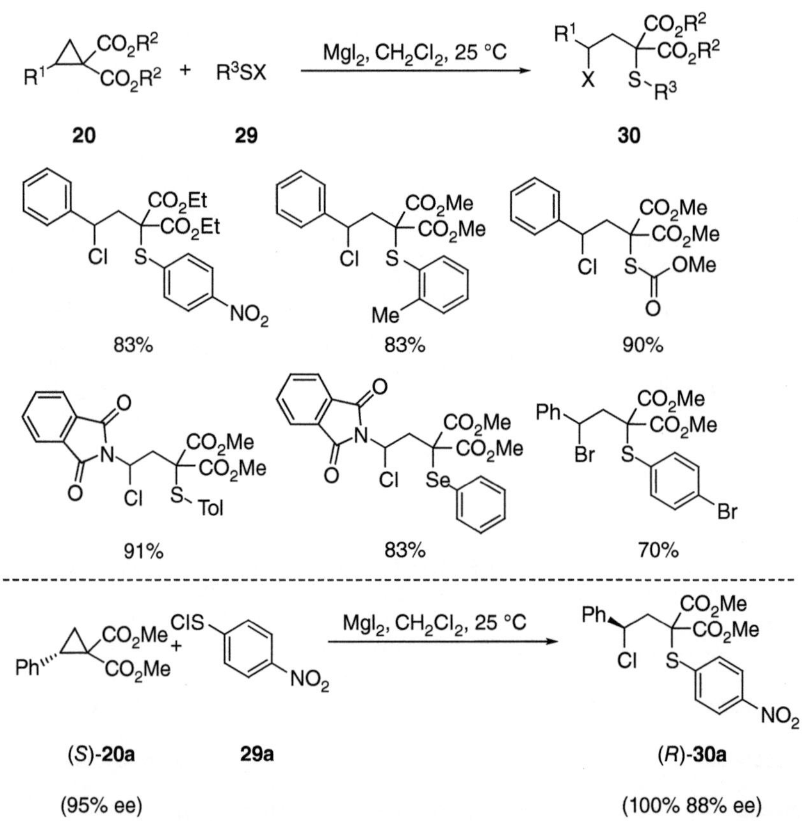

Scheme 5.6 1,3-Halochalcogenation of D–A cyclopropanes [14].

5.8 1,3-Aminobromination of D–A Cyclopropanes

In 2017, Studer and coworkers reported the $Sn(OTf)_2$-catalyzed 1,3-aminobromination of D–A cyclopropanes with different sulfonyl amides or electron-poor anilines **31** and *N*-bromosuccinimide **32** for the synthesis of γ-aminated α-brominated malonic diesters **33** [15]. The reaction proceeds under mild conditions and these reactions occurred with complete regio- and stereospecificity (Scheme 5.7). This methodology can tolerate various functional groups and in most cases moderate-to-good yields of γ-aminated α-brominated malonic diesters were obtained. These products functioned as valuable substrates for the subsequent reactions to afford substituted γ-lactams and azetidines in good yields. The reaction with anilines **31a** proceeded with complete stereospecificity. Simple experimentation, ready access to starting materials, and good yields of products are the notable features of this transformation.

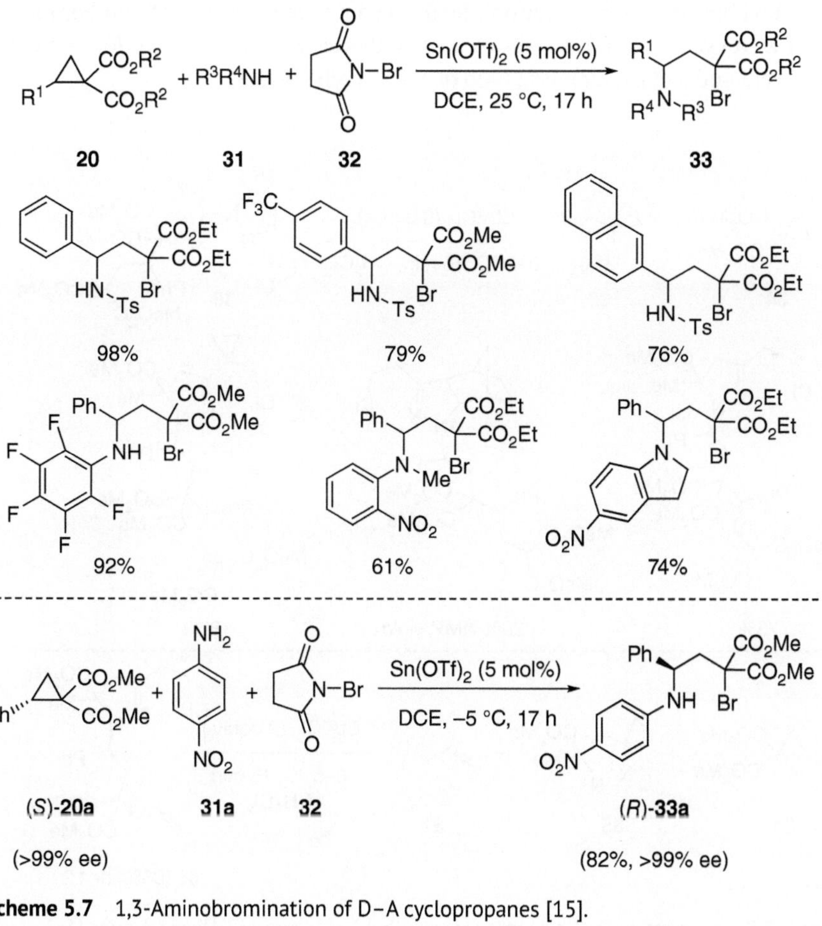

Scheme 5.7 1,3-Aminobromination of D–A cyclopropanes [15].

5.9 Reaction of D–A Cyclopropanes with 4,5-Diazaspiro[2.4]hept-4-enes

In 2018, Tomilov and coworkers uncovered a unique 1,3-bisfunctionalization reaction of D–A cyclopropanes and 4,5-diazaspiro[2.4]hept-4-enes for the synthesis of highly functionalized pyrazolines [16]. $EtAlCl_2$ has dual roles in this reaction. In the presence of $EtAlCl_2$, activation of D–A cyclopropane takes place. Also, the positively charged intermediate adds to the azocyclopropane system of the pyrazoline, leading to the opening of the second three-membered ring, along with the addition of a halide anion from the $EtAlCl_2$. Depending upon the ratio of the D–A cyclopropane or pyrazoline, a twofold addition can take place for both components. Product **36** was formed when a 5 : 1 ratio of D–A cyclopropane **34** and pyrazoline **35** were used, and in this case, 1,3-bisfunctionalization was achieved for one D–A cyclopropane moiety and the other D–A cyclopropane monofunctionalization was attained. The use of allyl bromide as the electrophilic third component, instead of another molecule of D–A cyclopropane, was also tested; in this case, product **38** was formed with a 63% yield and a diastereomeric ratio of 1.2 : 1 (Scheme 5.8).

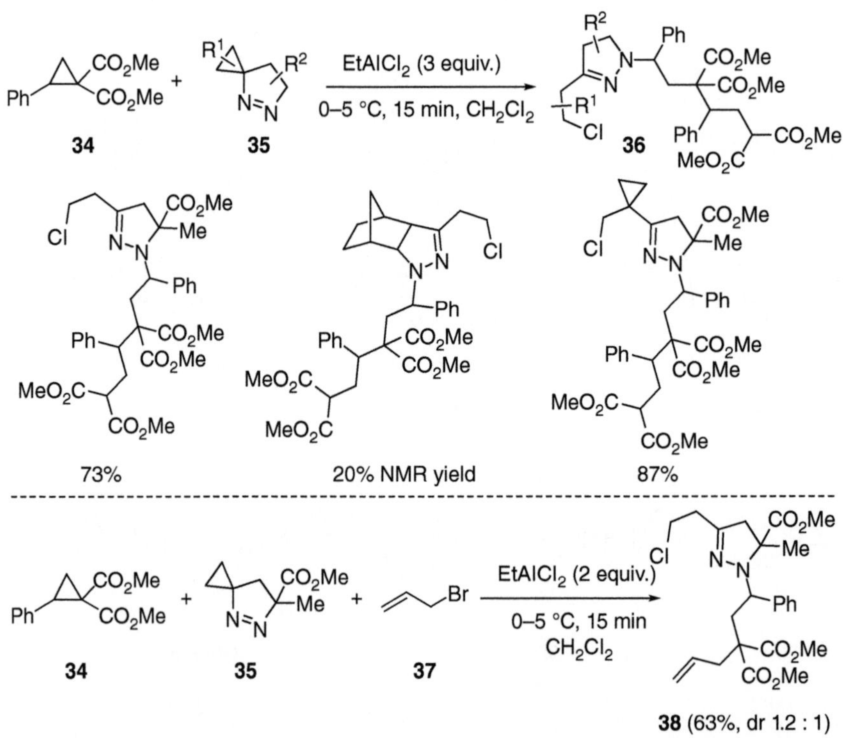

Scheme 5.8 Reaction of D–A cyclopropanes with 4,5-diazaspiro[2.4]hept-4-enes [16].

5.10 Four-Component Coupling of D–A Cyclopropanes

For the first time, Studer and his coworkers disclosed a four-component coupling reaction of D–A cyclopropanes **20** involving indolyl boron ate complexes and alkyl halide **41** [17]. Scandium triflate was found to be the optimum Lewis acid catalyst, with THF being the best solvent for this transformation. The reaction progressed with complete diastereoselectivity, to afford indolines. The formation of three C—C bonds and the construction of three contiguous stereocenters are notable features of this transformation. The reaction performed using enantiomerically enriched D–A cyclopropane **20a**, showed complete stereospecificity, to deliver highly enantiomerically enriched indoles and indolines in a single flask (Scheme 5.9). The present transformation can tolerate several valuable functional groups, and in most cases,

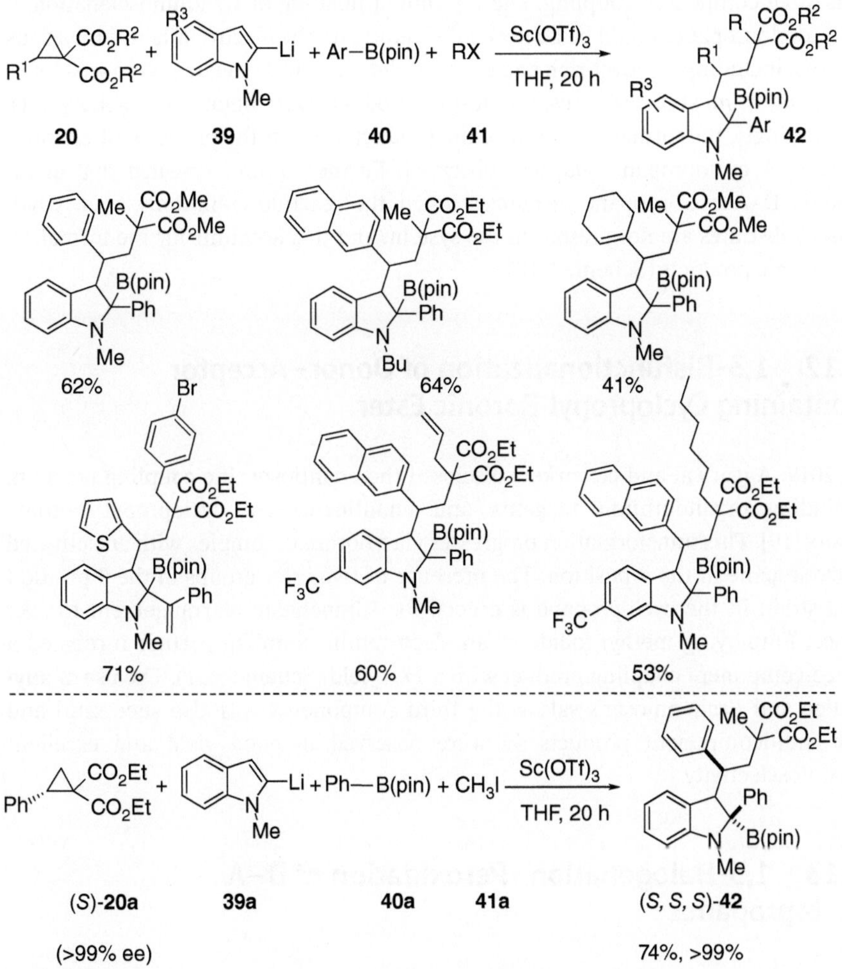

Scheme 5.9 Four-component coupling of D–A cyclopropanes with indolylboron ate complexes and alkyl halide [17].

the expected product is formed in moderate-to-good yields. The beneficial boron moiety is retained in the product, thus offering useful options for further transformation.

5.11 1,3-Aminochalcogenation of Donor–Acceptor Cyclopropanes

Inspired by the 1,3-aminobromination of donor–acceptor cyclopropanes demonstrated by Studer and his coworkers, Werz and coworkers later disclosed the 1,3-aminochalcogenation of donor–acceptor cyclopropanes using the multicomponent strategy [18]. $Sn(OTf)_2$ was established to be the Lewis acid of choice for this three-component coupling strategy. Sulfonamide **43** was found to be the nucleophile for this three-component coupling. The 1,3-aminothiolation or 1,3-aminoselenation of D–A cyclopropanes could be achieved depending on the electrophile used. Various donors, including electron-rich and electron-deficient aryl moieties, were well tolerated, and in most of the cases, the desired products were obtained in good yields. Interestingly, the formation of a racemic product, even in the presence of enantiopure D–A cyclopropane **20a**, was observed. Further studies revealed that enantiopure D–A cyclopropane racemizes under the reaction conditions, even when other substrates are not present in the system, and this accounts for the formation of racemic products (Scheme 5.10).

5.12 1,3-Bisfunctionalization of Donor–Acceptor Containing Cyclopropyl Boronic Ester

In 2019, Aggarwal and coworkers disclosed the enantiospecific coupling reaction, including organolithium reagents and enantioenriched cyclopropyl boronic esters [19]. This transformation progresses via a boronate complex with an activated cyclopropane in the α position. The presence of two ester groups in the β position and strain in the cyclopropane is crucial for 1,2-metalate rearrangement to take place. Employing methyl iodide as an electrophilic coupling partner furnished a three-component coupling product with a 77% yield (Scheme 5.11). The use of allyl iodide and Eschenmoser's salt as the third components was also successful and the multicomponent products **49** were observed in good yield and excellent enantioselectivity.

5.13 1,3-Halogenation–Peroxidation of D–A Cyclopropanes

Saha and coworkers revealed the use of hydroperoxides as the nucleophilic trigger in the multicomponent coupling reaction of D–A cyclopropanes [20]. This is the

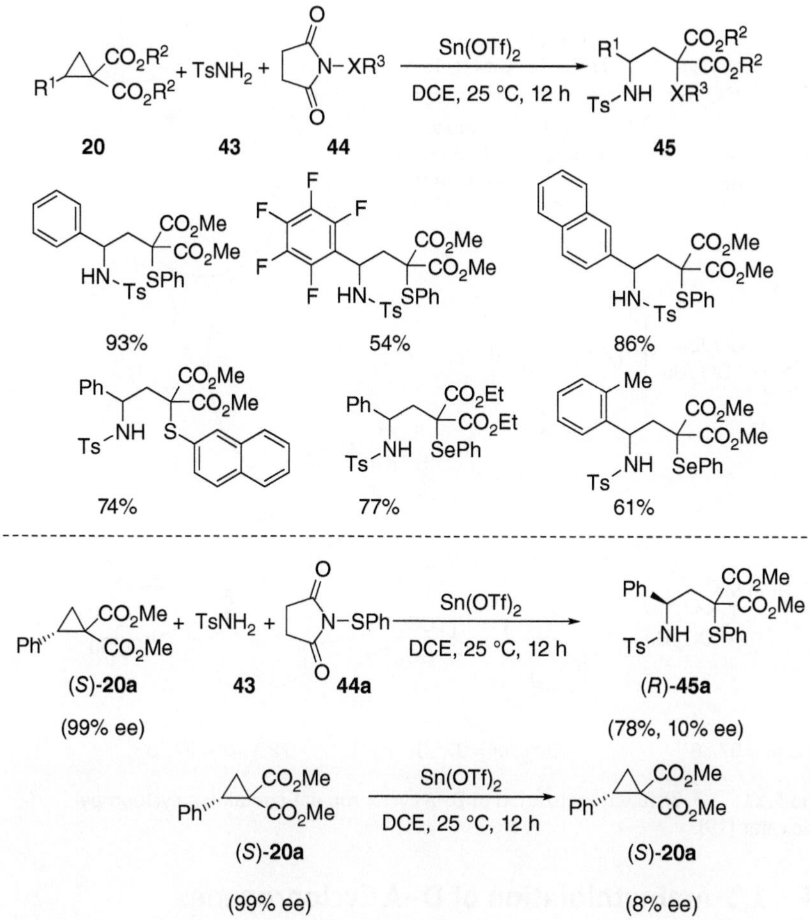

Scheme 5.10 1,3-Aminochalcogenation of donor–acceptor cyclopropanes [18].

first example of constructing highly functionalized γ-peroxy carbonyls through an ionic process and is distinctive from the previously reported radical methodologies. Operationally, a simple ring-opening 1,3-halogenation–peroxidation was achieved in the presence of hydroperoxide and an appropriate halogenating agent. $Sc(OTf)_3$ was found to be the choice of Lewis acid for the desired three-component coupling. The reaction was conducted at 0 °C in the presence of dichloromethane as the solvent. This multicomponent approach can tolerate several functional groups and shows variations for all three components, with products formed in moderate-to-good yields (Scheme 5.12). Enantiopure products were obtained when chiral D–A cyclopropane was used, which indicates the S_N2 nature of the addition of hydroperoxides to D–A cyclopropanes. Various synthetic transformations of 1,3-halogenation–peroxidation products were performed, which makes this methodology more attractive to synthetic organic chemists.

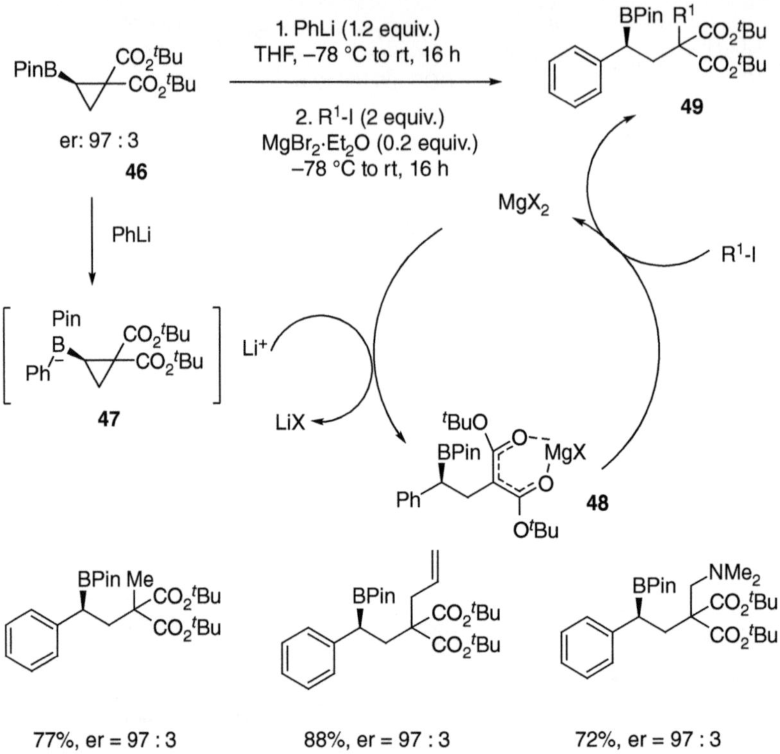

Scheme 5.11 1,3-Bisfunctionalization of D–A cyclopropane-containing cyclopropyl boronic ester [19].

5.14 1,3-Aminothiolation of D–A Cyclopropanes Using Sulfenamides

Considering the efficiency of 1,3-bisfunctionalization of D–A cyclopropanes, Biju and coworkers uncovered the $Yb(OTf)_3$-catalyzed mild and regioselective ring-opening 1,3-aminothiolation of D–A cyclopropanes, utilizing sulfenamides for the construction of γ-aminated α-thiolated malonic diester derivatives **54** [21]. The Lewis acid-catalyzed formation of C—N and C—S bonds was the key feature of this methodology. DCE was found to be the optimum solvent for this transformation. Under optimized conditions, various D–A cyclopropanes and sulfenamides were tested, and in all cases, the desired products were obtained in moderate-to-good yields (Scheme 5.13). Notably, in this case, enantiopure D–A cyclopropane yielded a chiral product with an inversion of configuration, which indicates that the nucleophilic attack of the sulfenamide proceeds in an S_N2-like fashion. Mechanistic studies revealed the role of 4-substitution on the *N*-methyl aniline moiety of **53** for the insertion to take place. In the absence of the 4-substitution on the *N*-methyl aniline moiety, sulfenamide can rearrange in the presence of $Yb(OTf)_3$, and thus N–H insertion of the rearranged product was observed as the sole product. The cross-over

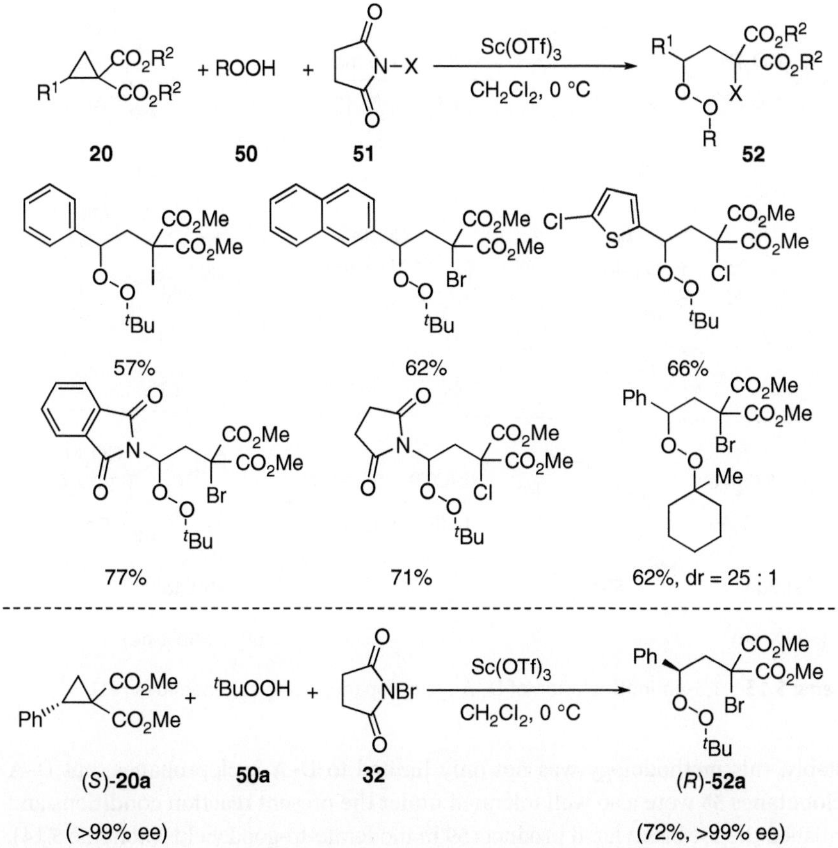

Scheme 5.12 1,3-Halogenation–peroxidation of D–A cyclopropanes [20].

experiments employing differently substituted sulfenamides revealed the formation of four γ-aminated α-thiolated malonic diester products, shedding light on the stepwise pathway for insertion and the intermolecular nature of the sulfur group migration.

5.15 1,3-Bisarylation of D–A Cyclopropanes with Electron-Rich Arenes and Hypervalent Arylbismuth Reagents

In 2020, Saha and coworkers disclosed a new catalytic procedure for the tandem 1,3-bisarylation of D–A cyclopropanes **20**, employing electron-rich arenes **55** as the nucleophile and hypervalent aryl bismuth components **56** as the electrophilic source of arene [22]. The formation of Lewis acid-catalyzed C—C bond using a multicomponent approach was the key highlight for this efficient transformation. $Sc(OTf)_3$ was found to be the choice of Lewis acid, and diethyl ether was the optimal solvent.

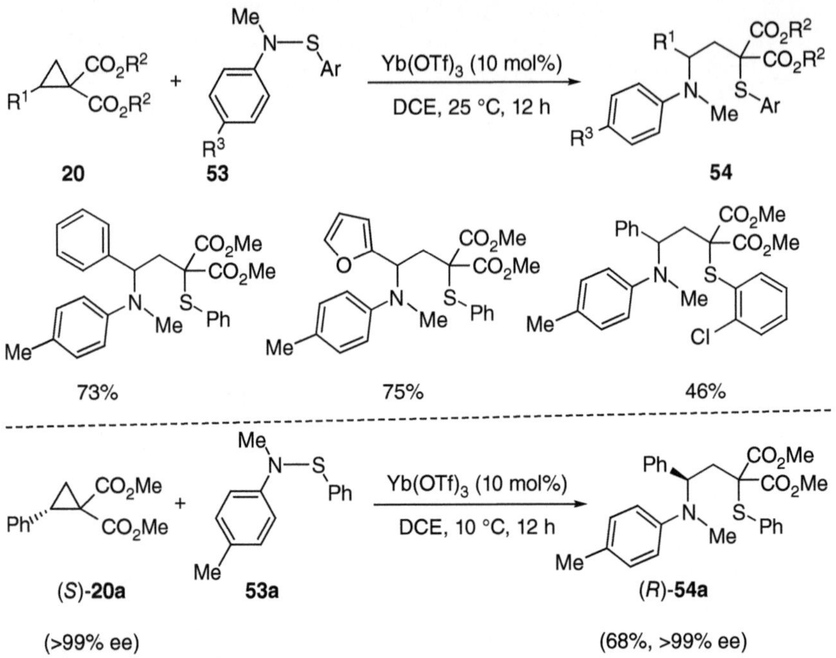

Scheme 5.13 1,3-Aminothiolation of D–A cyclopropanes using sulfenamides [21].

Notably, this methodology was not only limited to D–A cyclopropanes, but D–A cyclobutanes **58** were also well tolerated under the present reaction conditions and furnished the 1,4-bisarylated products **59** in moderate-to-good yields (Scheme 5.14). Mechanistic studies revealed that this transformation is stereospecific.

5.16 Conversion of D–A Cyclopropanes to β-Hydroxy Ketones

Werz and coworkers demonstrated an electrochemical strategy for the ring-opening of D–A cyclopropanes by single electron transfer (SET) initiation [23]. Although there are many ways to break the C—C bond of D–A cyclopropanes, including the use of Lewis acids, Brønsted acids, and breaking by radical addition, but there are not many reports for the electrocatalytic opening of D–A cyclopropanes by anodic oxidation. The use of triplet oxygen as the reaction partner was considered because of its abundance, cheapness, and diradical character. Interestingly, under the optimized conditions, the expected β-hydroxy ketones **60** were formed in moderate-to-good yields (Scheme 5.15). Remarkably, this methodology could be further extended toward D–A cyclobutanes and in this case, the expected γ-hydroxy ketones were formed in good-to-excellent yields. The reaction proceeds with the oxidation of the arene ring of the D–A cyclopropane at the anode, and the formation of radical cation **62** via the breaking of the C—C bond of cyclopropane. Radical combination with

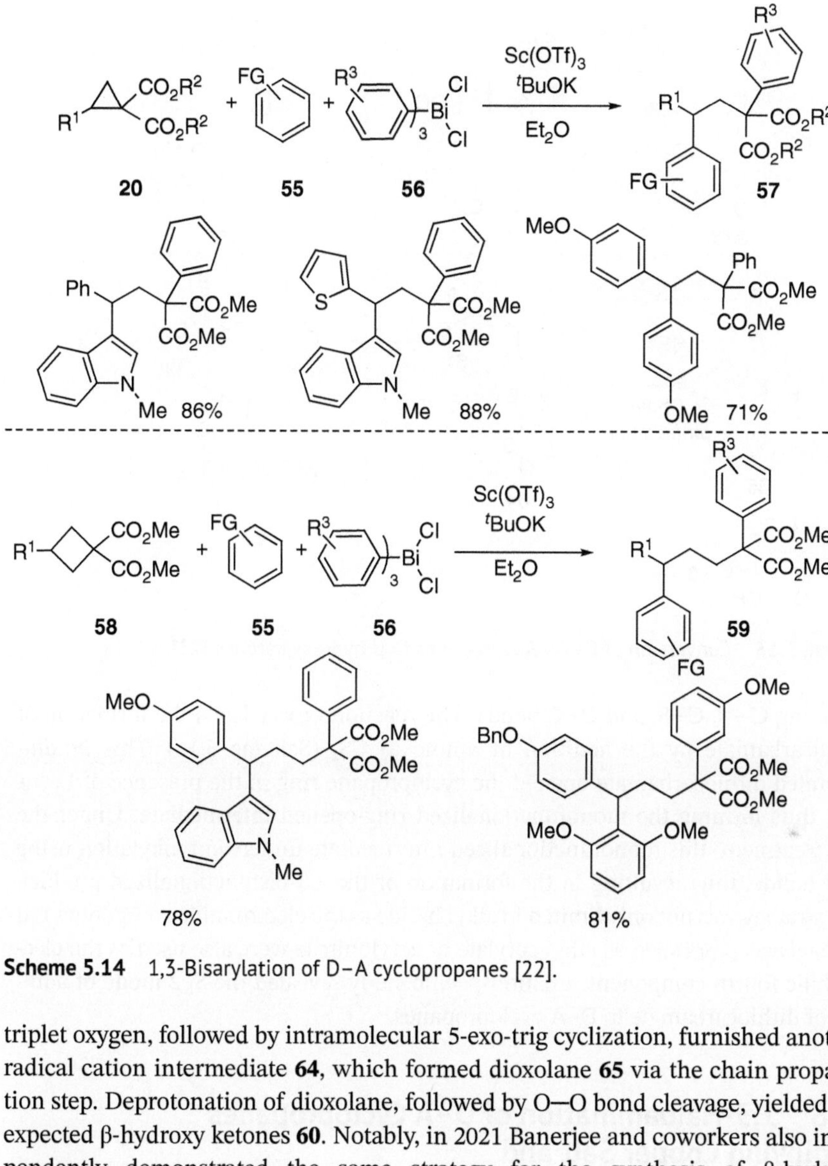

Scheme 5.14 1,3-Bisarylation of D–A cyclopropanes [22].

triplet oxygen, followed by intramolecular 5-exo-trig cyclization, furnished another radical cation intermediate **64**, which formed dioxolane **65** via the chain propagation step. Deprotonation of dioxolane, followed by O—O bond cleavage, yielded the expected β-hydroxy ketones **60**. Notably, in 2021 Banerjee and coworkers also independently demonstrated the same strategy for the synthesis of β-hydroxy ketones [24].

5.17 1,3-Carbothiolation of D–A Cyclopropanes

Biju and coworkers disclosed a two-step ring-opening 1,3-carbothiolation of D–A cyclopropanes utilizing alkyl halides and *in situ* generated dithiocarbamates (from amines and CS_2) [25]. This reaction is easy to operate and works with good functional group compatibility. The key highlight was the formation of three new bonds,

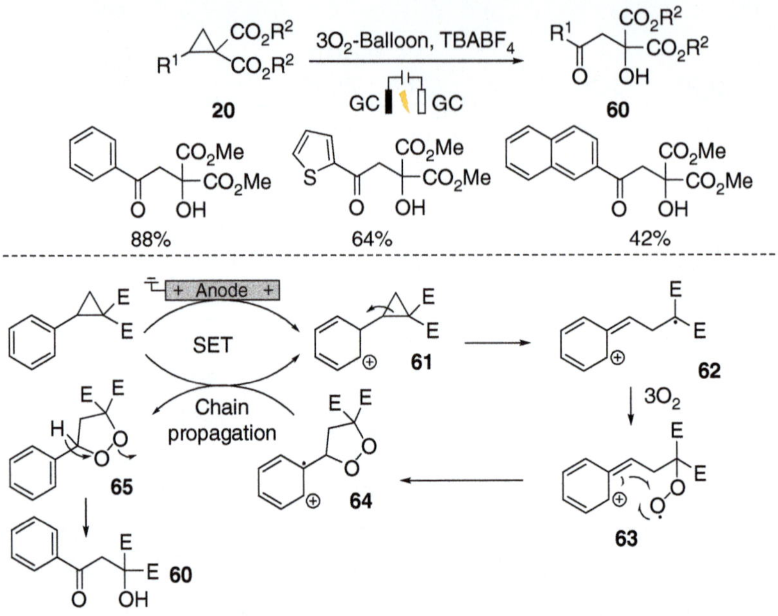

Scheme 5.15 Conversion of D–A cyclopropanes to β-hydroxy ketones [23].

including C—N, C—S, and C—C bonds. The reaction proceeds via the formation of dithiocarbamate by the addition of amine to CS_2 (Scheme 5.16). This *in situ*-generated dithiocarbamate opened the cyclopropane ring in the presence of Lewis acid, thus forming the monofunctionalized ring-opened intermediate. Under the base treatment, this monofunctionalized intermediate underwent alkylation using alkyl halide, thus resulting in the formation of the 1,3-bisfunctionalized product. This strategy was not only limited to alkyl halide as the electrophilic component but Michael acceptors such as ethyl acrylate or acrylonitrile were also used as the electrophilic fourth component. Enantiospecific study revealed the S_N2 mode of addition of dithiocarbamate to D–A cyclopropanes.

5.18 1,3-Haloamination of D–A Cyclopropanes Employing Copper Salt and *N*-Fluorobenzenesulfonimide

Li and coworkers uncovered a protocol for the divergent copper salt-controlled reactions of D–A cyclopropanes **20** and *N*-fluorobenzenesulfonimide (NFSI) **72** [26]. In the presence of CuX_2 (X = Cl, Br), which acts as both a mediator and a halide source, the cyclopropanes underwent 1,3-aminohalogenation via a free radical-mediated ring-opening process, to yield 1,3-aminochlorination and 1,3-aminobromination products **73** in moderate-to-good yields (Scheme 5.17). Interestingly, 1,3-aminohalogenation was a formal umpolung reaction, with the halogen anion directly connected to an electron-rich carbon center with two strongly electron-withdrawing groups. Mechanistically, the reaction proceeds with the generation of Cu(I) intermediate from

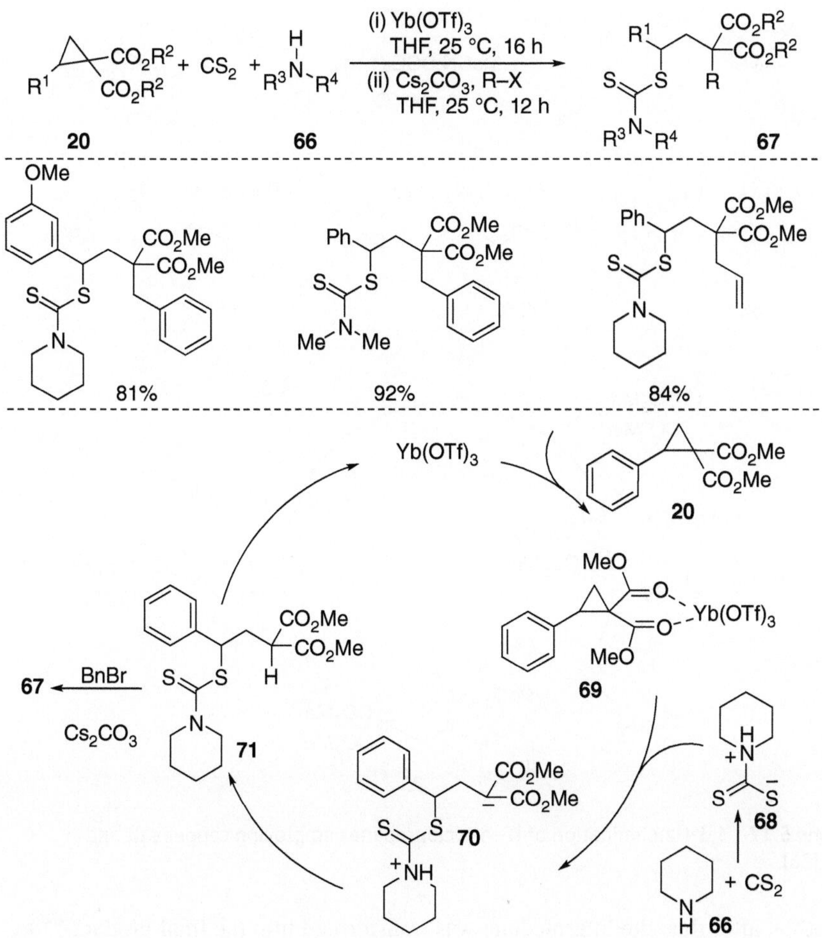

Scheme 5.16 1,3-Carbothiolation of donor–acceptor cyclopropanes [25].

Cu(II) species via the disproportionation of Cu(II) salt. Next, the NFSI underwent homolysis to produce phenylsulfamide nitrogen radical **74**. This radical **74** adds to D–A cyclopropane **20** to furnish intermediate **75**, which combines with CuX_2 via oxidative addition to generate metal–carbon species **76**. In the next step, product **73** was generated from intermediate **76** via the reductive elimination and simultaneously the active catalyst (Cu(I)) was regenerated. Thus, this methodology offers an alternative strategy for 1,3-haloamination proceeding via a radical pathway.

5.19 Ring-Opening 1,3-Carbocarbonation of D–A Cyclopropanes

Very recently, Werz and coworkers demonstrated the 1,3-carbocarbonation of D–A cyclopropanes by employing Grignard reagent as the nucleophilic trigger [27]. Using copper catalysis, the ring-opening addition with a Grignard reagent progressed

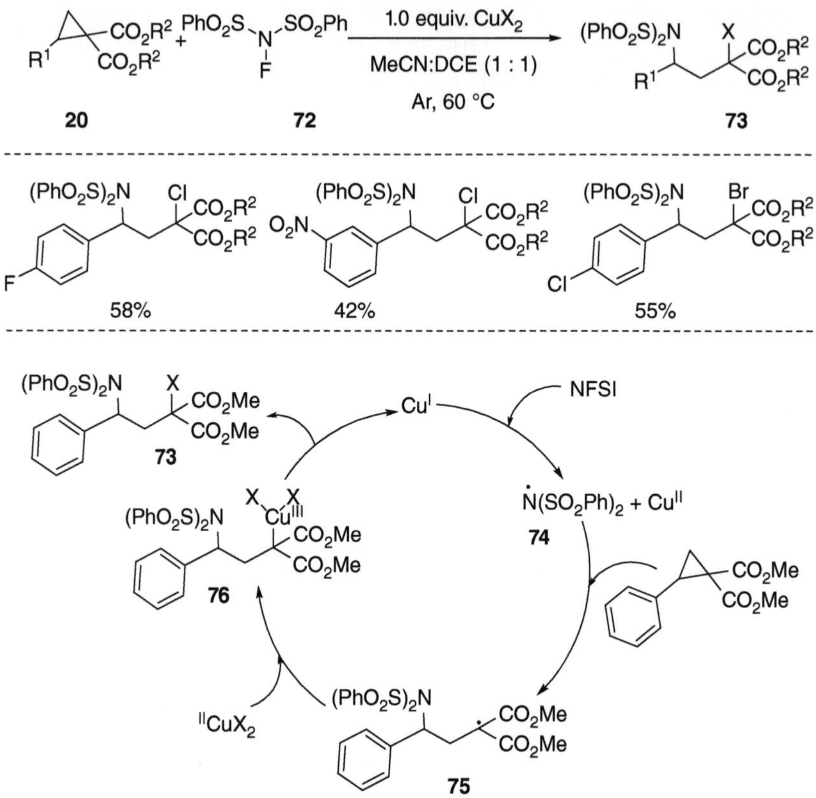

Scheme 5.17 1,3-Haloamination of D–A cyclopropanes employing copper salt and NFSI [26].

smoothly; after this, the intermediate was transformed into the final product **77** by the reaction with appropriate carbon-based electrophiles under basic conditions. Grignard reagents originated from sp^3-, sp^2-, and sp-hybridized carbon residues were successfully utilized as nucleophiles, whereas various aliphatic bromides and ethynylbenziodoxolone (EBX) derivatives (for sp moieties) functioned as electrophiles (Scheme 5.18).

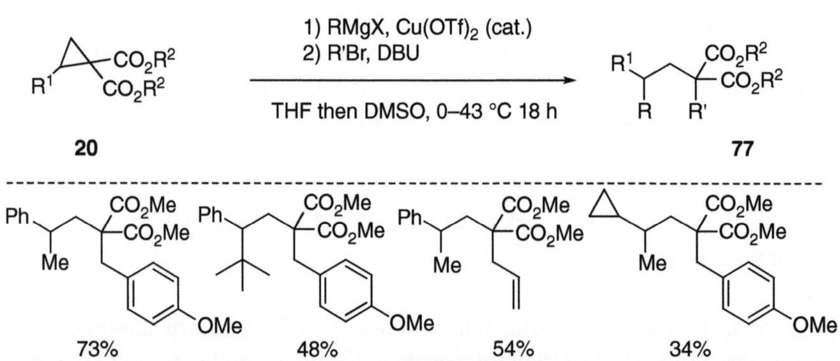

Scheme 5.18 Ring-opening 1,3-carbocarbonation of D–A cyclopropanes [27].

5.20 1,3-Aminofunctionalization of D–A Cyclopropanes

Very recently, Biju and coworkers devised 1,3-aminofunctionalization of D–A cyclopropanes via three-component coupling reactions using benzotriazoles as the nucleophilic trigger [28]. Under the optimized reaction conditions, benzotriazoles **78** were added to D–A cyclopropanes **20** in the presence of catalytic $Yb(OTf)_3$, and the *in situ* generated intermediate was trapped with various types of electrophiles. First, the scope of 1,3-aminohalogenation leading to the product **79** (employing *N*-halosuccinimides as electrophiles) was investigated and in each case, the expected products were formed in good-to-excellent yields. Later, the scope of the reaction was further expanded using different carbon-based electrophiles. Although the direct three-component coupling was not successful in that case, the one-pot strategy was successful. Several electrophiles, including alkyl/benzyl halides, Michael acceptors, EBX derivatives, and even Selectfluor, were used under the optimized conditions. The synthetic utility of the products was shown via some selected functional group interconversions (Scheme 5.19).

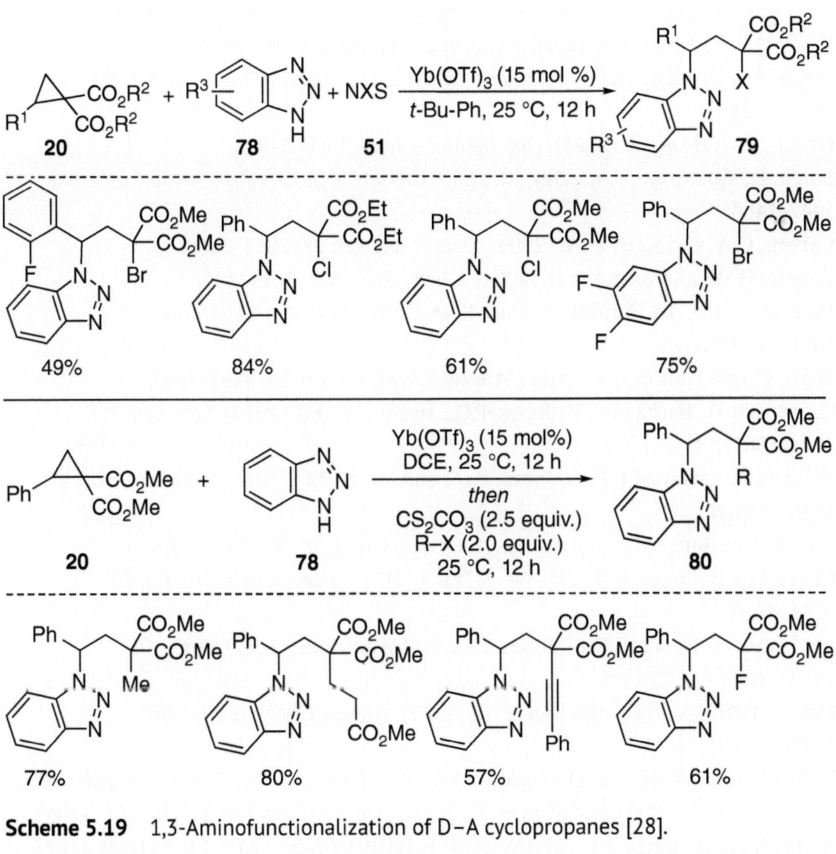

Scheme 5.19 1,3-Aminofunctionalization of D–A cyclopropanes [28].

5.21 Conclusion

In this chapter, the application of D–A cyclopropanes in ring-opening 1,3-bisfunctionalization reactions has been summarized. Ring-opening 1,3-bisfunctionalization reactions can be achieved in two ways. The first approach is the multicomponent approach, and the second type is the insertion reaction, which is not well explored in the D–A cyclopropane field. Various nucleophiles and electrophiles can combine with D–A cyclopropanes and produce several useful organic compounds by using this strategy. It is expected that the efficiency of 1,3-bisfunctionalization reactions in D–A cyclopropanes will continue and will result in remarkable outcomes for innovative 1,3-bisfunctionalizations in the future. It is realistic to believe that the potential of this 1,3-bisfunctionalization reaction is still not completely exposed, and more significant developments will be disclosed in the upcoming future.

References

1 Reissig, H.-U. and Hirsch, E. (1980). *Angew. Chem. Int. Ed. Engl.* 19: 813–814.
2 Werz, D.B. and Biju, A.T. (2020). *Angew. Chem. Int. Ed.* 59: 3385–3398.
3 Schneider, T.F., Kaschel, J., and Werz, D.B. (2014). *Angew. Chem. Int. Ed.* 53: 5504–5523.
4 Ghosh, K. and Das, S. (2021). *Org. Biomol. Chem.* 19: 965–998.
5 Singh, P., Varshnaya, R.K., Dey, R., and Banerjee, P. (2020). *Adv.Synth. Catal.* 362: 1447–1484.
6 Carson, C.A. and Kerr, M.A. (2009). *Chem. Soc. Rev.* 38: 3051–3060.
7 Reissig, H.-U. and Zimmer, R. (2003). *Chem. Rev.* 103: 1151–1196.
8 De Nanteuil, F., De Simone, F., Frei, R. et al. (2014). *Chem. Commun.* 50: 10912–10928.
9 Sparr, C. and Gilmour, R. (2011). *Angew. Chem. Int. Ed.* 50: 8391–8395.
10 Garve, L.K.B., Barkawitz, P., Jones, P.G., and Werz, D.B. (2014). *Org. Lett.* 16: 5804–5807.
11 Wallbaum, J., Garve, L.K.B., Jones, P.G., and Werz, D.B. (2016). *Chem. - Eur. J.* 22: 18756–18759.
12 Das, S., Daniliuc, C.G., and Studer, A. (2016). *Org. Lett.* 18: 5576–5579.
13 Garve, L.K.B., Jones, P.G., and Werz, D.B. (2017). *Angew. Chem. Int. Ed.* 56: 9226–9230.
14 Wallbaum, J., Garve, L.K.B., Jones, P.G., and Werz, D.B. (2017). *Org. Lett.* 19: 98–101.
15 Das, S., Daniliuc, C.G., and Studer, A. (2017). *Angew. Chem. Int. Ed.* 56: 11554–11558.
16 Novikov, R.A., Borisov, D.D., Zotova, M.A. et al. (2018). *J. Org. Chem.* 83: 7836–7851.
17 Das, S., Daniliuc, C.G., and Studer, A. (2018). *Angew. Chem. Int. Ed.* 57: 4053–4057.
18 Augustin, A.U., Jones, P.G., and Werz, D.B. (2019). *Chem. - Eur. J.* 25: 11620–11624.
19 Gregson, C.H.U., Ganesh, V., and Aggarwal, V.K. (2019). *Org. Lett.* 21: 3412–3416.

20 Singh, K., Bera, T., Jaiswal, V. et al. (2019). *J. Org. Chem.* 84: 710–725.

21 Guin, A., Rathod, T., Gaykar, R.N. et al. (2020). *Org. Lett.* 22: 2276–2280.

22 Mondal, B., Das, D., and Saha, J. (2020). *Org. Lett.* 22: 5115–5120.

23 Kolb, S., Petzold, M., Brandt, F. et al. (2021). *Angew. Chem. Int. Ed.* 133: 16064–16070.

24 Saha, D., Maajid Taily, I., and Banerjee, P. (2021). *Eur. J. Org. Chem.* 5053–5057.

25 Guin, A., Deswal, S., and Biju, A.T. (2022). *J. Org. Chem.* 87: 6504–6513.

26 Liu, L., Wang, X., Xiao, W. et al. (2022). *Chem. Eur. J.* e202202544.

27 von Köller, H.F., Jones, P.G., and Werz, D.B. (2022). *Chem. Eur. J.* e202203986.

28 Deswal, S., Guin, A., and Biju, A.T. (2023). *Org. Lett.* 25: 1643–1648.

6

Molecular Rearrangements in Donor–Acceptor Cyclopropanes

Igor V. Trushkov[1] *and Olga A. Ivanova*[2]

[1]*N.D. Zelinsky Institute of Organic Chemistry RAS, Laboratory of Directed Functionalization of Organic Molecular Systems, Leninsky Av., 47, Moscow, 119991, Russia*
[2]*M.V. Lomonosov Moscow State University, Department of Chemistry, Leninskie Gory, 1-3, Moscow, 119991, Russia*

CHAPTER MENU

6.1 Introduction

An amazing feature of donor–acceptor (D–A) cyclopropanes is their ability to produce a broad diversity of products in reactions involving no other educts. The fine tuning of reaction conditions allows each of the products to be obtained in high yields and selectivity. Among such processes are: (1) *cis-trans* isomerization [1–4] as well as racemization of the optically active substrates [4–8]; (2) various dimerizations that were reviewed recently [9, 10]; (3) oligo- and polymerization [11–15]; and (4) various molecular rearrangements producing isomers of the starting compounds with different carbon skeletons. It is the molecular rearrangements of D–A cyclopropanes that are discussed in this chapter. Seven major types of these processes can be distinguished. The simplest one is the cyclopropane-to-alkene isomerization (Section 6.1). Another important process is vinylcyclopropane-to-cyclopentene

Donor–Acceptor Cyclopropanes in Organic Synthesis, First Edition.
Edited by Prabal Banerjee and Akkattu T. Biju.

isomerization (Section 6.2), its analogs with heteroatom in double bond, i.e. formation of 2,3-dihydrofuran and 2,3-dihydropyrroles from acylcyclopropanes and their imines, respectively, (the Cloke–Wilson rearrangement) as well as the related processes (Section 6.3), homo-Nazarov cyclization (Section 6.4), and homo-version of 3,3-sigmatropic shifts (Section 6.5). Intramolecular cycloadditions can also be considered as a type of molecular rearrangement, but these processes are discussed in Chapter 3; therefore, this subject is not covered here. At last, intramolecular nucleophilic ring opening–ring closure (INRORC) processes can be considered as molecular rearrangement of D–A cyclopropanes. Some examples of these processes are discussed in Section 6.6. In general, we do not discuss here domino reactions wherein molecular rearrangement is one of the steps and other steps represent the addition or elimination of some molecules (group of atoms) except for rare examples given to better understand the subject of this chapter.

6.2 Donor–Acceptor Cyclopropane Isomerizations to Alkenes (Cyclopropane–Propene Rearrangement)

The isomerization of cyclopropane-to-propene was reported for the first time 100 years ago [16]. This reaction was proven to proceed via homolysis of the C–C bond [17] and required heating at 500–600 °C. Similar isomerization of D–A cyclopropanes was found to proceed under milder conditions; however, still requires very high temperatures. Thus, heating cyclopropa[*b*]thiophene **1** in toluene-d_8 at 150 °C in a sealed NMR tube failed to induce any reaction. However, **1** was converted to 2-thienylacetate **2** under treatment with iodine at room temperature (Scheme 6.1) [18].

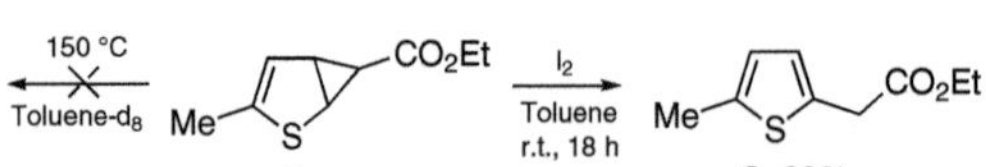

1 **2**, 69%

Scheme 6.1 Iodine-induced rearrangement of cyclopropa[*b*]thiophene **1**.

The isomerization of 2-alkoxycyclopropanecarboxylates **3** to the corresponding vinyl ethers **4** can be performed at relatively mild conditions in the presence of copper bronze or transition metal complexes, such as ($PtCl_2$•2PhCN) and $[Rh(CO)_2Cl]_2$ (Scheme 6.2) [19, 20].

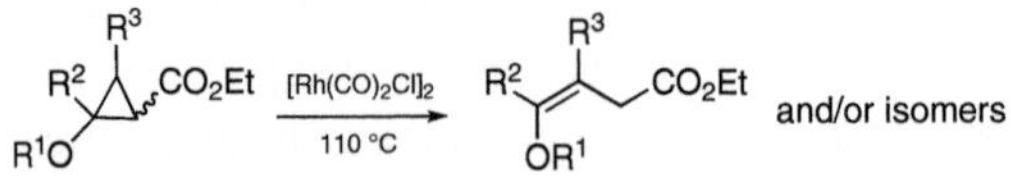

3 **4**, 46–98%

Scheme 6.2 Transition metal-catalyzed isomerizations of 2-alkoxycyclopropanecarboxylates.

However, these are Lewis and Brønsted acids that are most commonly used to initiate the isomerization of D–A cyclopropanes to the corresponding propenes. Thus, Reissig described the transformation of 2-siloxycyclopropanecarboxylates **5** to the corresponding alkenes **6** under catalysis with trimethylsilyl iodide (Scheme 6.3) [21].

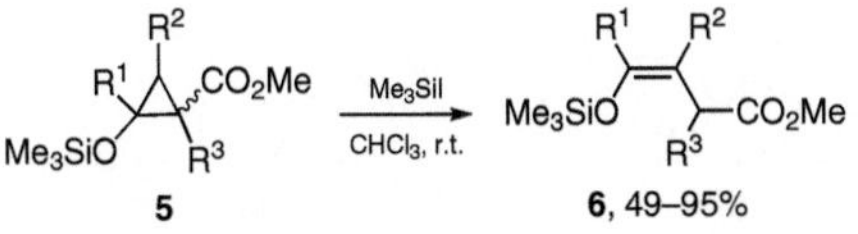

Scheme 6.3 Trimethylsilyl iodide-induced isomerization of 2-siloxycyclopropanecarboxylates **5**.

D–A cyclopropanes, bearing carbamates and related groups as donor substituents, behave similarly. Thus, cyclopropanes **8**, obtained from 2,3-dehydropiperidines **7** and vinyl acetates or silyl ethers, were rearranged to 3-acetonyl-2,3-dehydropiperidines **9** under treatment with 10-camphorsulfonic acid (10-CSA) in CH_2Cl_2 at 0 °C (Scheme 6.4). It should be noted that the rearrangement of corresponding cyclopropanecarboxylate (R^2 = OEt) required reflux in toluene [22].

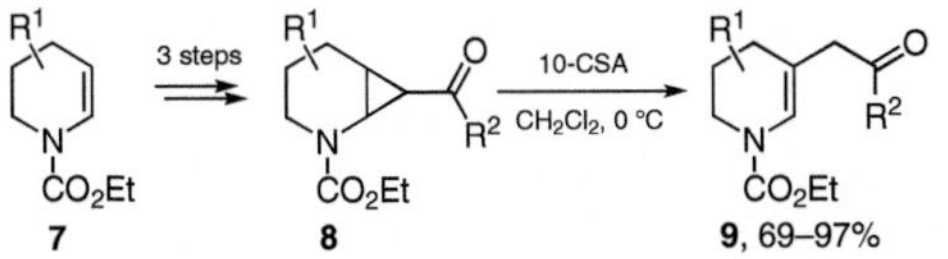

Scheme 6.4 Camphorsulfonic acid-initiated rearrangement of cyclopropapiperidines **8**.

Brönsted acids were also applied for the rearrangement of D–A cyclopropanes bearing arylsulfanyl groups as donors. Kulinkovich showed that 1-acyl-2,2-bis(phenylsulfanyl)cyclopropanes **10** isomerized to ketene dithioacetals **11** under treatment with *p*-toluenesulfonic acid (TsOH). The reaction should be carried out under anhydrous conditions to prevent hydrolysis of the products (Scheme 6.5) [23].

Scheme 6.5 TsOH-induced isomerization of cyclopropanes **10** to ketene dithioacetals **11**.

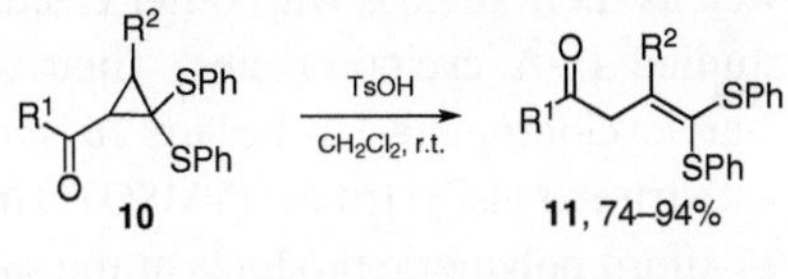

The important examples of D–A cyclopropanes with heteroatom-based functionality as donors are cyclopropanated heterocycles. In general, the cyclopropane-to-propene isomerization of these compounds can afford acyclic compounds, the corresponding heteroaromatic molecules, or products of ring expansion, depending on the nature of starting cyclopropanes. The trap of ring expansion products with alkenes and alkynes affording 8-aza- and 8-oxabicyclo[3.2.1]octane derivatives was reported by Reiser et al. [24]. These processes do not belong to the class of molecular rearrangements and are not discussed here in detail.

Indoles **12** cyclopropanation with ethyl diazoacetate **13** in the presence of Tp^{Br3}Cu(NCMe), where Tp^{Br3} is hydrotris(3,4,5-tribromopyrazolyl)borate, produced D–A cyclopropanes **14**, which isomerized to (indol-3-yl)acetates **15** during column chromatography on silica gel (Scheme 6.6) [25]. Heating the related *N*-acetylcyclopropa[*b*]indole with HCl in ethanol yielded predominantly 3-indolylacetate; this ring opening was accompanied by *N*-deacetylation [26].

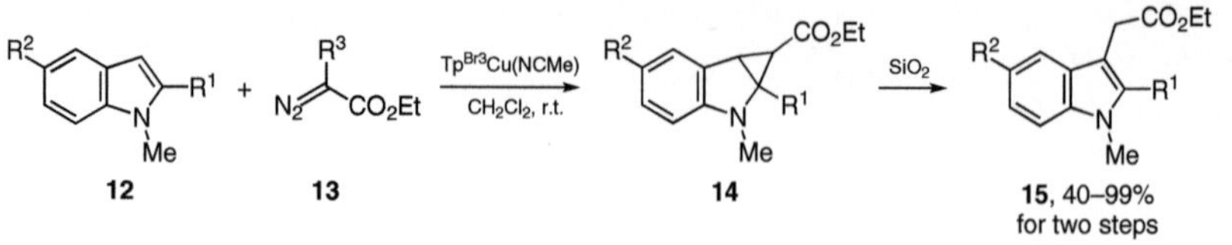

Scheme 6.6 Synthesis of 3-indolylacetic acids **15** by the rearrangement of cyclopropa[*b*]indoles **14**.

Nevertheless, 2-*H*-3-substituted indoles under the same conditions produced derivatives of (indol-2-yl)acetic acid, while *NH*-indoles afforded a mixture of 1,3-dialkylated and two monoalkylated products.

Similarly, under acidic conditions, cyclopropa[*b*]benzofuran **16** rearranged to (benzofuran-2-yl)acetate **17** [27], while substrate **18** produced (benzofuran-3-yl)acetic acid **19** [28] (Scheme 6.7).

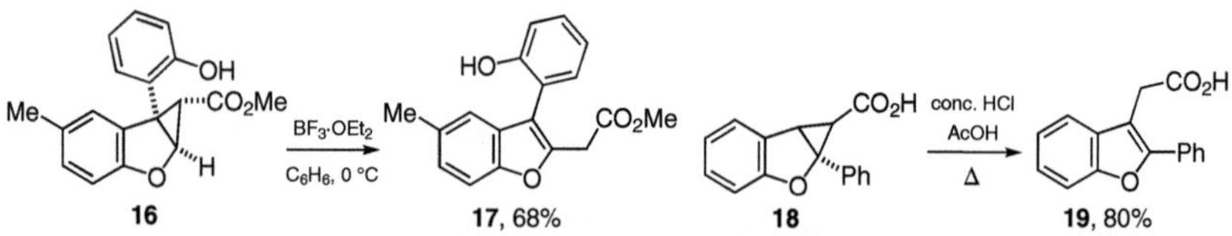

Scheme 6.7 Cyclopropane–propene isomerizations of cyclopropa[*b*]benzofurans **16** and **18**.

In the last few years, 2-aryl- and 2-alkenylcyclopropane-1,1-dicarboxylates, as well as their analogs with other electron-withdrawing groups, have been the most studied D–A cyclopropanes. Their studies showed that the moderately active 2-arylcyclopropane-1,1-diesters **20** isomerize to (*E*)-styrylmalonates **21** under reflux with trimethylsilyl triflate (TMSOTf) in chlorobenzene; the more reactive substrates **20** afford polymeric products at the same reaction conditions, while they yield styrylmalonates **21** in the presence of less activating Lewis acids such as $Sn(OTf)_2$. On the contrary to substrates studied in that work, 2,4,6-trimethylphenyl-derived cyclopropane **22** isomerizes mainly to alkylidenemalonate **23** (Scheme 6.8) [29]. It is noteworthy that without Lewis acid, the rearrangement of the active substrate **20** bearing a 4-methoxyphenyl group as a donor to the corresponding styrylmalonate required heating at 290 °C [30].

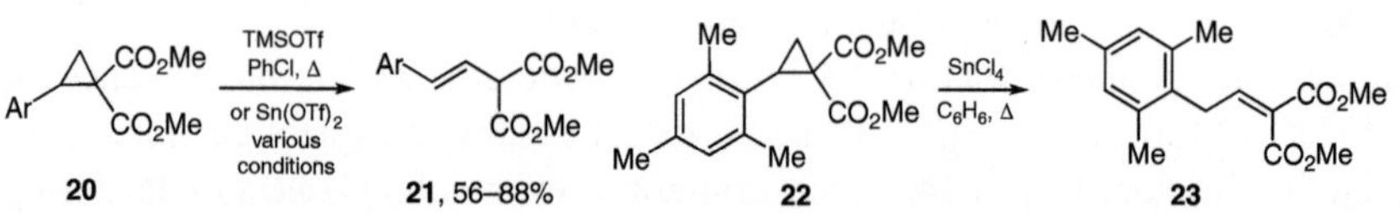

Scheme 6.8 Rearrangement of 2-arylcyclopropane-1,1-diesters **20** to styrylmalonate **21**; isomerization of 2,4,6-trimethylphenyl-substituted analog **22** to 2-arylethylidenemalonate **23**.

On the other hand, the broad variety of alkylidenemalonates **24** were synthesized by the treatment of cyclopropanes **20** with $GaCl_3$ in methylene chloride for 5–15 min at 0 °C (Scheme 6.9) [11, 31].

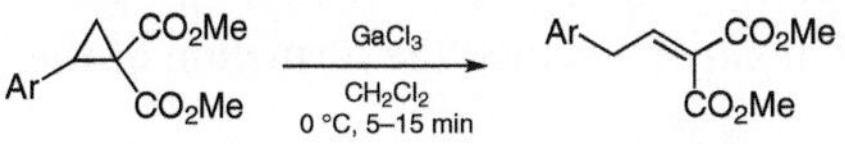

Scheme 6.9 Isomerization of cyclopropanes **20** to 2-arylethylidenemalonates **24** in the presence of $GaCl_3$.

The increase in the reaction time led to the decomposition of **24** and its partial isomerization to styrylmalonates **21**. This isomerization can be performed by stirring **24** with Lewis acids or bases, with pyridine being the most efficient additive for this goal (Scheme 6.10) [32]. Although for the preparation of styrylmalonates from substrates with highly donor substituents, this approach is less efficient than the direct isomerization described earlier, it allows one to involve cyclopropanes bearing acceptor group(s) in the aromatic ring in the desired transformation. Thus, the two approaches complement each other.

Scheme 6.10 Two-step isomerization of cyclopropanes **20** to styrylmalonates **21**.

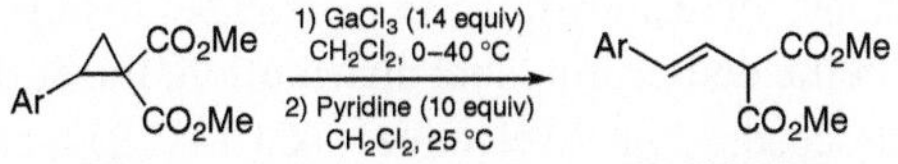

Vinilogous (2-phenylcyclopropyl)methylenemalonate **25** rearranged to diene **26** under stirring at room temperature with $GaCl_3$ while it gave cyclopentene **27** under heating with $Sc(OTf)_3$ (Scheme 6.11) [33].

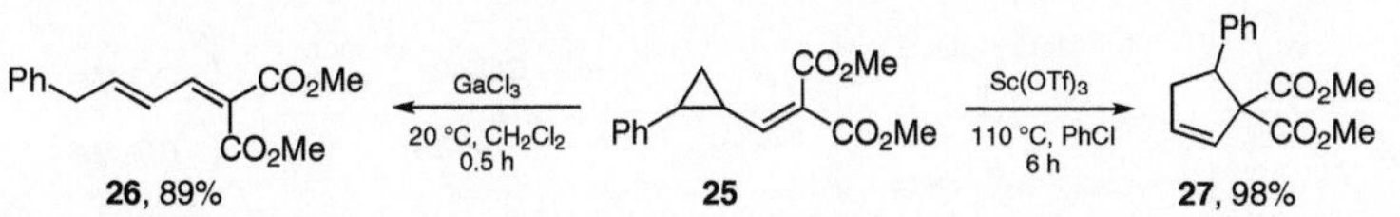

Scheme 6.11 Rearrangements of 2-phenylcyclopropylmethylenemalonate **25** to diene **26** and cyclopentene **27**.

Another type of vinylogues of **20**, i.e. 2-styrylcyclopropane-1,1-dicarboxylates, under treatment with Lewis acids isomerized usually to cyclopentenes (Section 6.2). However, compound **28**, bearing an *o*-nitrophenyl group, in the presence of $GaCl_3$, afforded diene **29**. Diene was also formed in the rearrangement of 2-vinylcyclopropane-1,1-diester **30**, albeit in low yield (Scheme 6.12) [34].

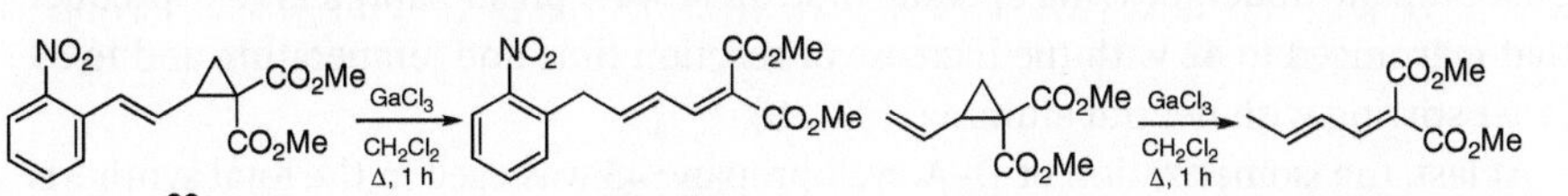

Scheme 6.12 Rearrangements of 2-alkenylcyclopropane-1,1-dicarboxylates **28** and **30** to dienes **29** and **31**.

The chemoselectivity of the isomerization of the related cyclopropanes **32** bearing an acyl group at the C(3) atom of the three-membered ring was found to depend on the Lewis acid applied. Under initiation with $TiCl_4$, they rearranged to dienones **33**; $SnCl_4$ induced their transformation to cyclopentenes **34** (Scheme 6.13) [35]. Again, the presence of the nitro group in the styryl moiety prevented the formation of the cyclopentene ring.

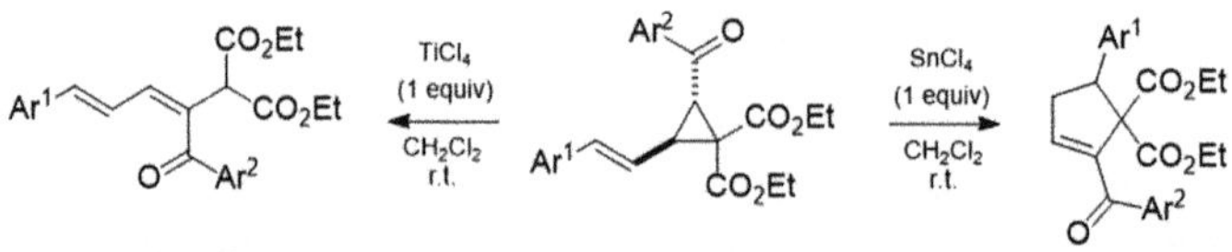

Scheme 6.13 Isomerization of 3-aroyl-2-styrylcyclopropane-1,1-diesters **32** to dienes **33** and cyclopentenes **34**.

The introduction of the phenyl group at the C(3) atom of styrylcyclopropanes also influences the reaction output; in the presence of titanium(IV) or iron(III) chlorides, these substrates **35** rearranged to but-3-enylidenemalonates **36** [36, 37]. Based on the deuterium label distribution, the authors concluded that this rearrangement proceeds via selective (C1)—C(2) bond heterolysis followed by aryl group 1,2-migration (Scheme 6.14). It is noteworthy that the formation of dienes **36** was proved to be reversible; the extension of the reaction time led to the formation of cyclopentenes **37**. Among other things, these experiments demonstrate that the styryl moiety stabilizes carbocation more efficiently than the related aryl group.

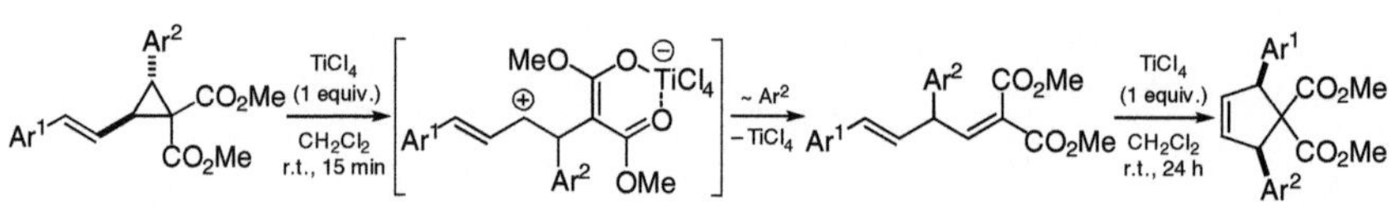

Scheme 6.14 Rearrangement of 3-aryl-2-styrylcyclopropane-1,1-dicarboxylates **35** to skipped dienes **36** and 2,5-diarylcyclopent-3-ene-1,1-dicarboxylates **37**.

2'-Phenylbicyclopropyl-1,1-dicarboxylate **38** was found to produce a broad diversity of isomerization products **39–44** depending on the reaction conditions (Scheme 6.15). According to these data, of the two three-membered rings, phenyl-bound cyclopropane undergoes ring opening first; diene **44** is presumably a kinetic product that rearranged to **42** with the increase of reaction time and temperature and to **39** (after stirring with neutral alumina) [33, 38].

At last, the isomerization of D–A cyclopropane **45** was used in the total synthesis of rhodomolleins XX and XXII. The treatment of ketoester **45** with TMSOTf and $(CH_2OTMS)_2$ led to both three-membered ring opening and acetal protection of the ketone moiety-producing compound **46** (Scheme 6.16) [39].

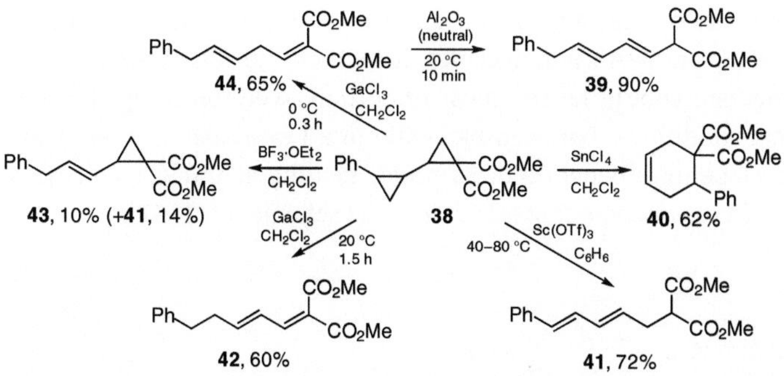

Scheme 6.15 A diversity of isomerization pathways for 2'-phenylbicyclopropyl-1,1-dicarboxylate **38**.

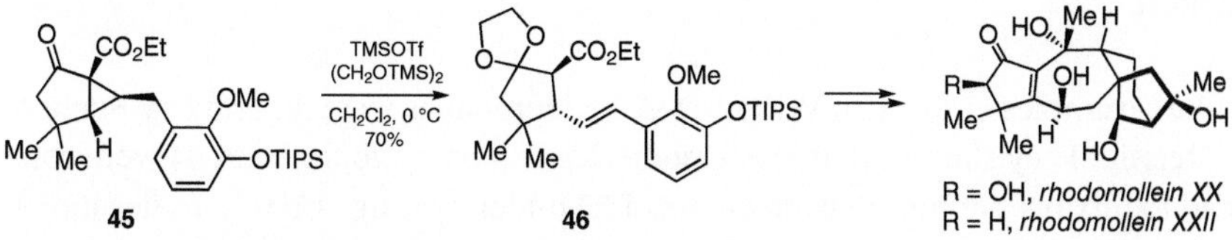

Scheme 6.16 The rearrangement of benzylcyclopropane **45** in the total synthesis of *rhodomolleins XX* and *XXII*.

6.3 Vinylcyclopropane–Cyclopentene Rearrangement

Vinylcyclopropane–cyclopentene rearrangement (VCR) was described for the first time in 1959 by Neureiter, who reported that the vacuum pyrolysis of 1,1-dichloro-2-vinylcyclopropane **47** at 500 °C (5 mm) produced 3,3-dichlorocyclopentene **48**, a product of its dehydrochlorination **49**, as well as 1,1-dichloropenta-1,3- and 1,4-dienes **50, 51** (Scheme 6.17) [40].

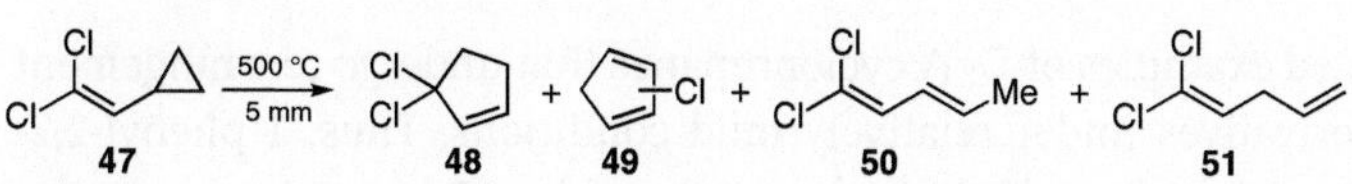

Scheme 6.17 The first example of vinylcyclopropane–cyclopentene rearrangement.

Already, this study demonstrated that thermal rearrangements of vinylcyclopropanes produce a complex mixture of products, the most common side products being dienes, resulting from the cyclopropane–propene rearrangement. Despite this, VCR began to be intensively studied immediately after its discovery; these reactions were considered as a very good approach to "dissonant" five-membered carbocycles, which are abundant in nature and have a broad application, including medicine. The results of these investigations were summarized in a series of reviews [41–47].

The introduction of donor and acceptor substituents into a three-membered ring decreases the energy barrier for VCR; moreover, this decrease is larger than expected based on the supposition of the stepwise mechanism with a biradical intermediate [48–50]. However, even D–A cyclopropanes, for which the stabilization of the zwitterionic intermediate should be the most important, were rearranged under rather harsh thermal conditions. For example, without a Lewis acid, cyclopropane **52** rearranged to cyclopenta[*b*[benzofuran **53** under heating at 150 °C in a sealed tube for four days, the conversion being less than 50% (Scheme 6.18) [51].

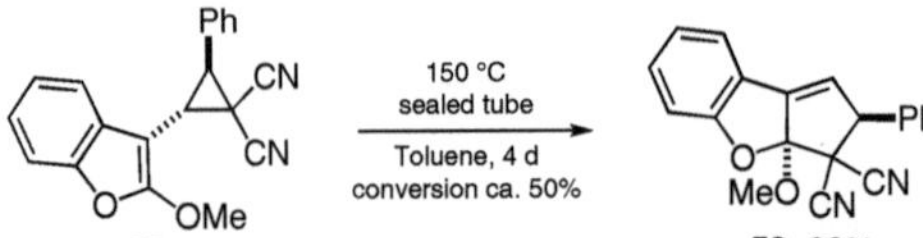

52 **53**, 33%

Scheme 6.18 Thermal rearrangement of 2-(benzofuran-3-yl)cyclopropane-1,1-dicarbonitrile **52**.

Other examples of thermal VCR of D–A cyclopropanes were reported by Buchert and Reissig. They showed that diastereomerically pure cyclopropanes **54** were converted to mixtures of four diastereomers of **55** under heating at 190 °C in decalin or benzonitrile, with the reaction in more polar solvent being significantly faster. Cyclopropanes **54** are stable under heating at 150 °C in DMF. Bicyclic analogs **56** demonstrated higher reactivity; their rearrangements at 150 °C also produced mixtures of isomers **57** (Scheme 6.19) [52].

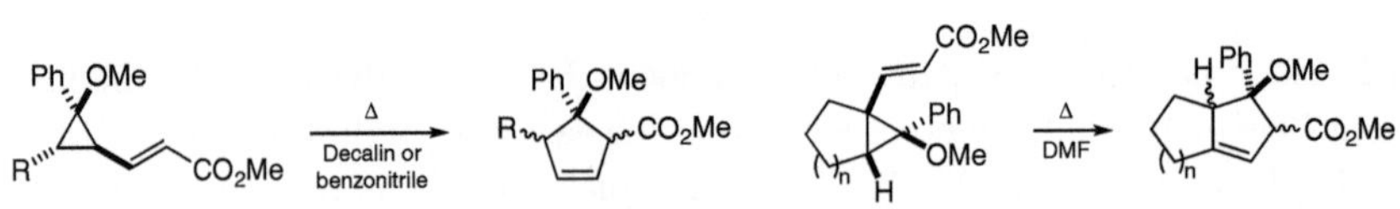

54 **55**, 71–95% **56**, *n* = 1, 2 **57**, quantitative

Scheme 6.19 Thermal rearrangement of cyclopropanes **54** and **56**.

There are restricted examples of D–A cyclopropanes that undergo rearrangement to cyclopentene derivatives under relatively mild conditions. Thus, 1-phenyl-2,2-divinylcyclopropanecarboxylic acid, its esters, and amides **58** isomerized to cyclopentenes **59** under reflux in toluene solution (Scheme 6.20) [53].

58 **59**, 73–91%

Scheme 6.20 Isomerization of 1-phenyl-2,2-divinylcyclopropanecarboxylic acid and its derivatives (**58**).

The use of transition metal catalysis was supposed to decrease the reaction temperature to provide reaction conditions more appropriate for the preparative synthesis. For example, butadienyl-substituted cyclopropane-1,1-dicarboxylate and their analogs **60** rearranged to cyclopentenes **61** in the presence of a catalytic amount of $Pd(PPh_3)_4$ at 25–60 °C (Scheme 6.21) [54, 55].

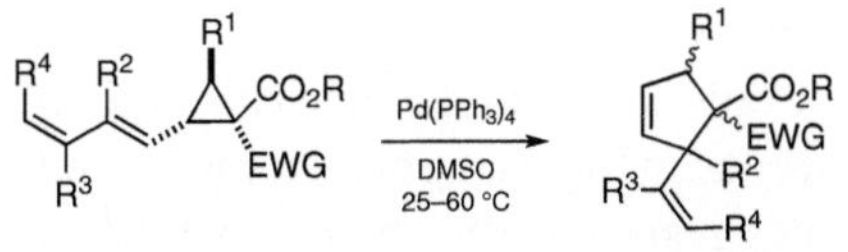

60 → 61, 28–96%, *dr* from 1:1 to 3:1

Scheme 6.21 $Pd(PPh_3)_4$-catalyzed isomerization of (butadienyl)cyclopropanes **60** to cyclopentenes **61**.

The VCR of similar substrates **62** was also carried out under radical conditions [56]. Refluxing a benzene solution of substrate with Ph_3SnH (Bu_3SnH, PhSH) and AIBN produced cyclopentenes **63** in reasonable to excellent yields (Scheme 6.22).

Scheme 6.22 Rearrangement of alkenylcyclopropanes **62** to cyclopentenes **63** under radical conditions.

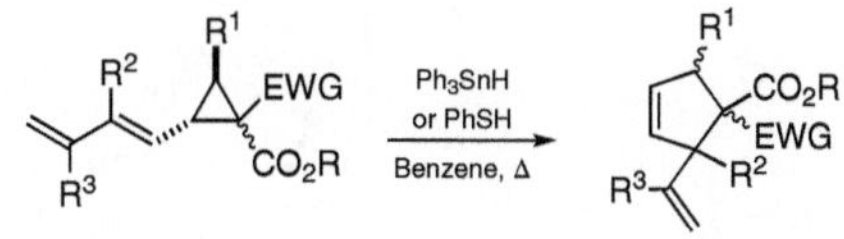

62 → **63**, 70–98%

It is noteworthy that the related isomerization of 1-acyl-2-alkenylcyclopropanes **64** produced the corresponding cyclopentens **65** in low yields, if at all, even though the rearrangement was carried out at 120 °C in the presence of nickel complexes with *N*-heterocyclic carbenes (NHC) (Scheme 6.23). This demonstrates that one acyl and one alkenyl group is not enough to provide a significant polarization of the C–C bond between endocyclic atoms bearing donor and acceptor substituents [57].

Scheme 6.23 Ni(NHC)-catalyzed isomerization of alkenylcyclopropanes **64** to cyclopentenes **65**.

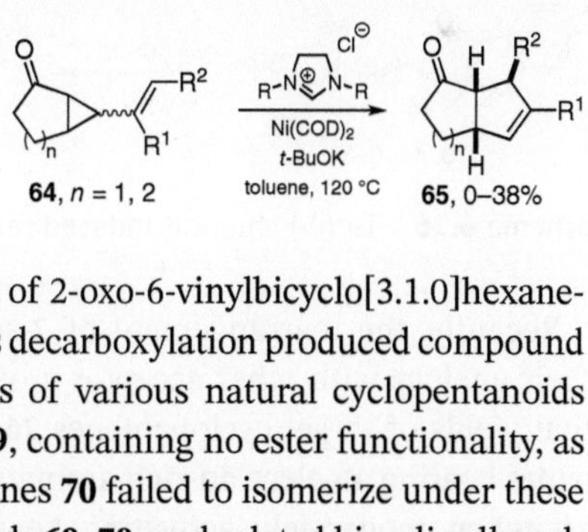

64, *n* = 1, 2 → **65**, 0–38%

Lithium iodide induced the rearrangement of 2-oxo-6-vinylbicyclo[3.1.0]hexane-1-carboxylate **66** to pentalene derivative **67**. Its decarboxylation produced compound **68**, a useful building block for the synthesis of various natural cyclopentanoids (Scheme 6.24). At the same time, substrate **69**, containing no ester functionality, as well as the corresponding bicyclo[4.1.0]heptanes **70** failed to isomerize under these conditions [58]. On the other hand, compounds **69**, **70**, and related bicyclic alkenylcyclopropanes were isomerized to bicyclo[3.3.0]octenones under thermolysis at 600 °C [59–61].

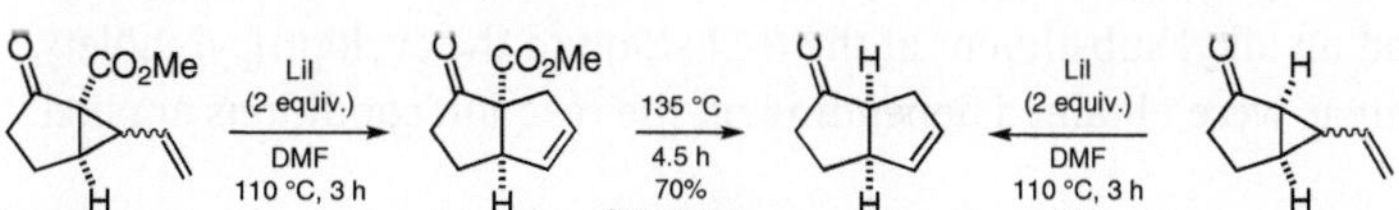

66 **67** **68** **69**

Scheme 6.24 LiI-initiated rearrangement of cyclopropane **66** to bicyclo[3.3.0]oct-6-ene-1-carboxylate **67**.

7-Alkenyl-2-oxobicyclo[4.1.0]heptanes **70** with *anti*-arrangement of the alkenyl group failed to form the product of VCR not only under treatment with lithium iodide but also under such standard procedures as pyrolysis, transition metal, and Lewis acid catalysis. However, these substrates rearranged to the corresponding cyclopentenes **72** under mild heating (40 °C) in acetonitrile in the presence of 1.5 equiv of MgI_2 (Scheme 6.25). The mechanism of this transformation, proposed by the authors, includes the reversible nucleophilic addition of iodide-ion to β-carbon atom of the double bond, followed by the cyclization of (*Z*)-isomer of intermediates **71** to products **72** and (*E*)-**71** to **70** [62]. LiI-induced rearrangements were supposed to proceed by the same mechanism; the different behavior of studied substrates in the presence of LiI and MgI_2 was not explained.

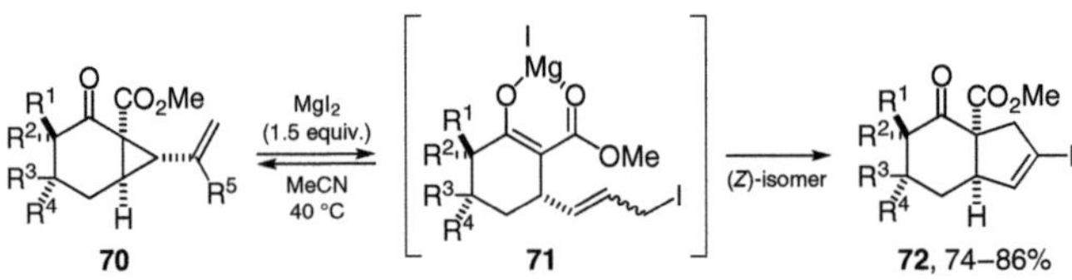

Scheme 6.25 MgI_2-induced isomerization of alkenylcyclopropanes **70** to compounds **72**.

Ila et al. described $SnCl_4$-initiated rearrangement of aryl 2-[4,4-bis(methylthio) butadienyl]cyclopropyl ketones **73** to the corresponding cyclopentenes **74** and their spirocyclic analogs to spirocyclopentenes (Scheme 6.26) [63].

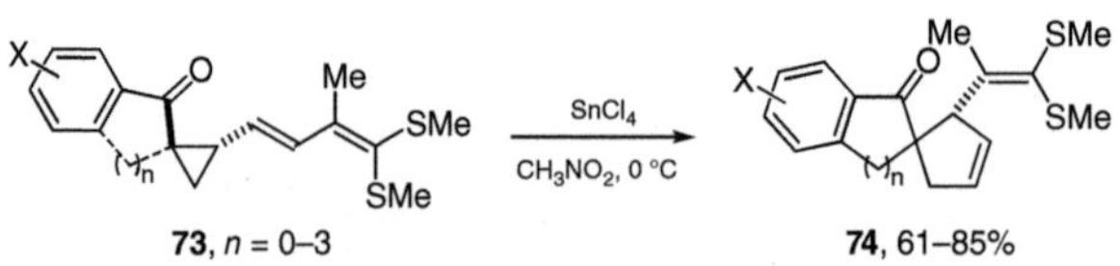

Scheme 6.26 Tin(IV) chloride-induced rearrangement of cyclopropanes **73**.

Recently, the rearrangement of 2-styrylcyclopropane-1,1-dicarboxylates **75** and their analogs with other acceptor groups was studied in detail [34]. It was shown that yields of target cyclopentenes **76** in the reaction of highly reactive cyclopropanes bearing an electron-rich aromatic group in the styryl substituent are the largest when moderately activating Lewis acids ($Ni(ClO_4)_2{\bullet}6H_2O$ or $Sc(OTf)_3$) are applied. On the other hand, for substrates with unsubstituted styryl or halogenostyryl groups, cyclopentenes were formed in the best yields using more activating $GaCl_3$, $SnCl_4$, or $TiCl_4$ (Scheme 6.27).

When styryl had an alkyl substituent at the α-position to the cyclopropyl moiety, two types of products were obtained depending on the reaction conditions applied.

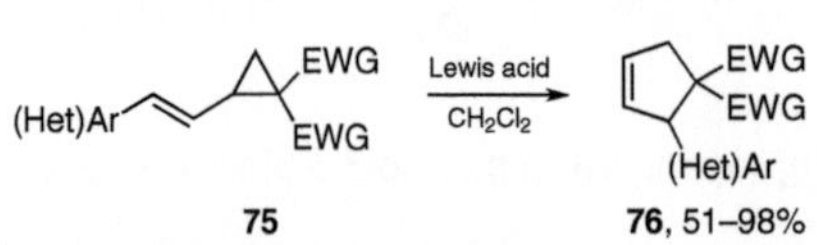

Scheme 6.27 The Lewis acid-initiated isomerization of styrylmalonates **75** to cyclopentenes **76**.

When methylene chloride solutions of substrates **77** were heated under reflux with tin(IV) chloride, 2,3-disubstituted cyclopent-3-ene-1,1-dicarboxylates **78** were obtained, while under heating with gallium(III) chloride, the isomeric cyclopent-2-ene-1,1-diesters **79** were prepared (Scheme 6.28).

Scheme 6.28 Rearrangement of 2-(1-alkyl-2-(het)arylvinyl) cyclopropane-1,1-diesters **77**.

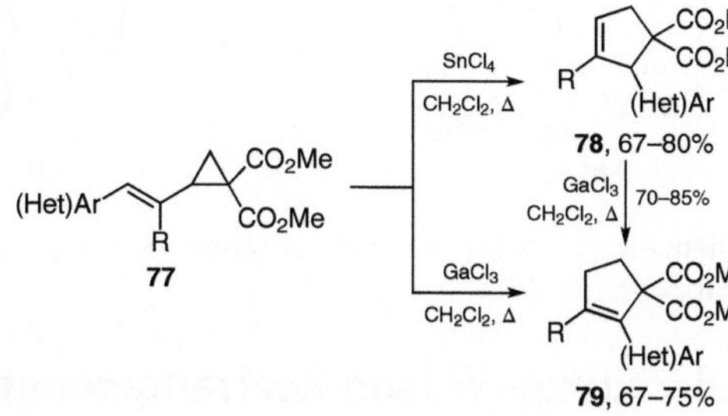

The investigation of the reaction stereochemistry revealed that Et_2AlCl-induced rearrangement of 1-alkenyl-2-alkoxycyclopropanecarboxylates **80** in CH_2Cl_2 proceeds with a significant loss of enantiomeric excess (Scheme 6.29), but the introduction of the additional group at the C(3) atom (compound **82**) was accompanied by a significant increase of the product *ee* [64].

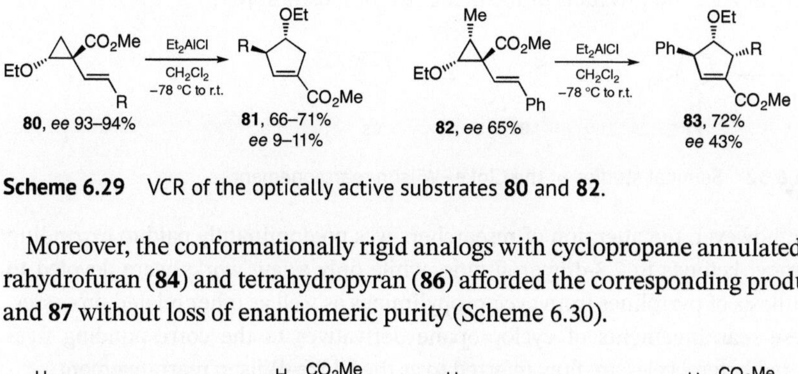

Scheme 6.29 VCR of the optically active substrates **80** and **82**.

Moreover, the conformationally rigid analogs with cyclopropane annulated to tetrahydrofuran (**84**) and tetrahydropyran (**86**) afforded the corresponding products **85** and **87** without loss of enantiomeric purity (Scheme 6.30).

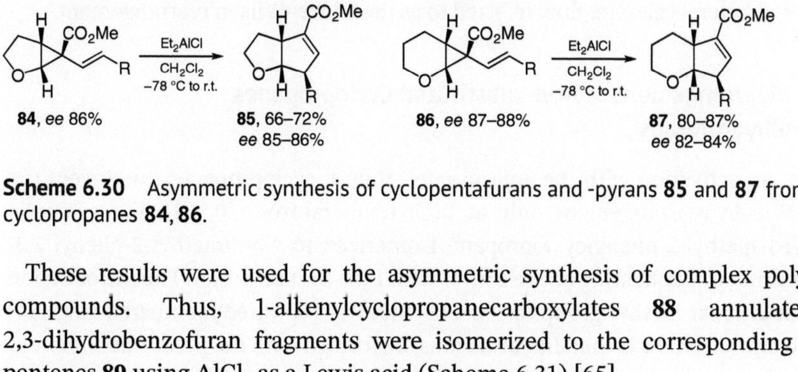

Scheme 6.30 Asymmetric synthesis of cyclopentafurans and -pyrans **85** and **87** from cyclopropanes **84, 86**.

These results were used for the asymmetric synthesis of complex polycyclic compounds. Thus, 1-alkenylcyclopropanecarboxylates **88** annulated to 2,3-dihydrobenzofuran fragments were isomerized to the corresponding cyclopentenes **89** using $AlCl_3$ as a Lewis acid (Scheme 6.31) [65].

Similar rearrangements were performed for the related compounds, wherein the same cyclopropane is annulated to the C(2)—C(3) bond of tetrahydrofuran, tetrahydropyran, and *N*-Boc-indoline using Et_2AlCl.

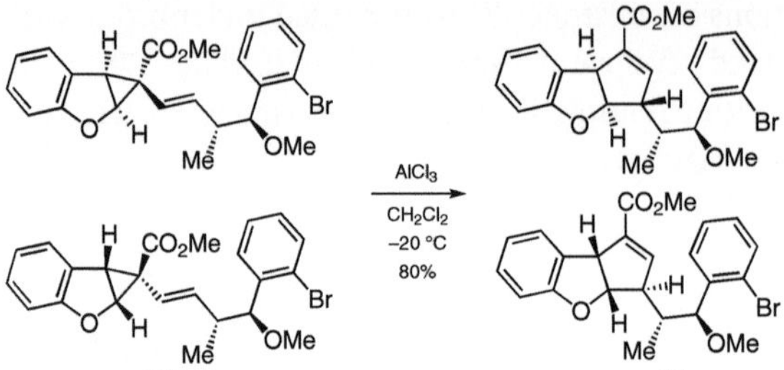

88 **89**

Scheme 6.31 $AlCl_3$-induced rearrangement of alkenylcyclopropanes **88** to cyclopentenes **89**.

6.4 Cloke–Wilson Rearrangement and Related Processes

The rearrangement of cyclopropylimines **90** or their hydrochloride to 2,3-dihydropyrroles **91** was disclosed by Cloke in 1929 [66]. In 1947, Wilson reported that cyclopropylcarbaldehyde **92** and 2,3-dihydrofuran **93** transform into each other upon heating (Scheme 6.32), with the initial focus on the transformation of **93** to cyclopropane **92** and products of its further decomposition [67].

90 **91** **92** **93**

Scheme 6.32 Seminal studies of the Cloke–Wilson rearrangement.

Later, however, the attention of researchers was predominantly paid to expanding cyclopropyl ketones to 2,3-dihydrofurans, while only a few works were devoted to the synthesis of pyrrolines from cyclopropylimines as well as other related processes. All these rearrangements of cyclopropane derivatives to the corresponding five-membered heterocycles are now referred to as the Cloke–Wilson rearrangement.

6.4.1 Rearrangement of Acyl-substituted Cyclopropanes to 2,3-dihydrofurans

Without an activation with the appropriate catalyst, cyclopropanes underwent the Cloke–Wilson rearrangement only at high temperatures [30, 68]. For example, 1-acetyl-1-methyl-2-phenylcyclopropane isomerized to 4,5-dimethyl-2-phenyl-2,3-dihydrofuran under heating above 250 °C with E_a of 48.1 kcal/mol. The introduction of the additional methyl group to the C(3) atom of the three-membered ring (**94**) allowed one to conclude that dihydrofurans are formed with the predominant retention of the relative stereochemistry of substituents at the C(2) and C(3) atoms (Scheme 6.33). Stereoisomers of the initial cyclopropane and isomeric enones were also formed under these conditions, demonstrating that under thermal conditions this process proceeds via a stepwise mechanism through 1,3-biradical intermediate,

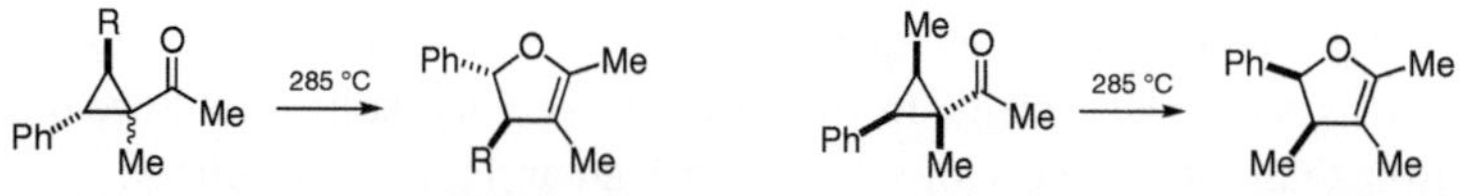

94a,b → **95a,b** 30–37%, *dr* 85:15 to 97:3; **94c** → **95c**, 40%, *dr* 70:30

Scheme 6.33 Thermal rearrangement of 1-acetyl-2-phenyl-1-methyl- and -1,3-dimethylcyclopropanes **94**.

wherein bond rotation competes with the ring closure step, and, as a result, it is non-chemoselective [68].

Similar results were also obtained under irradiation of compounds **94**. Moreover, Dauben and Shaffer found that the irradiation of the solution of chrysantemal **96a** and the corresponding methyl ketone **96b** in *t*-butanol led to both dihydrofurans **97** and a series of side products (Scheme 6.34) [69].

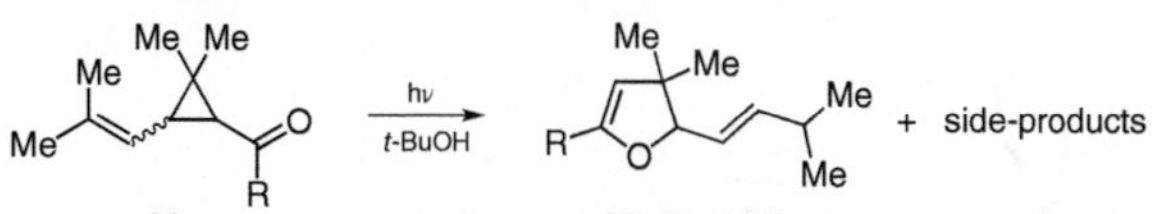

96 → **97**, 51–61%

Scheme 6.34 Photochemical rearrangement of chrysantemal **96a** and the related methyl ketone **96b**.

Similar to other transformations of D–A cyclopropanes, the Cloke–Wilson rearrangements proceed efficiently in the presence of Lewis or Brønsted acid. For example, Alonso and Morales reported that 2-(4-methoxyphenyl)-1-acetyl-3-R-cyclopropanecarboxylates **98** converted completely to 5-aryl-2-methyl-4-R-4,5-dihydrofuran-3-carboxylates **99** after stirring with neutral alumina (activity III) in chloroform for 24 h, the process proceeding with complete retention of cyclopropane stereochemistry (Scheme 6.35) [70].

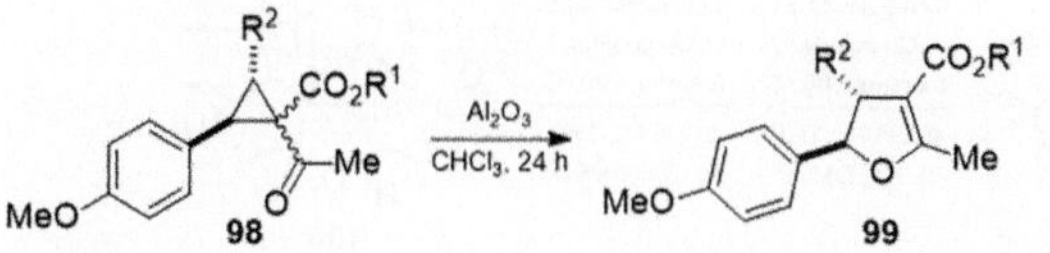

98 → **99**

Scheme 6.35 Isomerization of 1-acetyl-2-(4-methoxyphenyl)cyclopropanecarboxylates.

This study stimulated further investigations of the acid-induced Cloke–Wilson rearrangements. Thus, it was shown that spiro[cyclopropane-1,2'-cyclohexane-1',3'-diones] **100** rearranged efficiently to 2,3,6,7-tetrahydrobenzofuran-4(5*H*)-ones **101** in the presence of TMSI, TMSOTf, $Sc(OTf)_3$, MgI_2, $BF_3{\bullet}OEt_2$, or $TsOH{\bullet}H_2O$ (Scheme 6.36) [71, 72]. This approach was used by the authors in the total synthesis

Scheme 6.36 Acid-initiated rearrangement of spiro[cyclopropane-1,2'-cyclohexane-1',3'-diones] **100**.

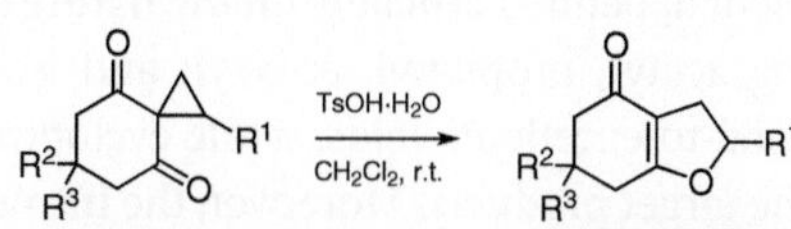

100 → **101**, 72–98%

of cuspidan B [72]. Even substrates having an alkyl group as a donor were found to be highly reactive. Nevertheless, yields in the reactions of alkyl-substituted cyclopropanes are lower than those for cyclopropanes with an aryl group as a donor.

The ease of this rearrangement was additionally confirmed by the isolation of tetrahydrobenzofuran-2-ones **101** during the synthesis of cyclopropanes **100** [73, 74]. At the same time, the efficiency of TsOH-induced rearrangement was low for spirocyclopropanes obtained from cyclopentane-1,3-dione, and various acyclic 1,3-diones [72].

The acid-induced isomerization was used by Werz et al. for the synthesis of tricyclic bisacetals and bishemiaminals **103** by double isomerization of dicyclopropanes **102,** wherein carbon atom of three-membered ring, connected to an acceptor group, is quaternary. Oppositely, if the second substituent at this atom is hydrogen, compounds **102** rearranged to 3,4-bis(acylmethyl)furans and -pyrroles **104**; these processes are examples of cyclopropane-to-propene rearrangement (Scheme 6.37) [75].

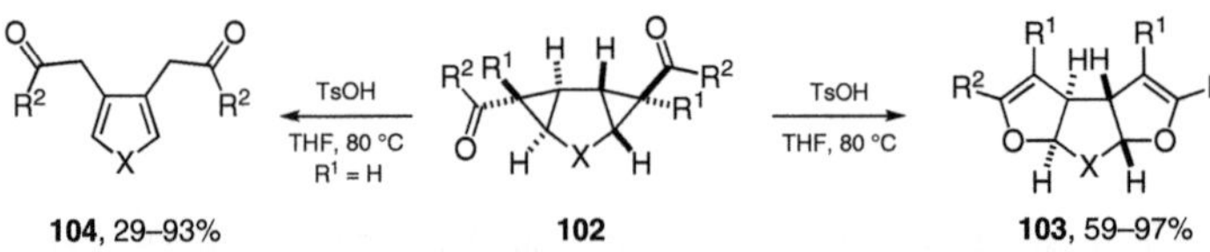

Scheme 6.37 TsOH-induced rearrangement of biscyclopropatetrahydrofurans and -pyrrolidines **102**.

The high efficiency of the rearrangement of cyclopropanecarbaldehyde to dihydrofuran was used for the synthesis of the related polycyclic compounds (**105, 106,** etc.), with both odd and even numbers of tetrahydrofurans annulated via C(2)—C(3) bonds containing up to nine rings (Scheme 6.38) [76, 77].

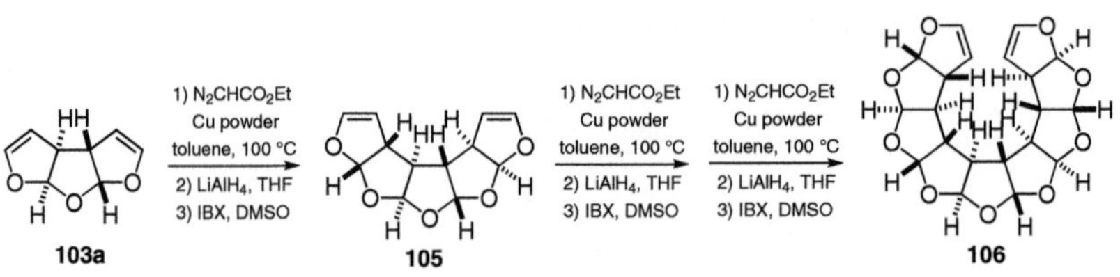

Scheme 6.38 Synthesis of annelated THF moieties by cyclopropanation-modification-isomerization sequence.

The use of BINOL-based phosphoric acids allowed to perform the asymmetric transformation of a broad series of *cis*-1-acyl-2-arylcyclopropanecarboxylates **107** to 2-aryl-2,3-dihydrofuran-4-carboxylates **108** in 48–95% yield and enantiomeric ratio (*er*) between 83:17 and 96:4 (Scheme 6.39) [78]. The efficiency of this rearrangement depended crucially on the nature of acyl functionality: cyclopropanes containing acetyl, propionyl, butyryl, and isobutyryl groups produced dihydrofurans in good-to-excellent yields, while cyclopropanes with benzoyl group failed to produce the target products. Moreover, the introduction of the second substituent to the C(2) atom of the three-membered ring led to a decrease in *er*.

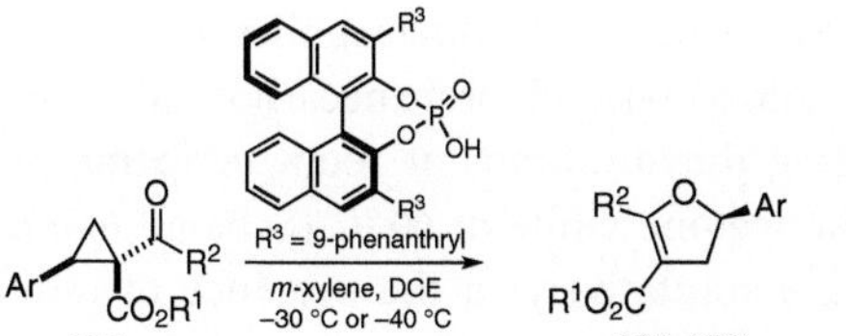

107 → **108**, 96%, *er* from 83:17 to 96:4

Scheme 6.39 Chiral phosphoric acid-catalyzed isomerization of *cis*-1-acyl-2-arylcyclopropanecarboxylates **107** to the optically active dihydrofurans **108**.

It is noteworthy that 1-acyl-2-arylcyclopropanes **109** also underwent an asymmetric transformation to the optically active 2-aryl-2,3-dihydrofurans **110** in 75–92% yields and good enantioselectivity (*er* from 90:10 to 95:5) (Scheme 6.40) [78]. The reversible cyclopropane ring opening under the reaction conditions is responsible for the success of this asymmetric transformation, as the stereochemical information of the starting substrates is lost during the process, and the stereochemical result of the reaction was determined exclusively by the chiral phosphoric acid used as a catalyst.

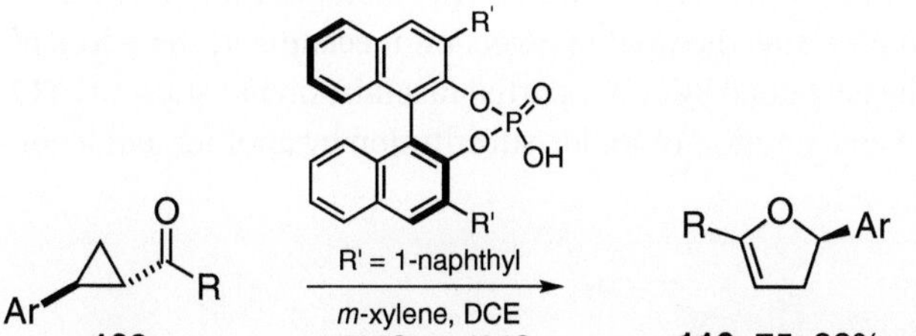

109 → **110**, 75–92%, *er* from 90:10 to 95:5

Scheme 6.40 Asymmetric synthesis of 2-aryl-2,3-dihydrofurans **110**.

The Cloke–Wilson rearrangement was carried out for a broad range of D–A cyclopropanes. For example, substrates **111**, wherein the silylmethyl group serves as a donor group, isomerize to dihydrofurans **112** (Scheme 6.41) [79]; spiro[cyclopropane-barbiturates] **113** rearranged to furo[2,3-*d*]pyrimidines **114** (Scheme 6.42) [80, 81], *etc.*

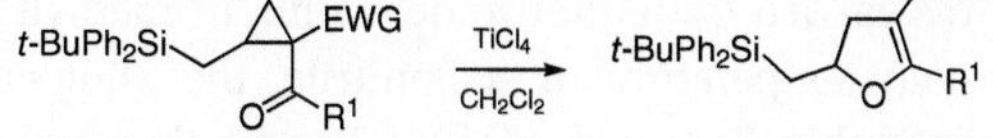

111 → **112**, 70–96%

Scheme 6.41 The rearrangement of 2-silylmethyl-1-EWG-substituted cyclopropyl ketones **111**.

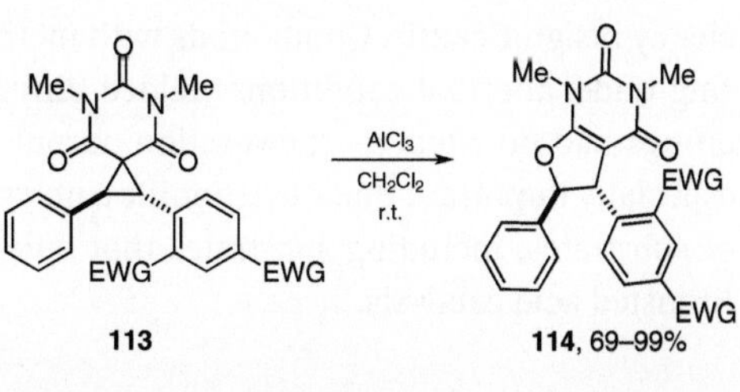

113 → **114**, 69–99%

Scheme 6.42 The Cloke-Wilson reaction of cyclopropane, spiro connected to barbituric acid moiety, **113** to furopyrimidine **114**.

The Cloke–Wilson rearrangement can also be induced by nucleophiles. For example, it was shown that 2-aryl-1-(tri- or difluoroacetyl)cyclopropanecarboxylates **115** isomerized efficiently to the corresponding dihydrofurans at room temperature under treatment with 10 equiv of NaI in acetone while at 60 °C (Scheme 6.43). Moreover, this rearrangement proceeded quantitatively in the presence of only 5 mol.% of NaI [82].

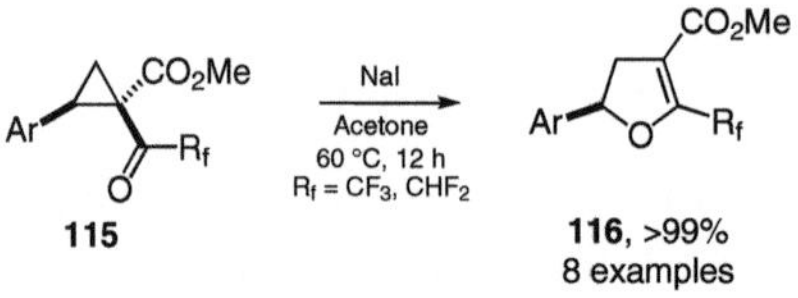

Scheme 6.43 NaI-induced isomerization of *trans*-1-acyl-2-arylcyclopropanecarboxylates **115** to **116**.

The stepwise mechanism was proposed for these processes. The initial nucleophilic attack of the iodide ion leads to the three-membered ring opening providing an acyclic intermediate containing an enolate anion. Then enolate serves as a nucleophile while iodide serves as a leaving group, affording dihydrofuran as a cyclization product. When this isomerization was performed for the optically active substrate **117**, product **118** was obtained with the retention of the configuration of the benzylic carbon atom, which confirms the double inversion mechanism; some loss of enantiomeric purity can be explained by either partial racemization of substrate **117** during the reaction or the contribution of iodide substitution by another iodide ion (Scheme 6.44) [83].

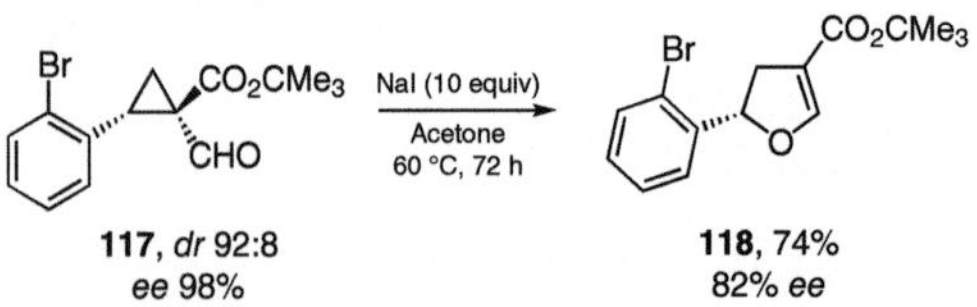

Scheme 6.44 The isomerization of the enantiomerically pure cyclopropane **117**.

A similar mechanism was proposed for $Bu_4N[Fe(CO)_3(NO)]$-catalyzed isomerization of 1,1-diacyl-2-vinylcyclopropanes **119** to the corresponding dihydrofurans **120** based on the results of DFT calculations [84]. Experimentally, this isomerization was carried out either under reflux in methylene chloride or under irradiation at room temperature in acetonitrile, the photochemical conditions being typically preferable (Scheme 6.45) [85]. The relative arrangement of substituents in the three-membered ring as well as substituents at the C=C bond influence the reaction efficiency insignificantly. Compounds with methyl group(s) at the C(3) atom of a small ring under thermal conditions yielded dihydrofurans with lower yields, but these groups had no effect on the reaction output under photochemical conditions. It is especially important that nucleophilic catalysis be equally efficient for a broad series of substrates, including substrates that failed to give dihydrofurans under Lewis/Brönsted acid catalysis.

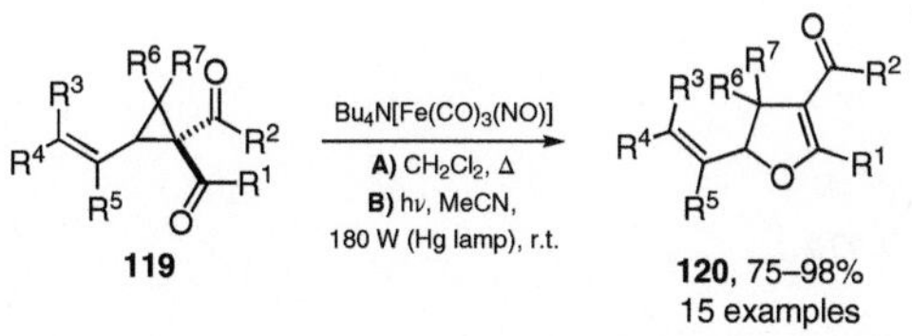

Scheme 6.45 $Bu_4N[Fe(CO)_3(NO)]$-catalyzed isomerization of 1,1-diacyl-2-vinylcyclopropanes **119**.

1,1-Diacyl-2-arylcyclopropanes **121** and the corresponding ketoesters behave similarly but require harsher reaction conditions. Under catalysis with $Bu_4N[Fe(CO)_3(NO)]$, they rearranged into the corresponding dihydrofurans **122** either under heating in DMF at 120 °C or under irradiation with UV light for 24 h. For these substrates, yields under thermal conditions are typically higher than under irradiation (Scheme 6.46) [85].

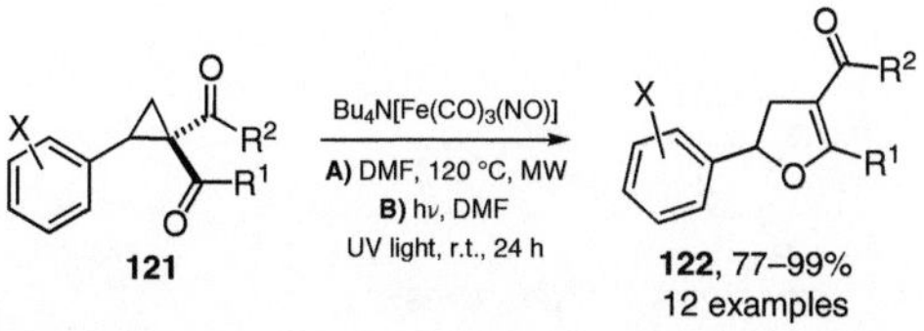

Scheme 6.46 $Bu_4N[Fe(CO)_3(NO)]$-catalyzed rearrangement of cyclopropanes **121**.

Tertiary amines can also induce the Cloke–Wilson rearrangement of D–A cyclopropanes, but the DABCO-initiated process requires heating in DMSO at 120 °C (Scheme 6.47) [86].

Scheme 6.47 DABCO-initiated rearrangement of 1,1-diacyl-2-aryl/alkenylcyclopropanes **119** and **121**.

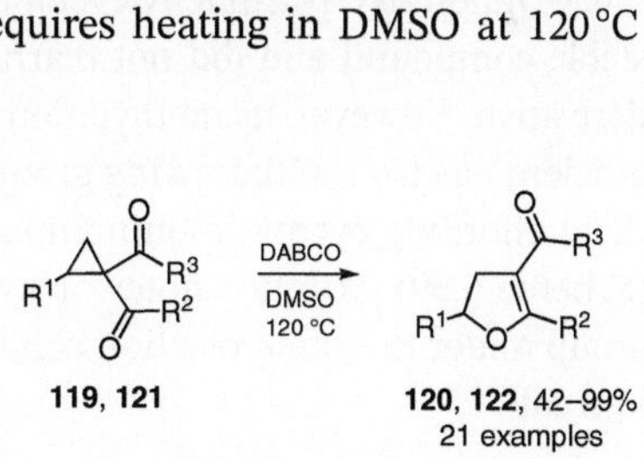

When the starting substrate contains two different acceptor substituents, the reaction chemoselectivity is determined by their nature. For example, the acyl group participates in the process while ester or amide groups remain intact. Cyclopropanes bearing acetyl and benzoyl groups as acceptors produced a mixture of two products with a small excess of 3-benzoyl-2-methyl-4,5-dihydrofurans.

In the reactions discussed above, the anionic Fe complex induced the Cloke–Wilson rearrangement, serving as a nucleophile. At the same time, alkenylcyclopropanes **123** bearing two acceptor groups at the vicinal position (one of them is an acyl) were shown to undergo the rearrangement to 5-vinyl-4,5-dihydrofurans **124** under common transition metal catalysis (Scheme 6.48) [87, 88].

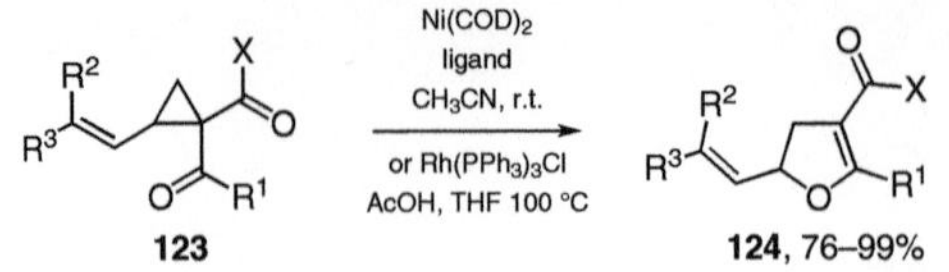

Scheme 6.48 Transition metal-catalyzed isomerization of 1-acyl-2-alkenylcyclopropanes **123**.

6.4.2 The Cloke–Wilson Rearrangements Affording Pyrrole Derivatives

There are only scarce examples of cyclopropylimine-to-pyrroline rearrangements for D–A cyclopropanes that can be related to the competition of imine hydrolysis with the Cloke–Wilson rearrangement. For example, highly active cyclopropane **125** during column chromatography was transformed into aldehyde **126**, pyrroline **127**, and dihydrofuran **128** (Scheme 6.49) [89]. In this reaction, silica gel served as an acidic catalyst, inducing both imine hydrolysis and rearrangement.

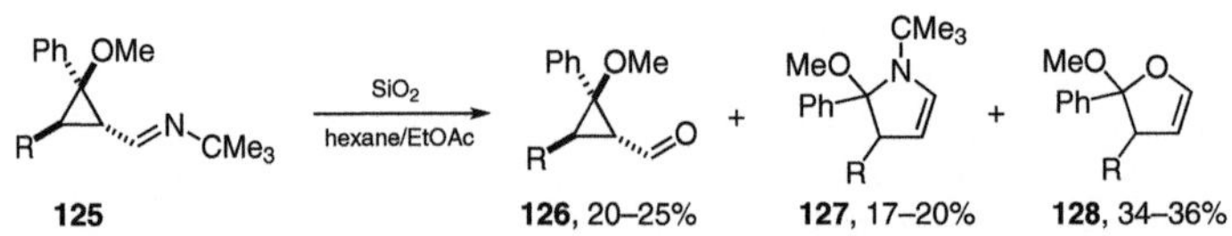

Scheme 6.49 The competition of imine hydrolysis and rearrangement for cyclopropyl imines **125**.

6,7-Dimethoxy-1-(2-phenylcyclopropyl)-3,4-dihydroisoquinoline **129** is a quite stable compound and did not rearrange to the corresponding pyrroloisoquinoline derivative. However, its methylation afforded salt **130**, which contains a much more efficient electron-withdrawing group and underwent the Cloke–Wilson rearrangement affording pyrroloisoquinolinium salt **131** even at the boiling point of acetone (Scheme 6.50) [90]. It is noteworthy that the related cyclopropane without phenyl group under the same reaction conditions did not isomerize to pyrroloisoquinolinium salt.

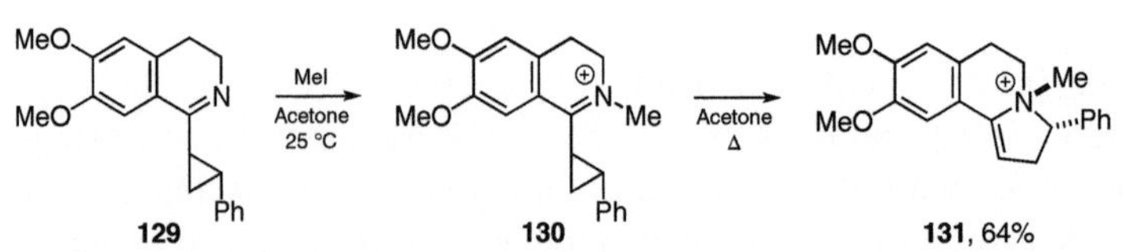

Scheme 6.50 Synthesis and isomerization of 1-cyclopropyl-2-methylisoquinolinum salt **130**.

On the other hand, 1-(2-pyridyl)-2-styrylcyclopropanecarboxylates **132** afforded dihydroindolizine derivatives **133** under catalysis with $Pd(PPh_3)_4$; most of the rearrangement products were isolated after aromatization [91]. A single example of the isolation of **133** was described (Scheme 6.51).

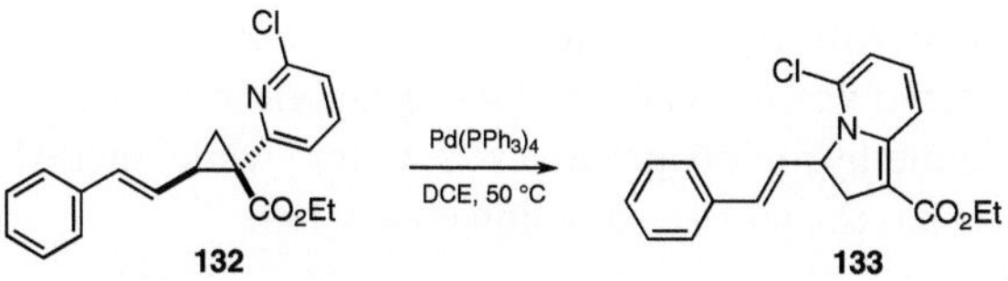

Scheme 6.51 Palladium-catalyzed rearrangement of 1-(pyridin-2-yl)-2-styrylcyclopropanecarboxylate **132**.

6.4.3 The Related Rearrangements Affording Other Heterocycles

Not only C=O and C=N groups can participate in the rearrangement of this type. For example, various nitrocyclopropanes were rearranged to the corresponding isoxazoline *N*-oxides [92–96]. An example of this reaction is given in Scheme 6.52.

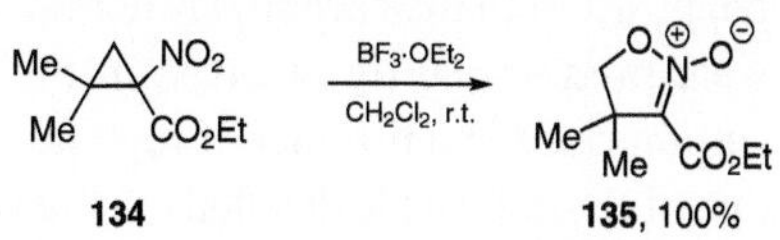

Scheme 6.52 Isomerization of ethyl 2,2-dimethyl-1-nitrocyclopropanecarboxylate **134**.

It is worth noting that the rearrangement of 2-aryl-1-nitro-1-(1-nitro-2-arylvinyl) cyclopropane **136** to the corresponding isoxazoline *N*-oxides **137** can be accompanied by an unusual shift of nitro group, affording 5-aryl-3-(2-aryl-2-nitrovinyl)isoxazoline *N*-oxides **138** (Scheme 6.53) [96].

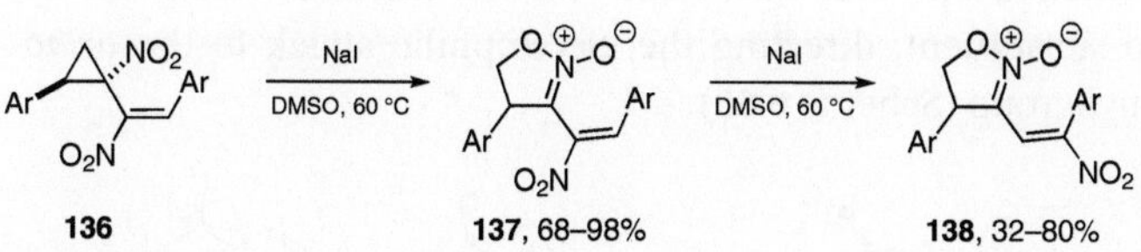

Scheme 6.53 Unusual shift of nitro group accompanying the rearrangement of cyclopropanes **136**.

It is also possible to point out that 1-nitroso-2-phenylcyclopropane **140**, formed as an intermediate in the 2-phenylcyclopropylamine **139** oxidation, undergoes in situ rearrangement to 5-phenyl-4,5-dihydroisoxazoline **141** (Scheme 6.54) [97].

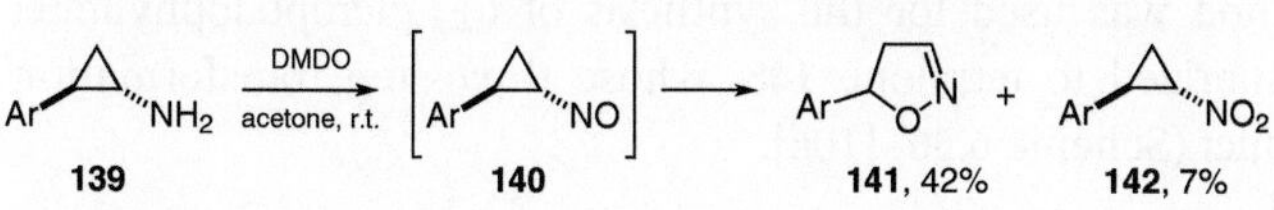

Scheme 6.54 The formation of 4,5-dihydroisoxazoline **141** and 1-nitro-2-phenylcyclopropane **142** in the oxidation of **139**.

The relative efficiency of various Cloke–Wilson-like rearrangements has been studied by the DFT method under the supposition of a concerted mechanism for this transformation. In general, the trend of reactivity for diverse donor groups was found to be $Me_2N > OMe > SMe$, SeMe, $Ph > Me$, PMe_2, Cl. The analogous trend for diverse acceptor groups is $N{=}O > CH{=}S > C(R){=}O$, CH=NR, $NO_2 > C(O)OR$, despite the fact that the relative reactivity in these sequences depends on the vicinal substituent. The lowest energy barrier was found for cyclopropane bearing a nitroso group as an acceptor and a dimethylamino group as a donor, while methyl 2-chlorocyclopropanecarboxylate has the largest activation energy [98].

6.5 Nazarov Reaction and its Homo-Version

D–A cyclopropanes bearing an aroyl group as an acceptor substituent can be involved not only in the Cloke–Wilson rearrangement but also in the homo-Nazarov cyclization, i.e. the acid-stimulated cyclization of alkenyl/aryl cyclopropyl ketones to cyclohexenones. This process proceeds by a stepwise mechanism with the intermediate cyclopropane ring opening, which facilitated by a donor group at the vicinal position to the carbonyl function. Therefore, these were D–A cyclopropanes that were predominantly involved in the homo-Nazarov cyclization. These studies are summarized in a number of reviews [99–101], although it should be noted that only one of them [101] is devoted exclusively to the homo-Nazarov cyclization.

The investigation of this process was started by Murphy and Wattanasin in the early 80s. In a series of articles [102–105], they reported that aryl 2-arylcyclopropyl ketones **143** rearrange to 4-aryltetralones **145** through the intermediate oxonium cation **144**, if aroyl moiety contains an electron-releasing group facilitating the Friedel–Crafts alkylation. 4-Aryl-4-hydroxybutyrophenones **146** were also formed after quenching the reaction mixtures before the reaction was over or if the aroyl moiety contained no substituent, directing the electrophilic attack to the *ortho*-position to the carbonyl group (Scheme 6.55).

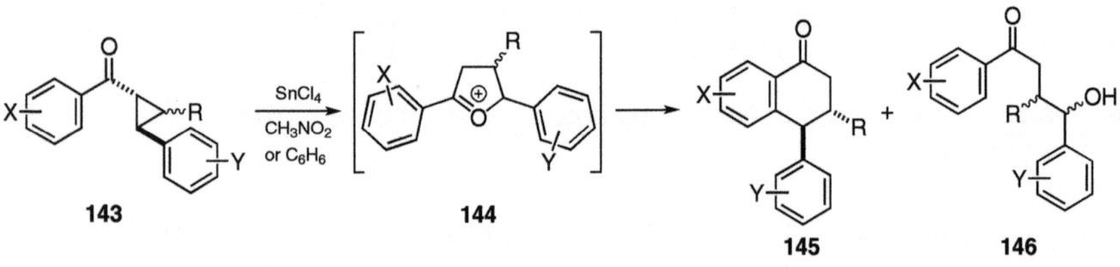

Scheme 6.55 Homo-Nazarov rearrangement of 1-aroy-2-arylcyclopropanes **143**.

The developed method was used for the synthesis of (±)-picropodophyllone: cyclopropane **147** isomerized to tetralone **148**, whose three-step transformation yielded the target product (Scheme 6.56) [106].

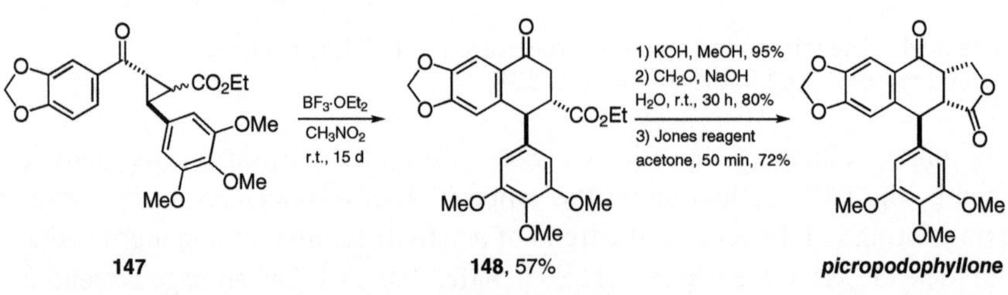

Scheme 6.56 Total synthesis of (±)-picropodophyllone.

At the same time, the belonging of cyclopropane **147** to the class of D–A cyclopropanes raises certain doubts as two acceptor groups at the vicinal carbon atoms decrease the mutual polarization of C—C bonds in a three-membered ring. Oppositely, cyclopropanes **149** can definitely be referred to as D–A ones. Under reflux of their dichloroethane (DCE) solution with titanium(IV) chloride, they underwent homo-Nazarov cyclization, affording 1-hydroxy-3,4-dihydronaphthalene derivatives **150** (Scheme 6.57) [107]. Optically active cyclopropanes **149** afforded the corresponding dihydronaphthalenes with retention of the absolute configuration of chiral atoms and without loss of enantiomeric excess.

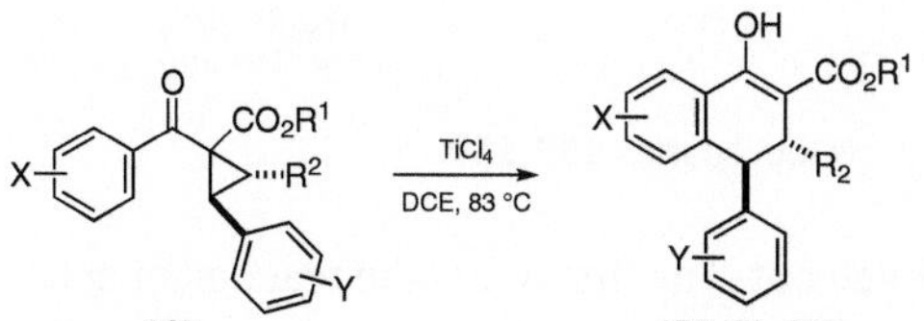

149 **150**, 31–92%

Scheme 6.57 Homo-Nazarov cyclization of 1-aroyl-2-arylcyclopropanecarboxylates **149**.

Cyclopropanes with heteroaroyl group as acceptor behave similarly to benzoyl-substituted cyclopropanes. Thus, under treatment with $SnCl_4$, 2-furyl- and 2-thienyl 2-arylcyclopropyl ketones **151** rearranged to 4-aryl-4,5,6,7-tetrahydrobenzofuran and -benzothiophene **152** (Scheme 6.58) [108, 109]. It should be noted that 4-methoxyphenyl derivative **151** isomerized at room temperature, while the cyclization of phenyl-substituted **151** required heating at 80 °C.

Scheme 6.58 $SnCl_4$-initiated isomerization of 2-arylcyclopropyl 2-furyl- and 2-thienyl ketones **151**.

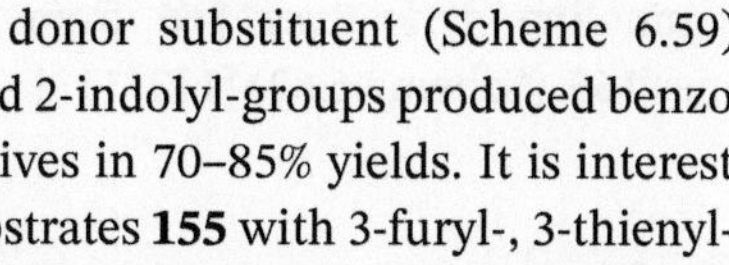

151 **152**, 54–94%

Heating was also needed for homo-Nazarov cyclization of cyclopropanes, wherein the silylmethyl group served as a donor substituent (Scheme 6.59). Substrates **153** with 5-R-2-furyl-, 2-thienyl-, and 2-indolyl-groups produced benzofuran, benzothiophene, and carbazole derivatives in 70–85% yields. It is interesting that for the corresponding reactions of substrates **155** with 3-furyl-, 3-thienyl-, and 3-indolyl-groups the reported yields were the same (80%) independently of the nature of heterocycles [109].

The reactivity of the related 1-(hetarylcarbonyl)cyclopropanecarboxylates **157** and **159** (hetaryl is 2- and 3-furyl, 2- and 3-thienyl, benzofuran-2- and 3-yl, 1-methylindol-2-

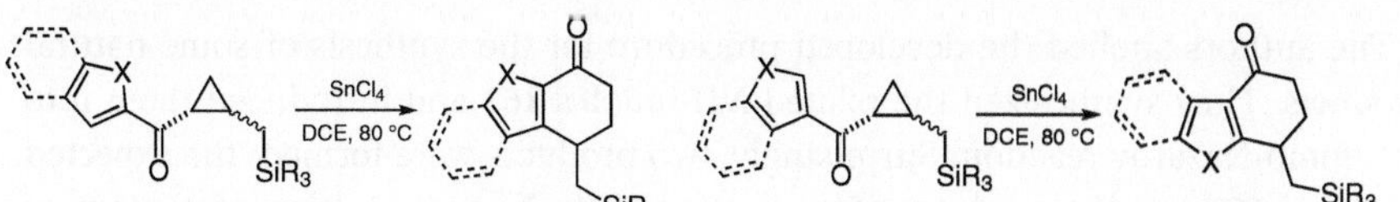

153 **154**, 70–85% **155** **156**, 80%

Scheme 6.59 Cyclization of 1-heteroaroyl-2-silylmethylcyclopropanes **153** and **155**.

and -3-yl) with diverse donor groups (aryl, silylmethyl) is much higher, which allowed performing their cyclizations to the corresponding annulated derivatives **158** and **160** under catalysis with milder Lewis acid, In(OTf)$_3$ (Scheme 6.60). At the same time, the presence of the second acceptor had some drawbacks: products were formed as mixtures of diastereomers [110, 111].

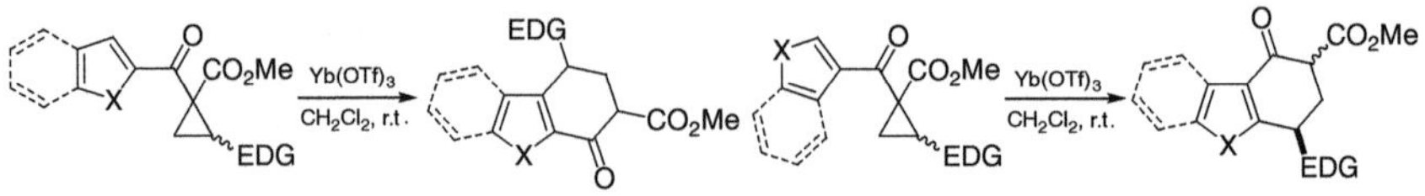

157 158, 63–91% *dr* from 1.1:1 to 2.4:1 159 160, 61–83% *dr* from 1.1:1 to 1.7:1

Scheme 6.60 Homo-Nazarov cyclization of cyclopropanes **157, 159**.

Two specific examples should be pointed out. The first one is the reaction of substrate **161**, bearing 2-bromo-3-thienylcarbonyl group as an acceptor group. In this compound, α-position was blocked by the bromine atom, and atom C(4) was included in the reaction providing benzo[*c*]thiophene derivative **162**. The second one is the reaction of cyclopropane **163** with an alkoxy group as a donor substituent; the elimination of alcohol from the initially formed ring leads to its aromatization furnishing 7-hydroxybenzothiophene **164** (Scheme 6.61) [110].

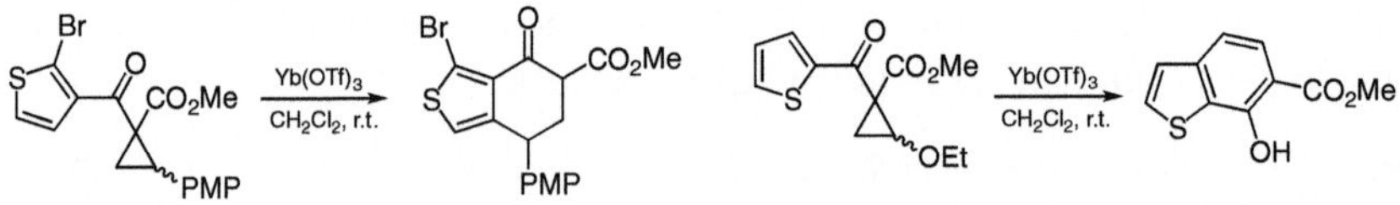

161 162, 56% 163 164, 51%

Scheme 6.61 Atypical examples of homo-Nazarov cyclization of thenoylcyclopropane derivatives **161, 163**.

Homo-Nazarov cyclization was used by Waser et al. for the synthesis of indolo[2,3-*h*]quinoline derivatives **166** from 7-(indol-2-ylcarbonyl)-1-azabicyclo[4.1.0]heptanes **165**. (Scheme 6.62) [112, 113].

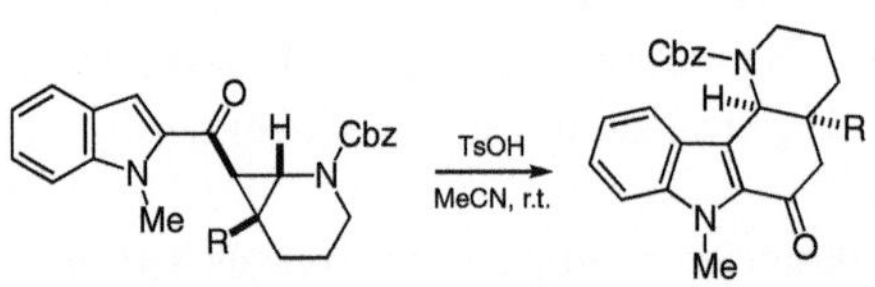

165 166, 90–100%

Scheme 6.62 Homo-Nazarov cyclization in the synthesis of octahydroindolo[2,3-*h*]quinolines **166**.

The authors applied the developed procedure for the synthesis of some natural products. They synthesized the related *NH*-indoles **167** and introduced them into the homo-Nazarov reaction. Surprisingly, two products were formed: the expected products **168** and the products of cyclization with the participation of *N*-atom of the indole ring **169** (Scheme 6.63). Optimization of reaction conditions allowed for preparing all of them in good yields [112–114]. Moreover, it was shown that

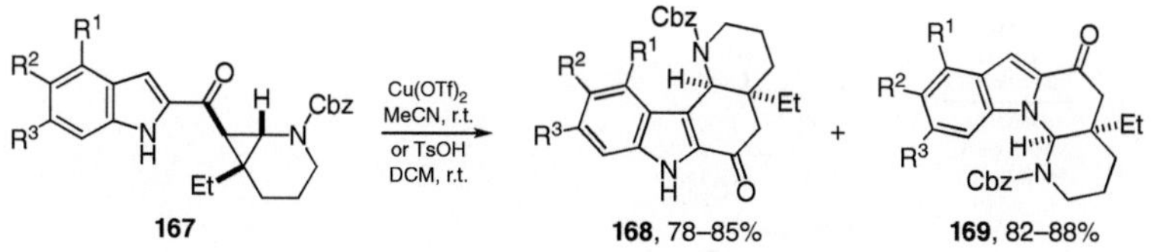

Scheme 6.63 Two ways for rearrangement of D–A cyclopropanes **167**.

compound **169** can rearrange to isomeric **168** in high yield under prolonged treatment with copper(II) triflate [113].

Deprotection of **168a** ($R^1 = R^2 = R^3 = H$) accomplished the formal synthesis of (±)-aspidospermidine. On the other hand, substrate **170** was converted to indolo[2,1-*h*]quinoline derivative **171**, which was transformed in 4 steps to (±)-goniomitine (Scheme 6.64) [112].

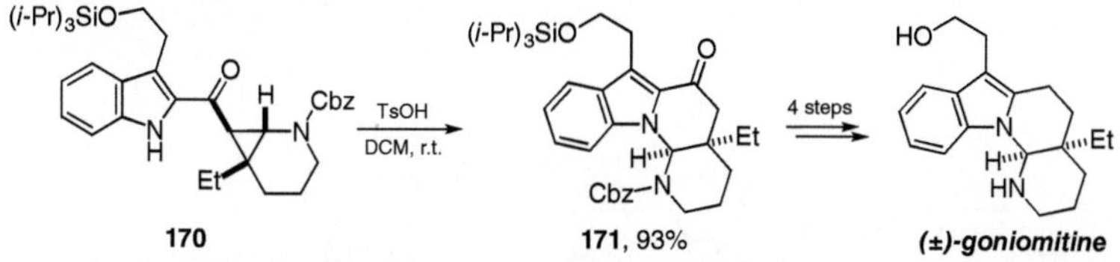

Scheme 6.64 Formal homo-Nazarov reaction in the total synthesis of (±)-goniomitine.

At last, homo-Nazarov cyclization of D–A cyclopropane **172** affording indoloquinoline derivative **173** was a key step in the total synthesis of jerantinine E (Scheme 6.65) [114].

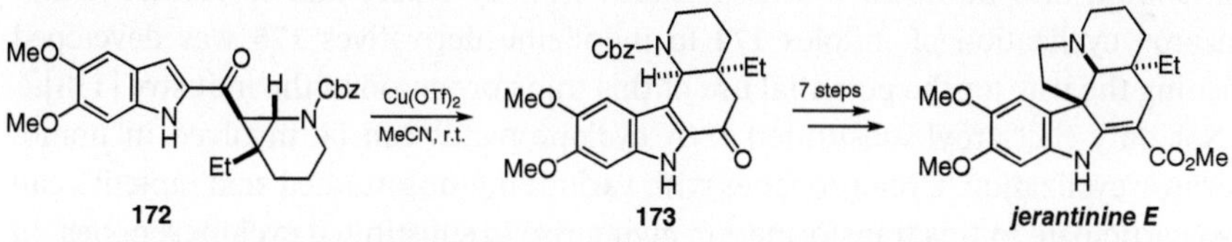

Scheme 6.65 Homo-Nazarov cyclization in the total synthesis of jerantinine E.

The transformation of 1-(indol-2-ylcarbonyl)-2-amidocyclopropane **168, 170** to indolo[2,1-*h*]quinolines **169, 171** can be classified as either nucleophilic ring opening/cyclization or as homo-Nazarov cyclization, accounting for the involvement of the N(1)—C(2) bond in the aromatic system of the indole ring that implies the partial double bond character of this bond. The same problem is faced when choosing the type of process, which includes the cyclizations of (indol-1-yl)carbonyl- and (pyrrol-1-yl)carbonylcyclopropanes **174** to the corresponding indolizine derivatives **175**. For ease of comparison, we included these reactions in this section (Scheme 6.66). It was found that for indole-derived cyclopropanes, the best yields were obtained using indium(III) triflate as a catalyst [115, 116], while for pyrrole-derived analogs, the best results were achieved with $Al(OTf)_3$ [117]. Various aryl groups, silylmethyl, amino, alkoxy, arylsulfanyl groups, and two alkyl substituents were used as donors; cyclopropanes with electron-abundant aromatic groups

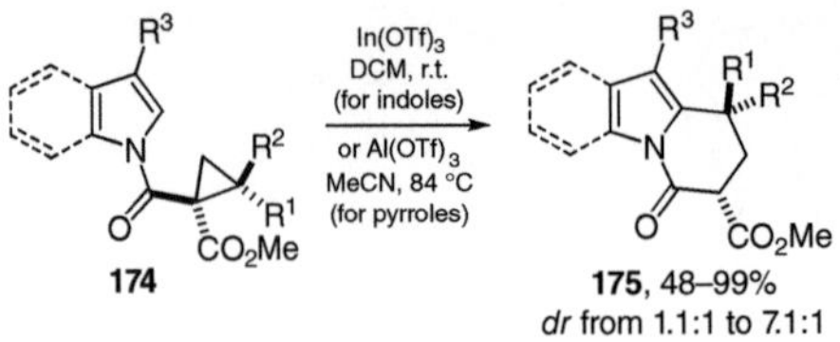

Scheme 6.66 Cyclization of (indol-1-yl)- and (pyrrol-1-yl)carbonylcyclopropanes **174** to indolizines **175**.

rearranged more efficiently than those with phenyl or 4-halophenyl substituents, while 4-nitrophenyl derivative was found to be unreactive.

The developed procedure was applied for the synthesis of (±)-deethyleburnamonine (Scheme 6.67). Formal homo-Nazarov cyclization of cyclopropane **176** afforded tetracyclic compound **177**, which was converted to the target product in two steps only [116].

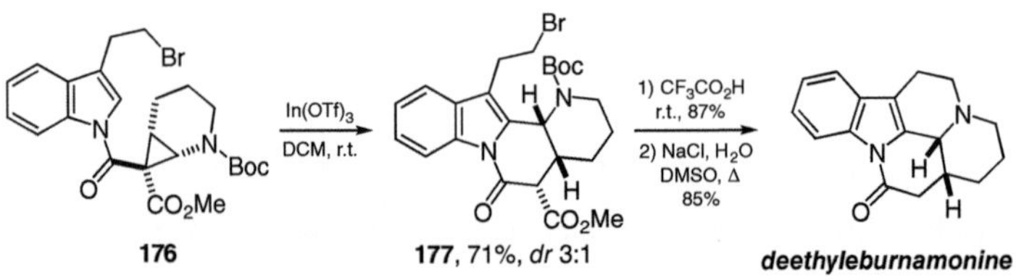

Scheme 6.67 Synthesis of (±)-deethyleburnamonine using formal homo-Nazarov cyclization.

It should also be noted that continuous flow cyclopropanation/formal homo-Nazarov cyclization of indoles **174** to indolizine derivatives **175** was developed opening the way for the potential use of this transformation in the industry [118].

Not only (het)aroyl-substituted D–A cyclopropanes can be involved in homo-Nazarov cyclization. Cyclopropanes with various α,β-unsaturated acid moieties can also participate in this transformation giving rise to substituted cyclohexenones. In the first study of this cyclization polyphosphoric acid in benzene was used for the reaction initiation, however under these conditions some side processes proceeded also, and yields of the corresponding cyclohexenones were from low to moderate [119]. More recently, TsOH was used for the cyclization of substrates **178**, bearing dihydropyran and dihydrofuran as fragments containing nucleophilic C—C bonds, to the corresponding bicyclic compounds **179** (Scheme 6.68) [120]. Simple alkenes **180** failed to give product of homo-Nazarov cyclization under these conditions. They reacted in the presence of In(OTf)$_3$, however, the expected products **181**

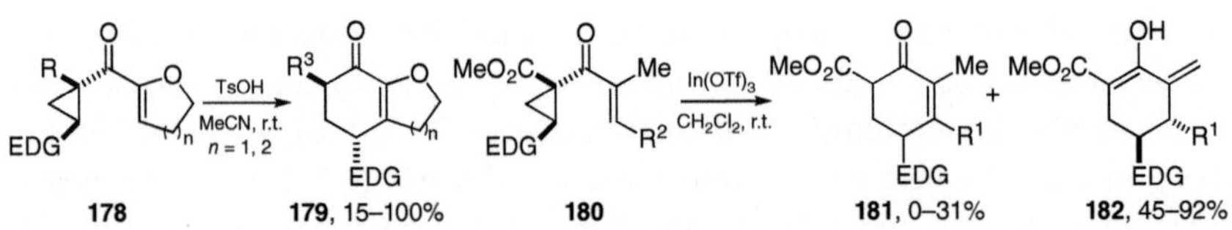

Scheme 6.68 Alkenyl cyclopropyl ketones **178** and **180** in homo-Nazarov cyclization.

were formed as minor products if at all. Their isomers **182** with exocyclic C—C double bonds were obtained as major or exclusive products [121].

All these processes demonstrate the large perspectives for using homo-Nazarov cyclization of D–A cyclopropanes. However, D–A cyclopropanes can also participate in the Nazarov cyclization itself. 2-Alkenyl-2-arylcyclopropane-1,1-diesters **183** under treatment with strongly activating Lewis acids underwent a three-membered ring opening, producing 1,3-zwitterionic intermediates, whose positively charged part is analogous to that in the common Nazarov cyclization. As a result, indenes or their heterocyclic analogs **184** were formed in good to excellent yields (Scheme 6.69). Similarly, 2,2-bis(β-methylstyryl)-substituted cyclopropane afforded the corresponding cyclopentadiene [122].

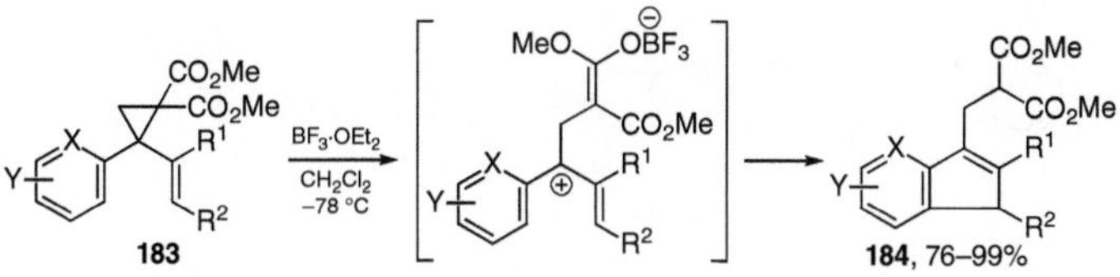

Scheme 6.69 Formal Nazarov cyclization of D–A cyclopropanes **183**.

Moreover, it was shown that $TiCl_4$ promoted rearrangement of 3-aroyl-2-arylcyclopropane-1,1-dicarboxylates **185** to (3-aryl-1-oxoindan-2-yl)malonates **186** via coordination of Lewis acid to both ester groups, leading to the C(1)–C(2) bond breaking, producing α-aroylstyrylmalonates (Section 6.1), and aroyl oxygen, inducing Nazarov cyclization (Scheme 6.70) [123].

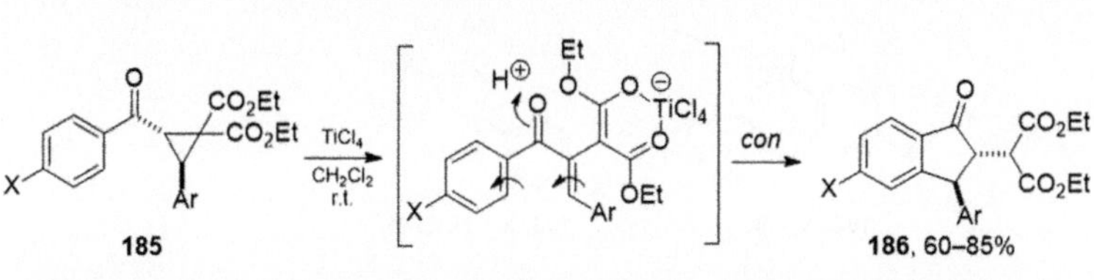

Scheme 6.70 $TiCl_4$-promoted rearrangement of cyclopropanes **185** to indanones **186**.

6.6 The Cope Rearrangement and Related Isomerizations of Donor–Acceptor Cyclopropanes

The rearrangement of 1,2-divinylcyclopropanes to cycloheptadienes has been studied since 1960 [124]; the results of these investigations were summarized in some reviews [125, 126]. It should be noted that *cis*-1,2-dialkenylcyclopropanes rearrange usually under very mild conditions, which complicates the study of this process. Nevertheless, some examples of the Cope rearrangement of D–A cyclopropanes are known.

Thus, compound **187**, a product of the furan cyclopropanation, can be isolated in a pure form, but it rearranged to 8-oxabicyclo[3.2.1]octadiene **188** as well as acyclic products **189** even at room temperature (Scheme 6.71) [127].

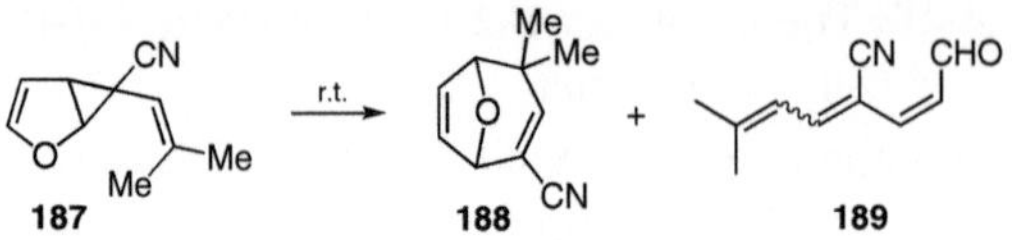

Scheme 6.71 Rearrangement of cyclopropa[*b*]furan **187**.

Another example of the Cope rearrangement was described relatively recently [128]. Namely, 2-vinylspiro[cyclopropane-1,3'-oxindoles] **190** rearranged to cyclohepta[*cd*]indoles **191** when vinyl and phenyl groups have *cis*-arrangement while another diastereomer was unreactive (Scheme 6.72) [128].

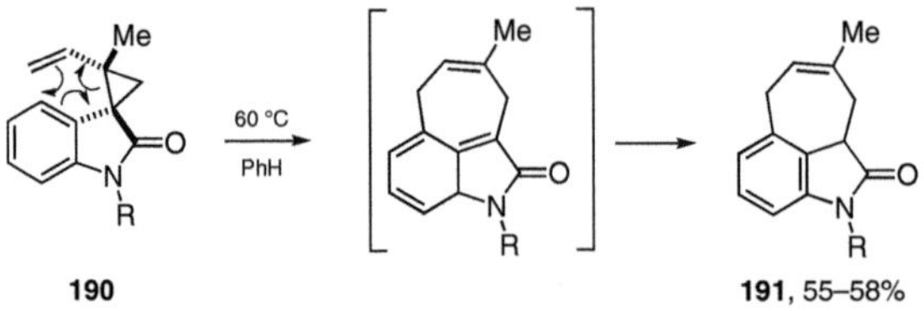

Scheme 6.72 The Cope rearrangement of 2-vinylspiro[cyclopropane-1,3'-oxindoles] **190**.

2-Alkenyl-1-arylcyclopropanecarboxylates, -phosphonates, and related compounds **191** can isomerize to either benzocycloheptenes **192** via 3,3-sigmatropic shifts or 4-arylcyclopentenes **193** via VCR. It was found that substrates, bearing two electron-withdrawing substituents in the vicinal position to the alkenyl group, underwent the Cope rearrangement; oppositely if the electron-withdrawing ability of these two substituents is not enough, VCR competes or predominates (Scheme 6.73) [129].

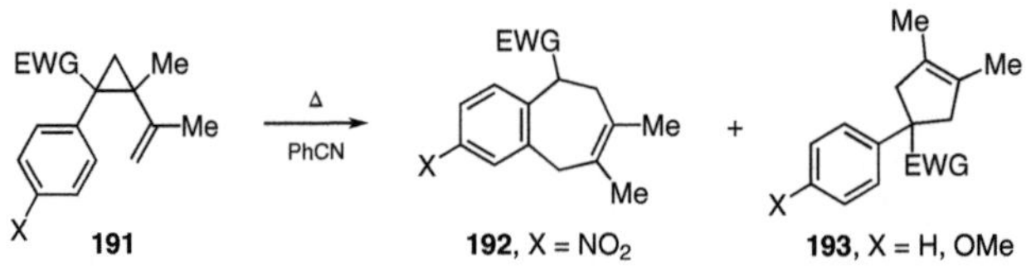

Scheme 6.73 The aromatic Cope rearrangement of D–A cyclopropanes **191**.

The phosphine-catalyzed isomerization of 2-alkenylcyclopropyl ketones **194** to cycloheptenones **195** was proposed by Xu et al. [130]. Despite this process is not a real 3,3-sigmatropic shift, the products correspond to the formal Cope rearrangement of the enol form of the starting cyclopropane (Scheme 6.74).

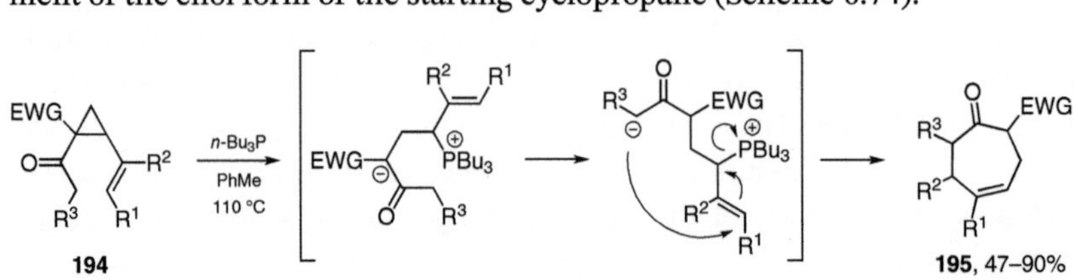

Scheme 6.74 Isomerization of 2-alkenylcyclopropyl ketones **194** to cycloheptenones **195**.

This approach was used in the total syntheses of daphenylline (Scheme 6.75) [131].

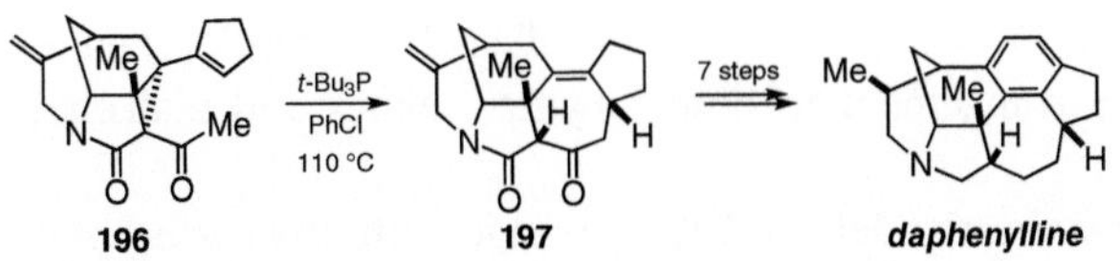

196 197 daphenylline

Scheme 6.75 Rearrangement of alkenylcyclopropyl ketone **196** in the total synthesis of daphenylline.

Heteroatom analogs of the Cope rearrangement have also been investigated. It was shown that *cis*-2-vinylcyclopropanecarbaldehyde **198** is in equilibrium with 2,5-dihydrooxepine **199**; the ratio of the two isomers being 95:5 in favor of cyclopropane (Scheme 6.76) [132]. Similarly, *syn*-bicyclo[3.1.0]hex-2-ene-6-carbaldehyde **200** was found to be in equilibrium with 2-oxabicyclo[3.2.1]octa-3,6-diene **201** [133]. On the other hand, for 2-phenyl-2-vinylcyclopropanecarboxaldehyde **202,** this ratio is 28:72 [134]. Some other examples of the related equilibriums were reported [135, 136].

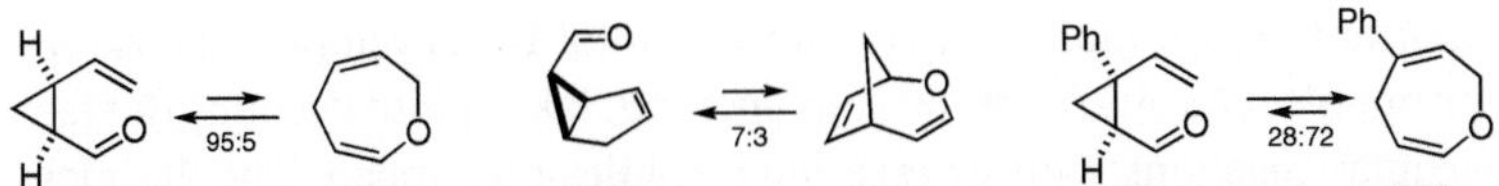

198 199 200 201 202 203

Scheme 6.76 Equilibrium between 2-alkenylcyclopropanecarbaldehydes and oxepines

When 2-alkenylcyclopropanecarbaldehydes are less stable than oxepines, they rearranged usually so fast that only oxepines can be isolated during attempts to synthesize aldehydes. Thus, Swern oxidation of 2-alkenyl-2-(*t*-butyldimethylsiloxy) cyclopropylmethanols **204** afforded oxepines **205** (Scheme 6.77). It should be noted that the formed 2,5-dihydrooxepines isomerized to 2,3-dihydrooxepines under acidic conditions and to 2-alkenyl-2-siloxy-2,3-dihydrofurans **206** under basic conditions. Moreover, some starting cyclopropylmethanols produced directly dihydrofurans [137, 138].

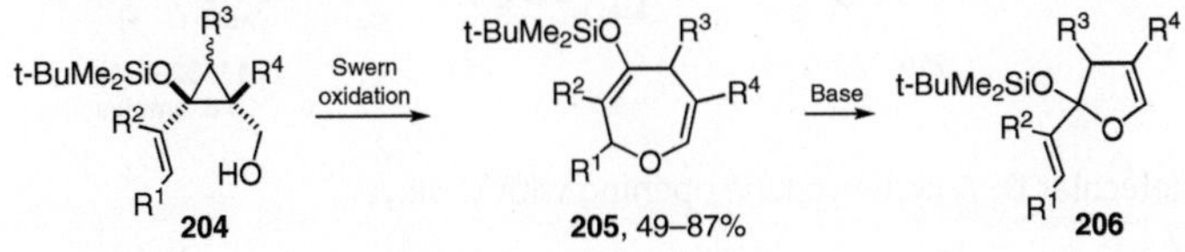

204 205, 49–87% 206

Scheme 6.77 Swern oxidation of cyclopropylmethanols **204** to 2,5-dihydrooxepines **205**.

On the other hand, oxepines, obtained by oxidation of cyclopropylmethanols, under treatment with Lewis acid rearranged to cyclopentenecarbaldehydes [132, 139]. In other words, 2-alkenylcyclopropanecarbaldehydes are in equilibrium with dihydrooxepines, 2-alkenyl-2,3-dihydrofurans, and cyclopentenecarbaldehydes; the predominant structure is determined by substituents in D–A cyclopropane and reaction conditions [140–144]. The competition of 3,3-sigmatropic shifts with other transformations of the starting D–A cyclopropanes was investigated by quantum chemical methods [138, 145–147].

Similarly, under heating at 140 °C, tosylimines of 2-alkenyl-1-arylcyclopropanecarbaldehydes underwent rearrangement to both dihydroazepines and 2-alkenylpyrrolines [148].

At last, 2-alkynylcyclopropanecarbaldehyde **207** rearranged into the corresponding alkylideneoxepine **208** under heating at 130–150 °C (Scheme 6.78) [149, 150].

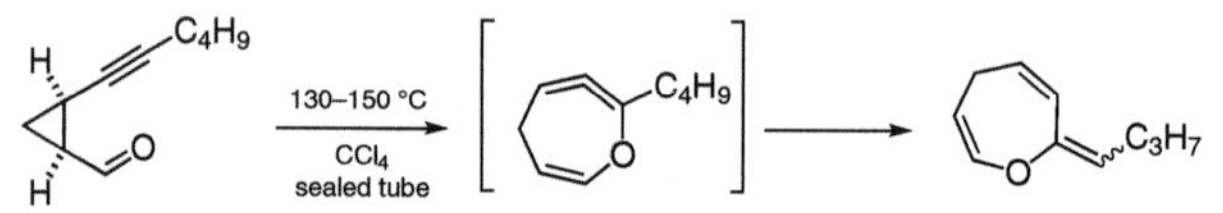

207 208

Scheme 6.78 Rearrangement of 2-alkynylcyclopropylcarbaldehyde **207**.

6.7 Intramolecular Nucleophilic Ring Opening/Ring Closure and Related Processes

One more important type of molecular rearrangement of D–A cyclopropane is represented by intramolecular nucleophilic ring opening, taking into account the fact that intermolecular reactions with diverse nucleophiles are, apparently, the most studied type of D–A cyclopropane reactions. Thus, Kerr et al. showed that the treatment of 2-aminooxyethyl-substituted cyclopropane **209** with $Yb(OTf)_3$ induced the cyclization affording isoxazolidine **210**. An interesting consequence of this process is the ability to control the diastereoselectivity of the transformation of cyclopropane **209** to pyrrolo[1,2-*b*]isoxazole derivatives **211**. The reaction of **210** with aldehydes afforded a new ring with *cis*-arrangement of substituent, while the initial formation of imine from aldehyde and amino group followed by (3+2)-cycloaddition furnished *trans*-**211** (Scheme 6.79) [151].

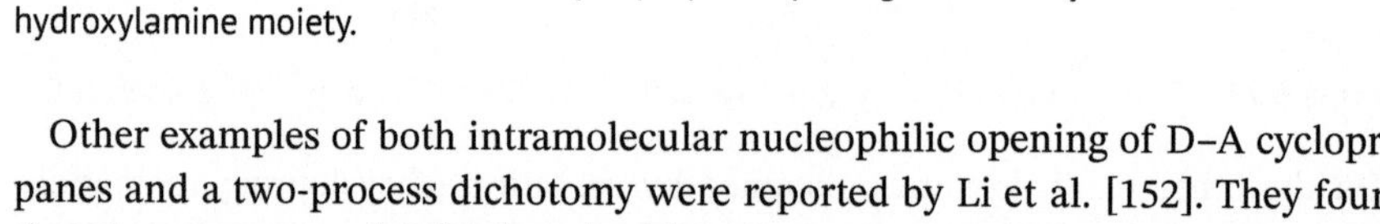

209 210 211, 32–85%
19 examples

Scheme 6.79 Intramolecular D–A cyclopropane opening with *O*-alkyl hydroxylamine moiety.

Other examples of both intramolecular nucleophilic opening of D–A cyclopropanes and a two-process dichotomy were reported by Li et al. [152]. They found that the treatment of 2-(biphenyl-2-yl)cyclopropane-1,1-diester **212** with a large excess of trifluoromethanesulfonic acid at room temperature or with catalytic quantities of this acid under heating produced (9*H*-fluoren-9-yl)methyl malonates **213** while at 0 °C in the presence of a catalytic quantity of TfOH dihydrophenanthrenes **214** were formed (Scheme 6.80). The authors explained the obtained results by the different intermediates formed during the reaction: protonation of cyclopropane **212** produced benzylic carbocation that alkylates the neighboring aromatic group; under milder conditions, this cation was transformed into alkylidene malonate; and products **214** were formed via intramolecular Michael addition (see also Scheme 6.83).

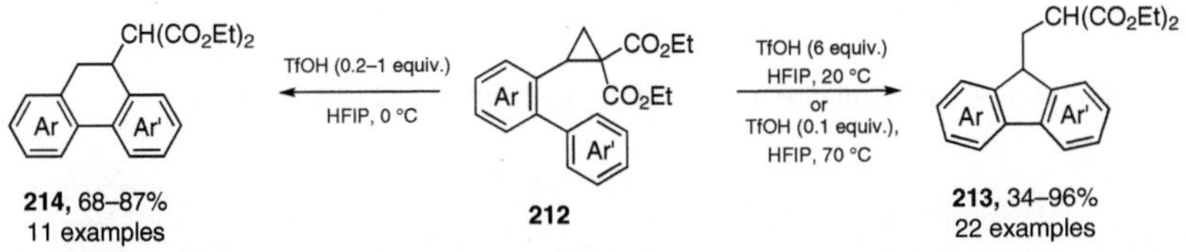

Scheme 6.80 Rearrangements of D–A cyclopropanes **212** to fluorenes **213** and dihydrophenanthrenes **214**.

A new class of potentially active β-lactam antibiotics, containing an azeto[2,3-*d*]isoxazolidine skeleton, was synthesized by intramolecular D–A cyclopropane ring opening by the amide functionality in compounds **215** (Scheme 6.81) [153]. This reaction is a rare example of a four-membered ring formation based on the D–A cyclopropanes reactivity.

Scheme 6.81 Annulated β-lactams **216** formation by base-induced rearrangement of D–A cyclopropanes **215**.

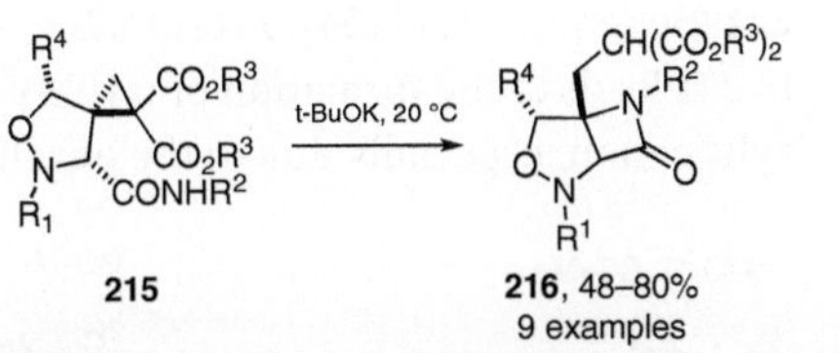

Nevertheless, the formation of four-membered rings is typically inefficient, and some concurrent processes proceed faster when furnishing other types of products. For example, $MgBr_2$-promoted rearrangement of 2-(2-hydroxyaryl)cyclopropane-1,1-diesters **217** produced 2,3-dihydrobenzofurans **218** (Scheme 6.82). *O*-Alkylated substrates can be also involved in this process. Moreover, dihydrobenzothiophene was synthesized by this method using thiophenol derivatives bearing 4-methoxybenzyl group at the sulfur atom [154].

Scheme 6.82 The rearrangement of 2-hydroxyaryl-substituted cyclopropanes **217** to dihydrobenzofurans **218**

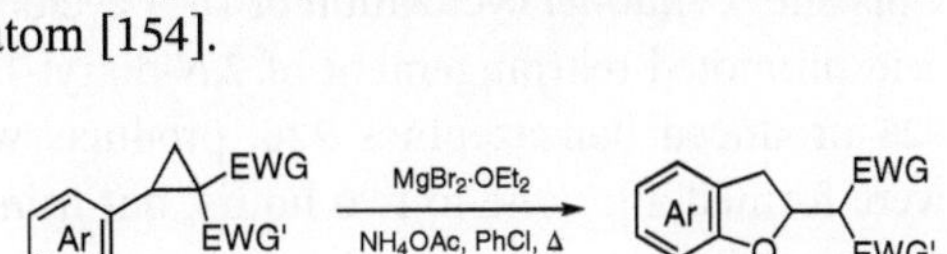

The proposed mechanism for dihydrobenzofurans **218** formation is shown in Scheme 6.83. The key steps of this rearrangement are the generation of magnesium dienolate and intramolecular Michael addition of phenolate ion to the electrophilic C=C bond.

Similar isomerizations were reported for 1-benzoyl-2-hydroxyphenyl-3,3-dimethylcyclopropane **219** affording dihydrobenzofuran **220** and related *o*-carboxamide **221** furnishing isoquinolone derivative **222** (Scheme 6.84) [155, 156].

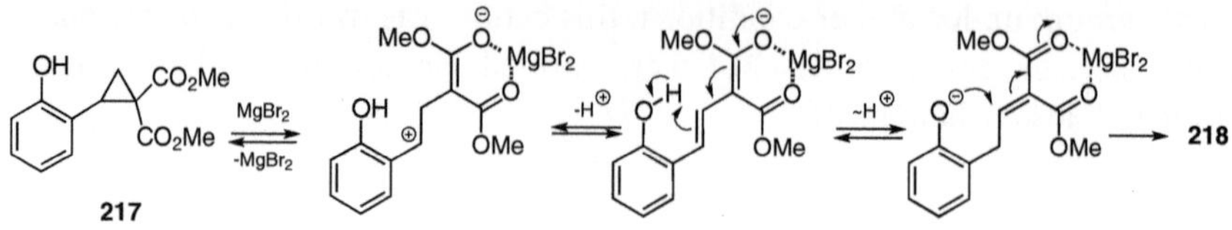

Scheme 6.83 Proposed mechanism for 2,3-dihydrobenzofurans **218** formation.

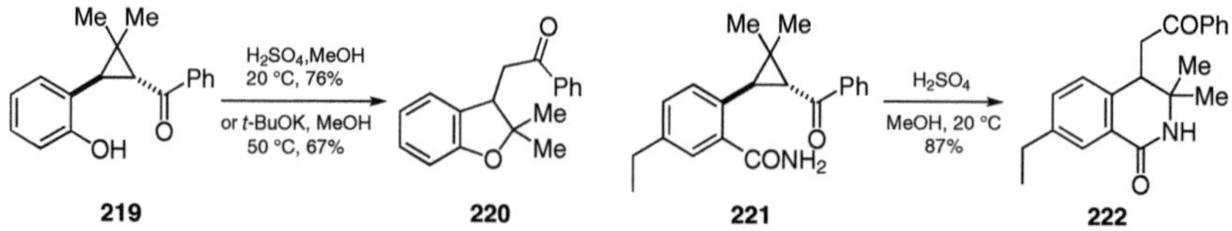

Scheme 6.84 Rearrangements of D–A cyclopropanes **219**, **221** bearing a nucleophilic group at the *ortho*-position of the aromatic donor substituent.

The last result seems to be unexpected as the nucleophilic attack on the benzylic carbon atom, providing a five-membered ring, is allowed here. For example, 2-hydroxymethyl-substituted substrate **223** rearranged to isobenzofuran derivative **224** (Scheme 6.85) [154]. Presumably, acid-induced three-membered ring opening in **221** leads to the formation of tertiary carbocation, which is more stable than benzylic cation, especially due to the negative effect of *ortho*-carbamoyl moiety.

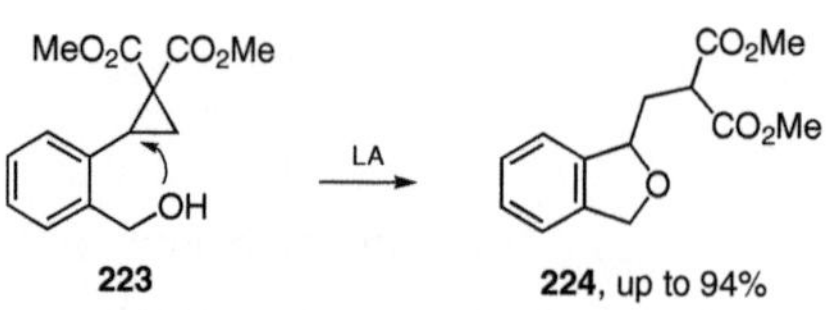

Scheme 6.85 Lewis acid-induced rearrangement of *ortho*-hydroxyphenyl-substituted cyclopropane **223**.

All these processes belong to the well-known type of n-*exo*-*tet* cyclizations. Oppositely, *endo*-*tet* cyclization of D–A cyclopropanes was reported recently. Lewis acid-promoted rearrangement of 2,*N*-diaryl-1-carbamoylcyclopropanecarboxylates **225** produced benzazepines **226**, products with *cis*-arrangement of substituents were formed after one to two hours, but *trans*-isomers were obtained after one to two days (Scheme 6.86) [157].

Evidently, molecular rearrangements of D–A cyclopropanes are not limited to the reactions discussed above; some interesting transformations are not included, but some other ones were disclosed.

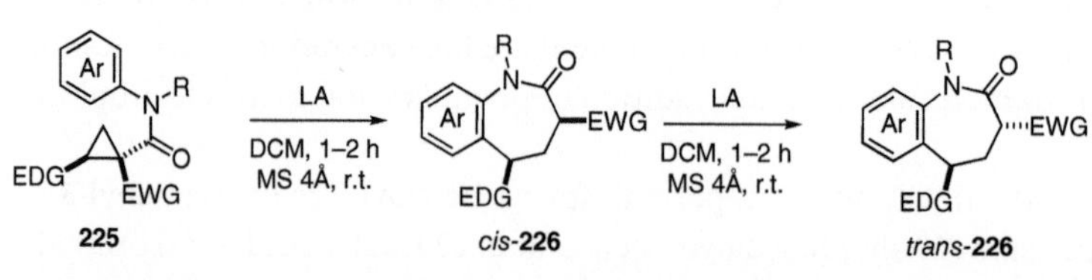

Scheme 6.86 *Endo*-*tet* cyclization of D–A cyclopropanes **225** yielding benzazepines **226**.

Acknowledgment

The preparation of this chapter was supported by the Russian Science Foundation (grant 21-13-00395).

References

1 Suzukamo, G., Fukao, M., and Tamura, M. (1984). *Tetrahedron Lett.* 25: 1595–1598.
2 Chmurny, A.B. and Cram, D.J. (1973). *J. Am. Chem. Soc.* 95: 4237–4244.
3 Li, C. and Vasella, A. (1993). *Helv. Chim. Acta* 76: 197–210.
4 Howe, N.E., Yankee, E.W., and Cram, D.J. (1973). *J. Am. Chem. Soc.* 95: 4230–4237.
5 Yankee, E.W., Badea, F.D., Howe, N.E., and Cram, D.J. (1973). *J. Am. Chem. Soc.* 95: 4210–4219.
6 Elling, G.R., Hahn, R.C., and Schwab, G. (1973). *J. Am. Chem. Soc.* 95: 5659–5662.
7 Pohlhaus, P.D., Sanders, S.D., Parsons, A.T. et al. (2008). *J. Am. Chem. Soc.* 130: 8642–8650.
8 Campbell, M.J., Johnson, J.S., Parsons, A.T. et al. (2010). *J. Org. Chem.* 75: 6317–6325.
9 Novikov, R.A. and Tomilov, Y.V. (2015). *Mendeleev Commun.* 25: 1–10.
10 Ivanova, O.A. and Trushkov, I.V. (2019). *Chem. Rec.* 19: 2189–2208.
11 Novikov, R.A., Korolev, V.A., Timofeev, V.P., and Tomilov, Y.V. (2011). *Tetrahedron Lett.* 52: 4996–4999.
12 Endo, T. and Sudo, A. (2015). Radical ring-opening polymerization: molecular designs, polymerization mechanisms, and living/controlled systems. In: *Controlled Radical Polymerization: Mechanism ACS Symposium Series*, vol. vol. 1187 (ed. K. Matyjaszewski, B.S. Sumerlin, N.V. Tsarevsky, and J. Chiefari), 19–50. Washington: Am. Chem. Soc. Chapter 2.
13 Chen, D.-F., Boyle, B.M., McCarthy, B.G. et al. (2019). *J. Am. Chem. Soc.* 141: 13268–13277.
14 Wu, S., Li, J.-F., Sun, X.-L. et al. (2020). *Polym. Chem.* 11: 5969–5973.
15 Schumacher, S., Pantawane, S., Gekle, S., and Agarwal, S. (2021). *Macromolecules* 54: 11–21.
16 Trautz, M. and Wikler, K. (1922). *J. Prakt. Chem.* 104: 53–79.
17 Corner, E.S. and Pease, R.N. (1945). *J. Am. Chem. Soc.* 67: 2067–2071.
18 Tranmer, G.K. and Capretta, A. (1998). *Tetrahedron* 54: 15499–15508.
19 Doyle, M.P. and van Leusen, D. (1981). *J. Am. Chem. Soc.* 103: 5917–5919.
20 Doyle, M.P. and van Leusen, D. (1982). *J. Org. Chem.* 47: 5326–5339.
21 Reissig, H.-U. (1985). *Tetrahedron Lett.* 26: 3943–3946.
22 Inokuchi, T., Takagishi, S., Akahoshi, F., and Torii, S. (1987). *Chem. Lett.* 16: 1553–1556.
23 Kulinkovich, O.G., Tishchenko, I.G., and Roslik, N.A. (1982). *Synthesis* 1982: 931–933.
24 Sonnleitner, C.M., Park, S., Eckl, R. et al. (2020). *Angew. Chem. Int. Ed.* 59: 18110–18115.

25 Delgado-Rebollo, M., Prieto, A., and Perez, P.J. (2014). *ChemCatChem* 6: 2047–2052.
26 Dutta, P.K., Chauhan, J., Ravva, M.K., and Sen, S. (2019). *Org. Lett.* 21: 2025–2028.
27 Fuerst, D.E., Stoltz, B.M., and Wood, J.L. (2000). *Org. Lett.* 2: 3521–3523.
28 Schmitz, E. and Sonnenschein, H. (1987). *Z. Chem.* 27: 171.
29 Chagarovskiy, A.O., Ivanova, O.A., Rakhmankulov, E.R. et al. (2010). *Adv. Synth. Catal.* 352: 3179–3184.
30 Alonso, M.E., Chitty, A.W., Pekerar, S., and Borgo, M.d.L. (1984). *J. Chem. Soc. Chem. Commun.* 1542–1543.
31 Novikov, R.A., Tarasova, A.V., and Tomilov, V.V. (2016). *Synlett* 27: 1367–1370.
32 Borisov, D.D., Chermashentsev, G.R., Novikov, R.A., and Tomilov, Y.V. (2021). *Synthesis* 53: 2253–2259.
33 Denisov, D.A., Novikov, R.A., Potapov, K.V. et al. (2016). *ChemistrySelect* 1: 6374–6381.
34 Ivanova, O.A., Chagarovskiy, A.O., Shumsky, A.N. et al. (2018). *J. Org. Chem.* 83: 543–560.
35 Thangamani, M. and Srinivasan, K. (2018). *J. Org. Chem.* 83: 571–577.
36 Richald, M., Delbrassinne, A., and Robiette, R. (2019). *Eur. J. Org. Chem.* 2019: 3779–3782.
37 Delbrassinne, A., Richald, M., Janssens, J., and Robiette, R. (2021). *Eur. J. Org. Chem.* 2021: 2862–2868.
38 Potapov, K.V., Denisov, D.A., Glushkova, V.V. et al. (2020). *J. Org. Chem.* 85: 15562–15576.
39 Yu, K., Yang, Z.-N., Liu, C.-H. et al. (2019). *Angew. Chem. Int. Ed.* 58: 8556–8560.
40 Neureiter, N.P. (1959). *J. Org. Chem.* 24: 2044–2046.
41 Mil'vitskaya, E.M., Tarakanova, A.V., and Plate, A.F. (1976). *Russ. Chem. Rev.* 45: 469–478.
42 Hudlicky, T., Kutchan, T.M., and Naqvi, S.M. (1985). *Org. React.* 33: 247–335.
43 Goldschmidt, Z. and Crammer, B. (1988). *Chem. Soc. Rev.* 17: 229–267.
44 Sonawane, H.R., Bellur, N.S., Kulkarni, D.G., and Ahuja, J.R. (1993). *Synlett* 1993: 875–884.
45 Baldwin, J.E. (2003). *Chem. Rev.* 103: 1197–1212.
46 Wang, S.C. and Tantillo, D.J. (2006). *J. Organomet. Chem.* 691: 4386–4392.
47 Hudlicky, T. and Reed, J.W. (2010). *Angew. Chem. Int. Ed.* 49: 4864–4876.
48 Wellington, C.A. (1962). *J. Phys. Chem.* 66: 1671–1674.
49 Simpson, J.M. and Richey, H.G. (1973). *Tetrahedron Lett.* 14: 2545–2548.
50 Richey, H.G. and Schull, D.W. (1976). *Tetrahedron Lett.* 17: 575–576.
51 Kassam, K., Venneri, P.C., and Warkentin, J. (1997). *Can. J. Chem.* 75: 1256–1263.
52 Buchert, M. and Reissig, H.-U. (1996). *Liebigs Ann. Chem.* 1996: 2007–2013.
53 Hay, E.B., Zhang, H., and Curran, D.P. (2015). *J. Am. Chem. Soc.* 137: 322–327.
54 Morizawa, Y., Oshima, K., and Nozaki, H. (1982). *Tetrahedron Lett.* 23: 2871–2874.
55 Morizawa, Y., Oshima, K., and Nozaki, H. (1984). *Isr. J. Chem.* 24: 149–152.
56 Miura, K., Fugami, K., Oshima, K., and Utimoto, K. (1988). *Tetrahedron Lett.* 29: 1543–1546.
57 Zens, A., Bauer, F., Kolb, B. et al. (2019). *Eur. J. Org. Chem.* 2019: 6285–6295.

58 Hashimoto, S., Shinoda, T., and Ikegami, S. (1986). *Tetrahedron Lett.* 27: 2885–2888.

59 Hudlicky, T., Koszyk, F.J., Kutchan, T.M., and Sheth, J.P. (1980). *J. Org. Chem.* 45: 5020–5027.

60 Hudlicky, T., Koszyk, F.J., Dochwat, D.M., and Cantrell, G.L. (1981). *J. Org. Chem.* 46: 2911–2915.

61 Hudlicky, T., Govindan, S.V., and Frazier, J.O. (1985). *J. Org. Chem.* 50: 4166–4171.

62 Coscia, R.W. and Lambert, T.H. (2009). *J. Am. Chem. Soc.* 131: 2496–2498.

63 Satyanarayana, J., Rao, M.V.B., Ila, H., and Junjappa, H. (1996). *Tetrahedron Lett.* 37: 3565–3568.

64 Davies, H.M.L., Kong, N., and Churchill, M.R. (1998). *J. Org. Chem.* 63: 6586–6589.

65 Wu, J., Becerril, J., Lian, Y. et al. (2011). *Angew. Chem. Int. Ed.* 50: 5938–5942.

66 Cloke, J.B. (1929). *J. Am. Chem. Soc.* 51: 1174–1187.

67 Wilson, C.L. (1947). *J. Am. Chem. Soc.* 69: 3002–3004.

68 McGreer, D.E. and McKinley, J.W. (1973). *Can. J. Chem.* 51: 1487–1493.

69 Dauben, W.G. and Shaffer, G.W. (1969). *J. Org. Chem.* 34: 2301–2306.

70 Alonso, M.E. and Morales, A. (1980). *J. Org. Chem.* 45: 4530–4532.

71 Nambu, H., Onuki, Y., Ono, N., and Yakura, T. (2018). *Adv. Synth. Catal.* 360: 2938–2944.

72 Nambu, H., Ono, N., and Yakura, T. (2016). *Synthesis* 48: 1892–1901.

73 Nambu, H., Fukumoto, M., Hirota, W. et al. (2015). *Tetrahedron Lett.* 56: 4312–4315.

74 Bosnidou, A.-E., Kalpogiannaki, D., Karanestora, S. et al. (2015). *J. Org. Chem.* 80: 1279–1283.

75 Kaschel, J., Schneider, T.F., Schirmer, P. et al. (2013). *Eur. J. Org. Chem.* 2013: 4539–4551.

76 Schneider, T.F., Kaschel, J., Dittrich, B., and Werz, D.B. (2009). *Org. Lett.* 11: 2317–2320.

77 Schneider, T.F., Kaschel, J., Awan, S.I. et al. (2010). *Chem. Eur. J.* 16: 11276–11288.

78 Ortega, A., Manzano, R., Uria, U. et al. (2018). *Angew. Chem. Int. Ed.* 57: 8225–8229.

79 Yadav, V.K. and Balamurugan, R. (2001). *Org. Lett.* 3: 2717–2719.

80 Zhu, Y., Zhao, S., Zhang, M. et al. (2019). *Synthesis* 51: 899–906.

81 Zhu, Y., Zhang, M., Li, T., and Song, X. (2019). *ChemistrySelect* 4: 10838–10842.

82 Wang, Y., Han, J., Chen, J., and Cao, W. (2016). *Chem. Commun.* 52: 6817–6820.

83 Xu, X., Wang, Y., Cui, X. et al. (2017). *Chem. Sci.* 8: 4347–4351.

84 Klein, J.E.M.N., Knizia, G., Miehlich, B. et al. (2014). *Chem. Eur. J.* 20: 7254–7257.

85 Lin, C.-H., Pursley, D., Klein, J.E.M.N. et al. (2015). *Chem. Sci.* 6: 7034–7043.

86 Zhang, J., Tang, Y., Wei, W. et al. (2017). *Org. Lett.* 19: 3043–3046.

87 Dzhemilev, U.M., Khusnutdinov, R.I., Shchadneva, N.A., and Tolstikov, G.A. (1990). *Russ. Chem. Bull.* 39: 1447–1450.

88 Bowman, R.K. and Johnson, J.S. (2006). *Org. Lett.* 8: 573–576.

89 Barluenga, J., Tomas, M., Lopez-Pelegrin, J.A., and Rubio, E. (1995). *J. Chem. Soc. Chem. Commun.* 665–666.

90 Bailey, T.S., Bremner, J.B., Pelosi, L. et al. (1995). *Aust. J. Chem.* 48: 1437–1445.

91 Kim, H., Kim, S., Kim, J. et al. (2017). *Org. Lett.* 19: 5677–5680.
92 Dean, F.M. and Johnson, R.S. (1980). *J. Chem. Soc. Perkin Trans. 1*: 2049–2053.
93 O'Bannon, P.E. and Dailey, W.P. (1990). *Tetrahedron* 46: 7341–7358.
94 Lifchits, O. and Charette, A.B. (2008). *Org. Lett.* 10: 2809–2812.
95 Varshnaya, R.K., Singh, P., Kaur, N., and Banerjee, P. (2021). *Org. Chem. Front.* 8: 1267–1274.
96 Bianchi, L., Ghelfi, F., Giorgi, G. et al. (2013). *Eur. J. Org. Chem.* 2013: 6298–6309.
97 Volkova, Y.A., Ivanova, O.A., Budynina, E.M. et al. (2009). *Tetrahedron Lett.* 50: 2793–2796.
98 Schneider, T.F. and Werz, D.B. (2011). *Org. Lett.* 13: 1848–1851.
99 Cavitt, M.A., Phun, L.H., and France, S. (2014). *Chem. Soc. Rev.* 43: 804–818.
100 Di Grandi, M.J. (2014). *Org. Biomol. Chem.* 12: 5331–5345.
101 Martin, M.C., Shenje, R., and France, S. (2016). *Isr. J. Chem.* 56: 499–511.
102 Murphy, W.S. and Wattanasin, S. (1980). *Tetrahedron Lett.* 21: 1887–1890.
103 Murphy, W.S. and Wattanasin, S. (1980). *Tetrahedron Lett.* 21: 3517–3520.
104 Murphy, W.S. and Wattanasin, S. (1981). *J. Chem. Soc. Perkin Trans. 1*: 2920–2926.
105 Murphy, W.S. and Wattanasin, S. (1982). *J. Chem. Soc. Perkin Trans. 1*: 1029–1035.
106 Murphy, W.S. and Wattanasin, S. (1982). *J. Chem. Soc. Perkin Trans. 1*: 271–276.
107 Takada, S., Takaki, N., Yamada, K., and Nishii, Y. (2017). *Org. Biomol. Chem.* 15: 2443–2449.
108 Greiner-Becher, L., Sprang, T., and Otto, H.-H. (2005). *Monatsch. Chem.* 136: 635–653.
109 Yadav, V.K. and Kumar, N.V. (2008). *Chem. Commun.* 3774–3776.
110 Phun, L.H., Patil, D.V., Cavitt, M.A., and France, S. (2011). *Org. Lett.* 13: 1952–1955.
111 Phun, L.H., Patil, D.V., Cavitt, M.A., and France, S. (2012). *Org. Lett.* 14: 6379–6380.
112 De Simone, F., Gertsch, J., and Waser, J. (2010). *Angew. Chem. Int. Ed.* 49: 5767–5770.
113 De Simone, F., Saget, T., Benfatti, F. et al. (2011). *Chem. Eur. J.* 17: 14527–14538.
114 Frei, R., Staedler, D., Raja, A. et al. (2013). *Angew. Chem. Int. Ed.* 52: 13373–13376.
115 Patil, D.V., Cavitt, M.A., Grzybowski, P., and France, S. (2011). *Chem. Commun.* 47: 10278–10280.
116 Patil, D.V., Cavitt, M.A., and France, S. (2012). *Heterocycles* 84: 1363–1373.
117 Cavitt, M.A. and France, S. (2016). *Synthesis* 48: 1910–1919.
118 Aponte-Guzman, J., Shenje, R., Huang, Y. et al. (2015). *Ind. Eng. Chem. Res.* 54: 9550–9558.
119 Tsuge, O., Kanemasa, S., Otsuka, T., and Suzuki, T. (1988). *Bull. Chem. Soc. Jpn.* 61: 2897–2908.
120 De Simone, F., Andres, J., Torosantucci, R., and Waser, J. (2009). *Org. Lett.* 11: 1023–1026.
121 Patil, D.V., Phun, L.H., and France, S. (2010). *Org. Lett.* 12: 5684–5687.
122 Sudhakar, G., Mahesh, S.K., Vemulapalli, S.P.B., and Nanubolu, J.B. (2017). *Org. Lett.* 19: 4500–4503.
123 Sathishkannan, G. and Srinivasan, K. (2014). *Adv. Synth. Catal.* 356: 729–735.

124 Vogel, E. (1960). *Angew. Chem.* 72: 4–26.
125 Hudlicky, T., Fan, R., Reed, J.W., and Gadamasetti, K. (1992). *Org. React.* 41: 1–133.
126 Krüger, S. and Gaich, T. (2014). *Beilstein J. Org. Chem.* 10: 163–193.
127 Franck-Neumann, M. and Buchecker, C. (1970). *Angew. Chem. Int. Ed.* 9: 526–527.
128 Schwarzer, D.D., Gritsch, P.J., and Gaich, T. (2012). *Angew. Chem. Int. Ed.* 51: 11514–11516.
129 Maas, G. (1979). *Chem. Ber.* 112: 3241–3272.
130 Wu, J., Tang, Y., Wei, W. et al. (2018). *Angew. Chem. Int. Ed.* 57: 6284–6288.
131 Wang, B., Xu, B., Xun, W. et al. (2021). *Angew. Chem. Int. Ed.* 60: 9439–9443.
132 Rhoads, S.J. and Cockroft, R.D. (1969). *J. Am. Chem. Soc.* 91: 2815–2816.
133 Rey, M. and Dreiding, A.S. (1965). *Helv. Chim. Acta* 48: 1985–1987.
134 Hofmann, B. and Reissig, H.-U. (1994). *Chem. Ber.* 127: 2327–2335.
135 Downing, W., Latouche, R., Pittol, C.A. et al. (1990). *J. Chem. Soc. Perkin Trans. 1*: 2613–2615.
136 Mahon, M.F., Molloy, K., Pittol, C.A. et al. (1991). *J. Chem. Soc. Perkin Trans. 1*: 1255–1263.
137 Boeckman, R.K., Shair, M.D., Vargas, J.R., and Stoltz, L.A. (1993). *J. Org. Chem.* 58: 1295–1297.
138 Zhang, W., Baudouin, E., Cordier, M. et al. (2019). *Chem. Eur. J.* 25: 8643–8648.
139 Nasveschuk, C.G. and Rovis, T. (2005). *Angew. Chem. Int. Ed.* 44: 3264–3267.
140 Allegre, K. and Tunge, J. (2019). *Tetrahedron* 75: 3319–3329.
141 Zhang, W. and Nay, B. (2020). *Eur. J. Org. Chem.* 2020: 3517–3525.
142 Hofmann, B. and Reissig, H.-U. (1993). *Synlett* 1993: 27–29.
143 Yamaguchi, S., Arisawa, A., Katoh, N. et al. (1997). *Bull. Chem. Soc. Jpn.* 70: 2215–2219.
144 Piotrowski, M.L. and Kerr, M.A. (2018). *Org. Lett.* 20: 7624–7627.
145 Sperling, D., Reissig, H.-U., and Fabian, J. (1997). *Liebigs Ann.* 1997: 2443–2449.
146 Sperling, D., Reissig, H.-U., and Fabian, J. (1999). *Eur. J. Org. Chem.* 1999: 1107–1114.
147 Wu, Y., Li, M., Jin, L., and Zhao, X. (2020). *ACS Omega* 5: 2957–2966.
148 Shang, H., Wang, Y., Tian, Y. et al. (2014). *Angew. Chem. Int. Ed.* 53: 5662–5666.
149 Bourelle-Wargnier, F., Vincent, M., and Chuche, J. (1979). *J. Chem. Soc. Chem. Commun.* 584–585.
150 Bourelle-Wargnier, F., Vincent, M., and Chuche, J. (1980). *J. Org. Chem.* 45: 428–435.
151 Jackson, S.K., Karadeolian, A., Driega, A.B., and Kerr, M.A. (2008). *J. Am. Chem. Soc.* 130: 4196–4201.
152 Wang, D., Zhao, J., Chen, J. et al. (2019). *Asian J. Org. Chem.* 8: 2032–2036.
153 Tran, T.Q., Savinkov, R.S., Diev, V.V. et al. (2013). *Tetrahedron* 69: 5173–5177.
154 Ivanova, O.A., Andronov, V.A., Vasin, V.S. et al. (2018). *Org. Lett.* 20: 7947–7952.
155 Zheng, G., Zhou, Z., Zhu, G. et al. (2020). *Angew. Chem. Int. Ed.* 59: 2890–2896.
156 Wu, M., Lin, S., Zhu, G. et al. (2020). *Org. Lett.* 22: 1295–1300.
157 Vartanova, A.E., Plodukhin, A.Y., Ratmanova, N.K. et al. (2021). *J. Am. Chem. Soc.* 143: 13952–13961.

7

Donor–Acceptor Cyclopropanes with an Amino Group as Donor

Ming-Ming Wang and Jerome Waser

Laboratory of Catalysis and Organic Synthesis, Institute of Chemical Sciences and Engineering, Ecole Polytechnique Fédérale de Lausanne, CH-1015, Lausanne, Switzerland

CHAPTER MENU

7.1 Introduction

The chemistry of donor–acceptor (DA)-substituted cyclopropanes (DA cyclopropanes) [1-4] has underwent an impressive renaissance in the last two decades because of their versatility either as building blocks in organic synthesis [5-14] or as substrates for asymmetric catalysis [15-18]. As a subclass of DA cyclopropanes, DA cyclopropanes with an amino group as donor (DA aminocyclopropanes) share many reactivity similarities with aryl- and alkoxy-substituted DA cyclopropanes. In addition, the importance of the nitrogen atom in biomolecules and pharmaceutical compounds makes aminocyclopropanes (acceptor-substituted or not) attractive starting materials for accessing more value-added products [19-21]. For example, with the formal [3 + 2] cycloaddition methods described in this chapter, cyclopentanes with an amino group can be synthesized. Such structures can be found in many bioactive compounds such as the anti-HIV drug Abacavir (**1**), and Ramipril (**2**), which is used to treat high blood pressure, heart failure, and diabetic kidney disease (Figure 7.1). Natural products can also be accessed by using DA aminocyclopropanes as key

Donor–Acceptor Cyclopropanes in Organic Synthesis, First Edition.
Edited by Prabal Banerjee and Akkattu T. Biju.

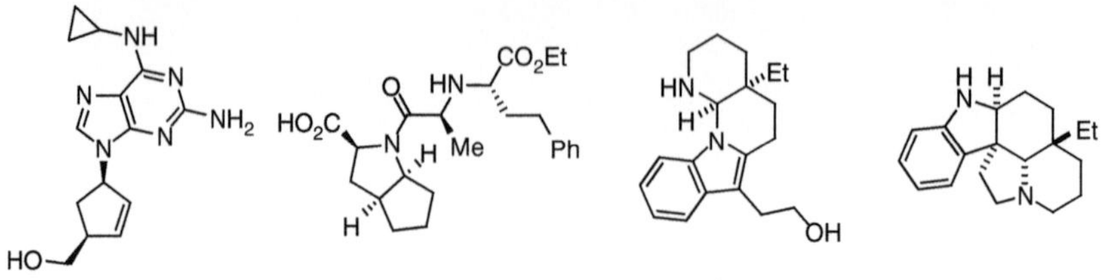

Figure 7.1 Examples of nitrogen-containing synthetic bioactive compounds and natural products.

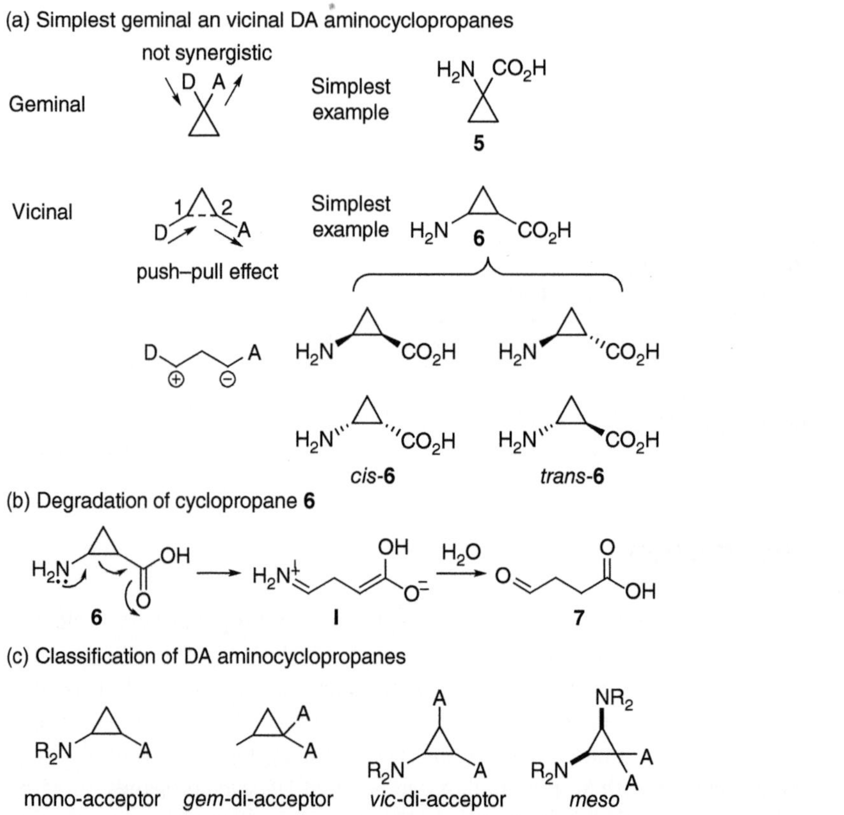

Scheme 7.1 Classification and reactivity of DA aminocyclopropanes.

intermediates in cyclization and annulation reactions, such as the alkaloids Goniomitine (**3**) and Aspidospermidine (**4**) [22].

In general, DA aminocyclopropanes can be divided into two types: geminal DA aminocyclopropanes and vicinal DA aminocyclopropanes (Scheme 7.1a) [1]. The former, such as acid **5**, lack synergistic activation for the C—C bond of the cyclopropyl ring and are of importance mostly as rigid nonnatural amino acid cores [23, 24], so they will not be covered in this chapter. Vicinal DA aminocyclopropanes can be considered

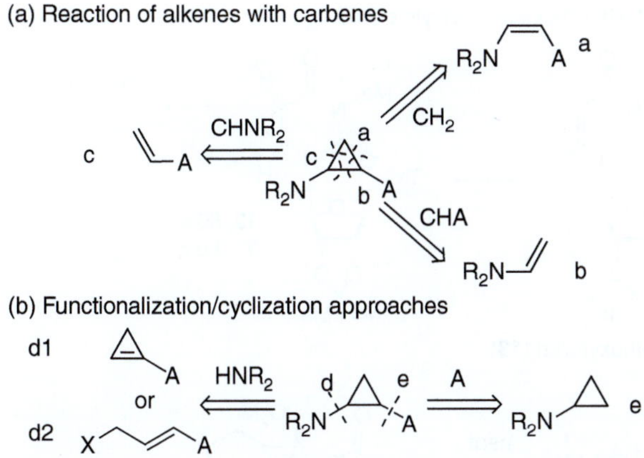

Scheme 7.2 Retro-synthetic analysis of DA aminocyclopropanes.

as 1,3-dipolar synthons because the C1—C2 bond is activated via an electronic push–pull effect. 2-Amino cyclopropanecarboxylic acid (**6**) [25], the simplest example of vicinal DA aminocyclopropanes, has two stereogenic centers and therefore four isomeric forms [26]. The high reactivity of vicinal DA aminocyclopropanes can compromise their stability. For example, **6** is prone to undergo ring opening, leading to aldehyde **7** in the presence of water via imine **I** (Scheme 7.1b) [27]. Therefore, protection of the nitrogen atom to reduce its electron density is necessary in order to have a good balance between reactivity and stability. Common protecting groups include succinimide, phthalimide, an acyl group, carbamates, sulfonamides, imines, and nitrogen-containing heterocycles. The acceptor is usually one or two electron-withdrawing groups, which can be an ester, an aldehyde/ketone, a cyanide, a nitro group, etc. Based on the number of donor/acceptor groups and their distribution on the cyclopropyl ring, DA aminocyclopropanes described in this chapter can be classified as mono-acceptor aminocyclopropanes, *gem*-di-acceptor aminocyclopropanes, *vic*-di-acceptor aminocyclopropanes, or *meso* aminocyclopropanes (Scheme 7.1c).

In this chapter, we will start by giving a brief summary of the synthetic methods to access DA aminocyclopropanes (Section 7.2). Their use as starting materials either in ring-opening reactions (Section 7.3) or in formal cycloadditions (Section 7.4) will then be described in detail.

7.2 Synthesis of DA Aminocyclopropanes

DA cyclopropanes are most commonly prepared by the addition of carbenes or their equivalents to alkenes, namely formal [2+1] cycloaddition or cyclopropanation (Scheme 7.2a). In this case, there are three different routes for constructing a cyclopropyl ring with the alkene partners being DA type alkenes such as β-dehydroamino acids (disconnection a), donor type alkene like enamines (disconnection b), or acceptor type alkenes such as acrylates (disconnection c), respectively [26]. It is also possible that all three carbon atoms come from one reaction partner, in which case

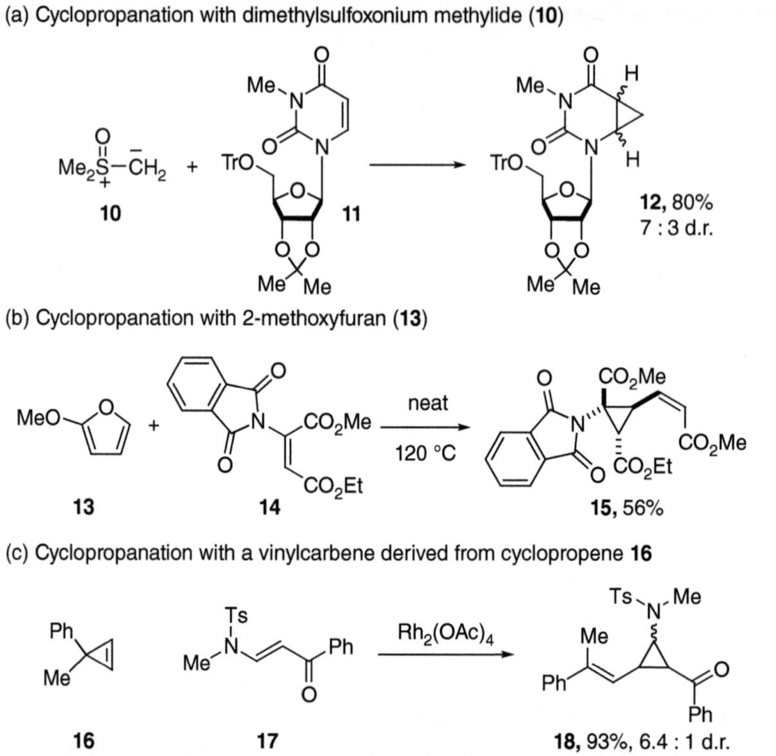

Scheme 7.3 Cyclopropanation from β-dehydroamino acid with different reagents.

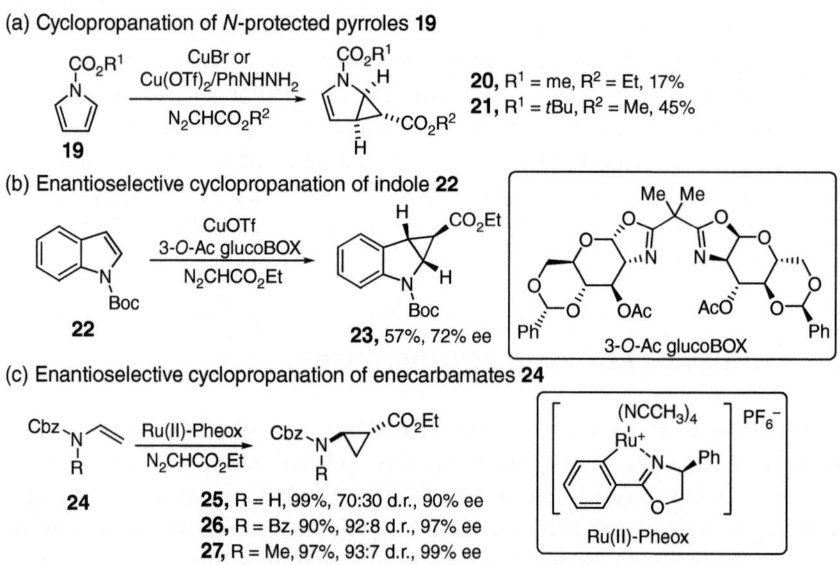

Scheme 7.4 Diastereoselective and enantioselective cyclopropanation of heterocycles and enamides.

there are two disconnections for introducing either an amino group (disconnection d) or an acceptor (disconnection e) (Scheme 7.2b). For the introduction of an amino group, there are two reported routes depending on the substrate: a cyclopropene bearing an acceptor group (disconnection d1) or 2-haloethylidene malonates (disconnection d2). Therefore, based on the starting materials, there are currently six straightforward routes for the synthesis of DA aminocyclopropanes. Each route will be briefly introduced below with selected examples.

7.2.1 Synthesis of DA Aminocyclopropanes from β-Dehydroamino Acids (Route A)

This cyclopropanation of DA-substituted olefins—β-dehydroamino acids—with a carbene gives DA aminocyclopropanes. The most convenient source of carbene equivalents is diazo compounds, which readily generate a carbenoid after reaction with a transition metal catalyst. Acceptor-substituted diazoalkanes were frequently used for the synthesis of DA cyclopropanes because they can be prepared easily and cyclopropanation with electron-rich alkenes is efficient. For example, *vic*-di-acceptor aminocyclopropane **9** can be synthesized by the copper-catalyzed cyclopropanation of **8** with diazoester in an intermolecular fashion (Eq. 7.1) [28]. Intramolecular cyclopropanation for this type of substrate has also been reported, with diazoalkane attached to the nitrogen atom to produce a bicyclic *vic*-di-acceptor aminocyclopropane [29].

*Copper-catalyzed cyclopropanation of alkene **8** to give vic-di-acceptor aminocyclopropane **9**.*

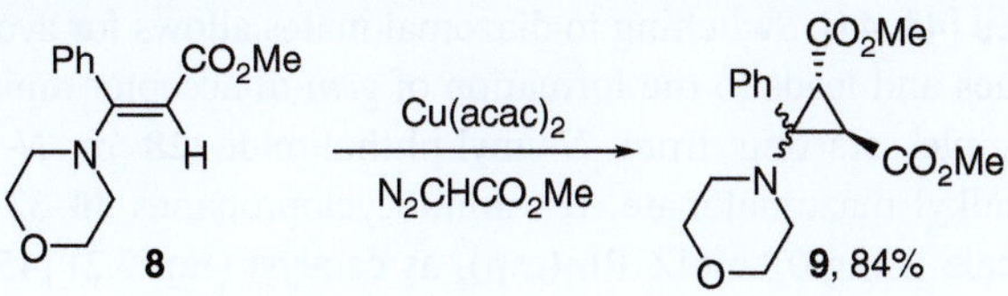

(7.1)

A Corey–Chaykovsky reaction with dimethylsulfoxonium methylide **10** can also be used to construct the cyclopropyl ring. For example, starting from protected uridine **11**, uridine analog **12** containing a cyclopropyl ring was isolated in 80% yield as a mixture of diastereoisomers in a ratio of 7 : 3 (Scheme 7.3a) [30]. 2-Methoxyfuran **13** has also been used for the cyclopropanation of electron-deficient olefins [31] such as alkene **14** [32], although no synthetic utility of the corresponding products has been disclosed so far (Scheme 7.3b). In 2019, the Loh group reported a cyclopropanation reaction of enaminone **17** with a vinylcarbene intermediate, which was generated *in situ* from cyclopropene **16** (Scheme 7.3c) [33]. The obtained product **18** was then subjected to hydrolysis under acidic conditions, and an α-vinyl aldehyde was formed after ring opening.

7.2.2 Synthesis of DA Aminocyclopropanes from Enamines (Route B)

Cyclopropanation of enamines with acceptor-substituted diazoalkanes is a straightforward way of synthesizing DA aminocyclopropanes. *N*-protected pyrroles **19** were versatile starting materials to react with diazoacetates for the synthesis of DA aminocyclopropanes **20–21**, and exclusive formation of the *exo* diastereoisomer was recorded with carbamate as protecting group on the nitrogen atom (Scheme 7.4a) [34, 35]. An asymmetric version of this reaction was reported by the Reiser group in 2003, with poor enantioselectivity obtained using a copper catalyst and a chiral bis(oxazoline) ligand [36]. In 2017, the Reiser group revisited this reaction, and good enantioselectivity was obtained by using a copper catalyst and an aza-BOX ligand [37]. In 2019, the Reiser group in collaboration with the Davies group further reported a Rh(II)-catalyzed cyclopropanation reaction of *N*-tosyl pyrroles [38] or cyclic enecarbamates [39], with monocyclopropanation products isolated in good yields and high enantioselectivity in both cases. In 2012, the Boysen group reported an enantioselective cyclopropanation of *N*-Boc indole **22** by using CuOTf as a catalyst and a carbohydrate-based ligand, 3-*O*-Ac glucoBOX, and the corresponding indole-derived DA aminocyclopropane **23** was obtained in good enantioselectivity (Scheme 7.4b) [40]. In 2017, an intramolecular enantioselective cyclopropanation of indoles was reported by the Zhu and Zhou groups [41]. Excellent yield and enantioselectivity were achieved by using copper or iron complexes with chiral spiro-BOX ligands as catalysts. The asymmetric cyclopropanation of acyclic vinylcarbamates **24** with diazoesters was also realized by the Iwasa group in 2013 by using a Ru(II)-Pheox catalyst, giving products such as **25–27** in excellent enantioselectivity (Scheme 7.4c) [42]. However, the diastereoselectivity of the reaction depended on the substrates, with higher diastereoselectivity observed for *N*,*N*-disubstituted vinylamines compared to *N*-monosubstituted vinylamines.

Indeed, achieving high diastereoselectivity can be difficult in the cyclopropanation of certain enamines [43, 44]. Switching to diazomalonates allows for avoiding diastereoselectivity issues and leads to the formation of *gem*-di-acceptor aminocyclopropanes. For example, starting from *N*-vinyl-phthalimide **28** or *N*-vinyl-succinimide **29** and dialkyl diazomalonate, DA aminocyclopropanes **30–32** were synthesized in gram scale using 0.1 mol% $Rh_2(esp)_2$ as catalyst (Eq. 7.2) [45, 46]. Using a phthalimide or a succinimide group, the electron density on the nitrogen atom is decreased, resulting in good stability for **30–32**.

*Synthesis of gem-di-acceptor aminocyclopropanes **30–32**.*

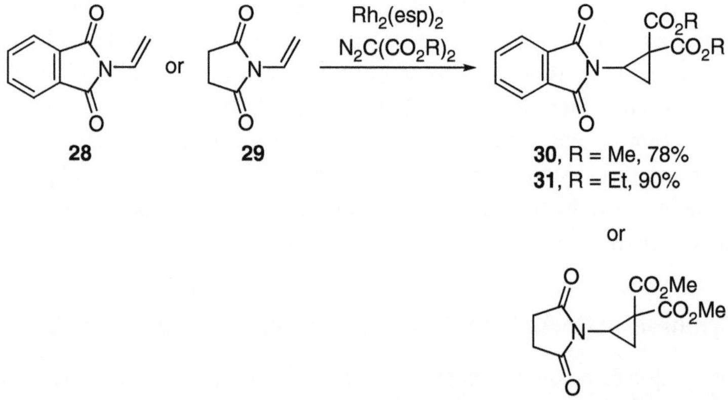

(7.2)

7.2.3 Synthesis of DA Aminocyclopropanes from Acrylates (Route C)

The synthesis of DA aminocyclopropanes starting from acrylates requires the use of aminocarbenes to assemble the cyclopropyl ring. As the donor-type diazo compound having an α-amino group does not exist [47], precedence is focused on Fischer carbenes. Amino-substituted Fischer carbenes can be easily prepared by replacing the alkoxy group of alkoxycarbene complexes with amines [48]. However, the reaction between most aminocarbene complexes and electron-deficient alkenes gave only byproducts, which arose from [2+2+1] cycloaddition [49] or ring-opening isomerization [50]. DA aminocyclopropane **35** was obtained only when using pyrrolocarbene complexes **33** or **34**, with pyrrole acting as a masked amino group (Scheme 7.5) [51]. Conversion of the pyrrole into other nitrogen-containing functionalities has been demonstrated by an ozonolysis-reduction sequence, which provided a formamide group in **36**. Other synthetic routes starting from simple acrylates involved several functional group interconversions, such as cyclopropanation followed by reduction of a nitro group [52] or Curtius rearrangement of a carboxylic group [53]. These examples of multi-step syntheses from simple acrylates are less efficient and will not be discussed.

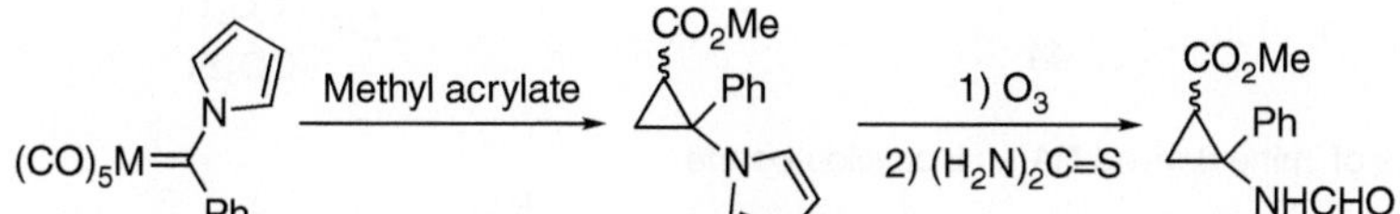

M = Mo (**33**) or W (**34**) **35**, 59–62% **36**, 73%

Scheme 7.5 Synthesis of cyclopropanes **35** and **36** from Fischer carbenes **33** or **34**.

7.2.4 Synthesis of DA Aminocyclopropanes from Cyclopropene (Route D1)

In 1988, it was reported that 1,4-addition of amines such as morpholine to cyclopropene carboxylates **37** occurred readily under mild conditions to form DA aminocyclopropanes such as **38–40** (Scheme 7.6a) [54]. Cyclopropene **I** can also be formed *in situ* from bromocyclopropane **41** after 1,2-elimination, and the addition of *N*-methylacetamide furnishes DA aminocyclopropane **42** in good yield and high diastereoselectivity, as reported by the Rubin group in 2013 (Scheme 7.6b) [55]. The scope of amides was limited to secondary amides, while products from primary amides were not stable, leading to the formation of aldehydes after ring opening and hydrolysis. Besides, nitrogen-containing heterocycles, such as pyrrole, indole, pyrazole, and anilines, were well tolerated in this formal substitution reaction.

7.2.5 Synthesis of DA Aminocyclopropanes from 2-Haloethylidene Malonates (Route D2)

2-Haloethylidene malonates could be used for a cyclopropanation reaction via an addition-substitution mechanism under basic conditions with amine nucleophiles.

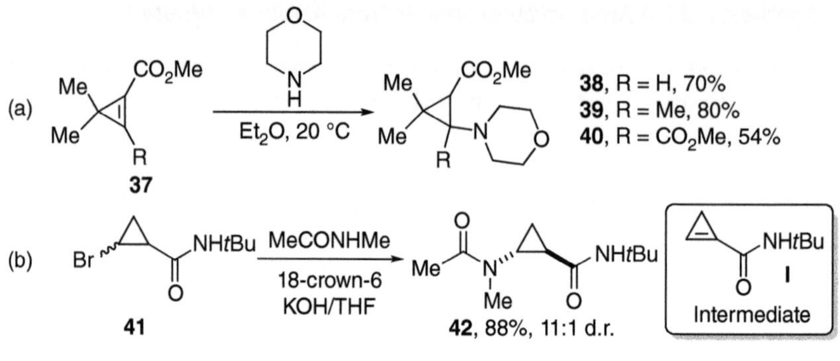

Scheme 7.6 Addition of amines to (*in situ* formed) cyclopropenes.

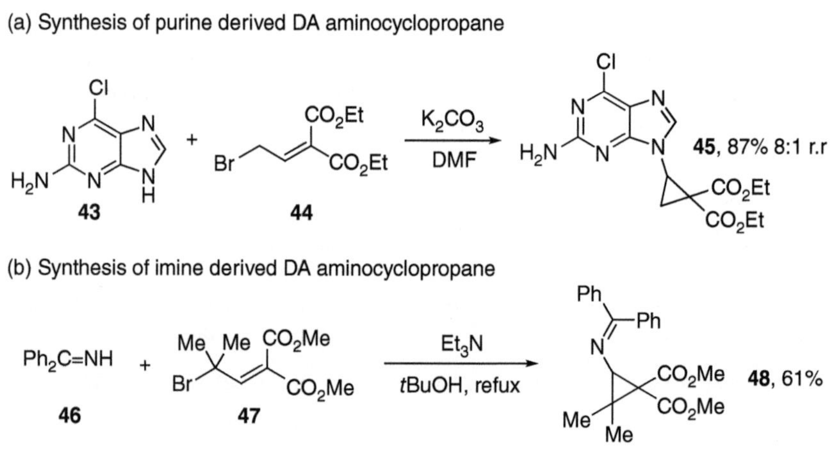

Scheme 7.7 Synthesis of DA aminocyclopropanes from 2-bromoethylidene malonates.

An elegant synthesis of purine-derived DA aminocyclopropane **45** was reported by the Geen group in 1992 by mixing 2-amino-6-chloropurine **43** with 2-bromoethylidene malonate **44** under basic conditions (Scheme 7.7a) [56]. Similarly, treating diphenylmethylidenamine **46** with 2-bromoethylidene malonate **47** in the presence of triethylamine provided **48** as a different class of DA aminocyclopropanes, as reported by the De Kimpe group in 2005 (Scheme 7.7b) [57].

7.2.6 Synthesis of DA Aminocyclopropanes from Cyclopropylamines (Route E)

Lithiation at the β position of cyclopropylcarbamate **50**, which was synthesized via an α lithiation cyclization sequence form **49**, has been observed, and further treatment with dimethyl carbonate for introducing an ester group as acceptor produced DA aminocyclopropane **51**, as reported by the Beak group in 1996 (Scheme 7.8) [58]. In contrast to other routes discussed above, *cis* DA aminocyclopropanes were

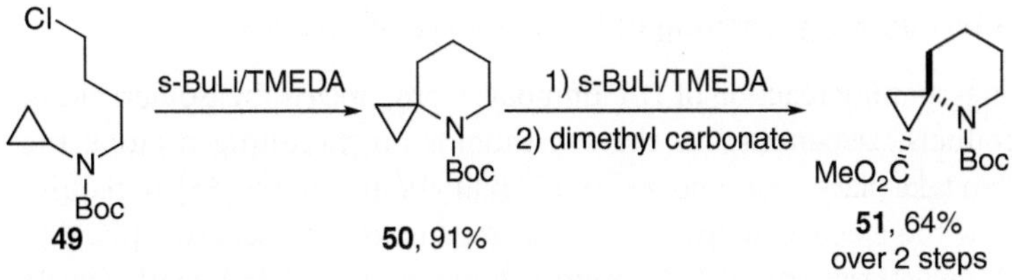

Scheme 7.8 Synthesis of DA aminocyclopropane **51** by two subsequent lithiations-alkylations from **49**.

obtained rather than *trans* products. The deprotonation is directed by the carbamate group leading to *cis* stereospecific β-lithiation.

To summarize this section, synthetic routes from six types of substrates have been well established for the efficient synthesis of DA aminocyclopropanes. Nevertheless, it would be still attractive to develop alternative synthetic routes in the future starting from other commercially available compounds, especially when compared with the more mature synthesis of aryl-substituted DA cyclopropanes [59]. For example, DA aminocyclopropanes could be accessed if a C(sp^3)–H amination reaction can be developed by using electrophilic aminating reagent [60], inspired by the recently reported C(sp^3)–H arylation of cyclopropyl carboxylic acids for accessing aryl-substituted DA cyclopropanes [61, 62]. Alternatively, the synthesis of DA aminocyclopropanes by aminocarbonylation of cyclopropenes might also be possible, based on a recent example of palladium-catalyzed intermolecular aminocarbonylation of alkenes [63].

7.3 Ring-Opening Reactions of DA Aminocyclopropanes

Similar to other types of DA cyclopropanes, heterolytic cleavage of the C1—C2 bond is a facile process, leading to the formation of 1,3-formal dipoles. Thermal epimerization of compound **52** indicated that the ring-opening step can be reversible [64] (Eq. 7.3).

*Thermal epimerization of DA aminocyclopropane **52**.*

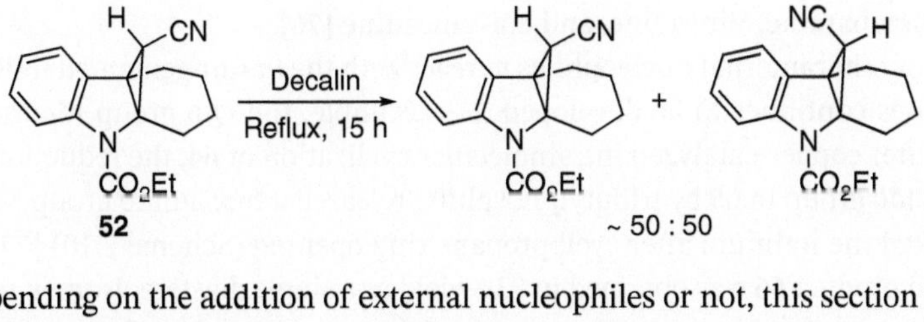

(7.3)

Depending on the addition of external nucleophiles or not, this section is divided into two parts describing the ring-opening reactions of DA aminocyclopropanes in an intramolecular fashion or an intermolecular fashion, respectively.

7.3.1 Intramolecular Ring-Opening of DA Aminocyclopropanes

The simplest ring-opening reaction of DA aminocyclopropanes is the isomerization of enamine products. Depending on the substituents on the nitrogen atom, the isomerization can take place spontaneously with dialkyl substituents [65], or require some activation if the electron density on the nitrogen atom is decreased [66, 67]. For example, the isomerization of DA aminocyclopropane **53** to indomethacin **54** was reported to occur efficiently at 160 °C (Eq. 7.4) [66].

*Thermal rearrangement of **53** into indomethacin **54**.*

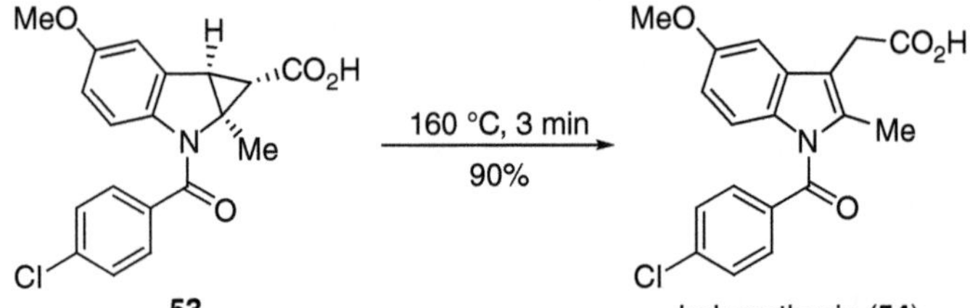

(7.4)

Although some ring-opening reactions took place under mild basic conditions, acidic conditions were more frequently used for the activation of DA aminocyclopropanes. In some cases of DA aminocyclopropane synthesis, the desired product was accompanied by the formation of ring-opened byproducts because of the presence of the catalyst, which could also act as a Lewis acid to activate the DA aminocyclopropanes [68, 69]. Apart from isomerization into enamine, the iminium species after ring opening can also undergo deprotonation to form an imine product [40] or can be trapped by nucleophiles. Nucleophiles can be internal functional groups, such as the oxygen atom of a carboxylic acid (**55**, Scheme 7.9a) [70]/ester (**56**, Scheme 7.9b) [71, 72]/carbonyl (**I** generated *in situ* from **57** via oxidation, Scheme 7.9c) [73, 74] group, or the nitrogen atom of cyanide (**58**, Scheme 7.9d) [75] group, leading to the formation of five-membered heterocycles **59–63**.

It is important to note that these ring-opening reactions can be potentially used in the synthesis of complex molecules. For example, based on the reaction illustrated in Scheme 7.9b, the Wenkert group reported first in 1978 the total synthesis of the natural products Eburnamonine and Aspidospermidine [71, 72]. By extending this reaction to the oxygen analog, the Wenkert group reported in 1981 the total synthesis of Quebrachamine, Vincadine, and epi-Vincadine [76].

The use of other internal nucleophiles to react with the *in situ*-generated iminium intermediates continues to be developed. For example, the Qin group reported in 2006 that after copper-catalyzed intramolecular cyclization of **64**, the reduction of a pendant azido group in **65** by tributylphosphine released a free amino group, which then attacked the iminium after cyclopropane ring opening (Scheme 7.10) [77, 78]. The desired product **66** was obtained in 83% yield as a single diastereoisomer, which was a key intermediate in the total synthesis of the natural product Communesin F (**67**). This application showcased the potential of DA aminocyclopropanes for the synthesis of complex molecules. Since 2020, this strategy has also been utilized by the Sen group for the synthesis of spirocyclic indole derivatives [79, 80].

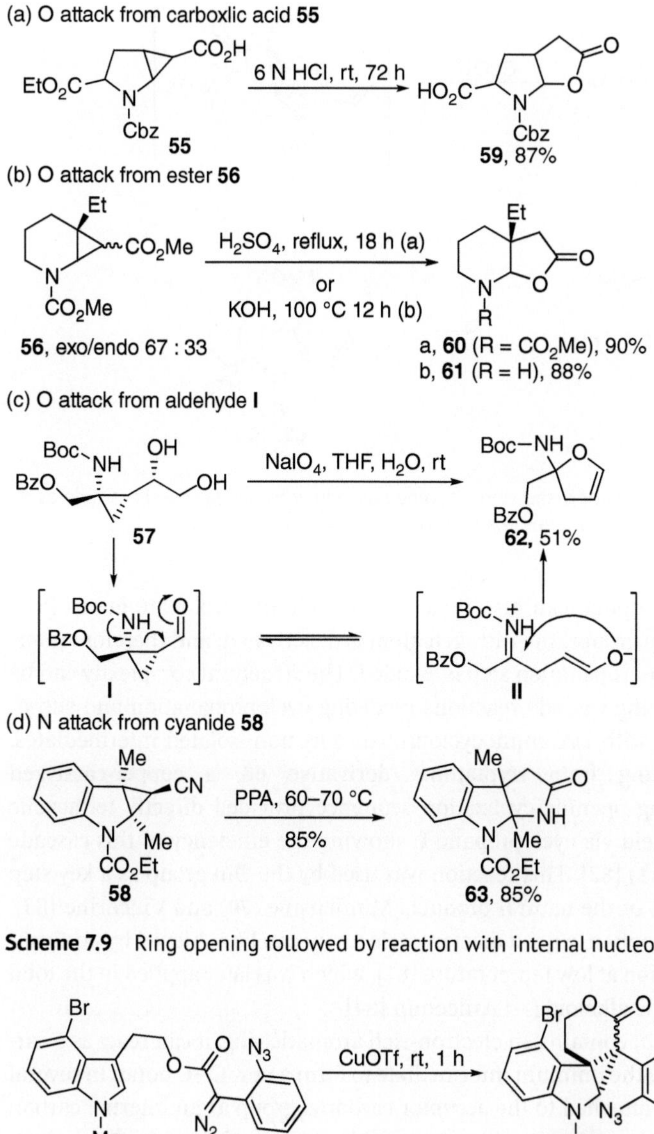

Scheme 7.9 Ring opening followed by reaction with internal nucleophiles.

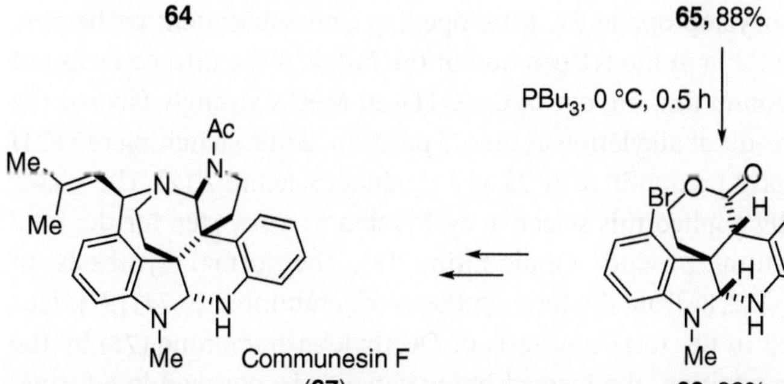

Scheme 7.10 Ring-opening cyclization with azide as a masked amino group for the synthesis of Communesin F (**67**).

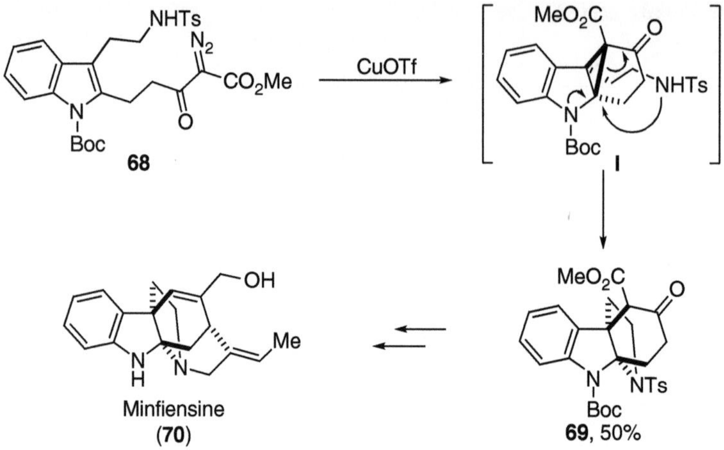

Scheme 7.11 Cyclopropanation and ring-opening cascade with pendant sulfonamide as a nucleophile for the synthesis of Minfiensine (**70**).

The pendant nucleophile can also be a sulfonamide or carbamate group [81]. With these protecting groups, the nitrogen atom is deactivated, and therefore interference with the cyclopropanation step is avoided. The structural complexity can be increased by developing cascade reactions involving cyclopropanation and subsequent ring opening, with DA aminocyclopropanes as non-isolated intermediates. For example, starting from tryptamine derivative **68**, a copper-catalyzed cyclopropanation-ring opening-cyclization sequence provided directly tetracyclic product **69** in 50% yield via cyclopropane **I**, showing the efficiency of this cascade reaction (Scheme 7.11) [82]. This reaction was used by the Qin group as a key step in the total synthesis of the natural products Minfiensine (**70**) and Vincorine [83]. Starting from L-tryptophan, a good diastereoselectivity can be achieved by performing the cascade reaction at low temperature [81], which was later applied in the total synthesis of the indole alkaloid (−)-Ardeemin [84].

When the side chain contains an electron-rich aromatic ring, it can react as a carbon nucleophile with the iminium intermediate to form a new C—C bond. In several cases, an indole was attached to the acceptor carbonyl group as an internal carbon nucleophile. The Waser group studied thoroughly this reaction with acyl indole-substituted DA aminocyclopropane **71**. Ring opening and subsequent cyclization occurred at either the C3 or at the N1 position of the indole if the nitrogen was not protected [85]. They found that the use of $Cu(OTf)_2$ in MeCN strongly favored the formation of **72** as a result of alkylation at the C3 position, while switching to TfOH in dichloromethane gave predominantly **73** as a product (Scheme 7.12). The Waser group also successfully applied this selective cyclization as a key step for the total synthesis of the natural product Goniomitine (**3**), the formal synthesis of Aspidospermidine (**4**) [22, 86] and the total synthesis of Jerantinine E (**74**) [87]. This strategy was also used in the total synthesis of Deethyleburnamonine (**75**) by the France group [88]. In addition, the formed imine can also be engaged in a formal

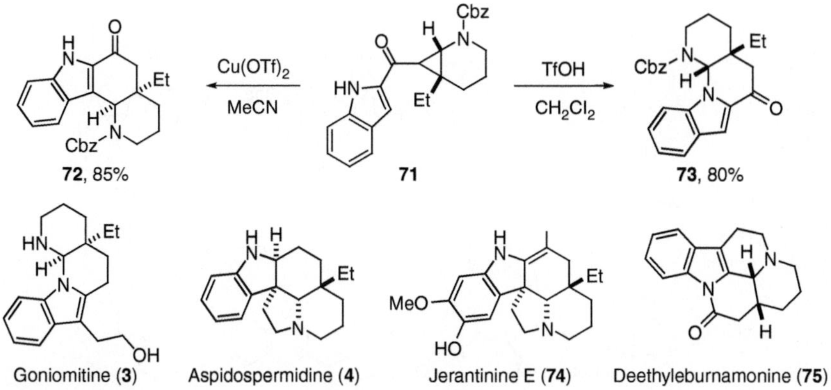

Scheme 7.12 Ring opening with internal indole and its application in total synthesis of indole alkaloids.

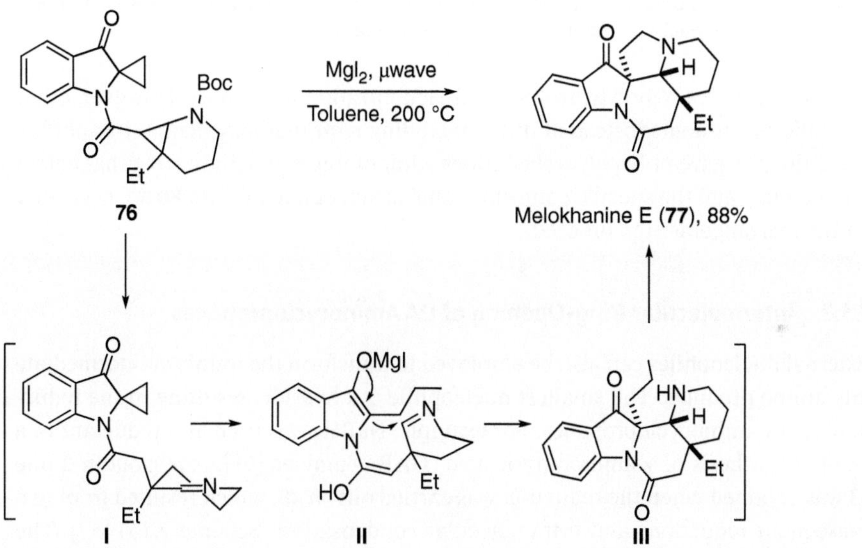

Scheme 7.13 Ring-opening cascade with enolate for the synthesis of melokhanine E (**77**).

[3 + 2] cycloaddition with a cyclopropane, as demonstrated by the Pierce group in the total synthesis of melokhanine E (Scheme 7.13) [89].

In a rare example reported in 2012 by the Reiser group, a furan migration product **81** was observed for *endo*-**80** after a Povarov reaction between **78** and **79**, while only isomerization into polycyclic imine **82** was observed for *exo*-**80** (Scheme 7.14) [90]. The authors proposed a plausible mechanism that involves the formation of intermediate **I** in the presence of $Sc(OTf)_3$ and subsequent furan migration to intermediate **II** via a spiroannulated intermediate. The unusual C-N bond cleavage is compensated by the rearomatization of a quinoline ring in intermediate **III**, which

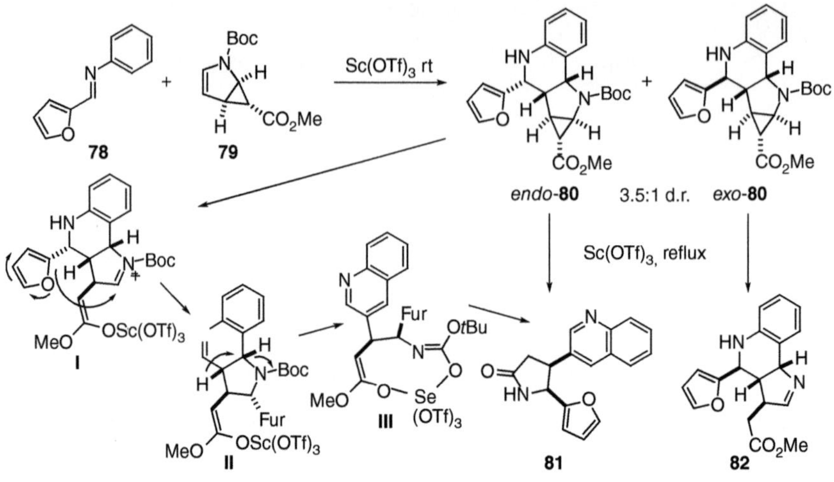

Scheme 7.14 Ring opening with or without furan migration of cyclopropane **80** obtained from the Povarov reaction of aryl imine **78** and enamide **79**.

further undergoes *N*-Boc hydrolysis and lactamization to form **81**. However, other aromatic substituents instead of furan, including *para*-methoxyphenyl, 1-naphthyl, and 2-thionyl, gave only polycyclic imines without migration, indicating that both a furan moiety and the specific conformational arrangement of *endo*-**80** are necessary for the rearrangement to proceed.

7.3.2 Intermolecular Ring-Opening of DA Aminocyclopropanes

External nucleophiles can also be employed to transform the iminium intermediate into amine products. The smallest nucleophile is a hydride, resulting in the reduction of DA aminocyclopropanes. For example, $NaBH_4$ was used as a reductant in a one-pot synthesis of γ-aminobutyric acid (GABA) amides [91]. A pyrrolidin-2-one **83** was obtained when the reduction was carried out on **48**, which resulted from two subsequent reductions and intramolecular condensation (Scheme 7.15) [57]. The cyclopropanation/ring-opening/reduction cascade is useful to construct cyclic systems and can thereby find applications in the synthesis of complex molecules [92]. For example, a domino cyclopropanation-ring opening-reduction has been applied as a key step in the total synthesis of Lyconesidine B, which was accomplished by the Takemoto group [93].

When water was employed as a nucleophile, an aldehyde was usually formed as a result of hydrolysis of the iminium intermediate [33], which is most often an undesired side reaction because of the loss of nitrogen. With alcohols as nucleophiles, the *in situ*-formed iminium intermediate can be stabilized in the form of an *N,O*-acetal, and the Qin group used this strategy in 2017 for the total synthesis of *Kopsia* indole alkaloids [94]. The Gharpure group reported in 2014 that thiophenol was a good nucleophile as well, which afforded an *N,S*-acetal as a product after ring opening [29].

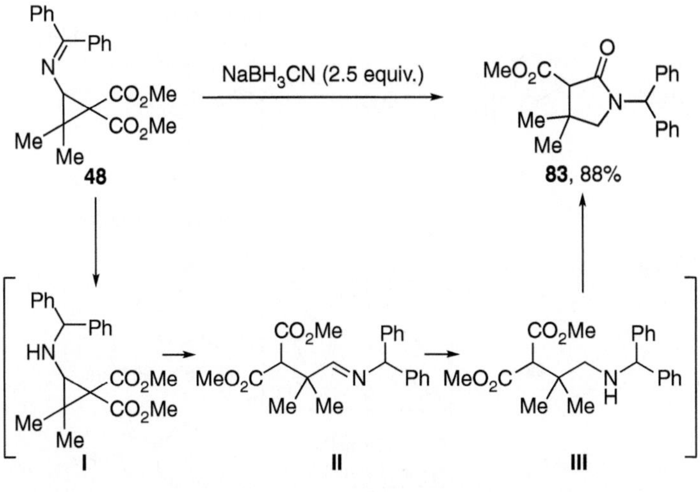

Scheme 7.15 Reduction of imine-substituted DA aminocyclopropane **48** followed by lactamization.

In 2019, the Saha group reported a $Sc(OTf)_3$-catalyzed ring opening of DA cyclopropanes by hydrogen peroxide [95]. Extension of this reaction to DA aminocyclopropanes was successful, resulting in the formation of α-amino peroxides as products with amino groups varying from phthalimide, succinimide to even maleimide. With peracid as a nucleophile, the α-amino peroxide formed from **84** was further fragmented to give pyrrolidin-2-one **85** (Eq. 7.5) [29].

*Formation of pyrrolidinone **85** by reaction of DA aminocyclopropane **84** with mCPBA.*

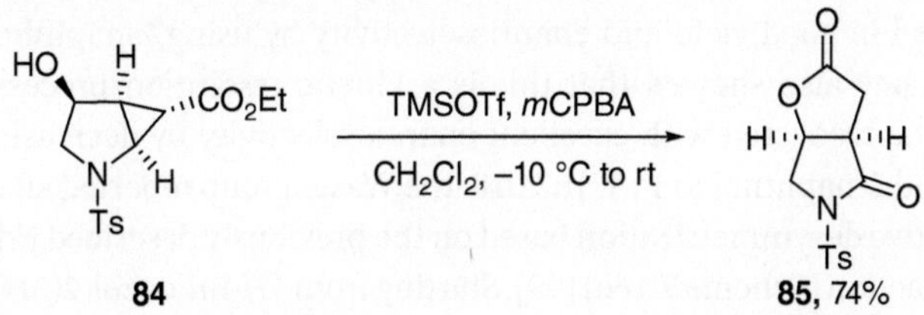

(7.5)

Carbon nucleophiles have also been successfully incorporated into the α-amino position in the presence of Lewis acid catalysts. In 2013, the Waser group reported a Friedel–Crafts reaction between DA aminocyclopropane **86** and indole **87**, leading to the formation of indole C3 alkylation product **88** (Scheme 7.16a) [96]. In the presence of C3-substituted indoles, C2 alkylation was observed. Pyrroles were also tested, and the regioselectivity between C2/C3 alkylation can be tuned by changing protecting groups on the pyrrole nitrogen atom. Other electron-rich aromatic compounds, such as anisole or phenol, can also be used but usually give a mixture of products with poorer regioselectivity. The Gharpure group reported in 2014 one example of nucleophilic attack by 1,3,5-trimethoxybenzene on **84** in the presence of the Lewis acid TMSOTf, forming a Friedel–Crafts alkylation product with high yield and diastereoselectivity [29]. In 2018, the Wang and Guo group developed the

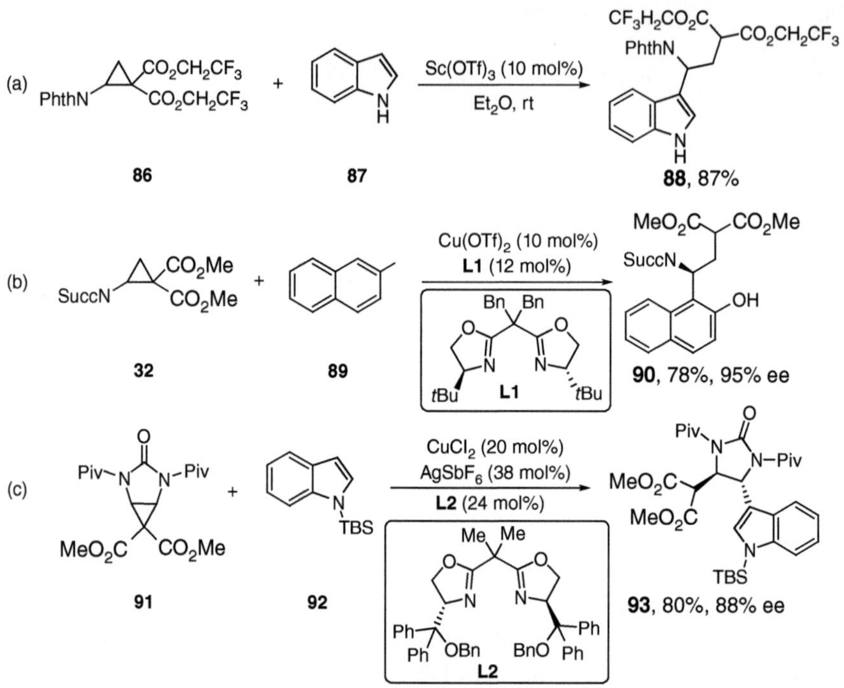

Scheme 7.16 Friedel–Crafts reaction between DA aminocyclopropanes and indoles or 2-naphthols.

Friedel–Crafts alkylation of DA aminocyclopropane **32** with 2-naphthol **89** as a nucleophile (Scheme 7.16b) [97]. With $Cu(OTf)_2$ and BOX ligand **L1**, GABA derivative **90** was obtained in good yield and enantioselectivity by using 2-naphthol as a limiting reagent. They also showed that this is a kinetic resolution process, and unreactive **32** can be recovered with excellent enantioselectivity by decreasing the ratio between **32** and 2-naphthol to 1 : 1. In 2018, the Waser group reported an example of enantioselective desymmetrization based on the previously described Friedel–Crafts alkylation reaction (Scheme 7.16c) [98]. Starting from 1*H*-imidazol-2(3*H*)-one, *meso*-diaminocyclopropane **91** was prepared by cyclopropanation with diazo malonate using $Rh_2(OAc)_4$ as catalyst. The combination of a copper catalyst and the fine-tuned BOX ligand **L2** enabled the conversion of **91** into an enantioenriched, diastereomerically pure urea derivative **93**.

In addition to aromatic compounds, organometallic reagents can also be used as nucleophiles. For instance, Me_2CuLi was used to introduce a methyl group by the Prieto and Pérez group in 2014 for C2, C3-difunctionalization of 1-methylindole **94** via an unstable indole-derived DA aminocyclopropane **95** (Eq. 7.6) [99].

Ring-opening reaction using Me2CuLi as nucleophile.

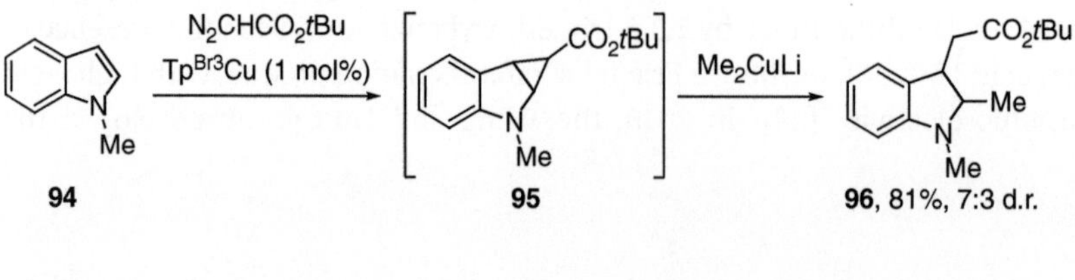

(7.6)

Apart from monofunctionalization of the DA aminocyclopropanes, 1,3-difunctionalization has also been reported in the presence of an electrophile. In 2014, the Werz group reported a 1,3-dichlorination of DA cyclopropanes by iodobenzene dichloride under mild conditions (Scheme 7.17a) [100]. The donor group could vary from succinimide, phthalimide, alkyl, aryl, etc., and the acceptors were usually diesters or dinitriles. A radical pathway, rather than an ionic process, was proposed to explain the mechanism of this reaction. Following this work, the Werz group reported a 1,3-halochalcogenation with magnesium iodide as a Lewis acid catalyst (Scheme 7.17b) [101]. The strongly polarized bonds of sulfenyl or selenyl halides were exploited for the incorporation of a chalcogen and a halogen atom into the nucleophilic and electrophilic sites of **30**, forming **100** or **101** in good yields. A similar strategy was utilized by the Studer group in 2017 for the ring-opening 1,3-aminobromination [102] of **31** as well as by the Saha group in 2019 for 1,3-haloperoxygenation [95] of **30** (Scheme 7.17c). In these cases, *p*-toluenesulfonamide or *tert*-butyl hydrogen peroxide were used as nucleophiles for the ring-opening reaction with the help of a Lewis acid catalyst, leading to the formation of **102** or **103**. In 2020, the Saha group also reported a bisarylation reaction of **30**, using 1-methylindole and an arylbismuth(V) reagent as nucleophile and electrophilic arylating agents, respectively (Scheme 7.17d) [103].

To summarize this section, ring-opening reactions of DA aminocyclopropanes have been discussed, including isomerization, cyclization, monofunctionalization

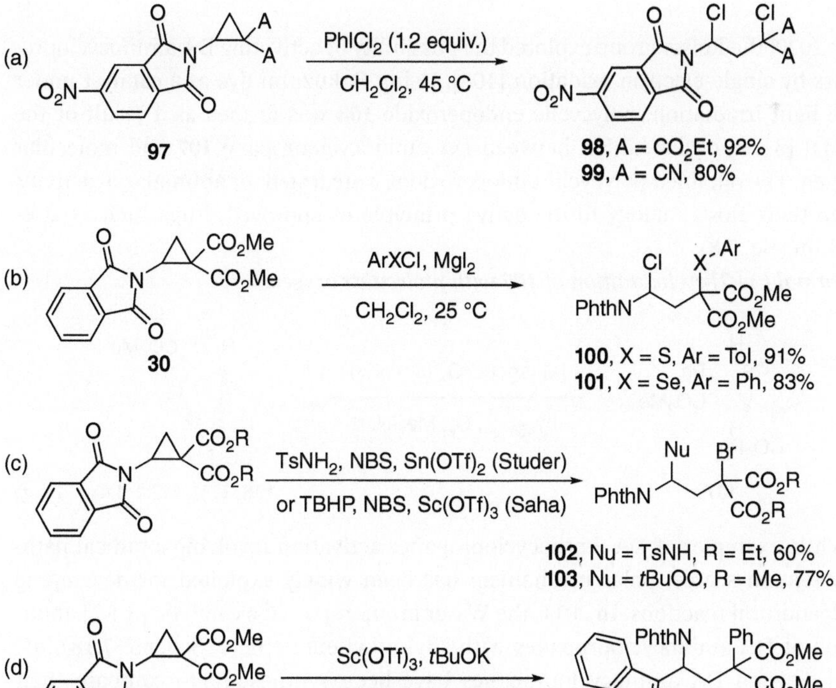

Scheme 7.17 Ring-opening 1,3-difunctionalization reactions with electrophiles.

as well as 1,3-difunctionalization. In all these cases, only classic bond cleavage involving the C1—C2 bond was observed. In addition to these reports, rare examples of ring-opening reactions involving non-classic bond cleavage were also disclosed [104-107].

7.4 Formal Cycloaddition of DA Aminocyclopropanes

In 1996, the Beak group reported a formal [3+2] cycloaddition between **51** and tetracyanoethylene (TCNE, **105**) (Eq. 7.7) [58]. In an attempt to broaden the synthetic utility of this reaction, less activated electrophiles including fumaronitrile, dimethyl maleate, and acrylonitrile were tested. However, no corresponding products were observed, and only the starting material was recovered. This may suggest that the first elementary step with tetracyanoethylene could be single electron transfer [20].

*Formal [3+2] cycloaddition of **51** with tetracyanoethylene.*

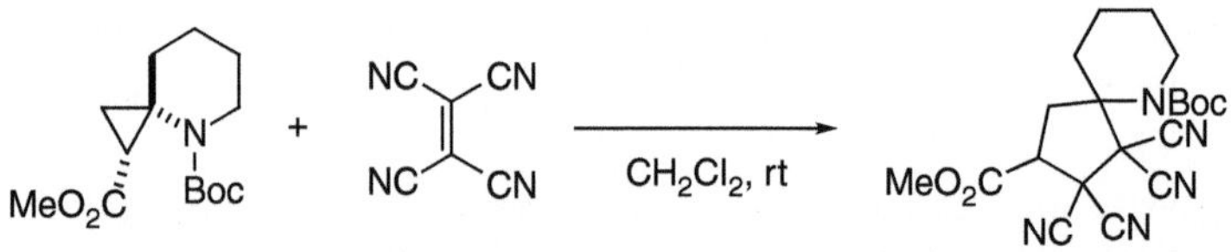

51 TCNE (**105**) **106**, 99%, single diastereoisomer (7.7)

In 2020, the Reiser group explored the possibility of activating DA aminocyclopropanes by single-electron oxidation [108]. Using Fukuzumi dye as a catalyst under blue light irradiation, polycyclic endoperoxide **108** was formed as a result of the formal [3+2] cycloaddition between DA aminocyclopropane **107** and molecular oxygen. The obtained polycyclic endoperoxides were tested for antimalarial activity, given their close analogy to the active principle of approved drugs such as artemisinin (Eq. 7.8).

*Formal [3+2] cycloaddition of **107** with molecular oxygen.*

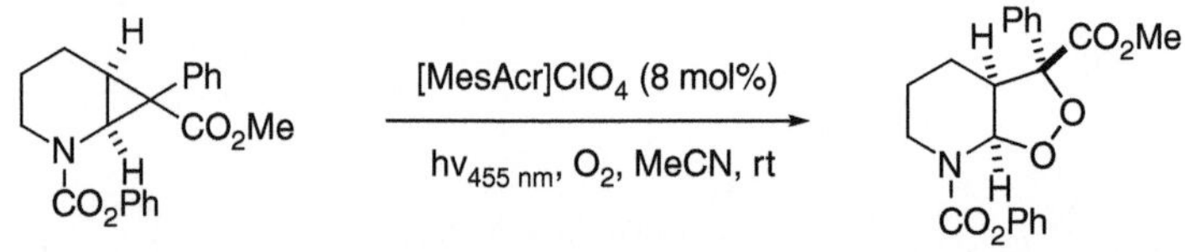

107 **108**, 37%, 5.3:1 d.r. (7.8)

While examples of DA aminocyclopropanes activation involving a radical pathway remain scarce, polar mechanism has been widely exploited for developing cycloaddition reactions. In 2011, the Waser group reported a catalytic [3+2] annulation of DA aminocyclopropanes with silyl/alkyl enol ethers (Scheme 7.18) [45, 109]. Several DA aminocyclopropanes have been synthesized to compare their reactivity and stability. It was found that mono-acceptor aminocyclopropanes **112–114** were not reactive under the examined conditions, while *gem*-di-acceptor

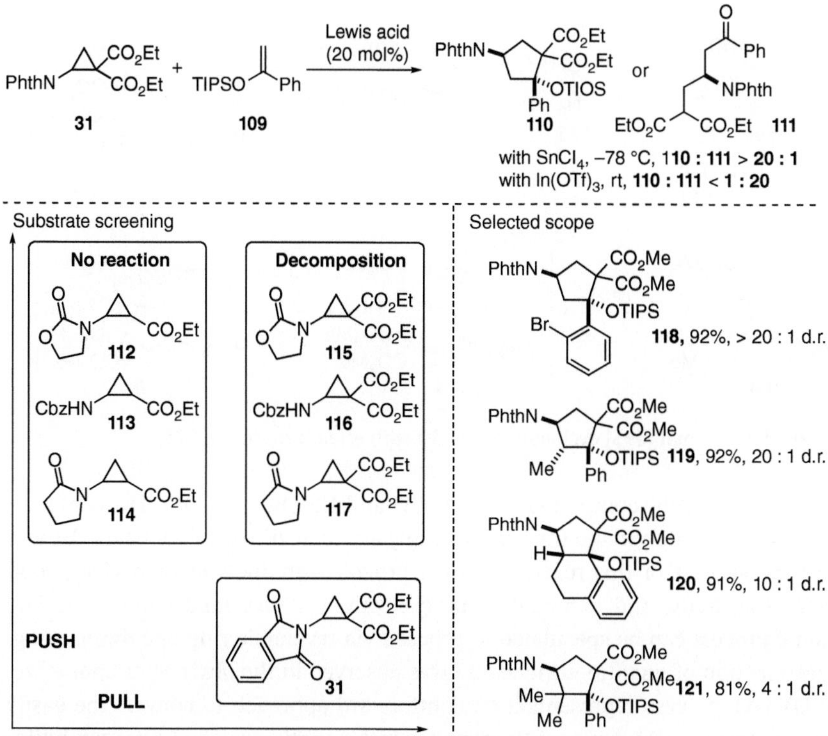

Scheme 7.18 Catalyst-tuned ring opening/cycloaddition with enol ether.

aminocyclopropanes **115–117** were not stable and decomposed directly after cyclopropanation. In the end, a balance between reactivity and stability was found when the nitrogen atom was protected as a phthalimide group to reduce its electron density (**31**), leading to the formation of cyclopentylamine **110**. Product **110** was obtained in 98% yield and with more than 20 : 1 diastereoselectivity using $SnCl_4$ as catalyst, while an acyclic product **111** arising from nucleophilic ring opening was formed when $In(OTf)_3$ was used as catalyst. This reaction was stereospecific in relation to the configuration of enol ether and enantiospecific. The size of the silyl group on the enol ether and the substituent on the benzene ring of the aminocyclopropane almost had no influence on the yield or selectivity of the products, as shown in the case of **118**. For trisubstituted or tetrasubstituted enol ethers, a drop in diastereoselectivity in the products **119–121** was observed. Interestingly, the opposite diastereoisomer was formed in the case of product **120**, while in other cases, products having nitrogen in *anti*-relationship with oxygen were determined as the major diastereoisomer.

In 2012, the Waser group also reported catalytic [3+2] annulations between DA aminocyclopropanes and carbonyl compounds such as aldehydes [110] or ketones [111], giving access to aminotetrahydrofurans. Built upon these results, the Waser group further reported in 2014 a dynamic kinetic asymmetric (DYKAT) [3+2] annulation reaction of **32** with enol ethers or aldehydes **122** to form

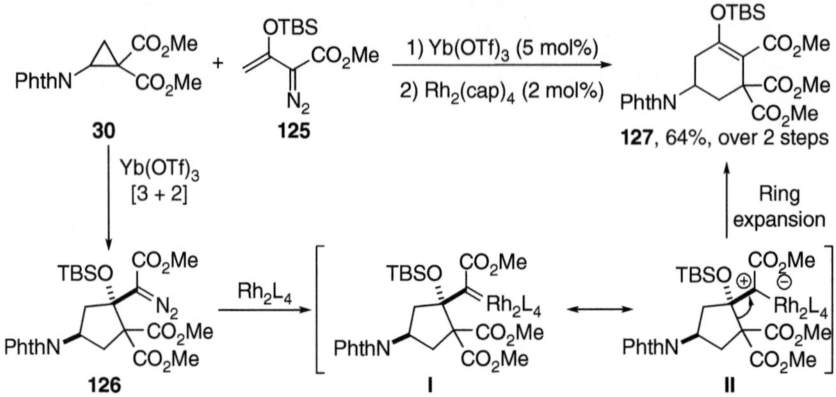

Scheme 7.19 Formal [3+3] cycloaddition of **30** with enoldiazoacetate **125**.

five-membered cyclic compounds such as **123** and **124** (Eq. 7.9) [46]. The combination of a copper catalyst and the commercially available bisoxazoline ligand **L1** was key to the success of this reaction, which provided products in high yield, good diastereoselectivity, and excellent enantioselectivity under mild conditions. The dynamic process can be speculated to proceed via reversible ring opening/closing, as racemization of enantio-enriched **32** was observed in the absence of enol ether. This DYKAT process represents a straightforward approach to convert the easily accessible racemic DA aminocyclopropanes into enantio-enriched nitrogen building blocks, which may be useful for the synthesis of bioactive compounds.

Dynamic kinetic asymmetric [3+2] annulation reaction of ***32****.*

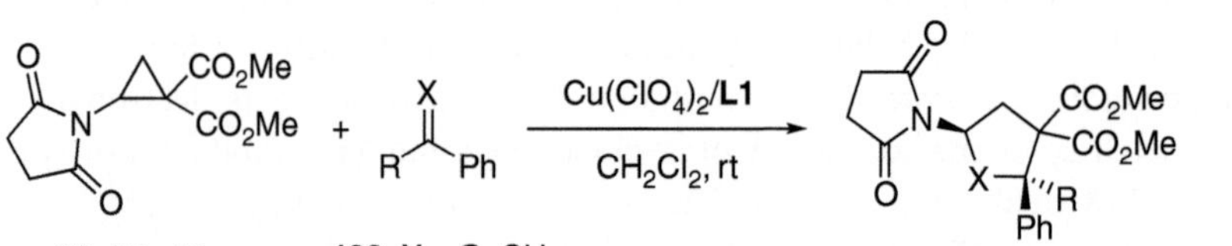

32, 50 : 50 e.r. **122**, X = O, CH_2 **123**, R = OBn, X = CH_2, 97%, 92% ee; **124**, R = H, X = O, 82%, 84% ee (7.9)

In 2015, the Doyle group reported a formal [3 + 3] cycloaddition of enoldiazoacetate **125** with DA cyclopropanes, which was extended successfully to DA aminocyclopropane **30** (Scheme 7.19) [112]. This reaction was realized by a tandem reaction, first with a Lewis acid-catalyzed [3 + 2] cycloaddition to form **126**, followed by a subsequent rhodium-catalyzed ring expansion to afford the six-membered cyclic compound **127** as the final product.

In addition to enol ethers/aldehydes/ketones, formal [3 + 2] cycloadditions with thioalkynes [113] and ynamides [114] were also reported. Based on previous studies from the Waser group [115], thioalkynes were easily synthesized from thiophenols and ethynylbenziodoxolones (EBX reagents). Annulation reactions of **30** and triethylsilyl substituted thioalkyne gave access to highly substituted cyclopentene **128** in high yield and regioselectivity (Scheme 7.20a), while a different product was

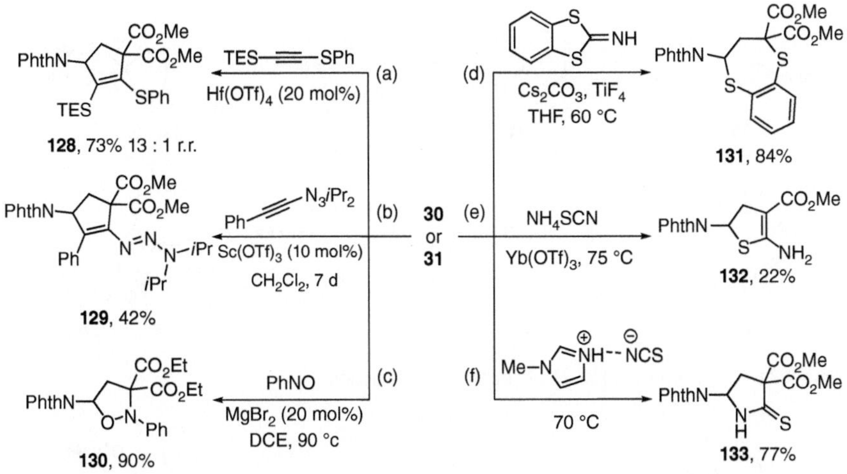

Scheme 7.20 Formal cycloaddition with thioalkynes, ynamides, nitrosobenzenes, benzodithioloimines, thiocyanates, or isothiocyanic acid.

formed with aliphatic thioalkynes [113]. 1-Alkynyltriazenes were synthesized by the Severin group, and their synthetic utility was demonstrated in several types of transformation, including the [3 + 2] annulation with **30** to form cyclopentene **129** (Scheme 7.20b) [114]. In 2014, the Studer group reported a formal [3 + 2] cycloaddition of **31** with nitrosobenzene to form **130** (Scheme 7.20c) [116]. The Werz group reported in 2016 a formal [4 + 3] cycloaddition of **30** with amphiphilic benzodithioloimine as surrogate for *ortho*-bisthioquinone to give **131** (Scheme 7.20d) [117], as well as a formal [3 + 2] cycloaddition of **30** with selenocyanate [118] or thiocyanate to form **132** (Scheme 7.20e) [119]. In 2021, the Trushkov group in collaboration with the Werz group reported a formal [3 + 2] cycloaddition of **30** with isothiocyanic acid to produce **133**, in which case an untypical *N*-attack took place (Scheme 7.20f) [120].

Apart from annulation of DA aminocyclopropanes with the dipolarophiles mentioned above, dearomative [3 + 2] annulation reactions have also been studied. In 2017, the Waser group described a dearomative [3 + 2] annulation of N-heterocycles with DA aminocyclopropanes (Scheme 7.21a) [121]. In this case, imido-substituted DA cyclopropanes exhibited superior reactivity compared to aryl or alkoxy-substituted DA cyclopropanes. Several different N-heterocycles, including pyridines, quinolines, and isoquinolines, were well tolerated. An asymmetric dearomative [3 + 2] cycloaddition reaction of DA aminocyclopropane **136** with benzazole **137** was later disclosed by the Guo and You group, forming tricyclic product **138** in good yield and enantioselectivity (Scheme 7.21b) [122]. In 2019, the Guo group successfully extended the asymmetric dearomative [3 + 2] cycloaddition to purines [123].

In 2019, the Wang group developed a cooperative Rh^{II}/In^{III} catalytic system for the cyclopentannulations of indole **139** to tetracyclic indolines (Scheme 7.22a) [124]. By tuning the reaction parameters carefully such as solvent and temperature, three structurally divergent tetracyclic indolines, cis-**140**, trans-**140,** and **141**, were synthesized with good yield and excellent selectivity. Mechanistic studies indicated an

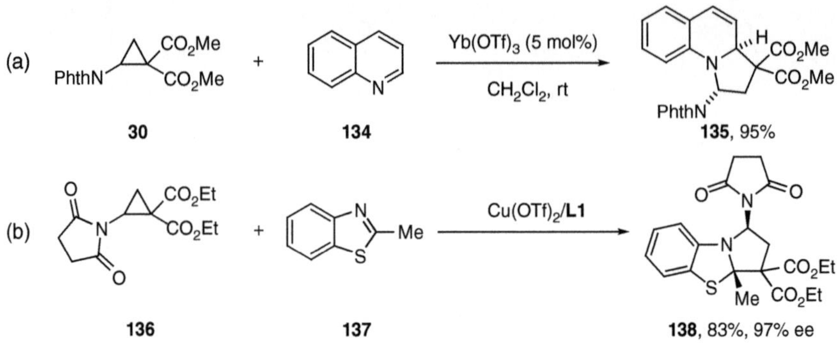

Scheme 7.21 Dearomative [3 + 2] cycloaddition reactions.

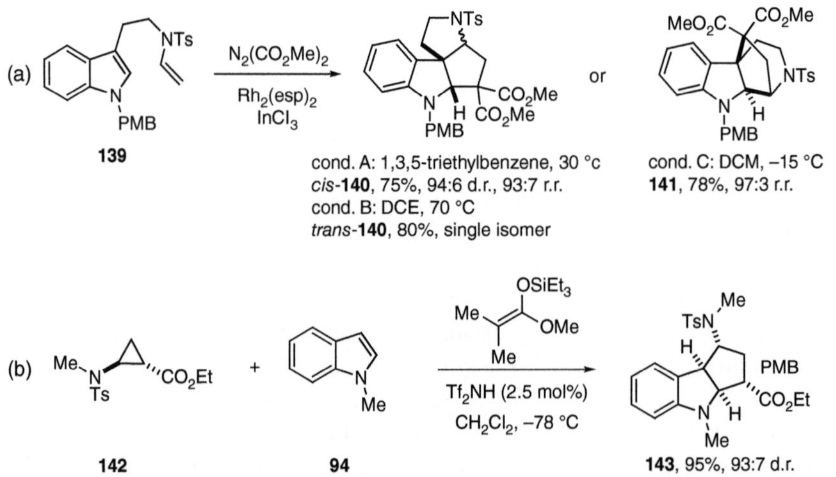

Scheme 7.22 Intra- and intermolecular dearomative [3 + 2] annulations of indoles.

intramolecular annulation of the indole with an *in situ*-formed aminocyclopropane. In 2021, the Waser group reported an intermolecular [3 + 2] dearomative annulation reaction between mono-acceptor aminocyclopropanes and indoles (Scheme 7.22b) [125]. Getting rid of one acceptor group was not easy, as a balance between reactivity and stability needed to be found. Protecting the amino group with a tosyl group and a methyl group as in **142** gave an optimal reactivity. The choice of the *in situ* formed triethylsilyl triflimide as catalyst was also crucial and led to the formation of tricyclic indoline product **143** with excellent yield and stereoselectivity.

Annulations with tropones were also explored. For example, the Sierra group reported in 2013 a formal [8 + 3] cycloaddition reaction between DA aminocyclopropane **31** and tropone, which produced amino-substituted tetrahydrocyclohepta[b]pyran **144** with complete regio- and diastereoselectivity (Eq. 7.10) [126]. Density functional theory (DFT) calculations revealed that the transformation proceeds stepwise through a zwitterionic intermediate, which is stabilized by some degree of π-aromaticity of the tropyl cation. The regio- and diastereoselectivity are controlled in the ring closure step. In the same year, the Adrio and Carretero group also reported a nickel-catalyzed formal [8 + 3] cycloaddition of tropones with 1,1-cyclopropanediesters, including DA aminocyclopropane **30** [127].

Formal [3 + 2] cycloaddition of ***31*** *with tropone.*

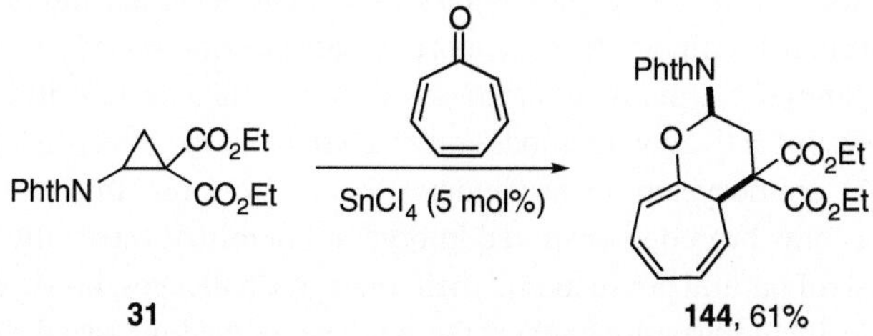

(7.10)

In addition to the commonly used imide-substituted DA cyclopropanes, the Waser group also developed an efficient synthesis of thymine/uracil-substituted DA cyclopropanes **145** and their subsequent formal [3 + 2] cycloaddition with enol ethers/aldehydes/ketones **146** for the synthesis of nucleoside analogs **147** (Scheme 7.23a) [128, 129]. Based on the recent progress in the synthesis of alkenyltriazene **148** [130], the Severin group in collaboration with the Waser group has synthesized triazene-derived DA cyclopropane **149** by cyclopropanation with diazomalonate. They illustrated the reactivity of **149** by developing ring-opening reaction with methanol to form **150** or by developing a formal [3 + 2] cycloaddition with TCNE or enol ethers to form five-membered cyclic products **151** or **152**, respectively (Scheme 7.23b) [131].

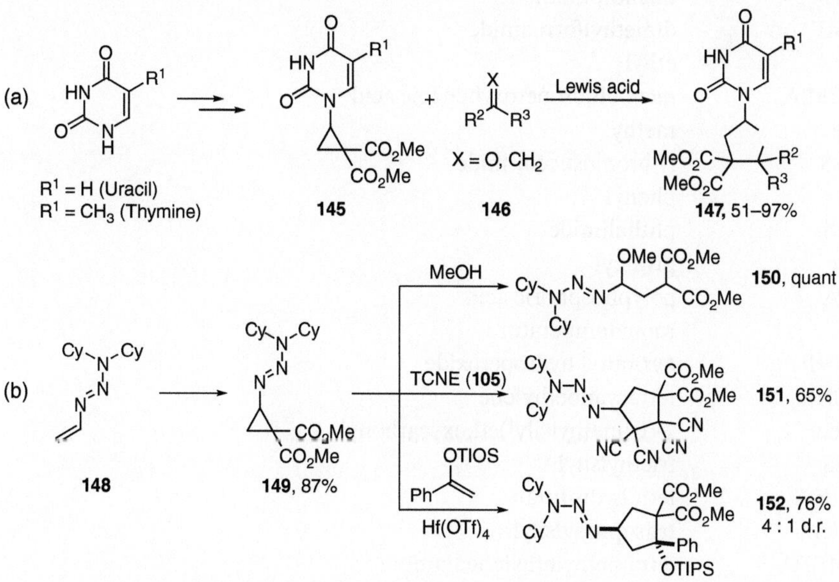

Scheme 7.23 Thymine/uracil- and triazene-substituted DA cyclopropanes **145** and **149**.

7.5 Conclusion

Activating the C—C bond of cyclopropane, in spite of its high ring strain, is challenging and usually requires the use of noble metal catalysts. Nevertheless, attaching an amino group as donor and one (or two) electron-withdrawing group(s) as acceptor to the vicinal carbon atoms of the C1—C2 bond makes it strongly polarized and prone to cleavage to give a formal 1,3-dipole. As an important subclass of DA cyclopropanes, DA aminocyclopropanes share many similarities with aryl- or alkoxy-substituted DA cyclopropanes in terms of activation methods and reaction patterns. Nevertheless, the nitrogen atom of DA aminocyclopropanes endows them with some unique properties. Their utility has now been demonstrated in organic chemistry, especially in the field of total synthesis of natural products and their analogs. Challenges, however, still remain for the future, including among others the synthesis of medium-sized rings by formal cycloaddition, the development of new diastereoselective and enantioselective reactions for mono-acceptor aminocyclopropanes, the use of other amino donor groups, and the further extension of the developed concepts to DA aminocyclobutanes [132-136], which have been only scarcely investigated so far.

Abbreviations

A	acceptor
BOX	bisoxazoline
D	donor
DCE	1,2-dichloroethane
DCM	dichloromethane
DMF	dimethylformamide
Et	ethyl
*m*CPBA	*meta*-chloroperoxybenzoic acid
Me	methyl
NBS	*N*-bromosuccinimide
Ph	phenyl
Phth	phthalimide
Piv	pivaloyl
PPA	polyphosphoric acid
rt	room temperature
TBHP	*tert*-butyl hydroperoxide
TCNE	tetracyanoethylene
Teoc	2-(trimethylsilyl)ethoxycarbonyl
TES	triethylsilyl
THF	tetrahydrofuran
TIPS	triisopropylsilyl
TMEDA	tetramethylethylenediamine
TMS	trimethylsilyl
TMSOTf	trimethylsilyltrifluoromethanesulfonate
Tr	triphenylmethyl

References

1 Reissig, H.-U. and Zimmer, R. (2003). *Chem. Rev. 103*: 1151–1196.
2 Yu, M. and Pagenkopf, B.L. (2005). *Tetrahedron 61*: 321–347.
3 Schneider, T.F., Kaschel, J., and Werz, D.B. (2014). *Angew. Chem. Int. Ed. 53*: 5504–5523.
4 O'Connor, N.R., Wood, J.L., and Stoltz, B.M. (2016). *Isr. J. Chem. 56*: 431–444.
5 Reissig, H.U. (1988). *Top. Curr. Chem. 144*: 73–135.
6 Wong, H.N.C., Hon, M.Y., Tse, C.W. et al. (1989). *Chem. Rev. 89*: 165–198.
7 Carson, C.A. and Kerr, M.A. (2009). *Chem. Soc. Rev. 38*: 3051–3060.
8 Zhang, D., Song, H., and Qin, Y. (2011). *Acc. Chem. Res. 44*: 447–457.
9 Tang, P. and Qin, Y. (2012). *Synthesis 44*: 2969–2984.
10 Wang, Z. (2012). *Synlett 23*: 2311–2327.
11 Grover, H.K., Emmett, M.R., and Kerr, M.A. (2015). *Org. Biomol. Chem. 13*: 655–671.
12 Reiser, O. (2016). *Isr. J. Chem. 56*: 531–539.
13 Ivanova, O.A. and Trushkov, I.V. (2019). *Chem. Rec. 19*: 2189–2208.
14 Werz, D.B. and Biju, A.T. (2020). *Angew. Chem. Int. Ed. 59*: 3385–3398.
15 Wang, L. and Tang, Y. (2016). *Isr. J. Chem. 56*: 463–475.
16 Pirenne, V., Muriel, B., and Waser, J. (2021). *Chem. Rev. 121*: 227–263.
17 Xia, Y., Liu, X., and Feng, X. (2021). *Angew. Chem. Int. Ed. 60*: 9192–9204.
18 Cohen, Y., Cohen, A., and Marek, I. (2021). *Chem. Rev. 121*: 140–161.
19 de Nanteuil, F., De Simone, F., Frei, R. et al. (2014). *Chem. Commun. 50*: 10912–10928.
20 Rassadin, V.A. and Six, Y. (2016). *Tetrahedron 72*: 4701–4757.
21 Sokolova, O.O. and Bower, J.F. (2021). *Chem. Rev. 121*: 80–109.
22 De Simone, F., Gertsch, J., and Waser, J. (2010). *Angew. Chem. Int. Ed. 49*: 5767–5770.
23 Brackmann, F. and de Meijere, A. (2007). *Chem. Rev. 107*: 4493–4537.
24 Cativiela, C. and Ordonez, M. (2009). *Tetrahedron Asymmetry 20*: 1–63.
25 Fulop, F. (2001). *Chem. Rev. 101*: 2181–2204.
26 Gnad, F. and Reiser, O. (2003). *Chem. Rev. 103*: 1603–1624.
27 Gu, P., Wu, X.-P., Su, Y. et al. (2013). *Tetrahedron Lett. 54*: 4957–4959.
28 Maas, G. and Mueller, A. (1998). *J. Prakt. Chem. 340*: 315–322.
29 Gharpure, S.J., Vijayasree, U., and Reddy, S.R.B. (2012). *Org. Biomol. Chem. 10*: 1735–1738.
30 Kunieda, T. and Witkop, B. (1969). *J. Am. Chem. Soc. 91*: 7751–7752.
31 Huisgen, R. and Mloston, G. (1994). *Tetrahedron Lett. 35*: 4981–4984.
32 Manabe, A., Matsumoto, R., and Shinada, T. (2015). *Synlett 26*: 1710–1714.
33 Chen, J., Guo, P., Zhang, J. et al. (2019). *Angew. Chem. Int. Ed. 58*: 12674–12679.
34 Tanny, S.R., Grossman, J., and Fowler, F.W. (1972). *J. Am. Chem. Soc. 94*: 6495–6501.
35 Bubert, C., Cabrele, C., and Reiser, O. (1997). *Synlett 1997*: 827–829.
36 Schinnerl, M., Böhm, C., Seitz, M., and Reiser, O. (2003). *Tetrahedron Asymmetry 14*: 765–771.
37 Pilsl, L.K.A., Ertl, T., and Reiser, O. (2017). *Org. Lett. 19*: 2754–2757.

38 Fu, J., Wurzer, N., Lehner, V. et al. (2019). *Org. Lett. 21*: 6102–6106.
39 Liu, W., Babl, T., Röther, A. et al. (2020). *Chem. Eur. J. 26*: 4236–4241.
40 Özüduru, G., Schubach, T., and Boysen, M.M.K. (2012). *Org. Lett. 14*: 4990–4993.
41 Xu, H., Li, Y.-P., Cai, Y. et al. (2017). *J. Am. Chem. Soc. 139*: 7697–7700.
42 Chanthamath, S., Nguyen, D.T., Shibatomi, K., and Iwasa, S. (2013). *Org. Lett. 15*: 772–775.
43 Kaufman, M.D. and Grieco, P.A. (1994). *J. Org. Chem. 59*: 7197–7198.
44 Dutta, P.K., Chauhan, J., Ravva, M.K., and Sen, S. (2019). *Org. Lett. 21*: 2025–2028.
45 de Nanteuil, F. and Waser, J. (2011). *Angew. Chem. Int. Ed. 50*: 12075–12079.
46 de Nanteuil, F., Serrano, E., Perrotta, D., and Waser, J. (2014). *J. Am. Chem. Soc. 136*: 6239–6242.
47 Zhu, D., Chen, L., Fan, H. et al. (2020). *Chem. Soc. Rev. 49*: 908–950.
48 Grotjahn, D.B. and Dötz, K.H. (1991). *Synlett 1991*: 381–390.
49 Sierra, M.A., Soderberg, B., Lander, P.A., and Hegedus, L.S. (1993). *Organometallics 12*: 3769–3771.
50 Sierra, M.A., Mancheño, M.J., Sáez, E., and del Amo, J.C. (1998). *J. Am. Chem. Soc. 120*: 6812–6813.
51 Merino, I. and Hegedus, L.S. (1995). *Organometallics 14*: 2522–2531.
52 Fan, R., Ye, Y., Li, W., and Wang, L. (2008). *Adv. Synth. Catal. 350*: 2488–2492.
53 Wheeler, J.W., Shroff, C.C., Stewart, W.S., and Uhm, S.J. (1971). *J. Org. Chem. 36*: 3356–3361.
54 Franck-Neumann, M., Miesch, M., and Kempf, H. (1988). *Tetrahedron 44*: 2933–2942.
55 Banning, J.E., Gentillon, J., Ryabchuk, P.G. et al. (2013). *J. Org. Chem. 78*: 7601–7616.
56 Geen, G.R., Kincey, P.M., and Choudary, B.M. (1992). *Tetrahedron Lett. 33*: 4609–4612.
57 Mangelinckx, S. and De Kimpe, N. (2005). *Synlett 2005*: 1521–1526.
58 Park, Y.S. and Beak, P. (1996). *Tetrahedron 52*: 12333–12350.
59 Tomilov, Y.V., Menchikov, L.G., Novikov, R.A. et al. (2018). *Russ. Chem. Rev. 87*: 201–250.
60 He, J., Shigenari, T., and Yu, J.-Q. (2015). *Angew. Chem. Int. Ed. 54*: 6545–6549.
61 Shen, P.-X., Hu, L., Shao, Q. et al. (2018). *J. Am. Chem. Soc. 140*: 6545–6549.
62 Hu, L., Shen, P.-X., Shao, Q. et al. (2019). *Angew. Chem. Int. Ed. 58*: 2134–2138.
63 Cheng, J., Qi, X., Li, M. et al. (2015). *J. Am. Chem. Soc. 137*: 2480–2483.
64 Ikeda, M., Matsugashita, S., Tabusa, F., and Tamura, Y. (1977). *J. Chem. Soc. Perkin Trans. I*: 1166–1171.
65 Martin, D., Canac, Y., Lavallo, V., and Bertrand, G. (2014). *J. Am. Chem. Soc. 136*: 5023–5030.
66 Welstead, W.J., Stauffer, H.F., and Sancilio, L.F. (1974). *J. Med. Chem. 17*: 544–547.
67 Novikov, M.S., Rostovskii, N.V., Koronatov, A.N. et al. (2017). *J. Org. Chem. 82*: 13396–13404.
68 Salim, M. and Capretta, A. (2000). *Tetrahedron 56*: 8063–8069.
69 Zhang, B. and Wee, A.G.H. (2012). *Org. Biomol. Chem. 10*: 4597–4608.
70 Arenare, L., De Caprariis, P., Marinozzi, M. et al. (1994). *Tetrahedron Lett. 35*: 1425–1426.

71 Wenkert, E., Hudlicky, T., and Showalter, H.D.H. (1978). *J. Am. Chem. Soc. 100*: 4893–4894.

72 Wenkert, E. and Hudlicky, T. (1988). *J. Org. Chem. 53*: 1953–1957.

73 Rifé, J. and Ortuño, R.M. (1999). *Tetrahedron Asymmetry 10*: 4245–4260.

74 Kaschel, J., Schneider, T.F., Schirmer, P. et al. (2013). *Eur. J. Org. Chem. 2013*: 4539–4551.

75 Ikeda, M., Matsugashita, S., and Tamura, Y. (1977). *J. Chem. Soc. Perkin Trans. I*: 1770–1772.

76 Wenkert, E., Halls, T.D.J., Kwart, L.D. et al. (1981). *Tetrahedron 37*: 4017–4025.

77 Yang, J., Song, H., Xiao, X. et al. (2006). *Org. Lett. 8*: 2187–2190.

78 Yang, J., Wu, H., Shen, L., and Qin, Y. (2007). *J. Am. Chem. Soc. 129*: 13794–13795.

79 Chauhan, J., Ravva, M.K., Gremaud, L., and Sen, S. (2020). *Org. Lett. 22*: 4537–4541.

80 Guha, S., Gadde, S., Kumar, N. et al. (2021). *J. Org. Chem. 86*: 5234–5244.

81 Song, H., Yang, J., Chen, W., and Qin, Y. (2006). *Org. Lett. 8*: 6011–6014.

82 Shen, L., Zhang, M., Wu, Y., and Qin, Y. (2008). *Angew. Chem. Int. Ed. 47*: 3618–3621.

83 Zhang, M., Huang, X., Shen, L., and Qin, Y. (2009). *J. Am. Chem. Soc. 131*: 6013–6020.

84 He, B., Song, H., Du, Y., and Qin, Y. (2009). *J. Org. Chem. 74*: 298–304.

85 De Simone, F., Saget, T., Benfatti, F. et al. (2011). *Chem. Eur. J. 17*: 14527–14538.

86 De Simone, F. and Waser, J. (2011). *Synlett 2011*: 589–593.

87 Frei, R., Staedler, D., Raja, A. et al. (2013). *Angew. Chem. Int. Ed. 52*: 13373–13376.

88 Patil, D.V., Cavitt, M.A., and France, S. (2012). *Heterocycles 84*: 1363–1373.

89 Cholewczynski, A.E., Williams, P.C., and Pierce, J.G. (2020). *Org. Lett. 22*: 714–717.

90 Roy, S. and Reiser, O. (2012). *Angew. Chem. Int. Ed. 51*: 4722–4725.

91 Maslivetc, V.A., Rubina, M., and Rubin, M. (2015). *Org. Biomol. Chem. 13*: 8993–8995.

92 Kurose, T., Tsukano, C., and Takemoto, Y. (2017). *Org. Lett. 19*: 4762–4765.

93 Kurose, T., Tsukano, C., Nanjo, T., and Takemoto, Y. (2021). *Org. Lett. 23*: 676–681.

94 Leng, L., Zhou, X., Liao, Q. et al. (2017). *Angew. Chem. Int. Ed. 56*: 3703–3707.

95 Singh, K., Bera, T., Jaiswal, V. et al. (2019). *J. Org. Chem. 84*: 710–725.

96 de Nanteuil, F., Loup, J., and Waser, J. (2013). *Org. Lett. 15*: 3738–3741.

97 Zhu, M., Wang, D.-C., Xie, M.-S. et al. (2018). *Chem. Eur. J. 24*: 15512–15516.

98 Perrotta, D., Wang, M.-M., and Waser, J. (2018). *Angew. Chem. Int. Ed. 57*: 5120–5123.

99 Delgado-Rebollo, M., Prieto, A., and Pérez, P.J. (2014). *ChemCatChem 6*: 2047–2052.

100 Garve, L.K.B., Barkawitz, P., Jones, P.G., and Werz, D.B. (2014). *Org. Lett. 16*: 5804–5807.

101 Wallbaum, J., Garve, L.K.B., Jones, P.G., and Werz, D.B. (2017). *Org. Lett. 19*: 98–101.

102 Das, S., Daniliuc, C.G., and Studer, A. (2017). *Angew. Chem. Int. Ed. 56*: 11554–11558.

103 Mondal, B., Das, D., and Saha, J. (2020). *Org. Lett. 22*: 5115–5120.

104 Gheorghe, A., Schulte, M., and Reiser, O. (2006). *J. Org. Chem. 71*: 2173–2176.

105 Yedoyan, J., Wurzer, N., Klimczak, U. et al. (2019). *Angew. Chem. Int. Ed. 58*: 3594–3598.

106 Sonnleitner, C.M., Park, S., Eckl, R. et al. (2020). *Angew. Chem. Int. Ed. 59*: 18110–18115.

107 McClure, C.K., Kiessling, A.J., and Link, J.S. (2003). *Org. Lett. 5*: 3811–3813.

108 Budde, S., Goerdeler, F., Floß, J. et al. (2020). *Org. Chem. Front. 7*: 1789–1795.

109 Serrano, E., de Nanteuil, F., and Waser, J. (2014). *Synlett 25*: 2285–2288.

110 Benfatti, F., de Nanteuil, F., and Waser, J. (2012). *Org. Lett. 14*: 386–389.

111 Benfatti, F., de Nanteuil, F., and Waser, J. (2012). *Chem. Eur. J. 18*: 4844–4849.

112 Cheng, Q.-Q., Qian, Y., Zavalij, P.Y., and Doyle, M.P. (2015). *Org. Lett. 17*: 3568–3571.

113 Racine, S., Hegedüs, B., Scopelliti, R., and Waser, J. (2016). *Chem. Eur. J. 22*: 11997–12001.

114 Perrin, F.G., Kiefer, G., Jeanbourquin, L. et al. (2015). *Angew. Chem. Int. Ed. 54*: 13393–13396.

115 Frei, R. and Waser, J. (2013). *J. Am. Chem. Soc. 135*: 9620–9623.

116 Chakrabarty, S., Chatterjee, I., Wibbeling, B. et al. (2014). *Angew. Chem. Int. Ed. 53*: 5964–5968.

117 Garve, L.K.B., Pawliczek, M., Wallbaum, J. et al. (2016). *Chem. Eur. J. 22*: 521–525.

118 Jacob, A., Jones, P.G., and Werz, D.B. (2020). *Org. Lett. 22*: 8720–8724.

119 Jacob, A., Barkawitz, P., Andreev, I.A. et al. (2021). *Synlett 32*: 901–904.

120 Andreev, I.A., Ratmanova, N.K., Augustin, A.U. et al. (2021). *Angew. Chem. Int. Ed. 60*: 7927–7934.

121 Preindl, J., Chakrabarty, S., and Waser, J. (2017). *Chem. Sci. 8*: 7112–7118.

122 Zhang, M.-C., Wang, D.-C., Xie, M.-S. et al. (2019). *Chem 5*: 156–167.

123 Hao, E.-J., Fu, D.-D., Wang, D.-C. et al. (2019). *Org. Chem. Front. 6*: 863–867.

124 Liu, H.-K., Wang, S.R., Song, X.-Y. et al. (2019). *Angew. Chem. Int. Ed. 58*: 4345–4349.

125 Pirenne, V., Robert, E.G.L., and Waser, J. (2021). *Chem. Sci. 12*: 8706–8712.

126 Rivero, A.R., Fernández, I., and Sierra, M.Á. (2013). *Org. Lett. 15*: 4928–4931.

127 Tejero, R., Ponce, A., Adrio, J., and Carretero, J.C. (2013). *Chem. Commun. 49*: 10406–10408.

128 Racine, S., de Nanteuil, F., Serrano, E., and Waser, J. (2014). *Angew. Chem. Int. Ed. 53*: 8484–8487.

129 Racine, S., Vuilleumier, J., and Waser, J. (2016). *Isr. J. Chem. 56*: 566–577.

130 Kiefer, G., Riedel, T., Dyson, P.J. et al. (2015). *Angew. Chem. Int. Ed. 54*: 302–305.

131 Suleymanov, A.A., Le Du, E., Dong, Z. et al. (2020). *Org. Lett. 22*: 4517–4522.

132 de Nanteuil, F. and Waser, J. (2013). *Angew. Chem. Int. Ed. 52*: 9009–9013.

133 Perrotta, D., Racine, S., Vuilleumier, J. et al. (2015). *Org. Lett. 17*: 1030–1033.

134 Zhong, X., Tang, Q., Zhou, P. et al. (2018). *Chem. Commun. 54*: 10511–10514.

135 Wei, S., Yin, L., Wang, S.R., and Tang, Y. (2019). *Org. Lett. 21*: 1458–1462.

136 Wei, S., Zheng, L., Wang, S.R., and Tang, Y. (2020). *Org. Lett. 22*: 1013–1017.

8

Reactivity of Cyclopropyl Monocarbonyls

Pankaj Kumar, Irshad Maajid Taily, Priyanka Singh, and Prabal Banerjee

Department of Chemistry, Indian Institute of Technology Ropar, Rupnagar, Punjab, 140001, India

CHAPTER OUTLINE

8.1 Introduction

Over the years, carbonyl groups have proven to be amazing acceptors in molecules where reactivity depends upon the resonance-induced electron-withdrawing nature of a suitably substituted functional group. Carbonyls work as useful withdrawing groups in aromatic systems, unsaturated molecules, and for activation of strained carbon—carbon bonds, like in small ring systems of cyclopropanes, aziridines, epoxides, and cyclobutanes. Considering cyclopropanes in particular, carbonyl compounds are among many other functional groups that have been used as acceptors to activate the cyclopropane ring. Some of the well-known acceptors are diesters, keto-esters, nitro, nitro-esters, nitriles, etc (Scheme 8.1a). Among all these, the most widely studied acceptor group for the cyclopropane chemistry is the diester group. Cyclopropane diesters are easy to prepare and are stable at room temperature. Cyclopropane diesters can be easily activated in the presence of Lewis acids, which particularly activate them through binding to the diester site of the cyclopropane [1]. This extensive study of the cyclopropane diesters has proven this molecule to be the

Donor–Acceptor Cyclopropanes in Organic Synthesis, First Edition.
Edited by Prabal Banerjee and Akkattu T. Biju.

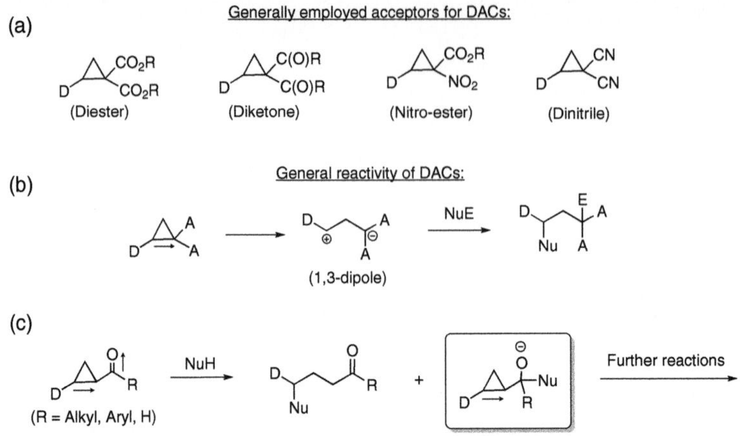

Scheme 8.1 (a) Generally employed acceptors for DACs; (b) common reactivity of DACs; and (c) reactivity of cyclopropane monocarbonyls.

best choice for the (3 + n) type of formal cycloaddition processes (Scheme 8.1b). This predictable reactivity of cyclopropane diesters being a three-carbon synthon, besides being an advantage, seems to saturate the versatile scope of this otherwise interesting molecule.

However, the removal of one of the two ester groups on the vicinally substituted cyclopropane molecule or having a ketone/aldehyde functionality in place of the diester can make the reactivity of these donor–acceptor cyclopropanes even more exciting. With cyclopropane carbaldehyde or ketone, the carbonyl group is also a pertinent reactive center that can undergo a number of reactions and, so, can make the donor–acceptor chemistry of the cyclopropane molecule even more fascinating (Scheme 8.1c).

8.2 Associated Challenges

Donor–acceptor cyclopropane monocarbonyls have come up with specific reactivity advantages, but there are also some challenges associated with these molecules. They are listed in the following section.

8.2.1 Reduced Reactivity

The first and major challenge that is associated with the cyclopropane monocarbonyls is their fairly reduced reactivity compared to other donor–acceptor cyclopropane (DAC), in particular the cyclopropane diesters. Clearly, the bond between the vicinal substituents will be more reactive when we have two ester groups to pull (with the alkyl/aryl group to push) rather than one ester (carbonyl) group for the pull/withdrawing effect (Figure 8.1). As a result, the cyclopropane ring-opening

Figure 8.1 Reduced polarization of reactive carbon—carbon bond in cyclopropane monocarbonyls.

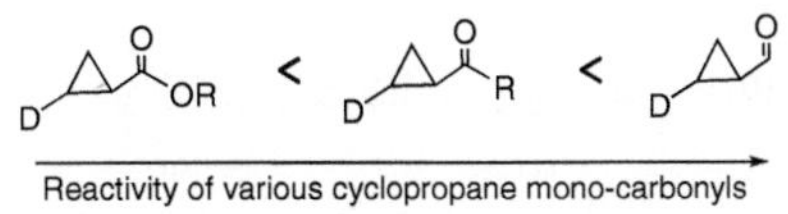

Figure 8.2 A general comparison of reactivity in various cyclopropane monocarbonyls.

becomes more difficult in the case of cyclopropane mono-carbonyls, and we may have to use harsh reaction conditions or modify the reaction conditions accordingly. However, the reduced reactivity of these cyclopropanes makes them more selective for specific reaction conditions in order to undergo ring-opening transformations.

On the other hand, the reactivity of DAC mono-carbonyls can be further improved by having a more electron-withdrawing carbonyl group as an acceptor. Evidently, aldehydes and ketones are significantly more electron-withdrawing than the esters [2]. Consequently, DA cyclopropane carbaldehydes/ketones would be more reactive than the DA cyclopropane monoesters (Figure 8.2). In this chapter, we will have a thorough discussion on the reactivity of DA cyclopropane carbaldehydes and DA cyclopropyl ketones.

8.2.2 Diastereomers and Controlled Reactivity

At first, cyclopropane monocarbonyls seem to be similar to the other DACs, but actually they come with a few practical challenges that need to be addressed before one starts to work with these molecules. In aryl substituted cyclopropane monocarbonyls, the first challenge resides in their diastereomeric nature. There are two stereo centers in the molecule, and the two substituents in the vicinal positions of the cyclopropane can be *cis* or *trans* to each other. Depending on the relative geometry of these two substituents that work in synergy, the molecule can be reactive for some transformation only in one geometry. So, tuning the geometry of these cyclopropane mono-carbonyls according to the designed transformation is the first challenge to be worked on before we move further with the reactivity.

Having this choice of diastereomers in the cyclopropane molecule can sometimes also be helpful. For instance, *trans*-geometry of the DA cyclopropane monocarbonyls would be preferred in the following scenario:

(i) Reaction with a binucleophile would prefer the *trans*-geometry for a sterically favored approach (Scheme 8.2a) [3].
(ii) If the nucleophilic species (approaching C—C bond of the cyclopropane) would have any interaction with the acceptor carbonyl group, or if the electrophilic center of the dipolar species (approaching C—C bond of the cyclopropane) would have any interaction with the donor group (Scheme 8.2a).

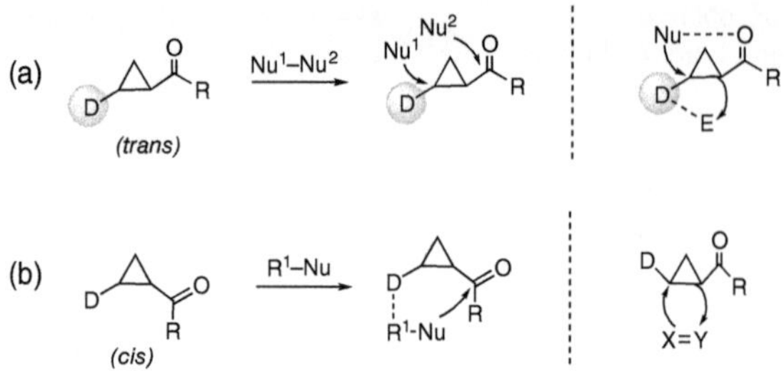

Scheme 8.2 (a) Steric factors would favor the *trans*-geometry, while (b) Donor group assistance and concerted cycloaddition would favor the *cis*-geometry of DA cyclopropane monocarbonyls.

On the other hand, *cis*-geometry of the DA cyclopropane monocarbonyls would be preferred in cases like;

(i) If the transformation is designed in such a way that the catalyst/reagent can activate/assist the carbonyl group as well as the donor substituent on the cyclopropane ring.
(ii) If there is any donor group assistance to the nucleophilic attack on the carbonyl center (Scheme 8.2b).
(iii) For a concerted cycloaddition process, where both nucleophilic and electrophilic ends of the dipolar species need to approach the cyclopropane from the same side (Scheme 8.2b).

8.3 Perks of Having a Monocarbonyl Substituent on Cyclopropane

8.3.1 DAC Monocarbonyls—Not Merely a Three-Carbon Synthon

Unlike Cyclopropane diesters, cyclopropane mono-carbonyls are not merely a three-carbon dipolar species; but the carbonyl group here is also reactive on its own. In many cases, it is seen that the carbonyl center, especially the aldehyde and ketone groups, is more reactive than the cyclopropane ring, and the nucleophilic center of a partner substrate first reacts on the carbonyl center rather than on the cyclopropane ring [4]. This unique property of the acceptor group being involved in reactions of cyclopropyl mono-carbonyls makes them rather interesting to work with.

8.3.2 Two Nucleophilic and Two Electrophilic Sites

As is known very clearly, DACs have an electrophilic (e1) and a nucleophilic center (n1) on the vicinal positions of the cyclopropane ring. In addition to these reactive centers, cyclopropane mono-carbonyls also have two more fairly reactive centers on the carbonyl group – an electrophilic carbonyl carbon (e2) and a nucleophilic carbonyl oxygen (n2). Thus, this class of cyclopropanes now possesses four highly reactive centers that can be used to design numerous transformations with these cyclopropanes. However, it is noteworthy that the vicinal positions of the cyclopropane are comparatively less reactive than the diester-based DACs as there is only one acceptor group here.

With all these structural properties, DA cyclopropane mono-carbonyls can be expected to undergo a number of transformations depending on the type of substrate partner/reagent used. Firstly, it can undergo the classical Cloke–Wilson type rearrangement on its own or in the presence of a nucleophilic species (Scheme 8.3a) [5]. This transformation can be catalyzed by a Bronsted acid, a Lewis acid, or by heating. Being a type of DAC, these cyclopropanes are also expected to undergo the most studied (3 + n) formal cycloaddition process of the DACs with a dipolarophile (Scheme 8.3b) [1]. These two processes are also well known with the other DACs, except for the fact that the cyclopropane C—C bond for DAC monocarbonyls is far less activated, and it should be challenging to make it undergo the requisite cleavage. The most fascinating part about the reactivity of DAC monocarbonyls is something that is not feasible with the other DACs. For instance, these cyclopropanes are also open to an unexpected (5 + n)-type formal cycloaddition process where the cyclopropane moiety could behave as a five-atom contributor to an even larger ring system (Scheme 8.3c). Intriguingly, these cyclopropanes are also apt to undergo annulation with a bi-nucleophilic species, making use of both of their electrophilic centers (Scheme 8.3d) [3]. Unlikely, in such a scenario, the cyclopropane would behave as a four-carbon contributor toward the annulation process.

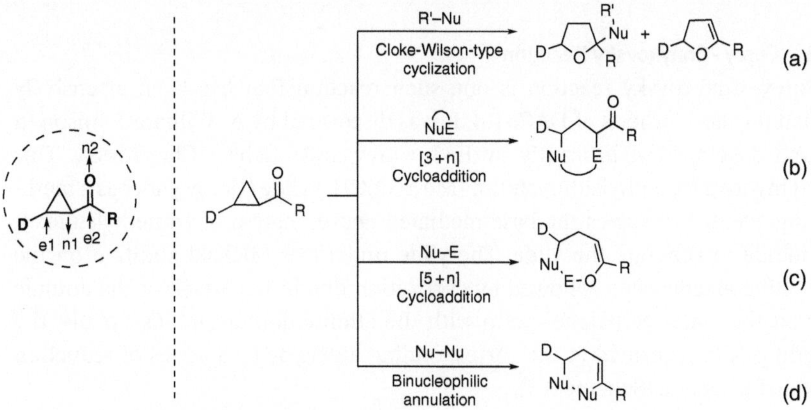

Scheme 8.3 Possible reactions of DA cyclopropane monocarbonyls.

8.3.3 Cyclopropane Mono-Carbonyls in Organocatalysis

There have been some great advancements in the field of chiral phosphoric acid as well as iminium ion catalysis with aldehydes and ketones for asymmetric synthesis, but unfortunately, iminium ion activation of the cyclopropane ring has not been explored much. With cyclopropyl carbaldehydes or with cyclopropyl ketones, this area of research has immense scope to work on (Scheme 8.4).

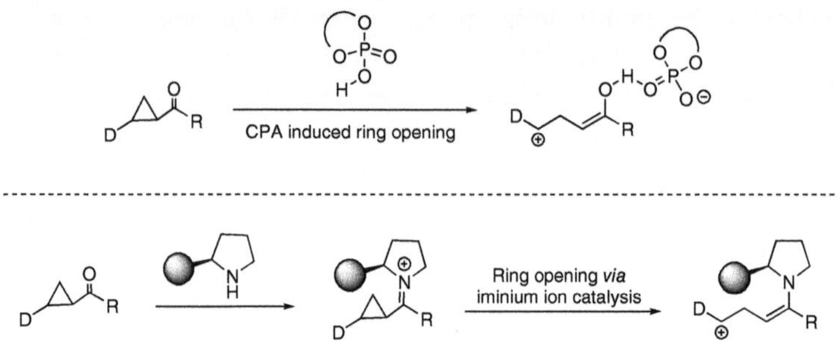

Scheme 8.4 Organocatalytic activation of DA cyclopropane monocarbonyls.

8.4 Methods for the Preparation of Cyclopropyl Monocarbonyls

There are a number of versatile methods available for the synthesis of cyclopropane carbaldehydes. However, only two approaches have become popular for the same—(i) those based on diazo compounds and (ii) those using sulfur ylides (Corey–Chaykovsky reaction). Nevertheless, a few other methods have their own strengths and applicability.

8.4.1 From Olefins

8.4.1.1 Corey–Chaykovsky Reaction

The Corey–Chaykovsky reaction is one such reaction that has been extensively exploited for the synthesis of DACs [6]. It was discovered by A. William Johnson in 1961 and developed significantly by E. J. Corey and Michael Chaykovsky. This method involves trimethylsulfoxonium [$Me_2S(O)CH_2$] ylide (Corey ylide) as a methylenating agent. It involves the base-mediated deprotonation of trimethylsulfoxonium iodide to generate the ylide. The ylide undergoes Michael addition on the electrophilic alkene ester, followed by cyclization due to the attack of the anionic center on the ylide methylene group with the elimination of DMSO. Finally, the ester group is transformed to the corresponding aldehyde by a series of reduction and oxidizing agents (Scheme 8.5).

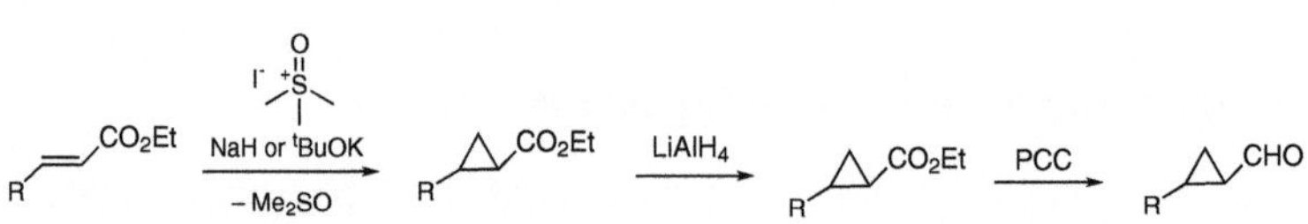

Scheme 8.5 Corey–Chaykovsky synthesis of cyclopropane carbaldehydes.

8.4.1.2 Hydroformylation of Cyclopropenes

The hydroformylation of cyclopropenes was first demonstrated by Orchin in 1981 by employing Mn- and Co-complexes, albeit in low yield [7]. Noyori in 1995 used $HMn(CO)_5$ complex, which somewhat improved the yield (Scheme 8.6) [8].

Scheme 8.6 Mn- and Co-catalyzed hydroformylation of cyclopropenes.

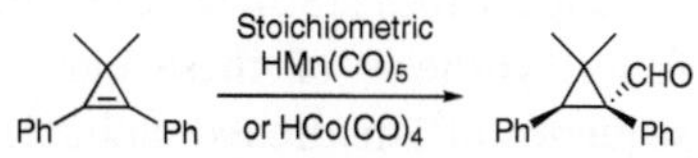

However, Rubin et al. reported the first catalytic diastereo- and enantioselective hydroformylation of prochiral cyclopropenes toward the synthesis of substituted cyclopropanecarbaldehydes under mild conditions and very low catalyst loading (Scheme 8.7) [9].

Scheme 8.7 Diastereo- and enantioselective hydroformylation of cyclopropenes.

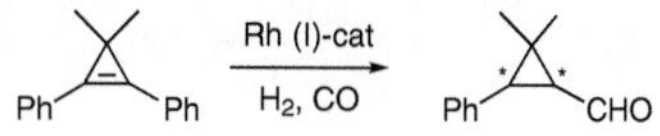

8.4.1.3 Ozonolysis of Vinyl Cyclopropanes

It involves the oxidative cleavage of vinyl cyclopropane to cyclopropylaldehydes by employing O_3/PPh_3 (Scheme 8.8) [10].

Scheme 8.8 Ozonolysis of vinyl cyclopropanes.

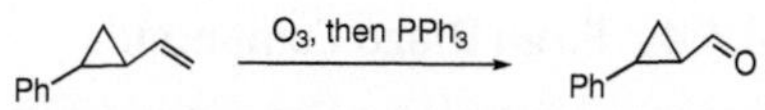

8.4.2 From Homoaldol Adducts

Taylor developed an efficient two-step homologation of aldehydes to cyclopropylaldehydes with excellent *trans*/*cis* selectivity in 2003 [11]. The protocol involves the activation of homoaldol adducts obtained from *O*-enecarbamates and *N*-enecarbamates to generate the heteroatom stabilized cyclopropylcarbinyl cationic intermediate to finally furnish the cyclopropane carbaldehydes (Scheme 8.9).

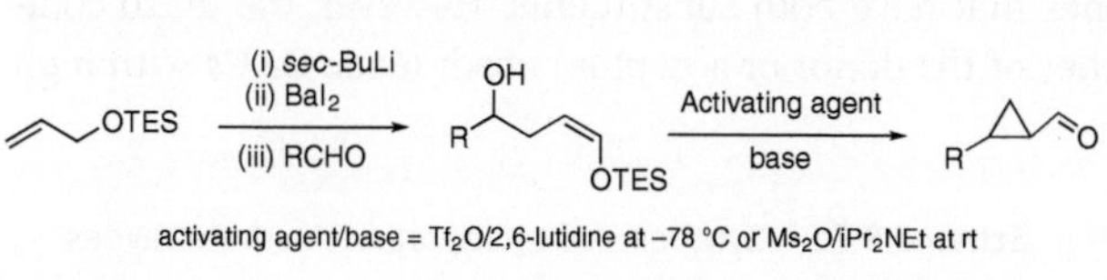

Scheme 8.9 Two-step homologation of aldehydes to cyclopropylaldehydes.

Moreover, the same group in 2004 disclosed the stereospecific cyclization of enantiomerically enriched homoaldol adducts to furnish enantiomerically enriched and diastereomerically pure 1,2,3-trisubstituted cyclopropyl aldehydes by asymmetric metalation [12].

8.4.3 From Arylthio Cyclopropyl Carbaldehydes

The stoichiometric metalation of arylthio ethers followed by trapping with formamides is the main route of their synthesis (Scheme 8.10a) [13–15]. The hydrolysis of cyclopropane carbonitriles could also be used to access these cyclopropanes (Scheme 8.10b) [13b]. In 2018, Secci et al. demonstrated the Bronsted acid-catalyzed facile synthesis of these carbaldehydes via addition/ring contraction reaction sequence of 2-hydroxycyclobutanones. The methodology exhibited wide substrate scope under mild reaction conditions (Scheme 8.10c) [16].

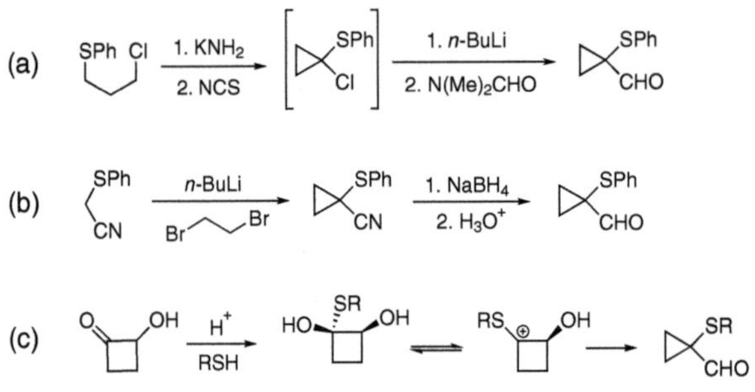

Scheme 8.10 Synthesis of arylthio cyclopropyl derivatives.

8.4.4 From Diazo Compounds

The synthesis of cyclopropane carbaldehydes has well exploited the carbene chemistry, as most of the methods for their synthesis are based on the addition of *in situ* generated carbenes or their equivalents to alkenes (Scheme 8.11) [17]. Diazo compounds (diazomethane, diazo carbonyls), which, in the presence of a metal catalyst (usually transition metals like Rh, Pd, Cu, Au, etc.) generate reactive carbenes, are utilized to bring out this transformation. The *in situ* generated carbenes undergo simultaneous [2+1] cycloaddition with the olefins to construct the three-carbon ring. Diazomethane is the easiest and most convenient source of carbenes for the cyclopropanation of alkenes that have both substituents. However, the diazo compound could also bear either of the donor or acceptor substituents. DACs with high

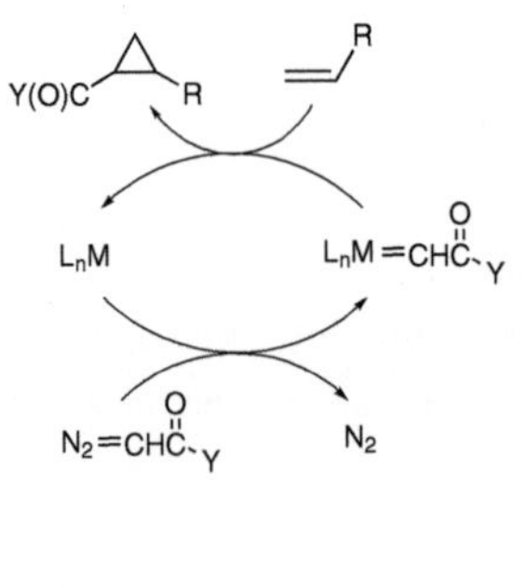

Scheme 8.11 Synthesis of cyclopropane carbaldehydes from diazo compounds.

yields and absolute stereoselectivity (*cis*/*trans*) could be obtained by designing the catalyst and the substituents in the reactants. Diazo compounds with electron-withdrawing groups (acetates, malonates) and alkenes with electron-donating groups are commonly employed as starting materials for the synthesis of DACs. This methodology has been found equally efficient in the case of alkyl vinyl ethers, vinyl sulfides, enamines, and other similar compounds where a heteroatom acts as a donor group. Moreover, this methodology has found its application at the industrial scale and in the flow chemistry. Alkyl diazoacetates are readily dediazotized in presence of the catalyst and are thus commonly utilized among alkyl diazocarbonyls.

This method has found its use in the synthesis of a number of chiral DACs by employing a chiral metal catalyst. Apart from these, the diazo compounds with two EWGs, like α-cyano- and α-nitro-diazocarbonyl compounds, have also been studied.

In 2016, the Lawasa group diastereo- and enantioselective Ru(II)–Amm–Pheox complex catalyzed stereoselective cyclopropanation of diazo Weinreb amides with olefins to furnish the corresponding chiral cyclopropyl Weinrebamides (Scheme 8.12) [18]. These Weinrebamides could be easily transformed to the corresponding aldehydes, ketones, and alcohols.

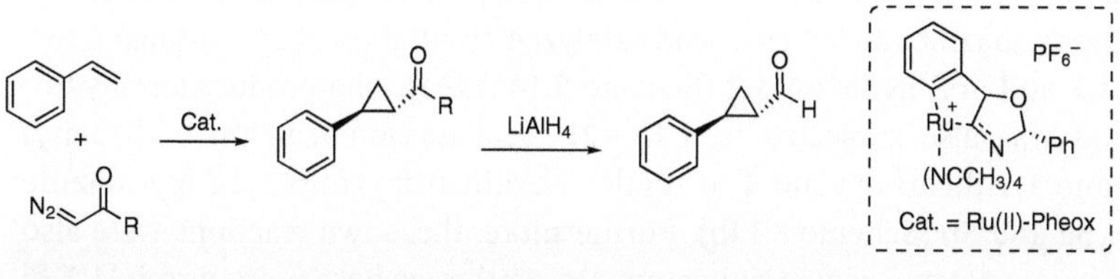

Scheme 8.12 Asymmetric cyclopropanation of various olefins.

8.4.5 From 1,2-Dicarbonyl Compounds

Keeping in view the potentially explosive nature and multi-step synthesis of diazo compounds, Zhuo et al. recently in 2021 reported molybdenum-catalyzed deoxygenative cyclopropanation of 1,2-dicarbonyl compounds (Scheme 8.13) [19]. The protocol elegantly presented the carbonyl group as the carbene precursor and furnished cyclopropanes with exclusive regioselectivity in up to 90% yield. Phosphines (or silanes) act as both a mild reductant and a good oxygen acceptor and bring about the reduction of Mo-oxo complexes to regenerate the Mo catalyst.

Scheme 8.13 Molybdenum-catalyzed deoxygenative cyclopropanation of 1,2-dicarbonyls.

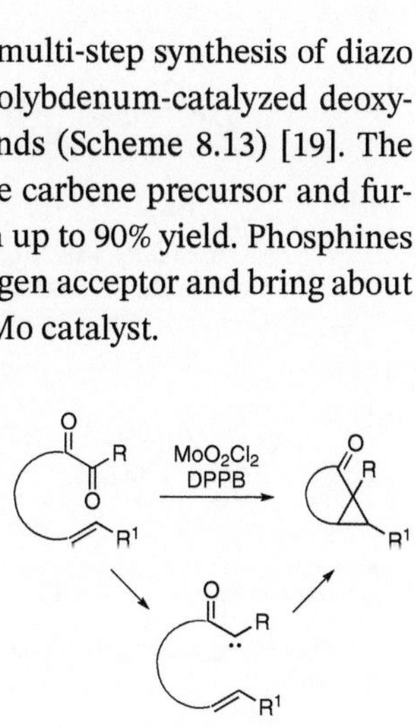

8.5 Cyclopropyl Monocarbonyls in Important Heterocyclic Synthesis

Similar to other types of DACs, aryl substituted cyclopropyl carbonyls, due to the vicinal arrangement of the functional groups, also undergo C—C bond cleavage, leading to the formation of 1,3-zwitterionic species. Various transformations such as cycloaddition, ring-opening, rearrangement, and ring-expansion reactions toward the formation of diverse high-value open-chain and cyclic compounds have been achieved through the formed 1,3-dipoles. This section of the chapter is mainly focused on the reactivity of cyclopropyl carbonyls toward different reactive partners.

8.5.1 Metal Catalyzed Annulation Reactions of Cyclopropyl Monocarbonyls

Among various cyclopropane derivatives, Lewis-acid catalyzed reactions of DACs have been studied widely [20], whereas the reactions of cyclopropane carbaldehydes are scarce. So in an attempt to explore the Lewis acid catalyzed reactivity of cyclopropane carbaldehydes, in 2018, Banerjee and group represented facile access to tetrahydropyridazines **3** via Lewis acid catalyzed annulation of cyclopropane carbaldehydes **1** and aryl hydrazines **2** (Scheme 8.14a) [3a]. The product tetrahydropyridazine **3** was also subjected to a [3+2] cycloaddition reaction with 2-aryl cyclopropane 1,1-dicarboxylate **4** to render hexahydropyrrolo[1,2-*b*]pyridazine derivatives **5a** and **5b** (Scheme 8.14b). Furthermore, these two reactions were also performed in one pot by a sequential manner to synthesize hexahydropyrrolo[1,2-*b*] pyridazine derivatives **5a** and **5b**. In the annulation step, $InCl_3$ was the most effective catalyst, whereas in cycloadditions step, $Yb(OTf)_3$ was the most efficient catalyst.

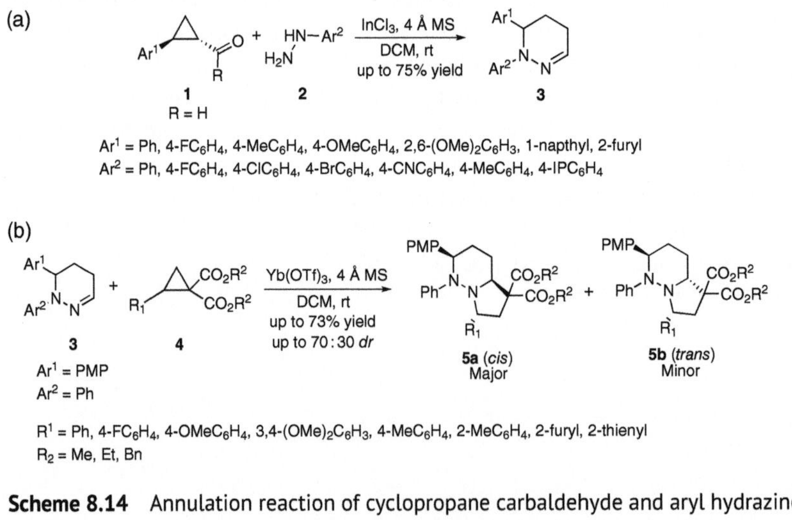

Scheme 8.14 Annulation reaction of cyclopropane carbaldehyde and aryl hydrazine.

Extending the Lewis acid catalyzed reactivity of cyclopropane carbaldehydes toward the Prins-type cyclization, Banerjee group subsequently in same year reported a versatile single-step TiX_4-mediated route to nine-membered ring (*E*)-hexahydrooxonines **8** or a fused bicyclic system octahydrocyclopenta[*b*]pyrans **9**, simply by picking up an appropriate alcohol. When cyclopropane carbaldehydes **1** were made to react with 3-buten-1-ol **6** through non classical Prins-type cyclization resulted in the highly stereoselective construction of relatively strained (*E*)-hexahydrooxonines **8**. While switching the alcohol to 3-butyn-1-ol **7** prompted a similar route, augmented by another classical Prins-type cyclization within a nine-membered ring to afford a bicyclized product (4,4-dihalo-5-aryloctahydrocyclopenta[*b*]pyran) **9** (Scheme 8.15) [3b]. Easy transformation of the resulting geminal dihalide to a vinyl halide and a ketone further supplemented the synthetic scope of this proposed approach.

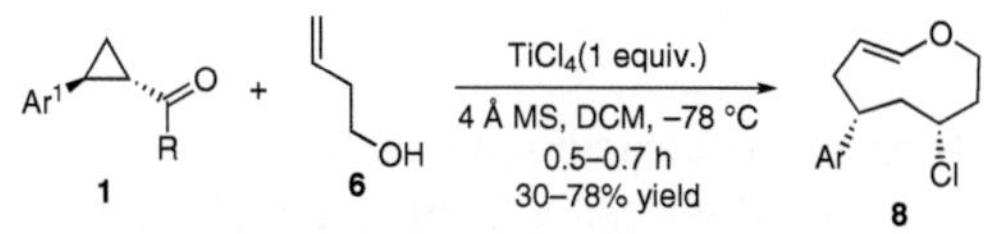

Ar^1 = 4-OMeC_6H_4, 4-OBnC_6H_4, 4-MeC_6H_4, 4-i-PrC_6H_4, 3,4-$(OMe)_2C_6H_3$, 2,4,6-$(Me)_3C_6H_2$

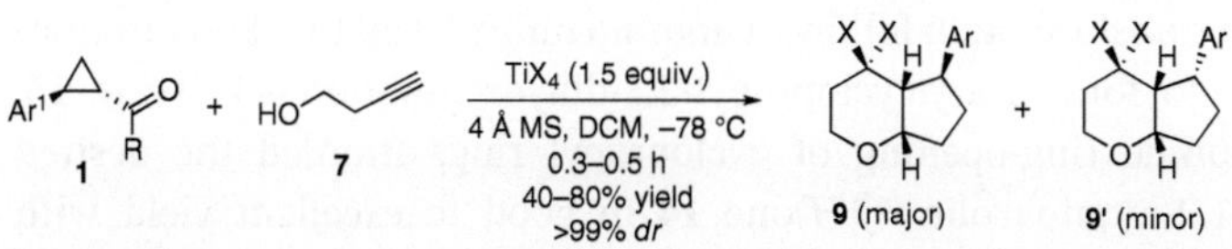

Ar^1 = Ph, 4-OMeC_6H_4, 4-OBnC_6H_4, 4-MeC_6H_4, 2,4,6-$(Me)_3C_6H_2$, 4-i-PrC_6H_4, 4-FC_6H_4, 4-ClC_6H_4, 4-BrC_6H_4, 2-napthyl
X = Cl, Br

Scheme 8.15 Prins-type annulation between cyclopropane carbaldehydes and alkenols/alkynols.

In contemporary time, the thought of exploring the metal-free reactivity of the cyclopropanes led us to test the ability of Bronsted acids to activate cyclopropane carbaldehydes. In 2019, our group reported a *p*-toluenesulphonic acid (*p*TSA) mediated mild and eco-friendly ring-opening/domino ring-opening cyclization reaction of cyclopropane aldehyde **1** and *N*-benzyl aniline **10** toward the construction of substituted 4-aminobutanal **11**/2,3-dihydro-1*H*-benzo[*b*]azepine **12** derivatives. The formation of 4-aminobutanal **11** or 2,3-dihydro-1*H*-benzo[*b*]azepines **12** depends upon the presence of methoxy substituent at the 3 position of phenyl ring of *N*-benzyl aniline **10**. The reaction proceeds through the formation of open chain 4-aminobutanal **11** via the nucleophilic ring-opening with *N*-benyl aniline **10** and cyclopropane aldehyde **1** in the presence of *p*TSA. Further, the aldehyde functionality of **11** gets activated by *p*TSA to undergo an intramolecular Friedel–Crafts type reaction to form the desired benzoazepines **12** (Scheme 8.16). A broad range of substrate scope was supported by the optimized reaction conditions, affording the designed compounds moderate to good yields [3c].

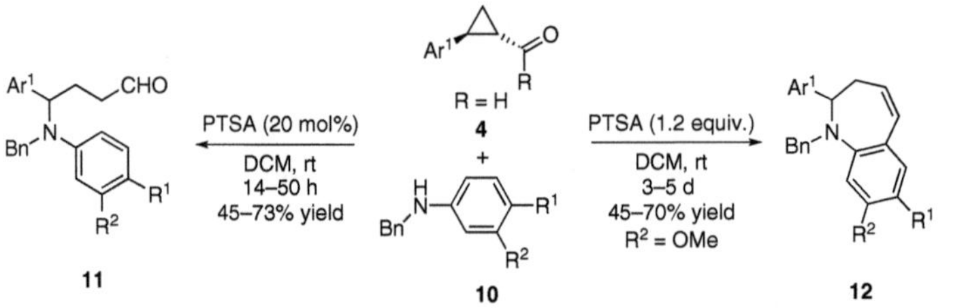

Ar^1 = Ph, 4-OMeC$_6$H$_4$, 4-OBnC$_6$H$_4$, 4-ClC$_6$H$_4$, 4-CH$_3$C$_6$H$_4$, C$_6$H$_4$CH = CH-, 2-napthyl
R^1 = H, OMe, Br, I, F, NO$_2$
R^2 = H, OMe

Scheme 8.16 Synthesis of 4-aminobutanal/2,3-dihydro-1*H*-benzo[*b*]azepine derivatives.

Further investigating the Bronsted acid-catalyzed reactivity of cyclopropane carbaldehydes, Banerjee and group also disclosed a highly regioselective synthesis of tetrahydropyrrolo[1,2-*a*]quinazolin-5(1*H*)one derivatives **14** by reacting cyclopropane carbaldehyde **1** with *N′*-aryl anthranil hydrazides **13** in the presence of *p*TSA. The reported transformation follows a domino imine formation and intramolecular cyclization to form 2-arylcyclopropyl-2,3-dihydroquinolin-4(1*H*)-one **15**, which, on nucleophilic ring-opening of cyclopropyl ring, afforded the desired tetrahydropyrrolo[1,2-*a*]quinazolin-5(1*H*)one **14** in good to excellent yield with complete regioselectivity (Scheme 8.17) [3d]. One-pot sequential and gram scale synthesis of the desired pyrroloquinazolinones also revealed the industrial utility of the reported protocol. The proposed methodology tolerates a great variety of functional groups and thus provides a simple and step-efficient method for the synthesis of pyrroloquinazolinones.

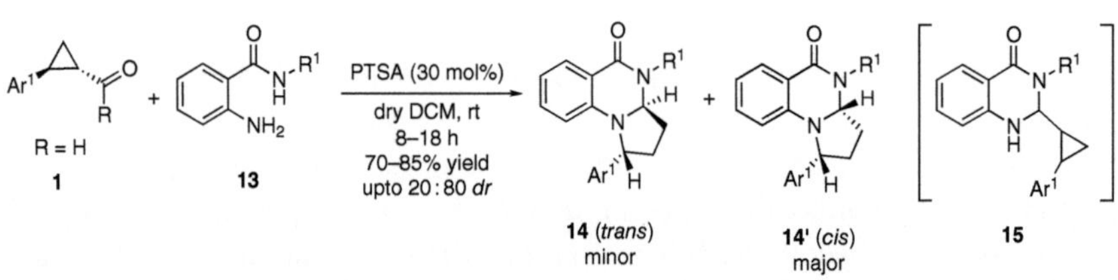

Ar^1 = Ph, 4-FC$_6$H$_4$, 4-ClC$_6$H$_4$, 4-BrC$_6$H$_4$, 4-MeC$_6$H$_4$, 4-OMeC$_6$H$_4$, 3,4-(OMe)$_2$C$_6$H$_3$, 4-Ch = CHC$_6$H$_4$, 2-napthyl, 2-furyl
R^1 = Ph, Bn, 4-FC$_6$H$_4$NH-, 4-ClC$_6$H$_4$NH-, 4-BrC$_6$H$_4$NH-, 4-MeC$_6$H$_4$NH-, 4-*i*-PrC$_6$H$_4$NH-, C$_6$H$_5$CH$_2$NH-

Scheme 8.17 PTSA catalyzed synthesis of tetrahydropyrrolo[1,2-*a*]quinazolin-5(1*H*)ones.

In continuation of this, Banerjee and group in 2020, put forward a convenient additive-free synthesis of dihydro-4*H*-1,2-oxazines **7** via a Cloke–Wilson-type ring expansion of aryl substituted cyclopropane carbaldehydes **1/1′** with the hydroxylamine salt **16** (Scheme 8.18a) [3e]. Here, hydroxylamine salt **16** acted as a precursor

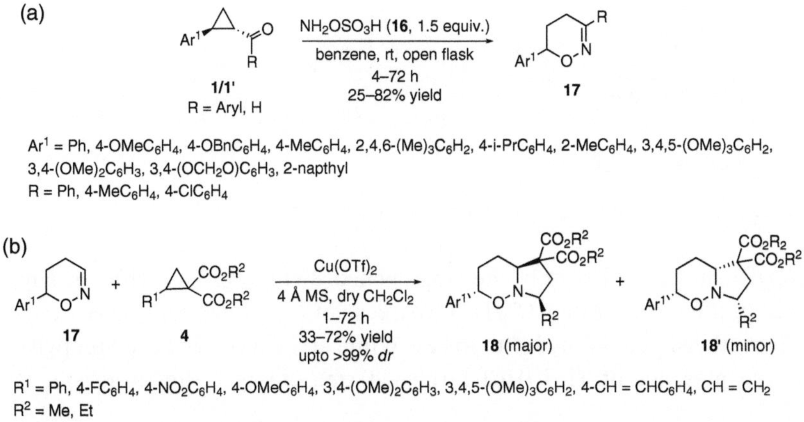

Scheme 8.18 Annulation reaction of cyclopropane carbaldehyde with hydroxylamine salt.

of cyclopropyl aldoximes and underwent rearrangement for the effortless construction of 6-aryl-5,6-dihydro-4*H*-1,2-oxazine **17**. Further to check the synthetic feasibility of the protocol, cyclopropyl ketones **1′** were also subjected to the standard conditions in the presence of catalytic *p*-toluene sulfonic acid monohydrate. The conversion rates were found to be very low as were the percentage yields, which could be due to the less reactivity of ketones as compared to the aldehydes. The transformation is performed in an open-air as it shows negligible sensitivity toward air/moisture. Dihydro-4*H*-1,2-oxazines **17** when subjected to cycloaddition with the cyclopropane diester **4** afforded a trouble-free formulation of the valued hexahydro-2*H*-pyrrolo[1,2-*b*][1, 2]oxazine derivatives **18** (Scheme 8.18b).

8.5.2 Ring Expansion and Ring-Opening Reactions of Cyclopropyl Monocarbonyls

Opening of the three-membered ring of cyclopropyl carbonyls under acidic conditions, followed by the attack of a nucleophile, is an area of considerable interest in organic synthesis. Several reports have been found in the literature discussing the ring-opening and ring expansion of the cyclopropane aldehydes.

Acylsilanes are a useful class of compounds in organic synthesis, and a number of methods for the synthesis of acylsilanes have been developed in the past [21]. Especially, cyclopropyl silyl ketones are expected to exhibit the specific reactivities of the silylcarbonyl group and the three-membered ring. However, the chemical behavior of cyclopropylacylsilanes has not yet been fully clarified because of their synthetic difficulty. In 1986, Nakajima and group reported for the first time the ring-opening reaction of cyclopropylacylsilanes **19** in the presence of Lewis acids or hydrogen chloride (HCl). The reactions were performed at a lower temperature and led to the formation of either 3-chloropropyl trimethylsilyl ketones **20** or 2-trimethylsilyl-4,5-dihydrofurans **21**, depending upon the substituents on the three-membered ring and the acids used (Scheme 8.19). The alkyl substituted

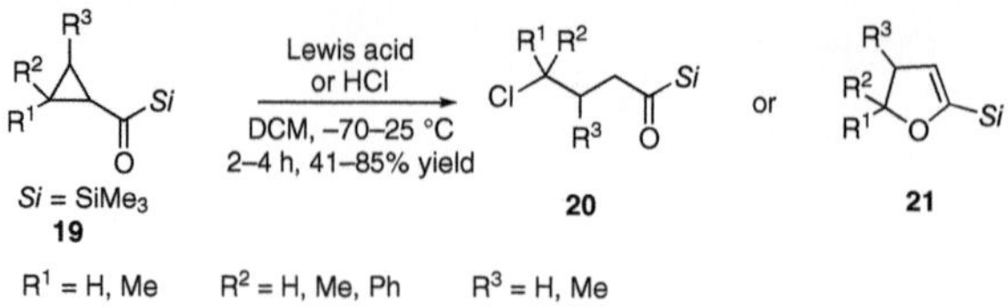

Scheme 8.19 Ring-opening and synthesis of silyl-dihydrofurans.

cyclopropylsilanes afforded only the ring-opened 3-chloropropyl trimethylsilyl ketones **20** in the presence of $TiCl_4$/HCl whereas $SnCl_4$ and BF_3. OEt_2 gave exclusively 2-trimethylsilyl-4,5-dihydrofurans **21**. While aryl substituted cyclopropylsilanes in the presence of $TiCl_4$/HCl furnished ring enlarged dihydrofurans **21** only [22].

In connection with the above findings, in 1995, Nakajima and group also reported that cyclopropylacylsilanes **19** react with strong acids having a low nucleophilic counter anion in aprotic solvents to yield two types of ring-enlargement products, cyclobutanones **23** or 2-silyl-4,5-dihydrofuran derivatives **21**, depending upon the substituents on the three-membered ring of **19** or the acids used. Aryl substituted cyclopropylsilanes **19** in the presence of H_2SO_4 in DME/THF gave cyclobutanones **23** only, while alkyl substituted cyclopropylsilanes **19** led to the formation of 2-silyl-4,5-dihydrofuran derivatives **21**. When solvent changed from DME/THF to MeOH, the formation of trimethylsilyl ketones **20** was observed (Scheme 8.20a) [23].

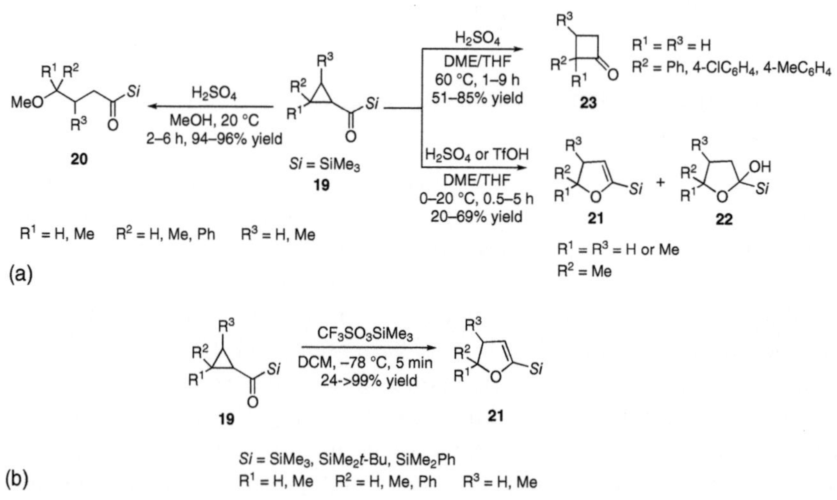

Scheme 8.20 Ring enlargement of cyclopropylacylsilanes to cyclobutanones and 2-silyl-4,5-dihydrofuran.

Futher, they have also put forward that the treatment of cyclopropyl silyl ketones **19** with trimethylsilyl trifluoromethanesulfonate ($CF_3SO_3SiMe_3$) as a strong acid with a low nucleophilic counter anion gives exclusively the corresponding 5-silyl-2,3-dihydrofuran derivatives **20** regardless of substituents on the cyclopropane ring or silicon atom (Scheme 8.20b). To check the synthetic utility of the reported compounds, subsequent reactions with electrophilic reagents or Heck-type reactions were also carried out [24].

In 2011, Sparr and Gilmour for the first time, reported an unprecedented enantioselective strategy for the synthesis of 1,3-dichlorides **24** via organocatalytic activation of the *meso*-cyclopropane carbaldehyde **22**. Not only does activation of cyclopropane carbaldehyde **22** by union with a secondary amine (MacMillan catalyst) **I** facilitate the ring-opening using pyridinium chloride **A** as a nucleophilic chlorinating source, it also generates a second reactive enamine **23** that can be intercepted by an electrophilic chlorinating reagent (perchlorinated quinone) **B**. The communicated protocol constitutes a formal 1,3-addition of Cl_2 across the C1—C2 bond of cyclopropyl carbaldehyde compounds **22** (Scheme 8.21) [25].

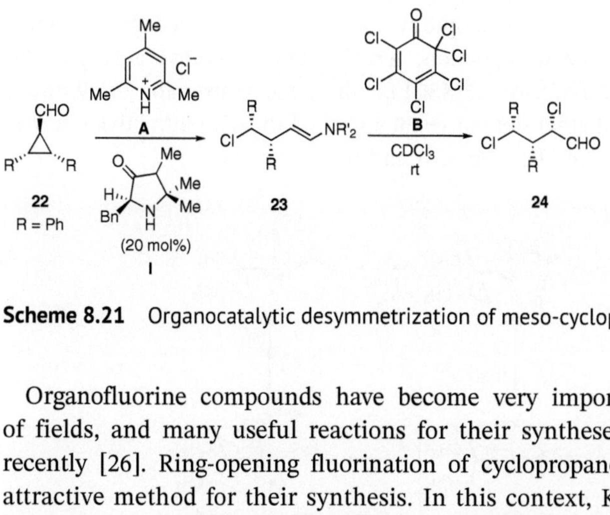

Scheme 8.21 Organocatalytic desymmetrization of meso-cyclopropane carbaldehydes.

Organofluorine compounds have become very important in a wide variety of fields, and many useful reactions for their syntheses have been developed recently [26]. Ring-opening fluorination of cyclopropane derivatives can be an attractive method for their synthesis. In this context, Kirihara et al. disclosed the reaction of arylated cyclopropane carbaldehydes **1** with diethylaminosulfur trifluoride (DAST), effectively causing the ring-opening fluorination to produce homoallylic fluorides **25**. The proposed reaction proceeded via a carbocation intermediate. DAST reacted with cyclopropane carbaldehydes **1** having electron-donating aromatic substituents to cause ring-opening fluorinations with the addition of two fluorine atoms to afford 1,4-difluorobut-1-enes **25**, while cyclopropane carbaldehyde having a *p*-nitrophenyl/alkyl group did not produce the ring-opening product but a (difluoromethyl)cyclopropane **26** was obtained (Scheme 8.22) [27]. An electron-donating aromatic substituent was found essential for the ring-opening fluorination to occur.

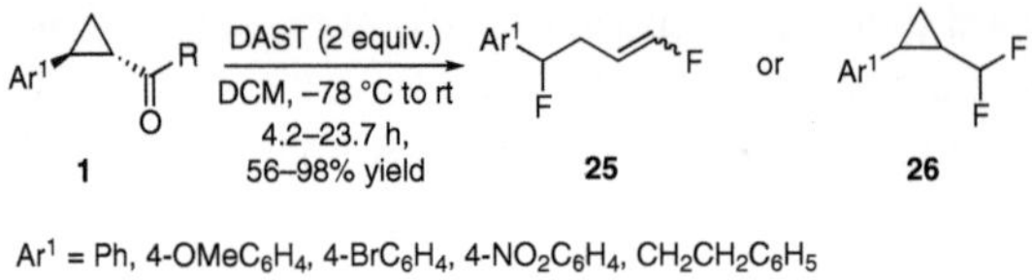

Scheme 8.22 Ring-opening fluorination of cyclopropane carbaldehydes with DAST.

8.6 Application in Total Synthesis

Functionally activated cyclopropanes are versatile synthetic intermediates in organic chemistry because they may undergo facile and predictable ring-opening reactions. Factors contributing to the stereo- and regioselectivity of cyclopropane carbonyls opening under reductive and nucleophilic conditions have been described for a number of systems. In 1997, Huang and Forsyth reported a trimethylsilyl halide induced ring-opening of spiro-fused cyclopropane carbaldehyde, which resulted in the total synthesis of the sesquiterpene natural products trifarienols A and B (Scheme 8.23) [28]. Oxidation to cyclopropyl aldehyde **28** activated the cyclopropane toward dissolving metal reduction, which gave 13-methyl aldehyde **29**. However, the potential use of TMSI/TMSCl for the direct conversion of **29** into a 13-(halo)methyl silyl enol ether derivative **30** appeared to be an attractive alternative towards the formation of Trifarienols A and B.

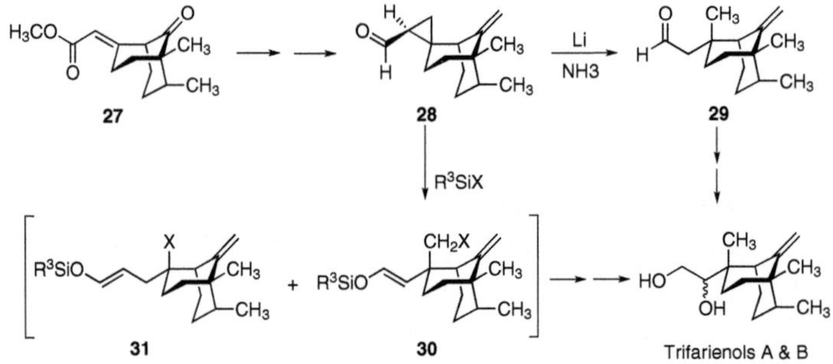

Scheme 8.23 Total synthesis of trifarienols A and B.

References

1 Selected reviews on donor-acceptor cyclopropanes: (a) Reissig, H.-U. and Zimmer, R. (2003). *Chem. Rev.* 103: 1151–1196. (b) Schneider, T.F., Kaschel, J., and Werz, D.B. (2014). *Angew. Chem. Int. Ed.* 53: 5504–5523. (c) Pandey, A.K., Ghosh, A., and Banerjee, P. (2016). *Isr. J. Chem.* 56: 512–521. (d) Budynina, E.M., Ivanov, K.L., Sorokin, I.D., and Melnikov, M.Y. (2017). *Synthesis* 49: 3035–3068. (e) Singh, P., Varshnaya, R.K., Dey, R., and Banerjee, P. (2020). *Adv. Synth. Catal.* 362: 1447–1484. (f) Ghosh, A., Dey, R., and Banerjee, P. (2021). *Chem. Commun.* 57: 5359–5373.

2 Kharasch, M.S. and Cooper, J.H. (1945). *J. Org. Chem.* 10: 46–54.

3 (a) Dey, R., Kumar, P., and Banerjee, P. (2018). *J. Org. Chem.* 83: 5438–5449. (b) Kumar, P., Dey, R., and Banerjee, P. (2018). *Org. Lett.* 20: 5163–5166. (c) Dey, R. and Banerjee, P. (2019). *Adv. Synth. Catal.* 361: 2849–2854. (d) Singh, P., Kaur, N., and Banerjee, P. (2020). *J. Org. Chem.* 85: 3393–3406. (e) Kumar, P., Kumar, R., and Banerjee, P. (2020). *J. Org. Chem.* 85: 6535–6550. (f) Dey, R., Rajput, S., and Banerjee, P. (2020). *Tetrahedron* 15: 131080.

4 (a) Burgi, H.B., Dunitz, J.D., Lehn, J.M., and Wipff, G. (1974). *Tetrahedron* 30: 1563–1572. (b) Burgi, H.B., Dunitz, J.D., and Shefter, E. (1973). *J. Am. Chem. Soc.* 95: 5065–5067.

5 (a) Cloke, J.B. (1929). *J. Am. Chem. Soc.* 51: 1174–1187. (b) Wilson, C.L. (1947). *J. Am. Chem. Soc.* 69: 3002–3004. (c) Hudlicky, T. and Reed, J.W. (2010). *Angew. Chem. Int. Ed.* 49: 4864–4876.

6 (a) Corey, E.J. and Chaykovsky, M. (1962). *J. Am. Chem. Soc.* 84: 867. (b) Corey, E.J. and Chaykovsky, M. (1965). *J. Am. Chem. Soc.* 87: 1353.

7 (a) Nalesnik, T.E. and Orchin, M.J. (1981). *Organomet. Chem.* 222: C5. (b) Nalesnik, T.E., Freudenberger, J.H., and Orchin, M.J. (1982). *Organomet. Chem.* 236: 95. (c) Matsui, Y. and Orchin, M.J. (1983). *Organomet. Chem.* 244: 369.

8 Jessop, P.G., Ikariya, T., and Noyori, R. (1995). *Organometallics* 14: 1510.

9 Sherrill, W.M. and Rubin, M. (2008). *J. Am. Chem. Soc.* 130: 13804–13809.

10 Melancon, B.J., Perl, N.R., and Taylor, R.E. (2007). *Org. Lett.* 9: 1425–1428.

11 Taylor, R.E., Risatti, C.A., Engelhardt, F.C., and Schmitt, M.J. (2003). *Org. Lett.* 5: 1377–1379.

12 Risatti, C.A. and Taylor, R.E. (2004). *Angew. Chem. Int. Ed.* 43: 6671–6672.

13 (a) Trost, B.M., Vladuchick, W.C., and Bridges, A.J. (1980). *J. Am. Chem. Soc.* 102: 3548. (b) Trost, B.M. and Jungheim, L. (1980). *J. Am. Chem. Soc.* 102: 7910.

14 (a) Cohen, T., Daniewski, W.M., and Weisenfeld, R.B. (1978). *Tetrahedron Lett.* 47: 4665. (b) Bernard, A.M., Frongia, A., Piras, P.P. et al. (2005). *Org. Lett.* 7: 4565.

15 (a) Tanaka, K., Uneme, H., Matsui, S. et al. (1980). *Chem. Lett.* 287. (b) Bernard, A.M., Cadoni, E., Frongia, A. et al. (2002). *Org. Lett.* 4: 2565. (c) Bernard, A.M., Frongia, A., Guillot, R. et al. (2007). *Org. Lett.* 9: 541.

16 Porcu, S., Luridiana, A., Martis, A. et al. (2018). *Chem. Commun.* 54: 13547–13550.

17 (a) Doyle, M.P., Mckervey, M.A., and Ye, T. (1998). *Modern Catalytic Methods for Organic Synthesis With Diazo Compounds: From Cyclopropanes to Ylides*. New York, Chichester, Weinheim, Brisbane, Singapore, Toronto: Springer. (b) Moss, R.A. and Doyle, M.P. (ed.) (2014). *Contemporary Carbene Chemistry*. Hoboken, NJ: Wiley.

18 Chanthamath, S.H., Mandour, S.A., Tong, T.M.T. et al. (2016). *Chem. Commun.* 52: 7814–7817.

19 Cao, L.Y., Luo, J.N., Yao, J.S. et al. (2021). *Angew. Chem. Int. Ed.* 60: 15254–15259.

20 (a) Zhang, J., Jiang, H., and Zhu, S. (2017). *Adv. Synth. Catal.* 359: 2924–2930. (b) Zhu, X., Hong, G., Hu, C. et al. (2017). *Eur. J. Org. Chem.* 2017: 1547–1551.

21 For recent reviews on the synthesis and reaction of acylsilanes, see: (a) Brook, M.A. (2000). *Silicon in Organic, Organometallic, and Polymer Chemistry. New York: John Wiley & Sons.* (b) Bonini, B.F., Comes Franchini, M., Fochi, M. et al. (1998). *Organomet. Chem.* 567: 181–189. (c) Qi, H. and Curran, D.P. (1995). Acyl silicon, germanium, or boron functions. In: *Comprehensive Organic Functional Group Transformations: Synthesis: Carbon with Two Attached Heteroatoms with at Least One Carbon-to-Heteroatom Multiple Link*, vol. 5, Chapter 5.09, (ed. A.R. Katrizky, O. Meth-Cohn, C.W. Rees, and C.J. Moody), 409–433. Oxford: Pergamon Press. (d) Cirillo, P.F. and Panek, J.S. (1992). *Org. Prep. Proced. Int.* 24: 553–582. (e) Page, P.C.B., Klair, S.S., and Rosenthal, S. (1990). *Chem. Soc. Rev.* 19: 147–195. (f) Ricci, A. and DeglInnocenti, A. (1989). *Synthesis* 647–660.

22 Nakajima, T., Miyaji, H., Segi, M., and Suga, S. (1986). *Chemistry Lett.* 15: 181–182.

23 Nakajima, T., Segi, M., Mituoka, T. et al. (1995). *Tetrahedron Lett.* 36: 667–1670.

24 Honda, M., Naitou, T., Hoshino, H. et al. (2005). *Tetrahedron Lett.* 46: 7345–7348.

25 Sparr, C. and Gilmour, R. (2011). *Angew. Chem. Int. Ed.* 50: 8391–8395.

26 (a) Kirsch, P. (2013). *Modern Fluoroorganic Chemistry. Weinheim: Wiley-VCH Verlag GmbH & Co KGaA.* (b) Bégué, J.-P. and Bonnet-Delpon, D. (2008). *Bioorganic and Medicinal Chemistry of Fluorine. Hoboken: John Wiely & Sons Inc.* (c) Uneyama, K. (2006). *Organofluorine Chemistry. Oxford: Blackwell Publishing Ltd.* (d) Inoue, M., Sumii, Y., and Shibata, N. (2020). *ACS Omega* 5: 10633.

27 Kirihara, M., Kikkawa, Y., Nakamura, R. et al. (2021). *Tetrahedron Lett.* 64: 152655.

28 Huang, H. and Forsyth, C.J. (1997). *Tetrahedron* 53: 16341–16348.

9

Chemistry of Aroyl- and Nitro-Substituted Donor–Acceptor Cyclopropanes

Thangavel Selvi and Kannupal Srinivasan

School of Chemistry, Bharathidasan University, Tiruchirappalli, Tamil Nadu, 620024, India

CHAPTER MENU

9.1 Introduction

The fascinating chemistry of donor—acceptor (D–A) cyclopropanes continues to attract the attention of numerous chemists around the globe [1–12]. The exploration of the reactivity of different D–A cyclopropanes has resulted in the discovery of novel synthetic routes to a wide variety of acyclic products, carbocycles, heterocycles, and natural products. The horizons of the field are ever expanding with the introduction of new D–A cyclopropanes, reaction partners, reagents, catalysts, and synthetic strategies in the field.

Aroyl- and nitro-substituted D–A cyclopropanes **1** and **2** (Figure 9.1) are relatively new entrants in the area of D–A cyclopropane chemistry [13]. Although few derivatives of **1** and **2** were reported many decades ago [14, 15], their synthetic potential has been intently being investigated in the past decade. Apart from the usual aryl donor and diester acceptor substituents in vicinal positions, the presence of an extra substituent (aroyl/nitro) in the third position of these cyclopropanes makes them unique building blocks in the field.

Donor–Acceptor Cyclopropanes in Organic Synthesis, First Edition.
Edited by Prabal Banerjee and Akkattu T. Biju.

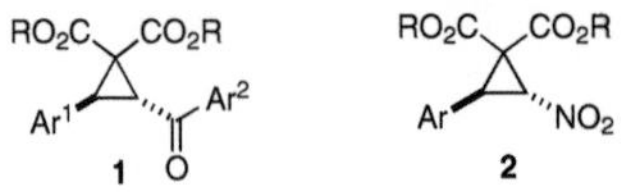

Figure 9.1 Structures of aroyl- and nitro-substituted D–A cyclopropanes.

9.2 Synthesis of Aroyl-Substituted D–A Cyclopropanes

Aroyl-substituted D–A cyclopropanes **1** have been invariably synthesized (in both racemic and chiral forms) by Michael initiated ring closure (MIRC) methodology, as summarized in Scheme 9.1. Accordingly, these cyclopropanes were synthesized by the reactions of (i) vinyl sulfonium salts **3** with dialkyl malonate **4** in the presence of NaOH in ethanol [14], (ii) arylidene malonates **5** with α-chloroacetophenones **6** in the presence of K_2CO_3 in xylene [16], (iii) α-bromochalcone **7** with dialkyl malonate **4** in the presence of K_2CO_3 in DMSO [17], and (iv) chalcones **8** with dialkyl bromomalonate (**9**) in the presence of cinchona alkaloid-derived chiral phase transfer catalysts and K_2CO_3 in mesitylene [18], or carbohydrate-derived chiral crown ethers and Na_2CO_3 in diethylether-THF [19]. They were also synthesized by the oxidative cyclization of Michael adducts of chalcones with malonates **10** using iodosobenzene and Bu_4NI in MeOH [20] or Bu_4NI and TBHP in THF [21].

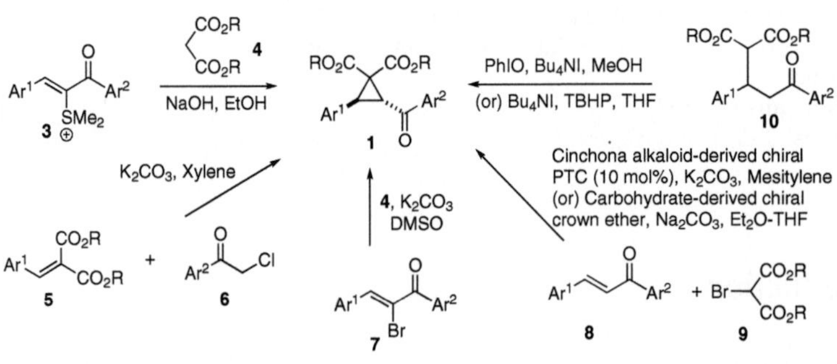

Scheme 9.1 Literature methods for the synthesis of aroyl-substituted D–A cyclopropanes.

When we started our research work in the area of D–A cyclopropanes in 2010, we required a variety of aroyl-substituted D–A cyclopropanes **1** for our work. So, we developed a convenient procedure for the synthesis of these cyclopropanes from Michael adducts of chalcones with dialkyl malonates **10** (Table 9.1) [22]. Accordingly, when Michael adducts **10** were treated with 2 equiv. each of iodine and DBU in toluene or DCM, D–A cyclopropanes **1** were obtained in yields of 86–95%. The reaction tolerated the presence of electron-donating, electron-withdrawing, and halogen substituents on both aromatic rings, and also heteroaromatic rings. Mechanistically, the base (DBU) first abstracts the more acidic malonate proton of **10**, and the resulting carbanion attacks iodine to give iodinated Michael adduct **11** as an intermediate. Next, the base removes the other acidic proton (adjacent to the keto group) in **11** and the anion so formed displaces iodide in an intramolecular S_N2 attack to give

Table 9.1 Synthesis of various aroyl-substituted D–A cyclopropanes.

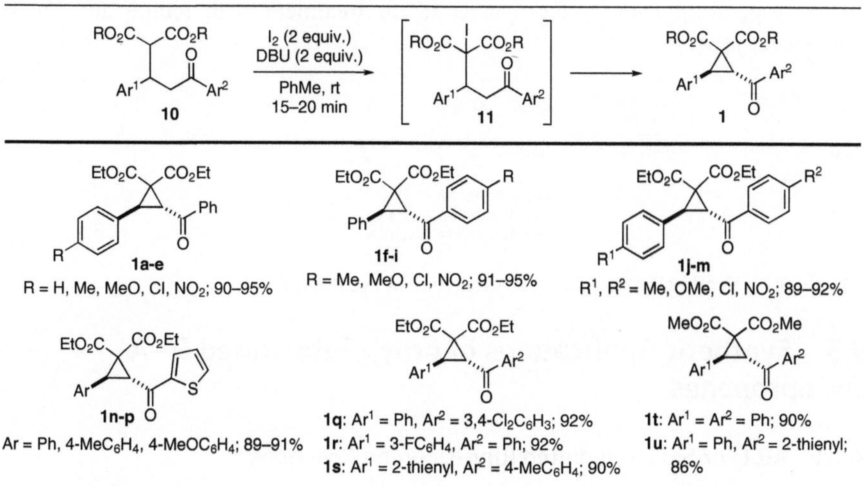

cyclopropanes **1**. The advantages of our procedure include the use of readily accessible Michael adducts as starting materials and the formation of products as single *trans*-diastereomers in good-to-excellent yields. The exclusive formation of *trans*-diastereomers in these reactions is attributed to DBU-induced isomerization of *cis*-isomers, formed if any, to more stable *trans*-isomers.

The above-mentioned methodology could also be extended to Michael adducts derived from other chalcone-like compounds. For example, Michael adducts of vinylogous chalcones with diethyl malonate **12** (obtained by the condensation of cinnamaldehydes with acetophenones followed by Michael addition of diethyl malonate) yielded a series of vinyl D–A cyclopropanes **13** under the same reaction conditions (Table 9.2) [23].

Table 9.2 Synthesis of aroyl-substituted vinyl D–A cyclopropanes.

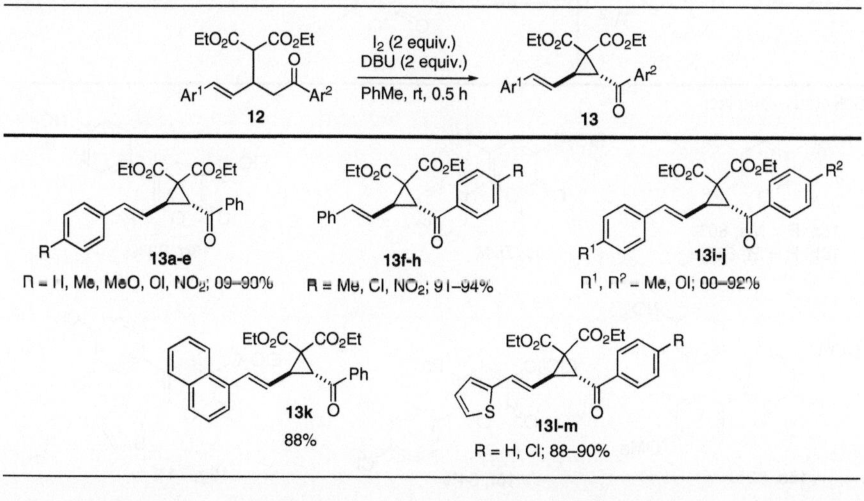

Similarly, Michael adducts **14**, obtained from diarylideneacetones, also furnished the corresponding D–A cyclopropanes **15** on treatment with iodine and DBU (Scheme 9.2) [24].

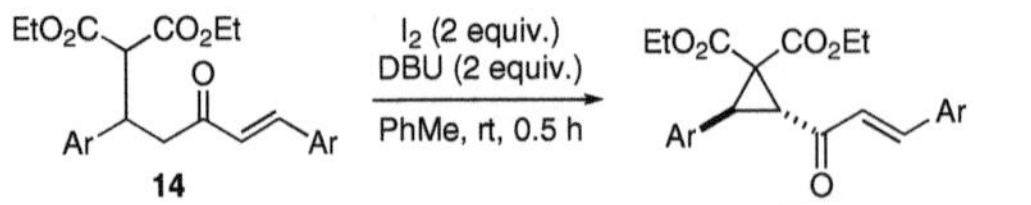

15a: Ar = Ph; 71%
15b: Ar = 4-MeC$_6$H$_4$; 74%
15c: Ar = 4-OMeC$_6$H$_4$; 67%

Scheme 9.2 Synthesis of acryloyl-substituted D–A cyclopropanes.

9.3 Synthetic Applications of Aroyl-Substituted D–A Cyclopropanes

9.3.1 $AlCl_3$ or $SnCl_4$-Mediated Ring-Opening Reactions

D–A cyclopropanes, when treated with Lewis acids, are capable of forming putative 1,3-zwitter ionic intermediates, which may undergo further transformations to yield useful products. With this view, aroyl-substituted D–A cyclopropanes **1** were treated with stoichiometric amounts of various Lewis acids. When these cyclopropanes were treated with 1 equiv. of $AlCl_3$ (or $SnCl_4$) in dichloromethane (or 1,2-DCE) at room temperature, they furnished 2-pyrones **16** (Table 9.3) [24]. A mechanism has

Table 9.3 Synthesis of 2-pyrones from aroyl-substituted D–A cyclopropanes.[a]

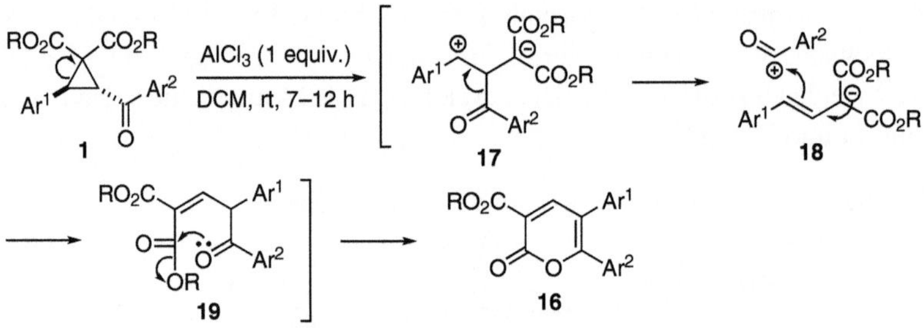

Selected examples:

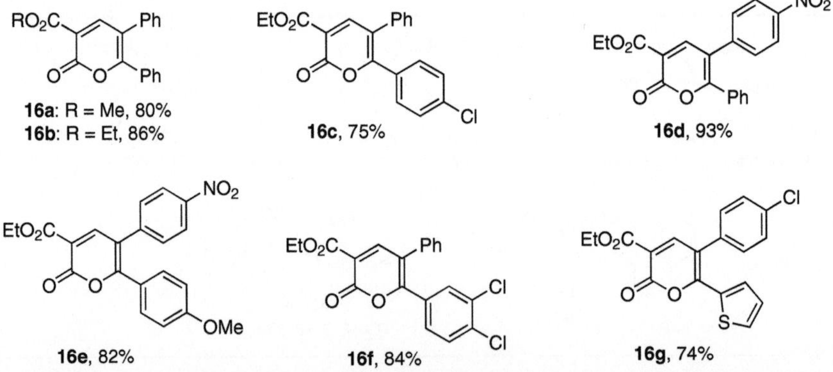

16a: R = Me, 80%
16b: R = Et, 86%

16c, 75%

16d, 93%

16e, 82%

16f, 84%

16g, 74%

a) In the mechanism, the Lewis acid coordination is not shown in the intermediate structures for clarity.

been formulated for the transformation, which involves the formation of 1,3-zwitterionic intermediate **17** through ring-opening, and this intermediate undergoes fragmentation to give ion pair **18**, followed by recombination of the fragments to form δ-keto ester **19** and lactonization of **19** to give 2-pyrone **16**. Except for a cyclopropane having *p*-nitrophenyl group as Ar^1 (because, *p*-nitrophenyl group is not capable of stabilizing the positive charge of the respective zwitterionic intermediate), all other cyclopropanes yielded 2-pyrones **16** in good-to-excellent yields.

Not only aroyl-substituted D–A cylopropanes **1** but also acryloyl-substituted D–A cyclopropanes **15** also gave the corresponding pyrones **19** in good yields (Scheme 9.3) [24]. The double bond of the acryloyl moiety remained unchanged during the transformation.

Scheme 9.3 Synthesis of 2-pyrones from acryloyl-substituted D–A cyclopropanes.

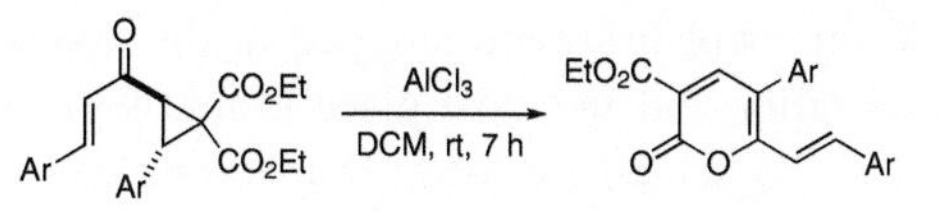

15

19a: Ar = Ph, 70%
19b: Ar = 4-MeC$_6$H$_4$, 71%
19c: Ar = 4-OMeC$_6$H$_4$, 80%

On the other hand, vinyl D–A cyclopropanes **13** upon treatment with $SnCl_4$ in dichloromethane, formed cyclopentenes **20** (Table 9.4) [23]. The transformation is basically a vinylcyclopropane–cyclopentene rearrangement, and it proceeds through the formation of 1,3-zwitterionic intermediate **21**, followed by cyclization to give cyclopentene **22**, which being a β,γ-unsaturated ketone, automatically rearranges to more stable α,β-unsaturated ketone, that is, cyclopentene **20**. However, the transformation does not work for a cyclopropane having a *p*-nitrophenyl group as Ar^1.

Table 9.4 Synthesis of cyclopentenes from aroyl-substituted vinyl D–A cyclopropanes.

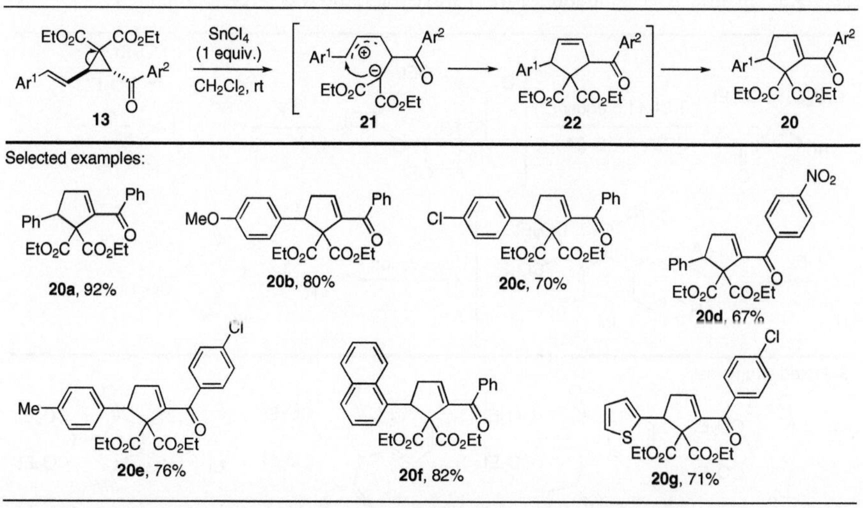

9.3.2 $TiCl_4$-Mediated Ring-Opening Reactions

During the course of the above-mentioned studies, we observed that the cyclopropanes yielded very different products when treated with $TiCl_4$ [24]. Thus, aroyl-substituted D–A cyclopropanes **1** when treated with 1.5 equiv. of $TiCl_4$ in dichloromethane at room temperature, they furnished 1-indanones **23** (Table 9.5). The transformation takes place through a partial cleavage of the C—C bond of **1** in between the Ar^1 and diester groups, S_Ni-like attack of chloride ion of $TiCl_4$ on the incipient carbocation to give an intimate ion-pair **24**, *E*2-like elimination of HCl from **24** to yield *E*-alkene **25** stereoselectively, and subsequent Nazarov cyclization to give the products **23**. Surprisingly, a cyclopropane having *p*-nitrophenyl group as Ar^1 was also compatible for the transformation.

When the above-mentioned procedure was extended to vinyl D–A cyclopropanes **13**, the transformation stopped at the alkene stage, and the subsequent Nazarov cyclization did not take place (Table 9.6) [23]. Nevertheless, a series of *E*,*E*-1,3-dienes **26**, which are versatile synthetic intermediates, were obtained in good yields in the transformations.

9.3.3 Ring-Opening Reactions with Hydrazines

Nucleophilic ring-opening reactions of D–A cyclopropanes with nitrogen nucleophiles provide access to various heterocyclic compounds [8, 10]. Generally, the nucleophile attacks the carbon to which the donor group is attached, resulting in the ring-opening of the cyclopropanes, and the intermediates so formed further undergo cyclization to give the products.

When aroyl-substituted D–A cyclopropanes **1** were heated under reflux conditions with aryl hydrazines **27** in EtOH, they gave 4,5-*trans*-substituted dihydropyrazoles **28** (Table 9.7). The products were obtained from the reaction mixture by

Table 9.5 Synthesis of 1-indanones from aroyl-substituted D–A cyclopropanes.

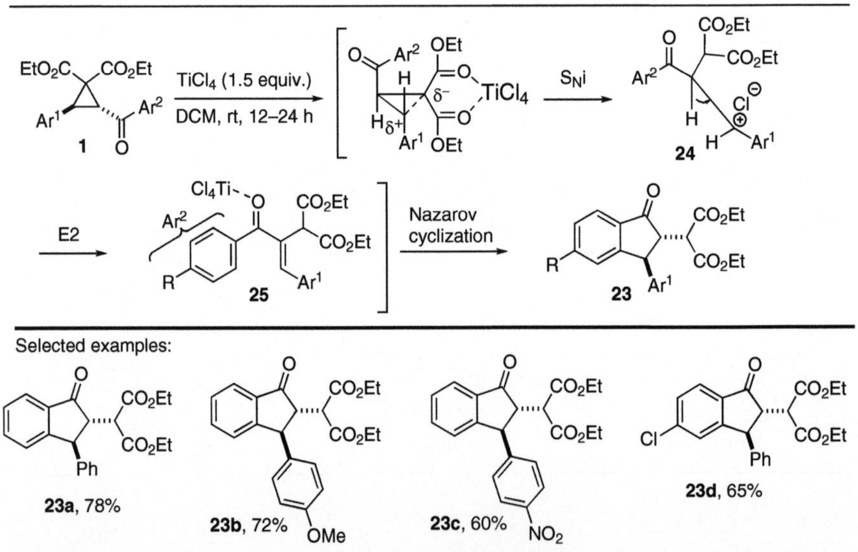

Table 9.6 Synthesis of *E,E*-1,3-dienes from aroyl-substituted vinyl D–A cyclopropanes.

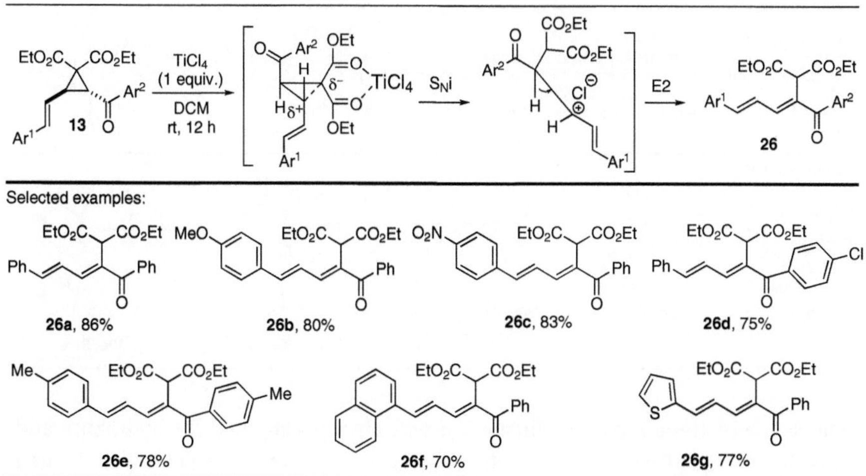

Table 9.7 Synthesis of dihydropyrazoles from aroyl-substituted D–A cyclopropanes.

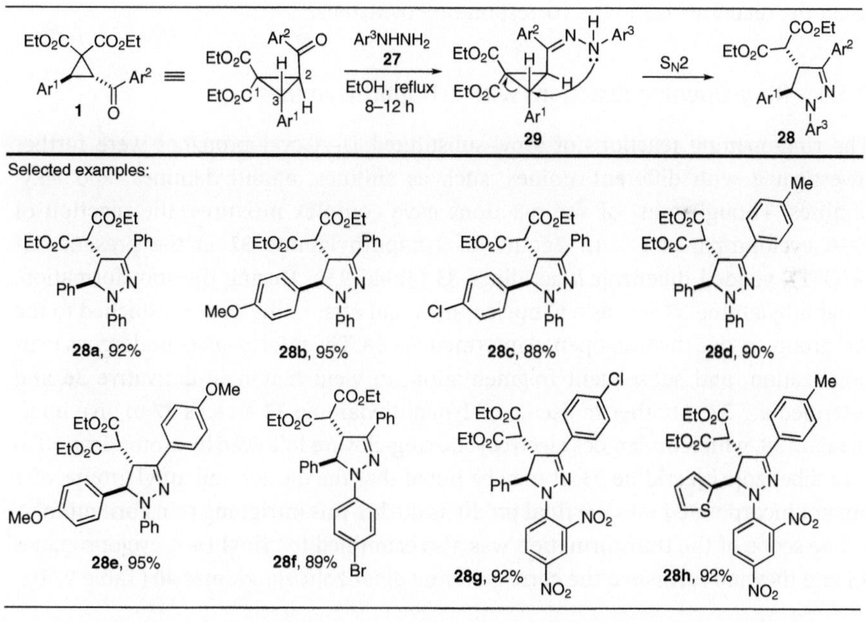

simple crystallization and did not require any column chromatographic purification. More importantly, the reaction did not require the assistance of any catalyst. The transformation proceeds through the formation of arylhydrazone **29** as an intermediate and subsequent intramolecular attack of the other nitrogen of hydrazone moiety on the carbon attached to the Ar^1 group in S_N2 fashion to give the products **28** diastereoselectively.

Table 9.8 Synthesis of pyridazinones from aroyl-substituted D–A cyclopropanes.

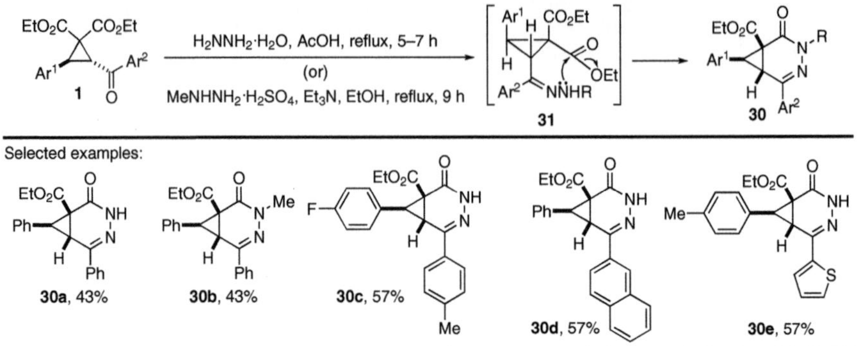

The scope of the above transformation was also investigated for hydrazine and methyl hydrazine (Table 9.8). Accordingly, when D–A cyclopropanes **1** were heated under reflux conditions with methyl hydrazine in ethanol or hydrazine in AcOH, cyclopropane-fused pyridazinones **30** were produced through the intermediates **31** in moderate-to-good yields. It may be noted that when hydrazine in ethanol was used, the reactions led to the corresponding pyrazoles.

9.3.4 Ring-Opening Reactions with 1-Naphthylamines

The ring-opening reactions of aroyl-substituted D–A cyclopropanes were further investigated with different amines, such as anilines, naphthylamines, and alkylamines. Though most of the reactions gave complex mixtures, the reaction of D–A cyclopropanes **1** with 2 equiv. of 1-naphthylamine **32** in the presence of $Sc(OTf)_3$ yielded dibenzo[*c,h*]acridines **33** (Table 9.9). During the transformation, 1-naphthylamine **32** acts as a C-nucleophile, and attacks the carbon attached to the Ar^1 group, to give the ring-opened intermediate **34**. The intermediate undergoes rearomatization, and subsequent fragmentation, to yield malonate derivative **36** and intermediate **37**. Another molecule of 1-naphthylamine **32** attacks **37** to give intermediate **38**, which undergoes electrocyclic ring-closure followed by aromatization to give dibenzo[*c,h*]acridine **33**. It may be noted that the diester and aroyl groups of **1** are not incorporated into the final products during this intriguing transformation.

The scope of the transformation was also examined for vinyl D–A cyclopropanes **13** and they too furnished the corresponding dibenzo[*c,h*]acridines **40** (Table 9.10).

9.3.5 (3 + 2) Annulations with Nitriles

The (3 + 2) annulation of D–A cyclopropanes with nitriles is an efficient strategy for the stereoselective synthesis of 1-pyrroline derivatives [7]. Other nitrogen heterocycles such as pyrroles and pyrrolidines, could also be easily accessed via 1-pyrrolines based on this strategy. The seminal work of Pagenkopf and coworkers on (3 + 2) annulation of carbohydrate-derived D–A cyclopropanes with various nitriles has acted as a stimulus for various works in this area [28].

Table 9.9 Synthesis of dibenzo[*c*,*h*]acridines from aroyl-substituted D–A cyclopropanes.

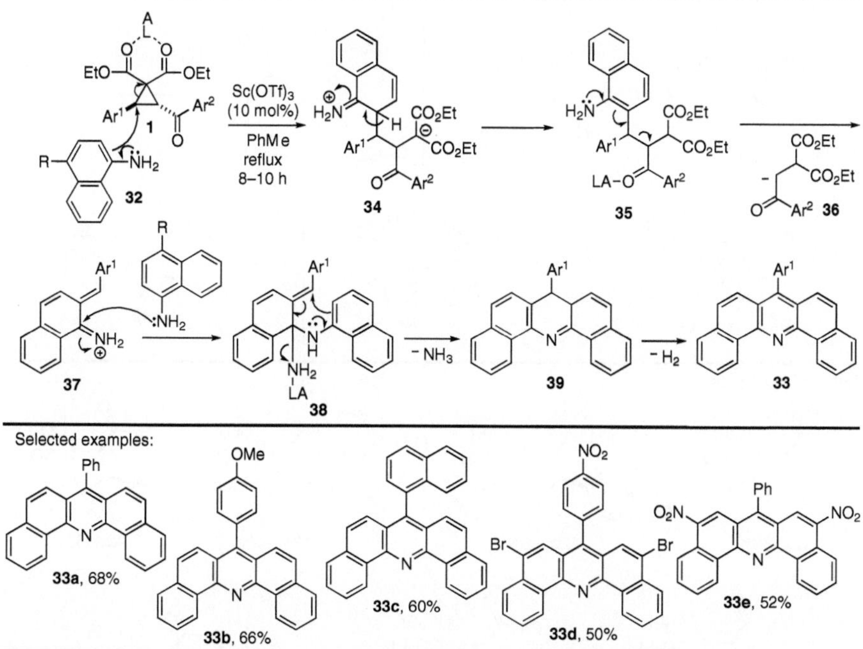

Table 9.10 Synthesis of dibenzo[*c*,*h*]acridines from aroyl-substituted vinyl D–A cyclopropanes.

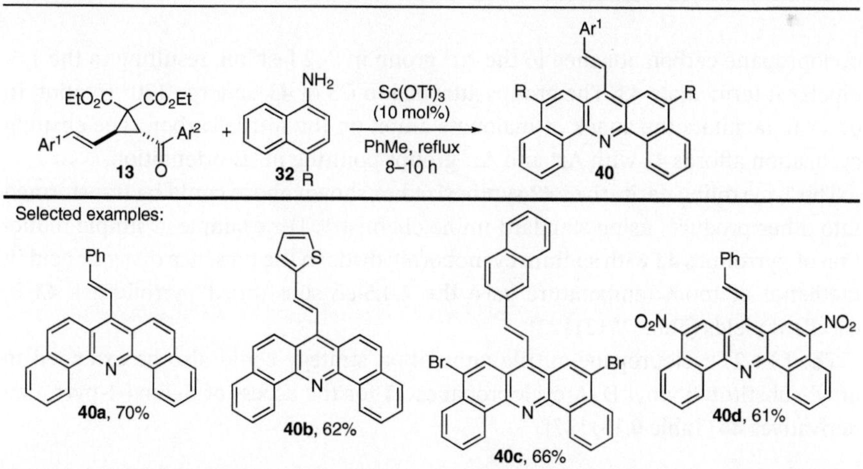

When aroyl-substituted D–A cyclopropanes **1** were subjected to (3 + 2) annulation with aliphatic and aromatic nitriles **41** in the presence of 1 equiv. of $SnCl_4$ in 1,2-DCE, they readily yielded the corresponding 1-pyrroline derivatives **42** in a highly diastereoselective manner (Table 9.11) [22]. The diastereoselectivity observed in the products could be explained as follows. The coordination of $SnCl_4$ to the diester moiety weakens the C1—C3 bond of the cyclopropane, and the nitrile **41** attacks the

Table 9.11 Synthesis of 1-pyrrolines from aroyl-substituted D–A cyclopropanes.

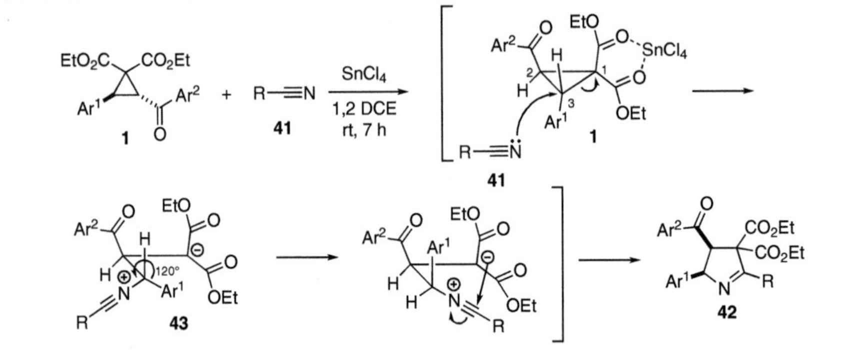

Selected examples:

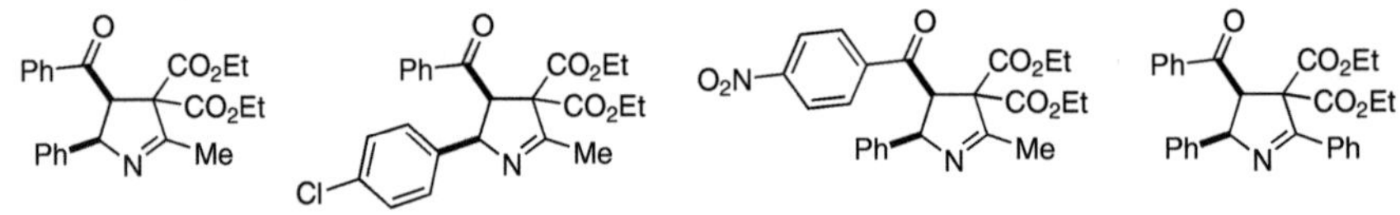

42a, 89% **42b**, 86% **42c**, 90% **42d**, 66%

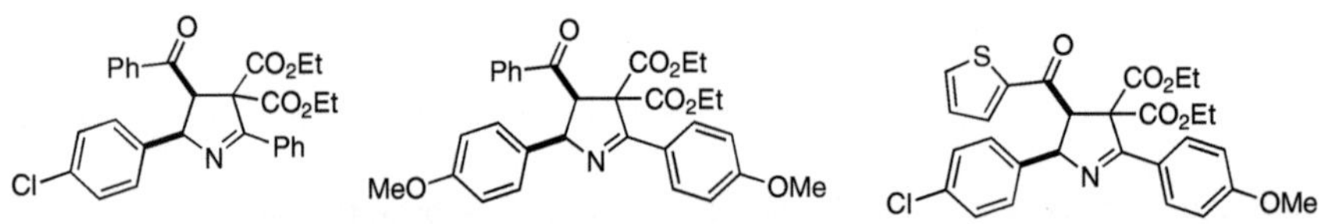

42e, 51% **42f**, 67% **42g**, 59%

cyclopropane carbon attached to the Ar[1] group in S_N2 fashion, resulting in the 1,5-dipolar intermediate **43**. The groups attached to C3 of **43** undergo 120° rotation in order to facilitate the attack of malonate anion on the nitrile carbon. The ensuing cyclization affords **42** with Ar[1] and Ar[2] groups pointing in *cis*-orientation.

The 1-pyrroline derivatives **42** synthesized as shown above could be transformed into other products using standard imine chemistry. For example, a simple reduction of pyrrolines **42** with sodium cyanoborohydride in the presence of acetic acid in methanol at room temperature gave the 2,4,5-*cis*-substituted pyrrolidines **43** in excellent yields (Table 9.12) [22].

The (3 + 2) cyclopropane–nitrile annulation strategy could also be extended to aroyl-substituted vinyl D–A cyclopropanes **13** for the access of 5-vinyl-1-pyrroline derivatives **44** (Table 9.13) [27].

9.3.6 (3 + 3) Annulation with Mercaptoacetaldehyde

As compared with (3+2) annulations of D–A cyclopropanes, only less number of (3 + 3) and other higher-order annulations have been reported in the literature [3]. Especially, for aroyl-substituted D–A cyclopropanes such higher-order annulations largely remain unexplored.

Table 9.12 Conversion of 1-pyrrolines to pyrrolidines.

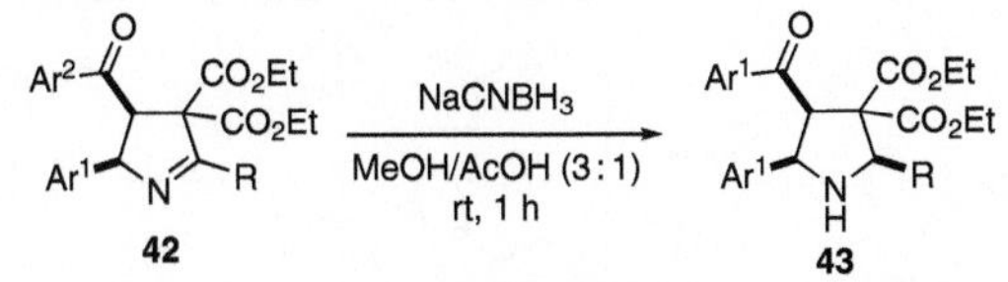

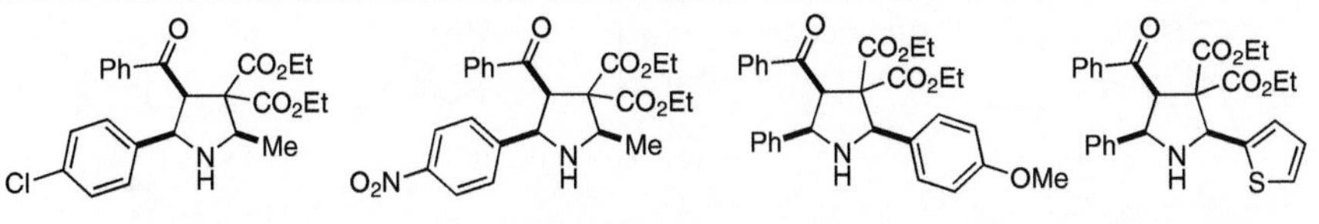

43a, 90% **43b**, 92% **43c**, 93% **43d**, 85%

Table 9.13 Synthesis of vinyl pyrrolines from vinyl D–A cyclopropanes.

44h, 65%

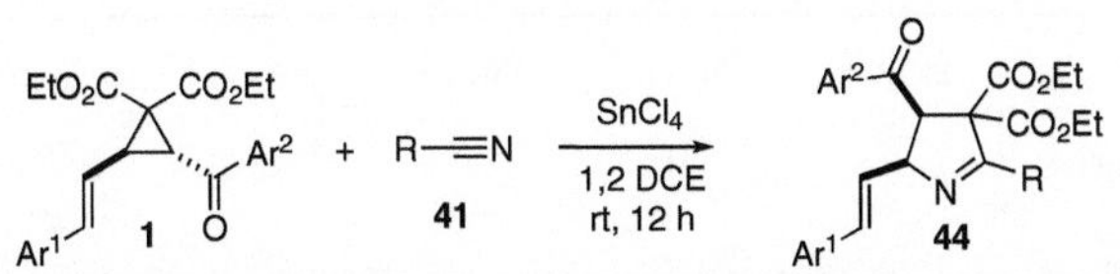

Selected examples:

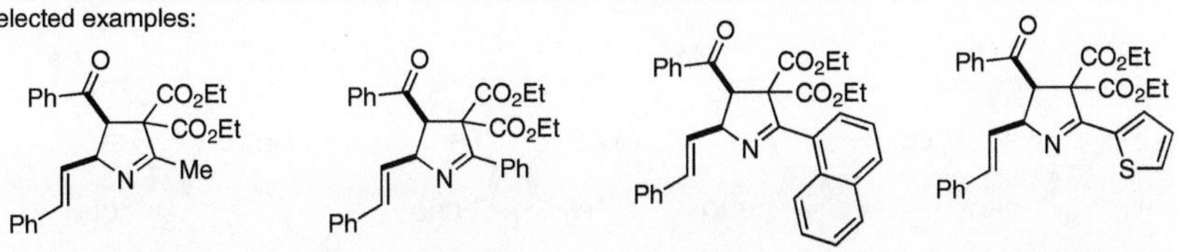

44a, 88% **44b**, 86% **44c**, 76% **44d**, 80%

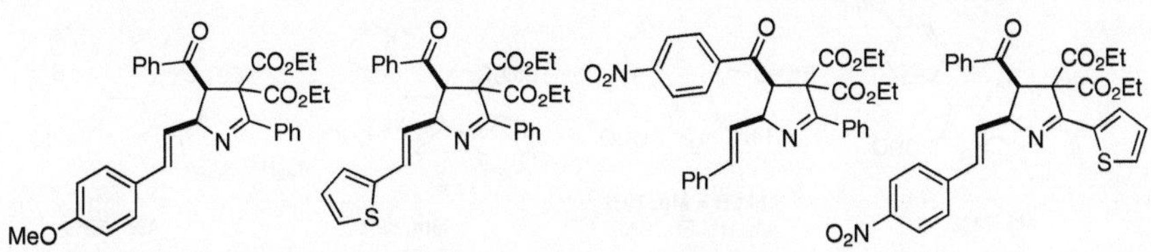

44e, 78% **44f**, 72% **44g**, 68%

1,4-Dithiane-2,5-diol (**45**), also called mercaptoacetaldehyde dimer, is an interesting synthetic precursor and capable of releasing the monomer, mercaptoacetaldehyde (**45′**) in situ when treated with tertiary amines or certain organocatalysts (Scheme 9.4) [30, 31].

Scheme 9.4 Generation of mercaptoacetaldehyde in situ.

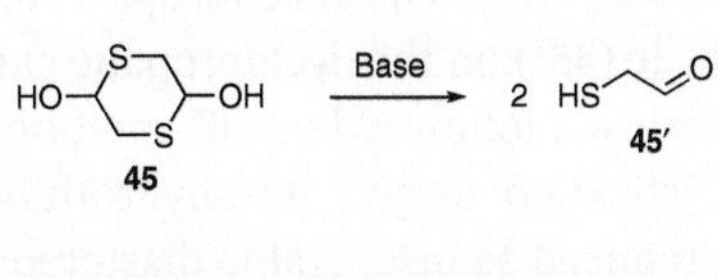

We envisaged that the treatment of **45** with a suitable Lewis acid would also form the monomer **45′** in situ (through coordination with sulfur), and the

Table 9.14 Synthesis of thiophenes from aroyl-substituted D–A cyclopropanes.[a]

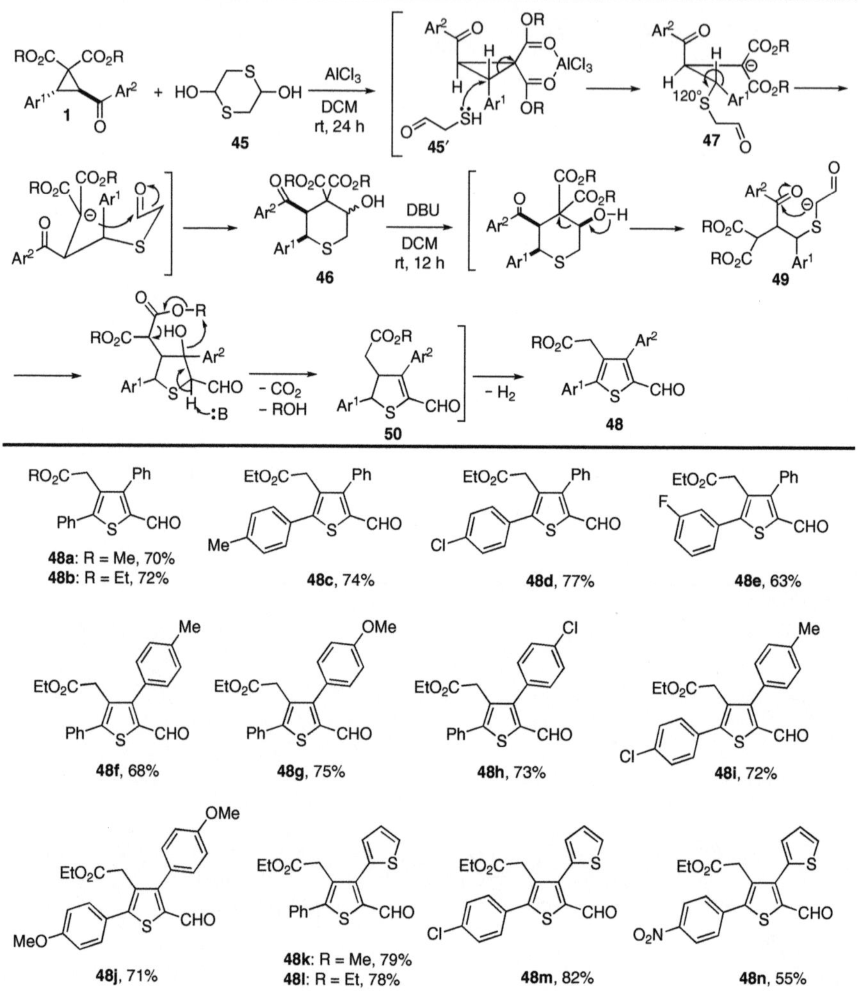

a) Isolated overall yield.

monomer could be used for (3 + 3) annulation with aroyl-substituted D–A cyclopropanes **1** under the same reaction conditions. Accordingly, when D–A cyclopropanes **1** were reacted with **45** in the presence of $AlCl_3$ in DCM, they smoothly underwent the desired (3 + 3) annulation and yielded tetrahydrothiopyranols **46** (Table 9.14). The reaction proceeds through an S_N2 attack of mercaptoacetaldehyde (**45′**) on the cyclopropane carbon attached to the Ar^1 group to give intermediate **47**, followed by 120° rotation of groups attached to C3 of **47** and subsequent cyclization to give tetrahydrothiopyranol **46**. Unfortunately, the products were produced as inseparable diastereomeric mixtures in most of the cases and hence, we decided to convert them into other useful compounds. Fortunately, when the diastereomeric mixtures of tetrahydrothiopyranols **46** were treated with DBU in

DCM, they yielded tetrasubstituted thiophenes **48** in excellent yields. The conversion takes place via DBU-induced ring-opening of **46** to give anionic intermediate **49**, followed by cyclization, elimination of carbon dioxide and alcohol, and subsequent aromatization to give thiophene **48**. It is interesting to note that *bis*-thiophenes **48k-n** were also accessible by this methodology by employing cyclopropanes having a thiophene ring as Ar^2.

9.3.7 Conversion of Aroyl-Substituted D–A Cyclopropanes into γ-Butyrolactone-Fused D–A Cyclopropanes and their Synthetic Applications

In order to expand the synthetic potential of aroyl-substituted D–A cyclopropanes **1** to new venues, we decided to convert them into γ-butyrolactone-fused D–A cyclopropanes **51** via reductive cyclization process. This kind of fusion of another ring to the D–A cyclopropane skeleton is expected to enhance or alter the reactivity of such cyclopropanes. Thus, when aroyl-substituted D–A cyclopropanes **1** were treated with $NaBH_4$ in MeOH at 0 °C, they underwent reductive cyclization to afford γ-butyrolactone-fused D–A cyclopropanes **51** as single diastereomers in good yields (Table 9.15) [32].

When γ-butyrolactone-fused D–A cyclopropanes **51** were reacted with nitriles **41** in the presence of $SnCl_4$ in 1,2-DCE at 40 °C, γ-butyrolactone-fused 1-pyrrolines **52** having four contiguous stereocenters, were produced in good yields (Table 9.16) [32]. The reaction takes place through a mechanism analogous to the one depicted in Table 9.11.

Table 9.15 Synthesis of γ-butyrolactone fused D–A cyclopropanes from aroyl-substituted D–A cyclopropanes.

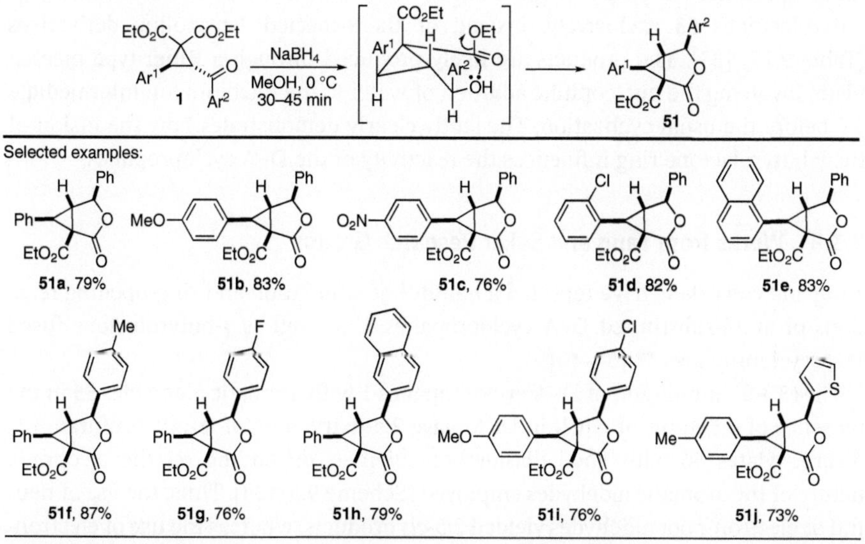

Table 9.16 Synthesis of γ-butyrolactone fused 1-pyrrolines from γ-butyrolactone fused D–A cyclopropanes.

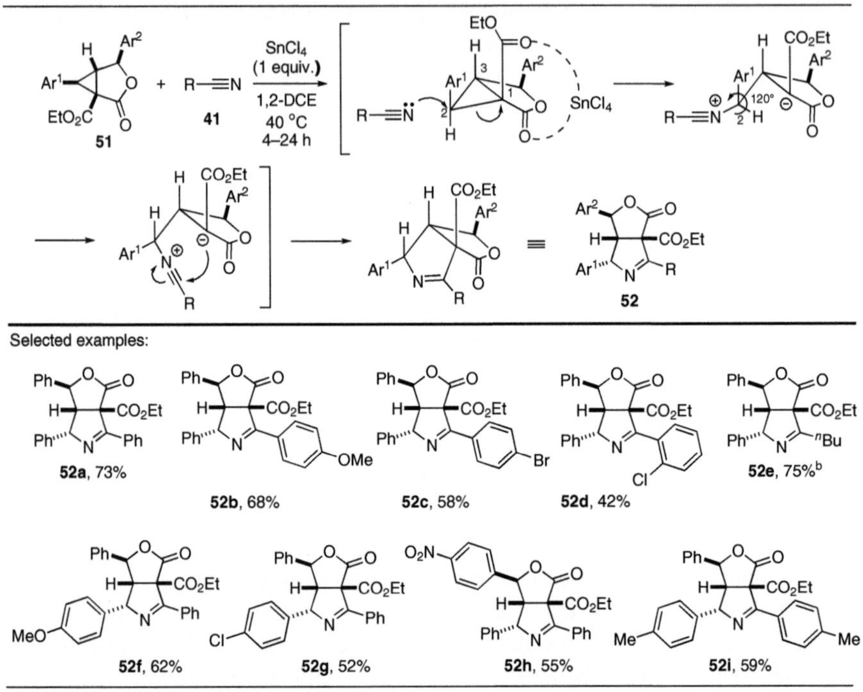

During the course of the above-mentioned investigation, we found that γ-butyrolactone-fused D–A cyclopropanes **51** when reacted with wet aliphatic nitriles in the presence of $AlCl_3$ in 1,2-DCE at 60 °C furnished 3,4,5-trisubstituted γ-butyrolactones **53** exclusively, instead of the expected 1-pyrroline derivatives (Table 9.17) [33]. The products are likely produced through a Ritter-type mechanism, involving the nucleophilic addition of water to the nitrilium ion intermediate **54**, before the usual cyclization. The study clearly demonstrates how the fusion of the γ-butyrolactone ring influences the reactivity of the D–A cyclopropanes.

9.3.8 Works from Yang and Sekar Research Groups

Yang and coworkers have reported a handful of annulation and ring-opening reactions of aroyl-substituted D–A cyclopropanes **1**, as well as γ-butyrolactone-fused D–A cyclopropanes **51** [34–40].

The (3 + 2) annulation of D–A cyclopropanes **1** with aromatic aldehydes **55** in the presence of 0.5 equiv. of $AlCl_3$ in DCM gave 2,5-diaryl-4-aroyl-tetrahydrofuran-3,3-dicarboxylates **56** with good diastereoselectivities, depending on the electronic nature of the aromatic aldehydes employed (Scheme 9.5) [34]. Thus, the use of neutral or electron-poor aldehydes yielded 2,5-*cis* products, whereas the use of electron-rich aldehydes furnished 2,5-*trans* products.

Table 9.17 Synthesis of 3,4,5-trisubstituted γ-butyrolactones from γ-butyrolactone fused D–A cyclopropanes.

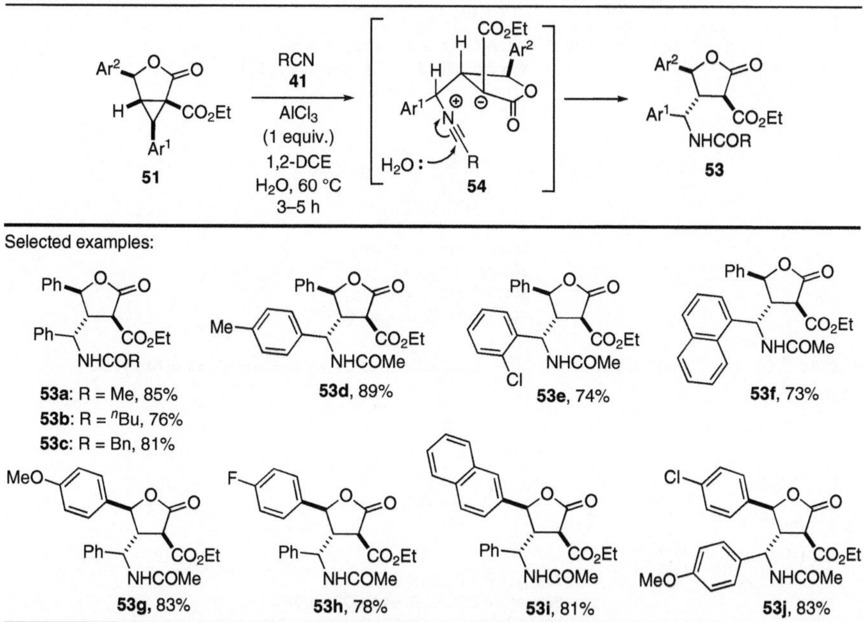

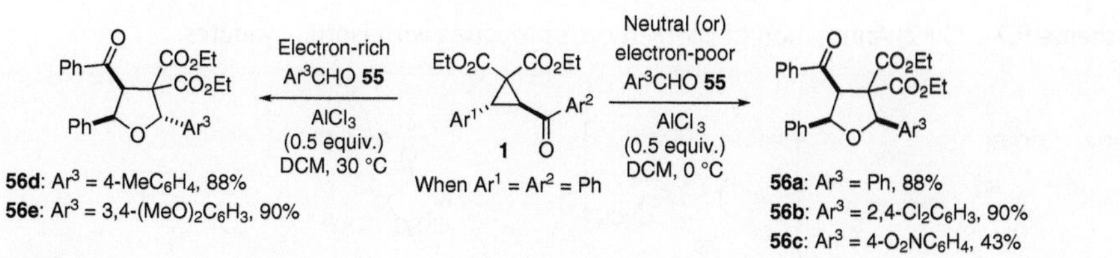

Scheme 9.5 (3 + 2) Annulation of aroyl-substituted D–A cyclopropanes with aldehydes.

The above-mentioned (3 + 2) cyclopropane–aldehyde annulation procedure has been extended to *cis*-diastereomers of D–A cyclopropanes **1** (Scheme 9.6) [35]. Unlike their *trans*-counter parts, these D–A cyclopropanes yielded only one type of diastereomers, irrespective of the electronic nature of the aromatic aldehydes used. Moreover, the products were produced comparatively in higher yields. The methodology was later expanded to cyclic ketones **57** to obtain spirotetrahydrofurans **58** in a highly diastereoselective manner [36].

The *cis*-diastereomers of D–A cyclopropanes **1** also underwent (3 + 2) annulation with aromatic and aliphatic isothiocyanates in the presence of 2 equiv. of $AlCl_3$ in nitromethane at 30 °C (Scheme 9.7) [37]. The products, 2-iminodihydrothiophenes, were produced as single *trans*-diastereomers in moderate-to-excellent yields.

The ring-opening reaction of aroyl-substituted D–A cyclopropanes **1** with PCl_5 has also been reported by the Yang group (Scheme 9.8) [38]. Accordingly, when D–A cyclopropanes **1** were treated with 5 equiv. of PCl_5 in chlorobenzene at 80 °C, they underwent ring-opening through the intermediate **59**, to give tri-substituted vinyl chlorides **60** stereoselectively.

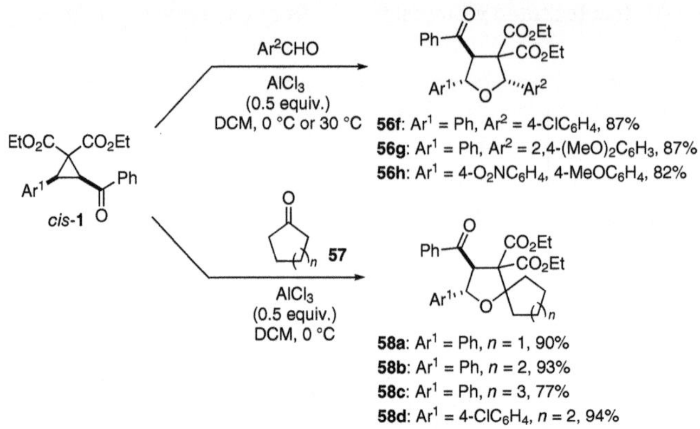

Scheme 9.6 (3 + 2) Annulation of *cis*-D–A cyclopropanes with aldehydes and cyclic ketones.

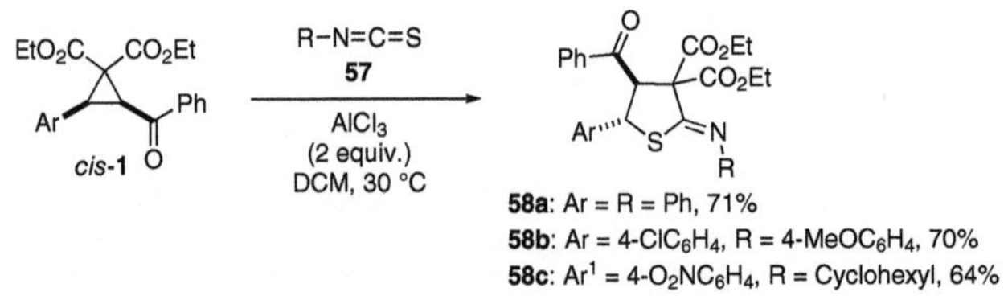

Scheme 9.7 (3 + 2) Annulation of *cis*-D–A cyclopropanes with isothiocyanates.

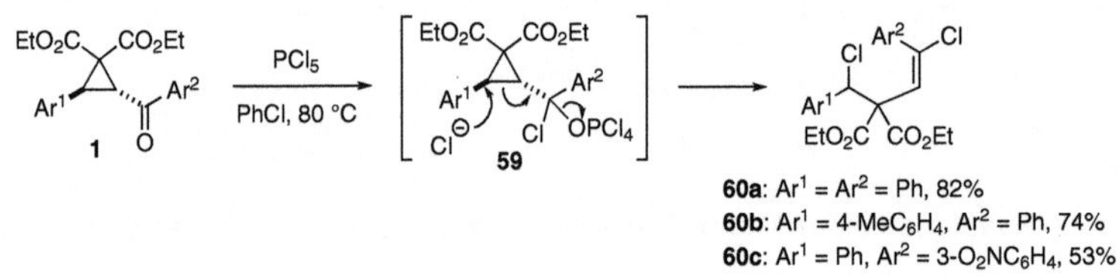

Scheme 9.8 Ring-opening of aroyl-substituted D–A cyclopropanes with PCl_5.

Yang and coworkers have also reported $FeCl_3$-promoted (3 + 2) annulation of γ-butyrolactone-fused D–A cyclopropanes **51** with isothiocyanates **57** and carbodiimides **61** (Scheme 9.9) [39]. A large number of bicyclic γ-butyrolactone-fused thioimidates **62** and γ-butyrolactone-fused amidines **63** having four contiguous stereocenters were obtained as single diastereomers in excellent yields in the reactions.

The Yang group has also explored the (3 + 2) annulation reaction of γ-butyrolactone-fused cyclopropanes **51** with aldehydes **55** and ketones **64**, in the presence of a catalytic amount of $Sn(OTf)_2$ in 1,2-DCE (Scheme 9.10) [40]. This reaction provided highly substituted γ-butyrolactone-fused tetrahydrofurans **65** or **66**, with excellent yields and diastereoselectivity.

Recently, Sekar and Sundarvelu have reported that 2-iodobenzoyl-substituted D–A cyclopropanes **67** upon treatment with 3 equiv. of potassium ethyl xanthate in the presence of a catalytic amount of $Cu(OAc)_2$ and 2 equiv. of AcOH in DMF

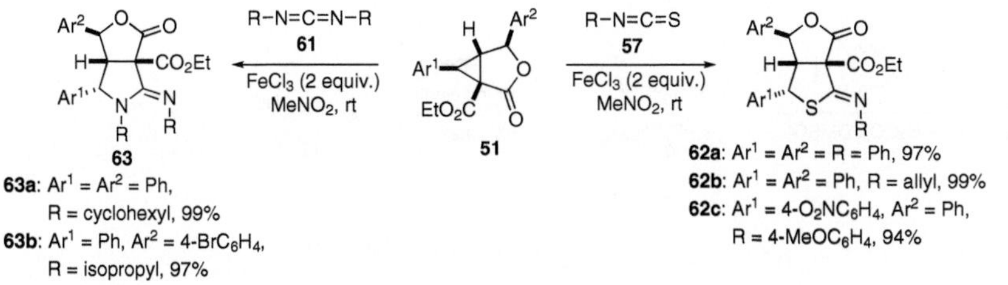

Scheme 9.9 (3 + 2) Annulation of γ-butyrolactone-fused D–A cyclopropanes with isothiocyanates and carbodiimides.

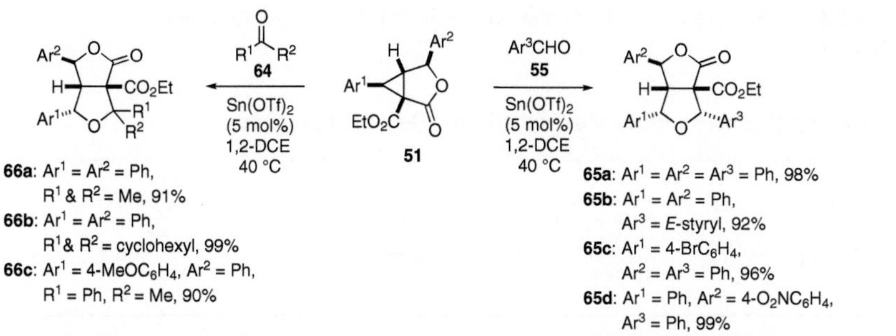

Scheme 9.10 (3 + 2) Annulation of γ-butyrolactone-fused D–A cyclopropanes with aldehydes and ketones.

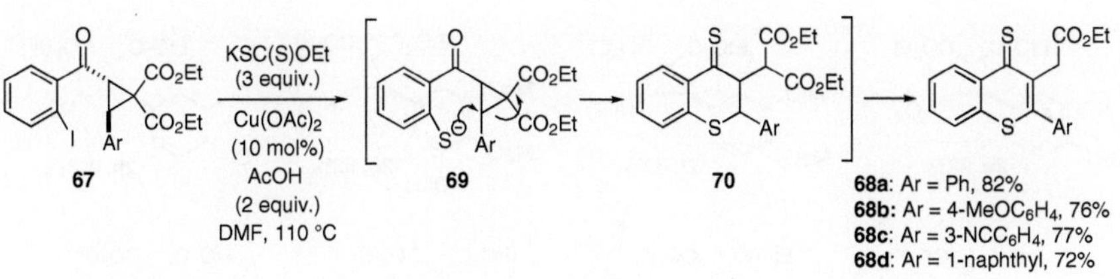

Scheme 9.11 Synthesis of thioflavothiones from 2-iodobenzoyl-substituted D–A cyclopropanes.

at 110 °C, yielded thioflavothiones **68** (Scheme 9.11) [41]. The reaction proceeds through in situ formation of thiolate intermediate **69**, followed by ring-opening to give thione **70**, and subsequent decarbethoxylation and oxidation to give **68**.

9.4 Synthesis of Nitro-Substituted D–A Cyclopropanes

Nitro-substituted D–A cyclopropanes **2** (both in racemic and chiral forms) were also routinely synthesized by the MIRC methodology (Scheme 9.12). Thus, they have been obtained by the following approaches: (i) the reaction of β-bromo-β-nitrostyrenes **71** with dialkyl malonates in the presence of K_2CO_3 in DMSO [15], (ii) the conjugate addition of halomalonates **9** to β-nitrostyrenes **72** in the presence

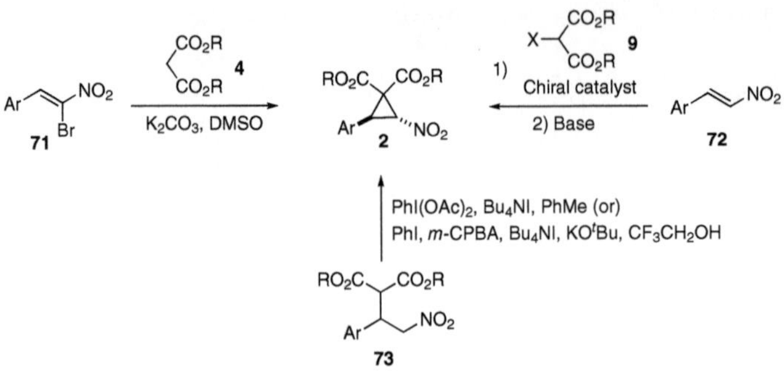

Scheme 9.12 Literature methods for the synthesis of nitro-substituted D–A cyclopropanes.

Table 9.18 Synthesis of various nitro-substituted D–A cyclopropanes.

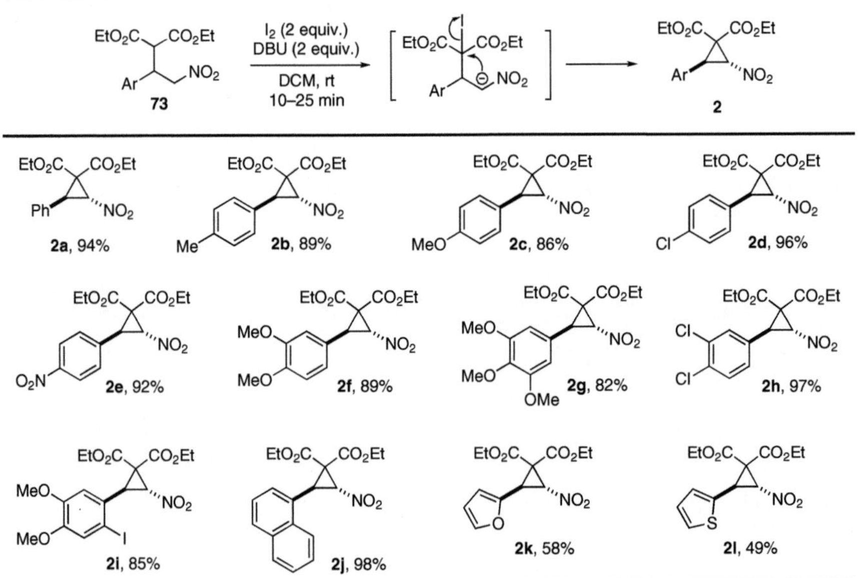

of various chiral catalysts, followed by base-mediated ring closure [42–44], and (iii) the oxidative cyclization of Michael adducts of nitrostyrenes with malonates **73** using iodobenzene diacetate and Bu_4NI in toluene [45] or iodobenzene, *m*-CPBA, Bu_4NI, and KO^tBu in trifluoroethanol [46].

In our research group, we have prepared nitro-substituted D–A cyclopropanes **2** by iodine/DBU-mediated cyclization of Michael adducts of nitrostyrenes with malonates **73** (Table 9.18) [47]. Accordingly, when the Michael adducts were treated with iodine (2 equiv.) and DBU (2 equiv.) in DCM at room temperature, D–A cyclopropanes **2** were produced as single *trans*-diastereomers in good yields. The reaction tolerated the presence of phenyl rings with neutral, electron-donating, and electron-withdrawing groups, naphthyl rings, and heteroaromatic rings.

9.5 Synthetic Applications of Nitro-Substituted D–A Cyclopropanes

In the realm of D–A cyclopropanes, nitro-substituted D–A cyclopropanes are known for their characteristic chemical reactivity. They normally undergo ring-opening reactions and a few ring-expansion reactions, but they seldom take part in annulation reactions.

9.5.1 BF_3-Mediated Ring-Opening Reactions

In order to understand the reactivity patterns of nitro-substituted D–A cyclopropanes **2** with Lewis acids, they were treated with a range of Lewis acids, such as $AlCl_3$, $SnCl_4$, $BF_3·OEt_2$, $FeCl_3$, $ZnCl_2$, $TiCl_4$, $InCl_3$, MgI, AgOTf, $Cu(OTf)_2$, $In(OTf)_3$, $Sc(OTf)_3$, and $Yb(OTf)_3$ in different solvents. Among these, $BF_3·OEt_2$ effectively transformed D–A cyclopropanes **2** into synthetically useful aroylmethylidene malonates **74** in DCM at room temperature, with yields up to 94% (Table 9.19). Other Lewis acids, such as $AlCl_3$, $SnCl_4$, and $FeCl_3$, were also capable of affording **74**, but the yields were poor. As expected, nitrocyclopropane (**2e**), having a *p*-nitrophenyl ring, did not yield the corresponding aroylmethylidene malonate. This transformation takes place through the formation of 1,3-dipolar intermediate **75**, fragmentation of **75** to form ion-pair **76**, recombination of **76** to give nitro compound **77**, and the Nef reaction of **77** finally furnished **74**.

The products of the above-mentioned transformation, aroylmethylidene malonates **74**, were found to be versatile synthetic precursors for the access of various heterocyclic compounds as depicted in Scheme 9.13 [47]. Thus, **74** reacted

Table 9.19 Synthesis of aroylmethylidene malonates from nitro-substituted D–A cyclopropanes.

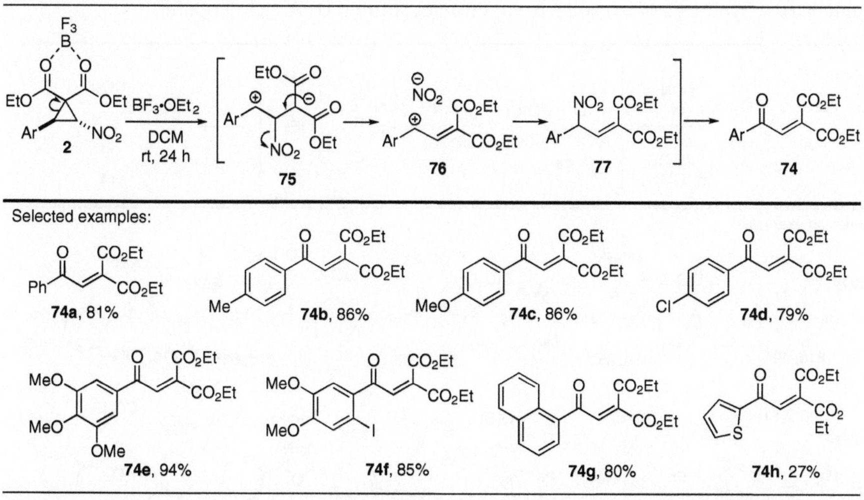

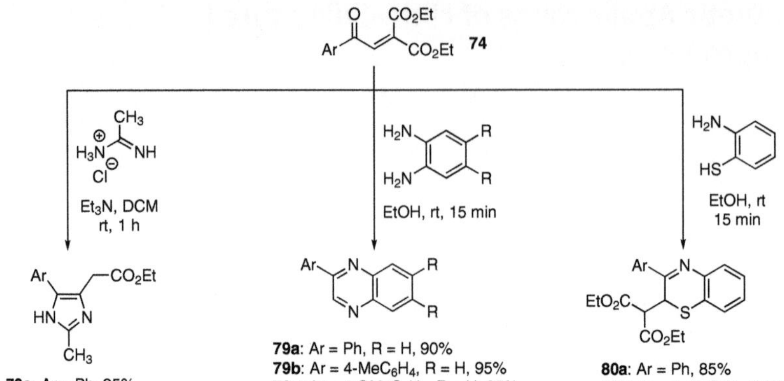

78a: Ar = Ph, 95%
78b: Ar = 4-ClC$_6$H$_4$, 93%

79a: Ar = Ph, R = H, 90%
79b: Ar = 4-MeC$_6$H$_4$, R = H, 95%
79c: Ar = 4-OMeC$_6$H$_4$, R = H, 97%
79d: Ar = Ph, R = Me, 92%

80a: Ar = Ph, 85%
80b: Ar = 4-MeC$_6$H$_4$, 96%

Scheme 9.13 Synthetic utility of aroylmethylidene malonates.

with (i) acetamidine hydrochloride in the presence of Et_3N in DCM to give imidazole derivatives **78**, (ii) *o*-phenylenediamines in ethanol to afford quinoxalines **79**, and (iii) *o*-aminothiophenol in ethanol to furnish benzo[1,4]thiazine derivatives **80** through a nucleophilic addition and cyclocondensation sequence.

9.5.2 Reactions with Nitriles

With a view to extend the (3+2) D–A cyclopropane–nitrile annulation strategy to nitro-substituted D–A cyclopropanes **2**, we treated them with various aliphatic and aromatic nitriles **41** in the presence of $SnCl_4$ in 1,2-DCE at room temperature. Instead of the expected 1-pyrrolines, the reactions yielded a wide variety of 2,4,5-trisubstituted oxazole derivatives **81** (Table 9.20) [48]. A mechanistic study revealed that D–A cyclopropanes **2** formed aroylmethylidene malonates **74** in situ, and **74** underwent

Table 9.20 Synthesis of oxazoles from nitro-substituted D–A cyclopropanes.

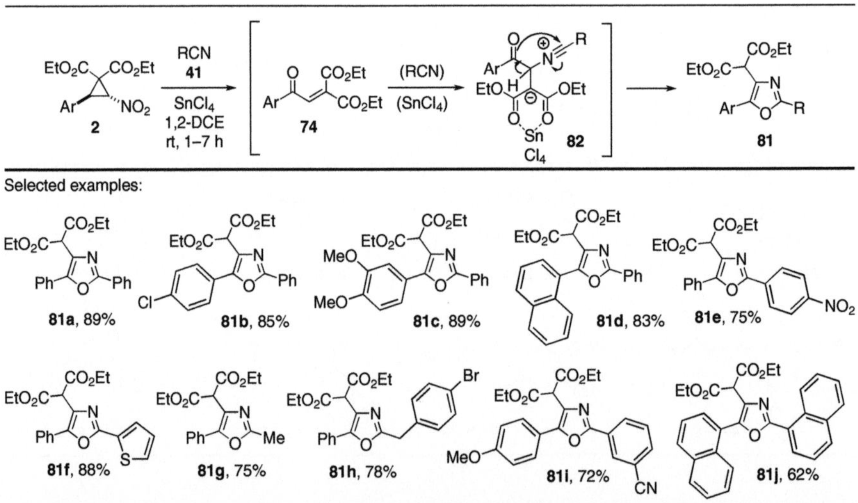

conjugate addition with nitriles **41** to give nitrilium ion intermediates **82**, which upon cyclization yielded oxazoles **81**. The literature methods reported for the generation of nitrilium ion intermediates involve the reaction of α-functionalized ketones with nitriles (the α-functionalization of ketones could be achieved only by using reagents based on transition metal or hypervalent iodine) [49–52] and, thus, the current method affords a hitherto unknown route to nitrilium ion intermediates.

9.5.3 Reactions with Activated Aromatics

The above work indicated that the treatment of nitro-substituted D–A cyclopropanes **2** with Lewis acids would result in the in situ formation of aroylmethylidene malonates **74**, which could be converted into other useful products by reaction with suitable reaction partners in the same flask. Such a sequential one-pot synthesis is better than the corresponding two-step synthesis, as it avoids the separation and purification of the intermediate products. We envisaged that the conjugate addition of carbon nucleophiles to in situ-generated aroylmethylidene malonates would give β-ketomalonates, which are potential synthetic intermediates [54]. Accordingly, nitro-substituted D–A cyclopropanes **2** were treated with 1 equiv. of $BF_3{\cdot}OEt_2$ in DCM at room temperature for 24 hours to form aroylmethylidene malonates in situ, and then the intermediates were reacted with activated aromatics **83** (indoles, carbazole, pyrrole, thiophenes, methoxybenzenes, and benzodioxole) by adding a catalytic amount of $In(OTf)_3$ in the same reaction vessel to obtain the respective adducts (β-ketomalonates) **84** (Table 9.21). It is interesting to note that both mono- and bis-Michael adducts of aroylmethylidine malonates with 2,2′-bithiophene, **84i** and **84k**, could also be achieved by this methodology.

9.5.4 Reaction with Mercaptoacetaldehyde Dimer

By applying a similar sequential one-pot synthetic strategy as a key step, we also devised a two-step synthesis for the access of 2,3-disubstituted thiophenes, which are difficult to access by direct electrophilic or metalation procedures [56]. Accordingly, in the first step, nitro-substituted D–A cyclopropanes **2** were treated with $BF_3{\cdot}OEt_2$ in DCM to form aroylmethylidene malonates in situ, and then reacted with mercaptoacetaldehyde (**45′**) generated in situ from 1,4-dithiane-2,5-diol (**45**) in the presence of Et_3N, to obtain tetrahydrothiophenes **85** in a sequential one-pot manner (Table 9.22). A minimum of 2 equiv. of Et_3N was necessary for the reaction in order to neutralize the Lewis acid as $BF_3{\cdot}OEt_2$-Et_3N complex and also for the generation of the monomer **45′** from mercaptoacetaldehyde dimer. The products **85** were formed as inseparable diastereomeric mixtures via the thia-Michael addition of **45′** to aroylmethylidene malonates followed by an aldol reaction.

In the second step, tetrahydrothiophenes **85** were treated with *p*-TsOH in toluene at 90 °C to obtain 2,3-disubstituted thiophenes **86** through dehydration, followed by monodecarbethoxylation and aromatization (Table 9.23).

Table 9.21 Synthesis of β-ketomalonates from nitro-substituted D–A cyclopropanes.

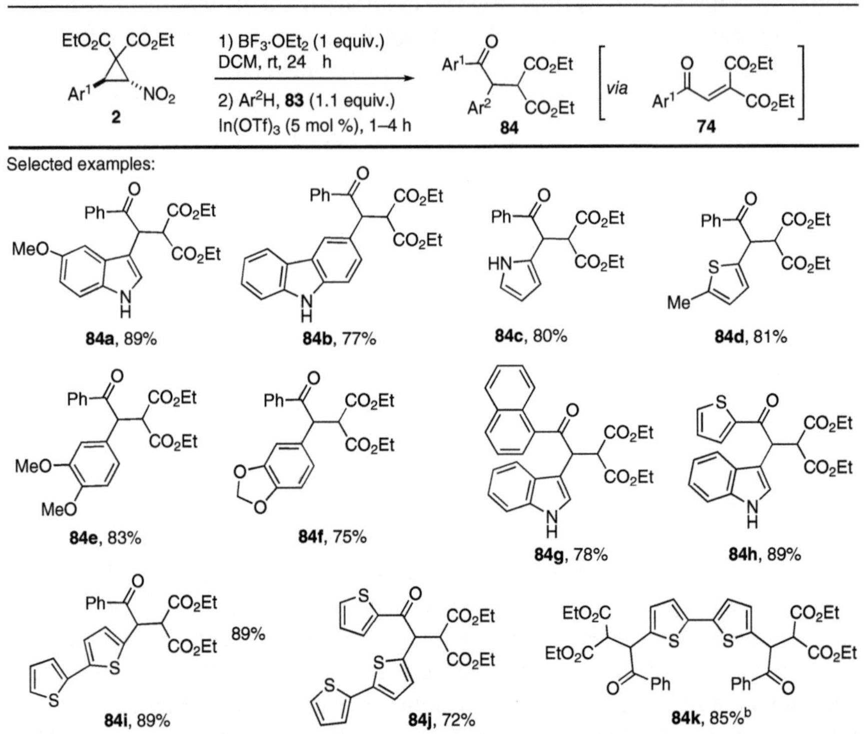

Table 9.22 Synthesis of tetrahydrothiophenes from nitro-substituted D–A cyclopropanes[a] (step-1).

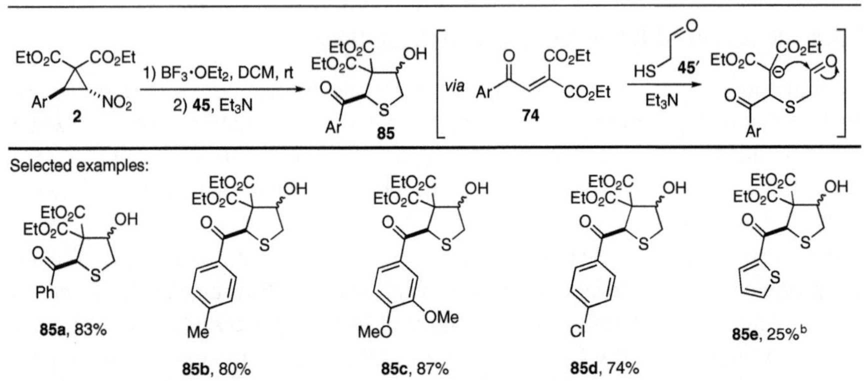

a) The *dr* values ranged from 1 : 1.1 to 1 : 2.2.
b) The reaction was carried out at −15 °C.

9.5.5 Ring-Opening Reactions with 2-Aminopyridines

While investigating the ring-opening reactions of nitro-substituted D–A cyclopropanes **2** with different amines, we observed that they undergo tandem ring-opening/cyclization with 2-aminopyridines **87** to yield pyrido[1,2-a]pyrimidin-4-one derivatives **88** (Table 9.24). Water was found to be a better solvent for this transformation

Table 9.23 Conversion of tetrahydrothiophenes into thiophenes (step-2).

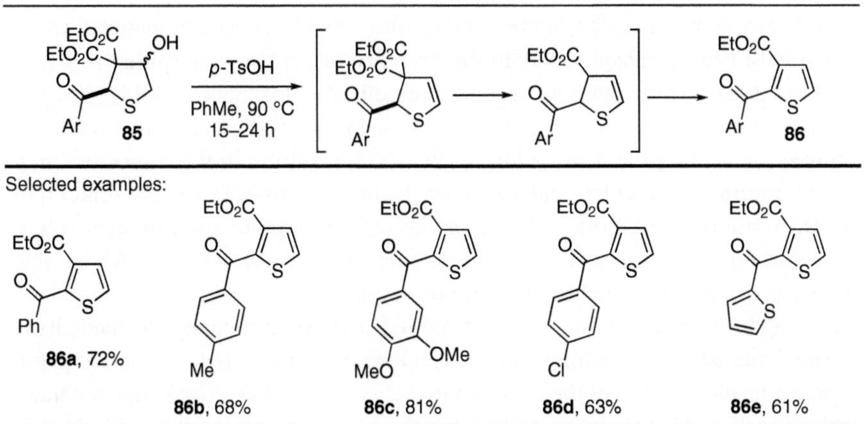

Table 9.24 Synthesis of pyridopyrimidinones from nitro-substituted D–A cyclopropanes.

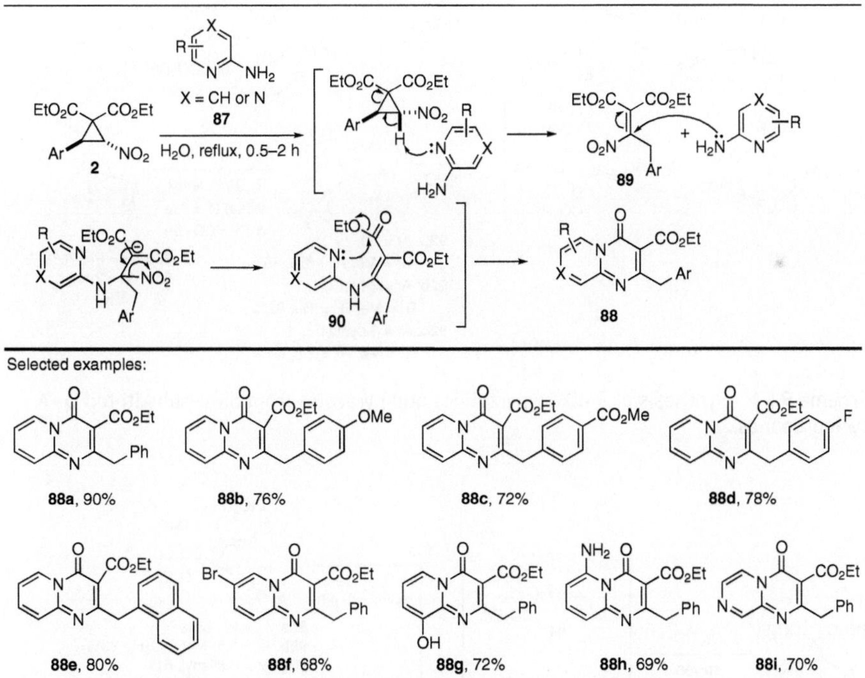

as compared to organic solvents such as MeOH, EtOH, AcOH, toluene, DCM, 1,2-DCE, THF, 1,4-dioxane, MeCN, $MeNO_2$, DMF, and DMSO. In addition, these reactions did not require any other reagent or catalyst to take place. Mechanistically, 2-aminopyridine initially removes the acidic proton of cyclopropane at C3 to give a ring-opened intermediate **89**. The addition of 2-aminopyridine to intermediate **89**, followed by the elimination of the nitro group, forms a new intermediate **90**, which upon cyclization, affords pyridopyrimidinone **88**.

9.5.6 Works from He, Xia, and Asahara Groups

He and coworkers have reported that nitro-substituted D–A cyclopropanes **2**, when treated with azodicarboxylates **91** in the presence of triphenylphosphine in DCM, furnished 3-alkoxypyrzolines **92** in good-to-excellent yields (Scheme 9.14) [58]. The Huisgen zwitterion, produced by the attack of PPh_3 on **91**, served as a base and removed the acidic proton of cyclopropane at C3, resulting in the respective ring-opened intermediate, which underwent cyclization, followed by the elimination of Ph_3PO, to afford the product. If one of the ester groups in the products was *t*-butoxycarbonyl, the products were further converted into pyrazoles **93** by acid hydrolysis using TFA, followed by decarboxylation.

Xia et al. have disclosed that nitro-substituted D–A cyclopropanes **2** formed allene intermediates **94** upon treatment with Cs_2CO_3 via base-mediated ring-opening, followed by the elimination of the nitro group (Scheme 9.15) [59]. These intermediates yielded the dimerization products **95** when the reactions were performed in H_2O in the presence of soybean lecithin (a surfactant), whereas they furnished the

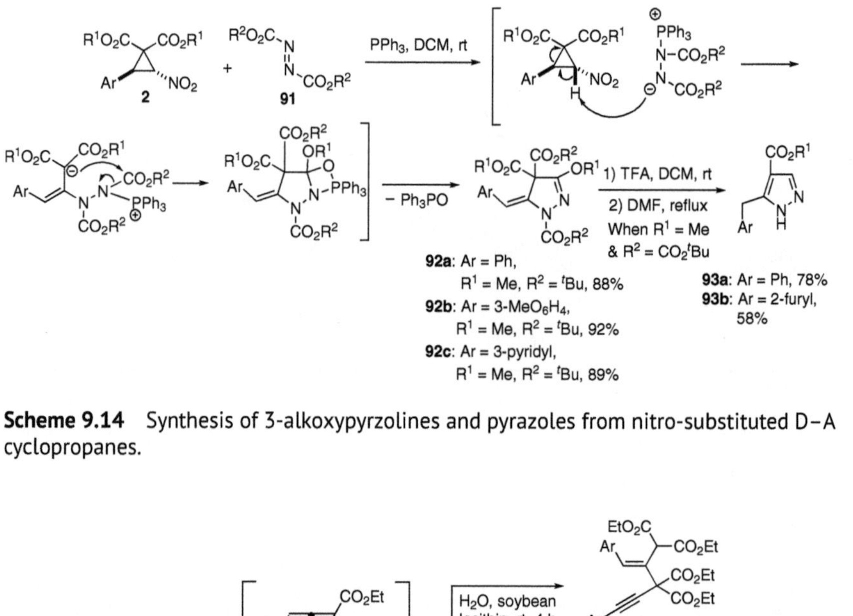

Scheme 9.14 Synthesis of 3-alkoxypyrzolines and pyrazoles from nitro-substituted D–A cyclopropanes.

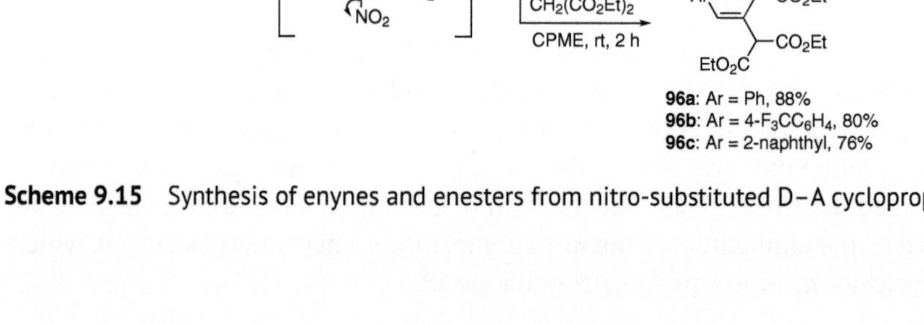

Scheme 9.15 Synthesis of enynes and enesters from nitro-substituted D–A cyclopropanes.

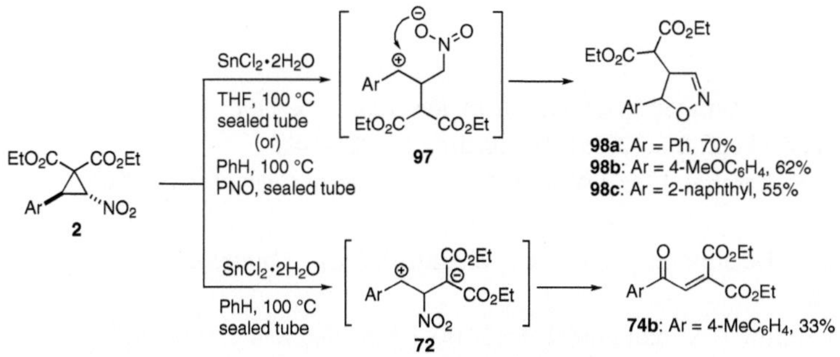

Scheme 9.16 Synthesis of isoxazolines and aroylmethylidene malonates from nitro-substituted D–A cyclopropanes.

nucleophilic addition products **96** when the reactions were carried out in cyclopentylmethyl ether (CPME) in the presence of diethylmalonate.

Recently, Asahara and coworkers have reported that nitro-substituted D–A cyclopropanes **2** could undergo two types of ring-opening with $SnCl_2{\cdot}2H_2O$, depending upon the solvent or ligand employed (Scheme 9.16) [60]. Thus, when the reactions were performed in THF or benzene in the presence of pyridine-*N*-oxide (PNO), the C2—C3 bond of **2** underwent cleavage to give 1,3-dipolar intermediates **97**, which upon cyclization, yielded isoxazolines **98**. On the other hand, the cyclopropanes underwent the usual C1—C2 bond cleavage to give aroylmethylidene malonates **74** through intermediates **72**, when the reactions were carried out in benzene in the absence of PNO.

9.6 Conclusion

The investigation of the reactivity of aroyl-and nitro-substituted D–A cyclopropanes has revealed that the former undergo all the routine reactions of D–A cyclopropanes, such as ring-opening, ring-opening/cyclization, and annulation reactions, while the later undergo only ring-opening and ring-opening/cyclization reactions. These studies have provided efficient synthetic routes to a range of acyclic, carbocyclic, and heterocyclic products, including medicinally important compounds [61, 62]. The presence of the extra aroyl- or nitro-groups in these cyclopropanes, along with regular D–A substituents, indeed make them unique and versatile building blocks in the field of D–A cyclopropane chemistry.

Acknowledgments

K.S. thanks the Science and Engineering Research Board (SERB) and Council of Scientific and Industrial Research (CSIR), India, for financial support. T.S. thanks the Department of Science and Technology (DST) for a Women Scientist Fellowship (WOS-A).

References

1 Reissig, H.-U. and Zimmer, R. (2003). *Chem. Rev.* 103: 1151–1196.
2 Yu, M. and Pagenkopf, B.L. (2005). *Tetrahedron* 61: 321–347.
3 Schneider, T.F., Kaschel, J., and Werz, D.B. (2014). *Angew. Chem. Int. Ed.* 53: 5504–5523.
4 de Nanteuil, F., De Simone, F., Frei, R. et al. (2014). *Chem. Commun.* 50: 10912–10928.
5 Cavitt, M.A., Phun, L.A., and France, S. (2014). *Chem. Soc. Rev.* 43: 804–818.
6 Grover, H.K., Emmett, M.R., and Kerr, M.A. (2015). *Org. Biomol. Chem.* 13: 655–671.
7 Pagenkopf, B.L. and Vemula, N. (2017). *Eur. J. Org. Chem*: 2561–2567.
8 Budynina, E.M., Ivanov, K.L., Sorokin, I.D., and Melnikov, M.Y. (2017). *Synthesis* 49: 3035–3068.
9 Ivanova, O.A. and Trushkov, I.V. (2019). *Chem. Rec.* 19: 2189–2208.
10 Singh, P., Varshnaya, R.K., Dey, R., and Banerjee, P. (2020). *Adv. Synth. Catal.* 362: 1447–1484.
11 Werz, D.B. and Biju, A.T. (2020). *Angew. Chem. Int. Ed.* 59: 3385–3398.
12 Xia, Y., Liu, X., and Feng, X. (2021). *Angew. Chem. Int. Ed.* 60: 9192–9204.
13 Selvi, T. and Srinivasan, K. (2016). *Isr. J. Chem.* 56: 454–462.
14 Schmidt, G. and Gosselck, J. (1969). *Tetrahedron Lett.* 31: 2623–2625.
15 Sopova, A.S., Perekalin, V.V., Yurchenko, O.L., and Arnautova, G.M. (1969). *Zh. Org. Khim.* 5: 858–863.
16 Gaosheng, Y., Yuanyuan, H., and Yue, S. (2009). *Chin. J. Chem.* 27: 1811–1819.
17 Sun, Y., Yang, G., Shen, Y. et al. (2013). *Tetrahedron* 69: 2733–2739.
18 Herchl, R. and Waser, M. (2013). *Tetrahedron Lett.* 54: 2472–2475.
19 Rapi, Z., Nemcsok, T., Grun, A. et al. (2018). *Tetrahedron* 74: 3512–3526.
20 Ye, Y., Zheng, C., and Fan, R. (2009). *Org. Lett.* 11: 3156–3159.
21 Gao, W.-C., Hu, F., Tian, J. et al. (2016). *Chem. Commun.* 52: 13097–13100.
22 Sathishkannan, G. and Srinivasan, K. (2011). *Org. Lett.* 13: 6002–6005.
23 Thangamani, M. and Srinivasan, K. (2018). *J. Org. Chem.* 83: 571–577.
24 Sathishkannan, G. and Srinivasan, K. (2014). *Adv. Synth. Catal.* 356: 729–735.
25 Sathishkannan, G., Tamilarasan, V.J., and Srinivasan, K. (2017). *Org. Biomol. Chem.* 15: 1400–1406.
26 Thangamani, M. and Srinivasan, K. (2021). *J. Org. Chem.* 86: 1172–1177.
27 Thangamani, M., Thangamalar, S., and Srinivasan, K. (2021). *RSC Adv.* 11: 14980–14985.
28 Yu, M. and Pagenkopf, B.L. (2003). *J. Am. Chem. Soc.* 125: 8122–8123.
29 Sathishkannan, G. and Srinivasan, K. (2014). *Chem.Commun.* 50: 4062–4064.
30 Barco, A., Baricordi, N., Benetti, S. et al. (2006). *Tetrahedron Lett.* 47: 8087–8090.
31 Tang, J., Xu, D.Q., Xia, A.B. et al. (2010). *Adv. Synth. Catal.* 352: 2121–2126.
32 Tamilarasan, V.J. and Srinivasan, K. (2019). *J. Org. Chem.* 84: 8782–8787.
33 Tamilarasan, V.J. and Srinivasan, K. (2010). *Eur. J. Org. Chem.* 2010: 593–598.
34 Yang, G., Shen, Y., Li, K. et al. (2011). *J. Org. Chem.* 76: 229–233.

35 Yang, G., Sun, Y., Shen, Y. et al. (2013). *J. Org. Chem.* 78: 5393–5400.
36 Yang, G., Wang, T., Chai, J., and Chai, Z. (2015). *Eur. J. Org. Chem.* 2015: 1040–1046.
37 Sun, Y., Yang, G., Chai, Z. et al. (2013). *Org. Biomol. Chem.* 11: 7859–7868.
38 Hua, D., Yang, P., Shen, Y. et al. (2017). *Tetrahedron Lett.* 73: 1205–1210.
39 Feng, M., Yang, P., Yang, G. et al. (2018). *J. Org. Chem.* 83: 174–184.
40 Yang, P., Shen, Y., Feng, M. et al. (2018). *Eur. J. Org. Chem.* 2018: 4103–4112.
41 Sundaravelu, N. and Sekar, G. (2019). *Org. Lett.* 21: 6648–6652.
42 McCooey, S.H., McCabe, T., and Connon, S.J. (2006). *J. Org. Chem.* 71: 7494–7497.
43 Xuan, Y., Nie, S., Dong, L. et al. (2009). *Org. Lett.* 11: 1583–1589.
44 Lee, H.J., Kim, H.J., and Kim, D.Y. (2012). *Tetrahedron Lett.* 53: 3437–3439.
45 Fan, R., Ye, Y., Li, W., and Wang, L. (2008). *Adv. Synth. Catal.* 350: 2488–2492.
46 Li, Y., Guo, H., and Fan, R. (2020). *Synthesis* 52: 928–932.
47 Selvi, T. and Srinivasan, K. (2014). *J. Org. Chem.* 79: 3653–3658.
48 Selvi, T. and Srinivasan, K. (2014). *Chem. Commun.* 50: 10845–10848.
49 Nagayoshi, K. and Sato, T. (1983). *Chem. Lett.* 1983: 1355–1356.
50 Lee, J.C. and Hong, T. (1997). *Tetrahedron Lett.* 38: 8959–8960.
51 Ishiwata, Y. and Togo, H. (2009). *Tetrahedron* 65: 10720–10724.
52 Saito, A., Hyodo, N., and Hanzawa, Y. (2012). *Molecules* 17: 11046–11055.
53 Selvi, T. and Srinivasan, K. (2015). *Adv. Synth. Catal.* 357: 2111–2118.
54 Liu, G., Shirley, M.E., Van, K.N. et al. (2013). *Nature Chem.* 5: 1049–1057.
55 Selvi, T., Vanmathi, G., and Srinivasan, K. (2015). *RSC Adv.* 5: 49326–49329.
56 O'Connor, C.J., Roydhouse, M.D., Przybyl, A.M. et al. (2010). *J. Org. Chem.* 75: 2534–2538.
57 Selvi, S. and Srinivasan, K. (2017). *Eur. J. Org. Chem.* 2017: 5644–5648.
58 Yang, G., Liu, W., He, Z., and He, Z. (2016). *Org. Lett.* 18: 4936–4939.
59 Zhou, Z.-Y., Xu, Z.-Y., Shen, Q.-Y. et al. (2021). *Chem. Commun.* 57: 6424–6427.
60 Asahara, H., Kamidate, R., and Nishiwaki, N. (2021). *Heterocycles* 103: 379–391.
61 Kandasamy, S., Subramani, P., Srinivasan, K. et al. (2020). *J. Mol. Struct.* 1214: 128177.
62 Meenakshi, M., Antojenifer, P., Karthikeyan, M. et al. (2022). *J. Heterocycl. Chem.* 58: 351–358.

10

Metal-Free Activation of the Donor–Acceptor Cyclopropanes: Protic Acids, Bases, and Thermal Reactions

Lijia Wang[1] *and Yong Tang*[2]

[1]*East China Normal University, Shanghai Engineering Research Center of Molecular Therapeutics and New Drug Development, Department of Chemistry, 3663 North Zhongshan Road, Shanghai, 200062, China*
[2]*The State Key Laboratory of Organometallic Chemistry, Shanghai Institute of Organic Chemistry, Chinese Academy of Sciences, 345 Lingling Road, Shanghai, 200032, China*

CHAPTER MENU

10.1 Introduction

Donor–acceptor (D–A) cyclopropanes are very useful three-carbon synthons in modern organic synthetic chemistry, which have been widely applied in the total synthesis of biologically active natural products as well as important pharmaceutical intermediates. Aiming at efficient transformations of D–A cyclopropanes, various activation strategies have been developed. In this chapter, metal-free activations of the D–A cyclopropanes by means of protic acids, non-metal Lewis acids, bases, thermal reactions, and so on, are discussed. Four main sections based on different modes of activation with abundant examples are reviewed, including electrophilic activation, nucleophilic activation, catalyst-free activation, and activation of D–A cyclopropanes via radical processes.

Donor–Acceptor Cyclopropanes in Organic Synthesis, First Edition.
Edited by Prabal Banerjee and Akkattu T. Biju.

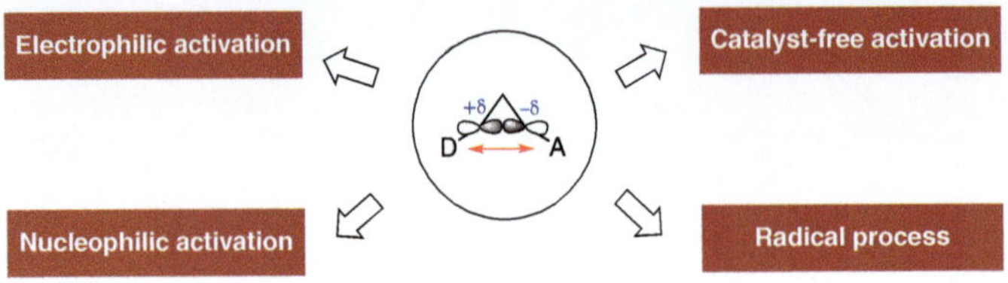

10.2 Metal-Free Electrophilic Activation of D–A Cyclopropanes

Electrophilic activations on the carbonyl groups of D–A cyclopropanes, employing catalytic protic acids such as trifluoromethanesulfonic acid (TfOH), could increase the C—C bond polarization of the cyclopropane, which allowed the following nucleophilic ring-opening in an S_N1 or S_N2 manner, depending on both the catalyst acidity and the electronic nature of substituents on the cyclopropanes (Scheme 10.1a). Besides, other non-metal catalysts or regents, such as trivalent boron species, TMSOTf, and so on, that promote electrophilic activations of D–A cyclopropanes, are also reviewed in this section. An alternative strategy of metal-free electrophilic activation of D–A cyclopropanes in the concept of iminium–enamine activation (Scheme 10.1b) is also included and discussed.

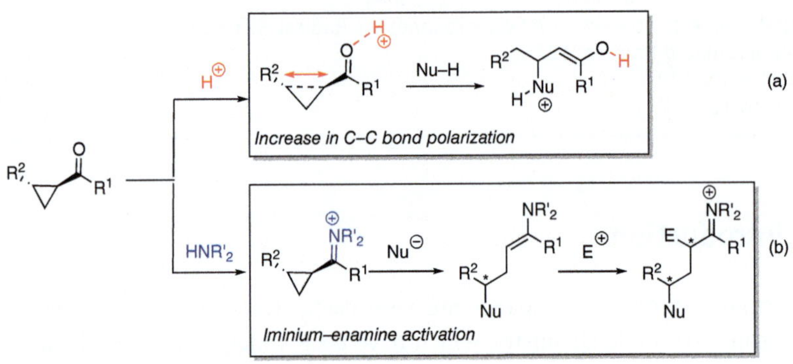

Scheme 10.1 Metal-free electrophilic activation of D–A cyclopropanes.

Moran and coworkers developed a TfOH-catalyzed nucleophilic ring-opening of D–A cyclopropanes with a wide range of nucleophiles, including arenes, indoles, azides, diketones, and even alcohols (Scheme 10.2) [1]. The reaction worked very efficiently in an expanded cyclopropane scope, where D–A cyclopropanes with diester groups (**1**) or single ketoacceptor groups (**2**), and those bearing electron-deficient aryl groups, could all react at room temperature within 3 hours in the presence of 10 mol% of TfOH as catalyst. Mechanistic study, employing an enantiopure cyclopropane under standard conditions, delivered a configuration-inverted ring-opening product, suggesting that the nucleophilic ring-opening via a predominantly S_N2-like pathway. The fluorinated alcohol solvent, hexafluoroisopropanol (HFIP),

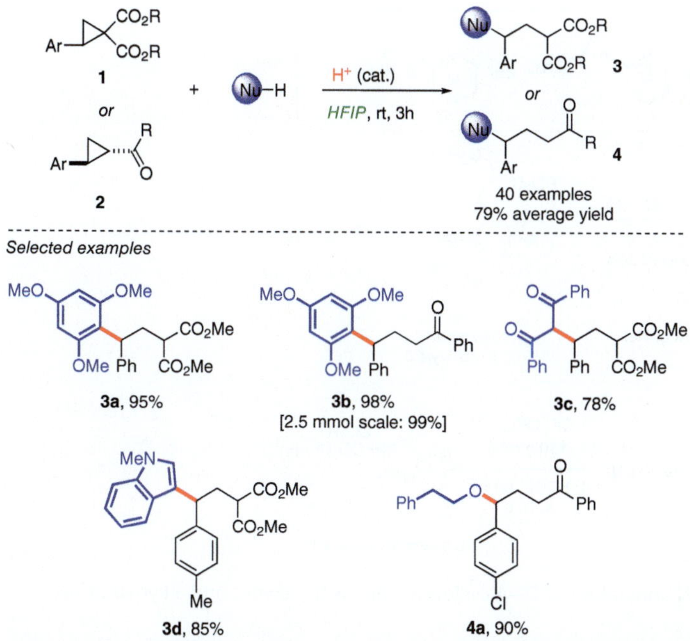

Scheme 10.2 TfOH-catalyzed nucleophilic ring-opening of D–A cyclopropanes.

was found to play an important role. Zhang and Feng performed density functional theory (DFT) calculations to better understand the detailed mechanisms of this reaction [2]. Computational studies indicate that the explicit solvation effect of HFIP significantly lowered the activation free energy from 29.5 kcal/mol to 20.5 kcal/mol, which suggested the cooperative role of TfOH and HFIP in facilitating the ring-opening step.

Similar to the metallic Lewis acid catalysis, in the case of Brønsted acids activated reactions of D–A cyclopropanes, the strained ring with highly polarized C—C bond could also be considered as an equivalent of 1,3-dipole (type I and II), which allowed [3+n] annulation of D–A cyclopropanes with various dipolarophiles. In 2015, Budynina and coworkers developed a TfOH-catalyzed [3+2] annulation of arylcyclopropane-1,1-diesters with alkynes, which led to facile access to various functionalized indenes (**5**) (16 examples) in yields up to 89% at room temperature (Scheme 10.3a) [3]. As a solvent with strong polarity, $MeNO_2$ was used in this reaction. In their studies, Lewis acids such as $BF_3 \cdot Et_2O$ could also promote this [3+2] annulation; however, stoichiometric Lewis acids were needed. In 2022, Goswami and coworkers reported a $BF_3 \cdot OEt_2$-mediated [3+2] annulation of cyclopropane-1,1-diesters with alkynylnitriles in a chemoselective manner for the synthesis of propargylic cyclic imines (**6**) (28 examples) in yields up to 93% (Scheme 10.3b) [4].

In 2021, Zhai and Zhao reported a Brønsted acid-catalyzed formal [3+2] dehydration cycloaddition of vinyl substituted D–A cyclopropanes (**7**) with 2-naphthols in CH_2Cl_2 (Scheme 10.4) [5]. TfOH was employed as the catalyst and delivered the naphthalene-fused cyclopentanes (**8**) in moderate-to-good yields (55–87% yields) with

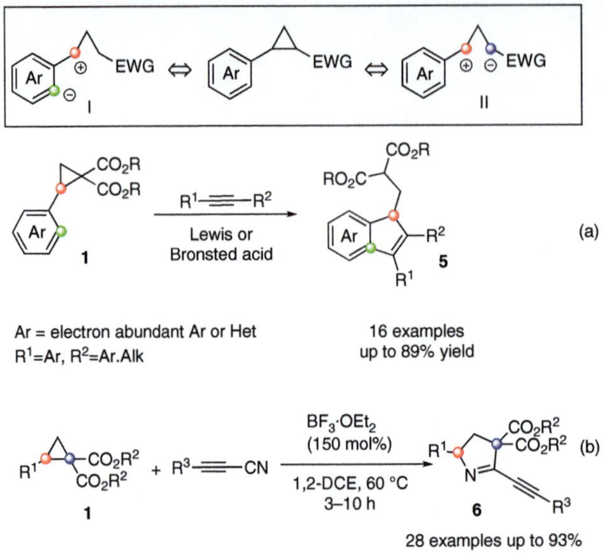

Scheme 10.3 [3+2] annulation of D–A cyclopropanes with alkynes and alkynylnitriles.

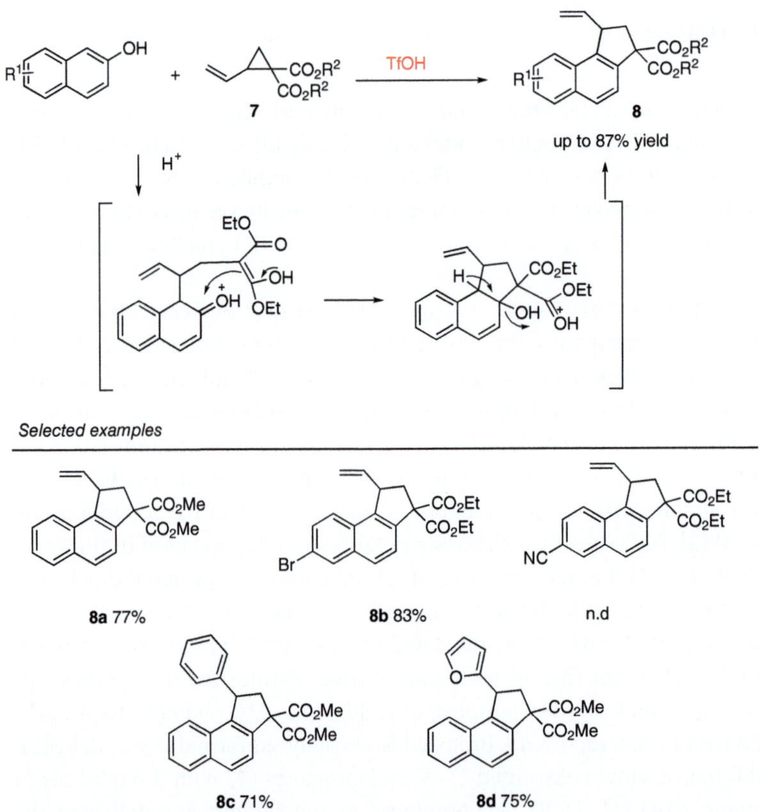

Scheme 10.4 TfOH promoted formal [3+2] dehydration cycloaddition of D–A cyclopropanes with 2-naphthols.

high regioselectivity in a metal-free catalysis. Both vinyl- and aryl-cyclopropane-1,1-diesters worked well with 2-naphthols bearing many different substituents, while electron-withdrawing cyano-substituted 2-naphthol decreased the performance of the reaction. In this reaction, the polarization of the C—C bond increased by TfOH enabled the nucleophilic ring-opening of the cyclopropane by 2-naphthol, followed by an intermolecular aldol reaction to form the cyclopentane intermediate, which eliminated H_2O to generate the naphthalene-fused cyclopentane.

In 2020, Baranov and Mikhaylov developed a Brønsted acid-promoted [3+2] annulation of D–A cyclopropanes (**9**) with aldehydes. In the reaction, stoichiometric TsOH was employed, and imidazol-5-one was introduced as an acceptor group of the spirocyclopropanes. Products **10** and **10′** in 51–92% yields with moderate diastereoselectivity were obtained over 25 examples (Scheme 10.5) [6]. Computational insight into the mechanism of the [3+2] annulation between D–A spirocyclopropane and different aldehydes was revealed by Xue and coworkers in 2022 [7], which showed that electron-rich aldehydes are more nucleophilic and have a relatively lower Gibbs free energy barrier in the rate-determining step during the reaction, which is more favorable for the reaction to occur.

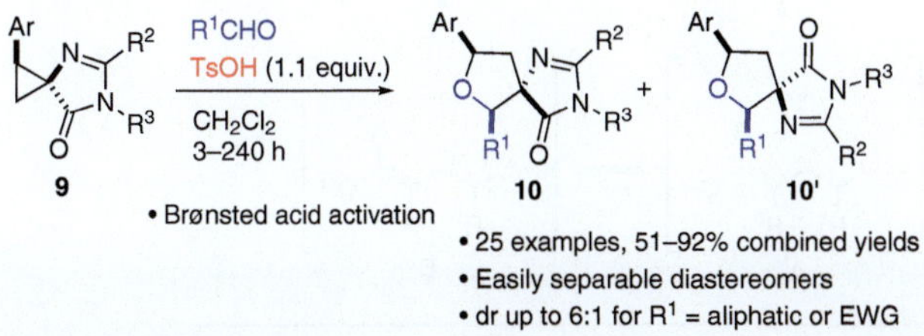

Scheme 10.5 TsOH promoted [3+2] annulation of D–A cyclopropanes with aldehydes.

In 2012, Werz and coworkers developed a TsOH-promoted domino approach of ring-enlargement of D–A cyclopropanes [8–10]. Ketone-substituted cyclopropanes (**11**) were transformed into corresponding imines, which underwent the intramolecular ring-opening rearrangement to build up 3,3′-linked oligopyrroles (**12**). In addition, the *in situ*-generated highly nucleophilic bispyrroles could react with a Michael acceptor to afford pyrrolo[3,2-e]indoles (**13**) (Scheme 10.6).

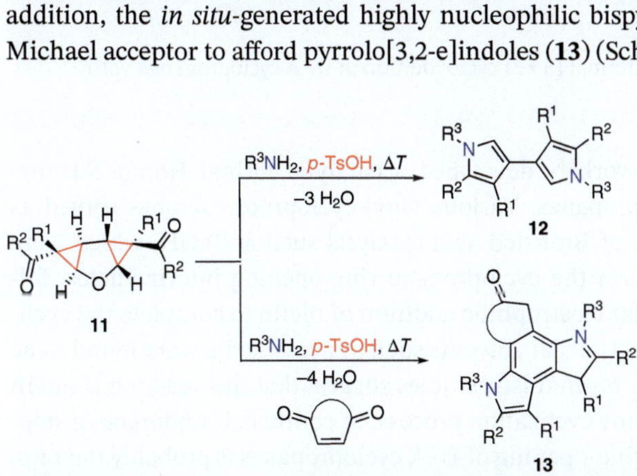

Scheme 10.6 TsOH-promoted domino approach to bispyrroles.

Protic ionic liquid (PIL) as a Brønsted acid was found to be useful to promote reactions of D–A cyclopropanes. In 2021, Werz, Trushkov, and coworkers developed a metal-free formal [3+2]-cycloaddition of D–A cyclopropanes with isothiocyanic acid by employing thiocyanate-based PILs as the sources of nucleophile, the solvent, and the Brønsted acid catalyst. The substrate scope of this reaction is quite broad that various D–A cyclopropanes (33 examples) could be attacked by the nucleophilic 1-methylimidazolium thiocyanate, resulting in the formation of corresponding pyrrolidine-2-thiones (**14**) in good yields. The PIL could be recovered and the reaction scale could be enlarged to 3 grams without the loss of efficiency and sustainability (Scheme 10.7) [11].

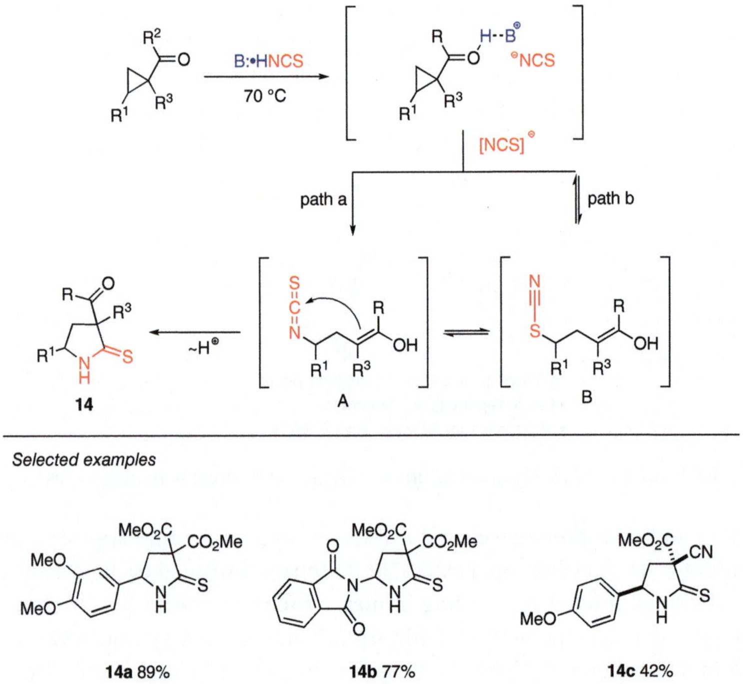

Scheme 10.7 PIL-promoted formal [3+2] cycloaddition of D–A cyclopropanes with isothiocyanic acid.

In 2009, Waser and coworkers developed a catalytic formal Homo-Nazarov cyclization of D–A cyclopropanes. Various vinyl-cyclopropyl ketones served as substrates in the presence of Brønsted acid catalysts such as TsOH, which were activated by protons to form the cyclopropane ring-opening intermediates, followed by the intramolecular electrophilic addition of olefin to complete the cyclization (Scheme 10.8) [12–14]. Polar solvents such as acetonitrile were found to be more appropriate solvents. Mechanistic studies suggest that this reaction is not in a classical concerted Nazarov cyclization process; in contrast, it undergoes a stepwise mechanism that the ring-opening of D–A cyclopropanes is probably the rate-limiting step.

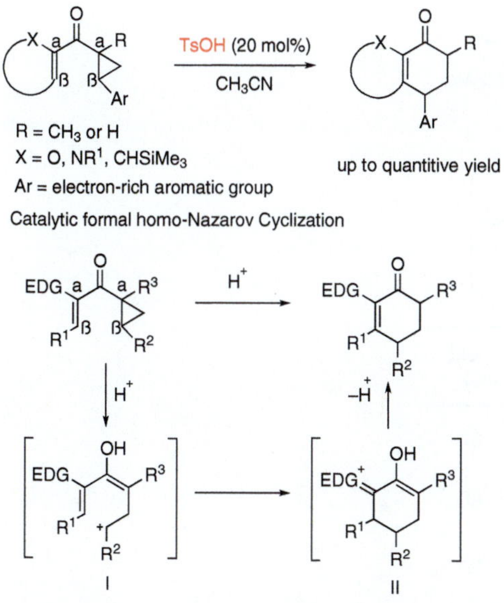

Scheme 10.8 TsOH-promoted formal Homo–Nazarov cyclization of D–A cyclopropanes.

In 2006, Pagenkopf and coworkers demonstrated an annulation of mono-acceptor D–A cyclopropanes with electron-deficient pyridines and quinolones, leading to the synthesis of indolizines (**15**) and benzoindolizines (Scheme 10.9) [15]. The ring-opening reaction of D–A cyclopropanoate esters with pyridines or quinolines was promoted by TMSOTf as a Lewis acid, followed by the intramolecular Mannich annulation to form 2,3-dihydroindolizines, which underwent rapid autooxidation to generate indolizines or benzoindolizines.

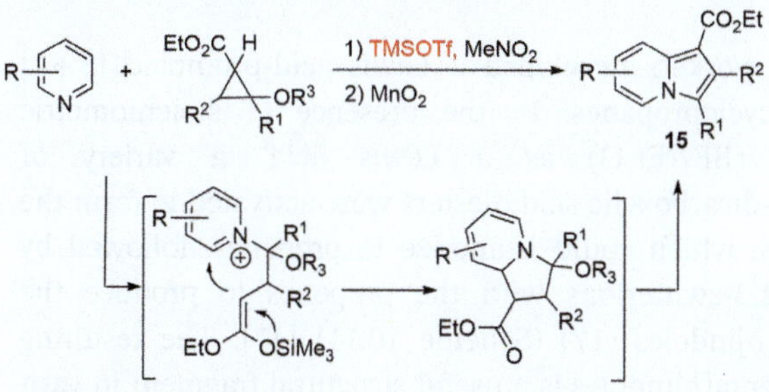

Scheme 10.9 TMSOTf-promoted annulation of mono-acceptor D–A cyclopropanes with pyridines.

In 2010, Melnikov and coworkers developed a TMSOTf-catalyzed isomerization of 2-arylcyclopropane-1,1-dicarboxylates, which provided a facile access to 2-styrylmalonates (**16**) (Scheme 10.10) [16]. A variety of D–A cyclopropanes bearing both electron-donating groups and electron-withdrawing groups on the aryl substituents were tolerated, leading to the corresponding products in moderate-to-good yields.

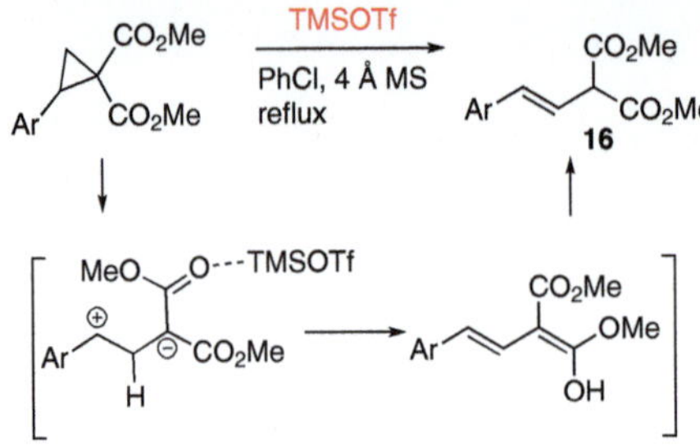

Entry	Ar	Time (h)	Yield (%)
1	C_6H_5	1	67
2	4-BrC_6H_4	3	80
3	4-FC_6H_4	1	63
4	4-MeC_6H_4	1	56
5	2-MeO-5-(O_2N)C_6H_3	1	59
6	α-naphthyl	1	78
7	β-naphthyl	1	86
8	biphenyl-4-yl	2	73

Scheme 10.10 TMSOTf-promoted isomerization of 2-arylcyclopropane-1,1-dicarboxylates.

In 2014, Budynina and coworkers developed a Lewis acid-promoted [3+2] cyclodimerization of D–A cyclopropanes. In the presence of stoichiometric boron trifluoride etherate ($BF_3 \cdot Et_2O$) as a Lewis acid, a variety of 2-(3-indolyl)cyclopropane-1,1-dicarboxylic acid diesters were activated to form the corresponding 1,3-zwitterions, which could isomerize to propenes, followed by the [3+2] annulations of 1,3-zwitterions with the propenes to produce the 3-indolyl-derived cyclopenta[*b*]indoles (**17**) (Scheme 10.11) [17]. The resulting tricyclic system of the cyclopenta[*b*]indoles is a useful structural fragment in various biologically active indole diterpenoids.

In 2016, Csákÿ and coworkers developed the ring-opening of cyclopropane-1,1-dicarboxylates with the *in situ*-generated electron-deficient trivalent boron species, which was promoted by trifluoroacetic anhydride (TFAA) from boronic acids and potassium organotrifluoroborates under metal-free conditions. Electrophilic activation formed by the *in situ*-formed boron species may occur by coordinating with one of the ester groups of the cyclopropane, followed by the internal nucleophilic attack to form the ring-opening product. The *in situ*-generated boron species could be considered as synthetic equivalents of alkenyl, alkynyl, and allyl carbon

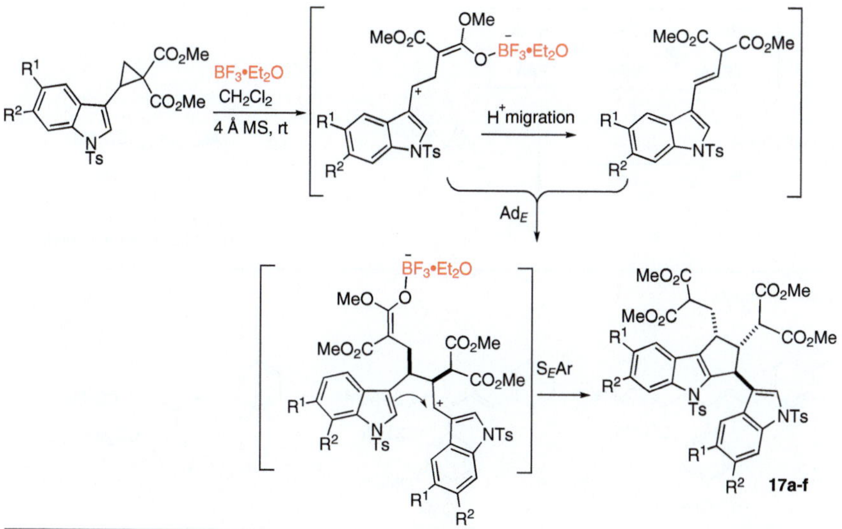

Entry	Product	R^1	R^2	Time (min)	Yield (%)
1	**17a**	H	H	100	85
2	**17b**	Cl	H	100	76
3	**17c**	Br	H	90	72
4	**17d**	CN	H	210	35
5	**17e**	H	F	120	74
6	**17f**	H	Cl	120	70

Scheme 10.11 $BF_3 \cdot Et_2O$-promoted [3+2] cyclodimerization of D–A cyclopropanes.

nucleophiles, producing the corresponding ring-opening products (**18**) in good yields (Scheme 10.12) [18].

Besides the direct electrophilic activation of D–A cyclopropane by Brønsted acid or Lewis acid, an alternative masked D–A strategy via electrophile-induced C—C bond activation of vinylcyclopropanes was developed by Chandrasekaran and coworkers (Scheme 10.13) [19]. An *in situ*-generated electrophilic bromine acted on the double bond of the vinylcyclopropane to form a π-complex, which could be considered as a D–A cyclopropane analog to accept the attack of a nucleophile leading to the ring-opening product. Treatment of chloramine-T with a catalytic amount of phenyltrimethylammonium tribromide (PTAB) resulted in the *in situ* generation of an interhalogen compound like BrCl, which was usually used as the electrophilic activation catalyst. By applying the masked D–A strategy, Chandrasekaran and coworkers explored several tandem annulation reactions of vinylcyclopropanes bearing a donor group for the synthesis of various bicyclic amidines and Z-alkylidenetetrahydrofurans.

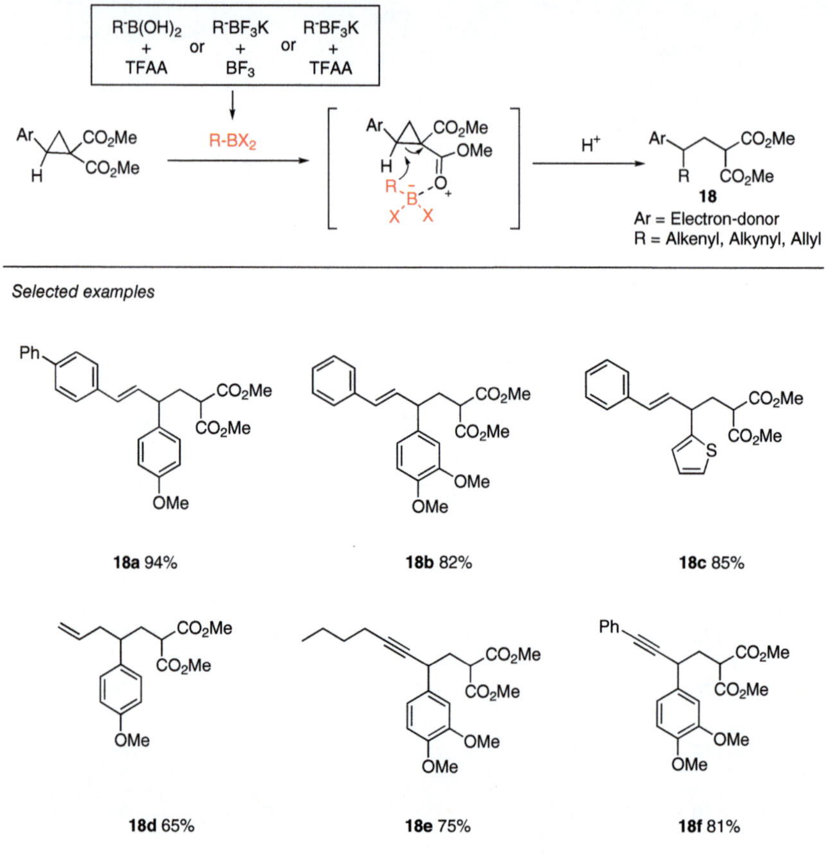

Scheme 10.12 The *in situ*-formed boron species promoted ring-opening of D–A cyclopropanes.

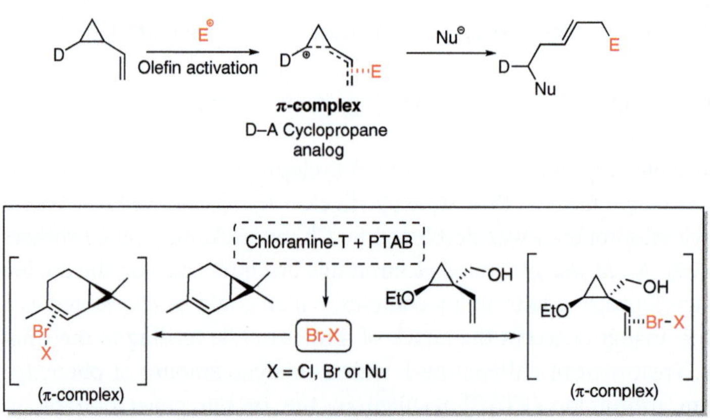

Scheme 10.13 Electrophile-induced C—C bond activation of vinylcyclopropanes.

In 2011, they reported a tandem ring-opening/cyclization of vinylcyclopropane derivatives with various cyanides through a Ritter-type rearrangement to give enantiomerically pure [4.4.0] and [4.3.0] bicyclic amidine derivatives (**19**) in good yield

(Scheme 10.14a) [19, 20]. In 2013, they demonstrated a highly diastereoselective synthesis of Z-alkylidenetetrahydrofurans (**20**) from a variety of vinylcyclopropanes bearing a hydroxyl group. A plausible mechanism has been presented explaining the Z-selectivity observed during the product formation (Scheme 10.14b) [21].

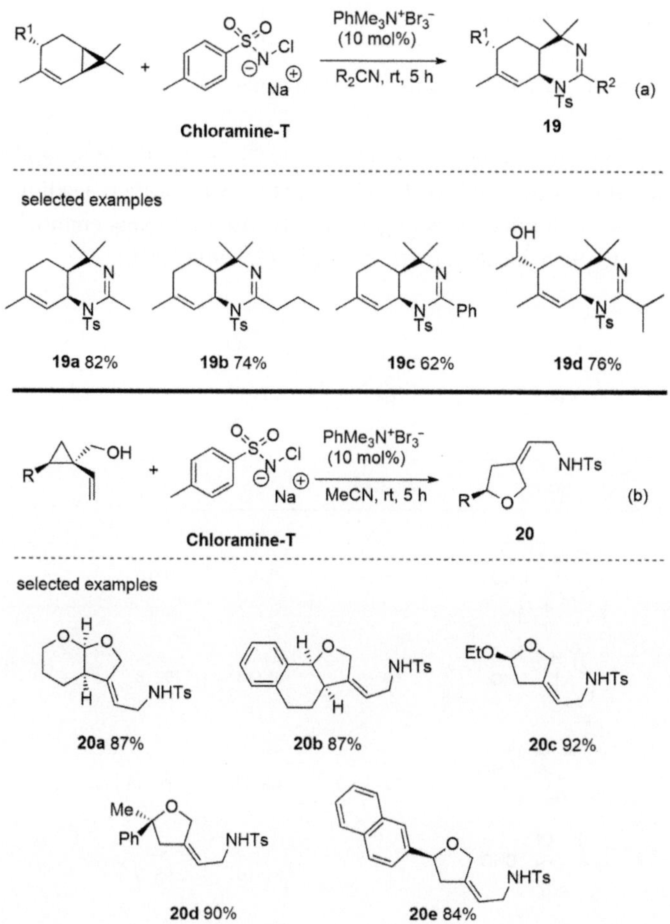

Scheme 10.14 Reaction of vinylcyclopropane with chloramine-T. Ts = *p*-toluenesulfonyl.

An alternative strategy of metal-free electrophilic activation of D–A cyclopropanes was developed in the concept of iminium–enamine activation to form the highly electrophilic three-membered ring intermediate, which was attacked by the nucleophilic reagent and then the electrophilic reagent, successively. In this type of reaction, cyclopropanes bearing an aldehyde group are essential, and chiral secondary amines served as catalysts to promote the ring-opening of cyclopropanes (Scheme 10.15).

In 2011, Gilmour and Sparr reported the first cyclopropyl iminium activation of D–A cyclopropanes for the asymmetric synthesis of 1,3-dichlorides (**22**) in an enantioselective desymmetrization of meso-cyclopropanecarbaldehyde compounds (**21**). A variety of secondary amine organocatalysts could promote this reaction, and the first-generation MacMillan catalyst (**Cat.-1**) was found to be the best one.

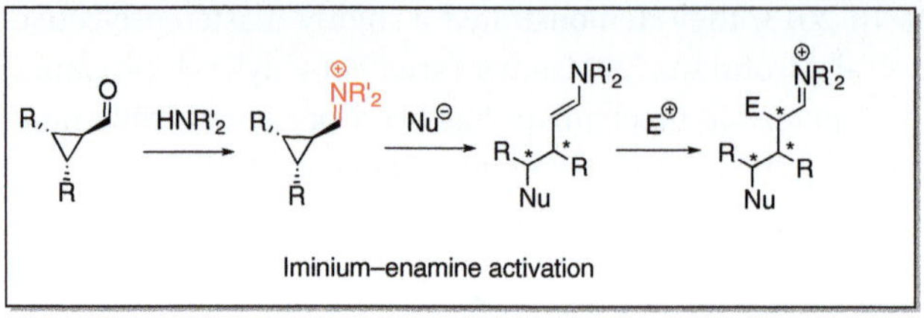

Scheme 10.15 Iminium–enamine activation of D–A cyclopropanes.

Reagents **A** and **B** were employed as the sources of Cl^- and Cl^+, respectively. Several meso-cyclopropanecarbaldehydes could work well in this reaction system, leading to the optically active 1,3-dichloraldehydes in good yields with up to 96:4 enantiomeric ratio (er) and up to > 95:5 diastereomeric ratio (dr) (Scheme 10.16) [22].

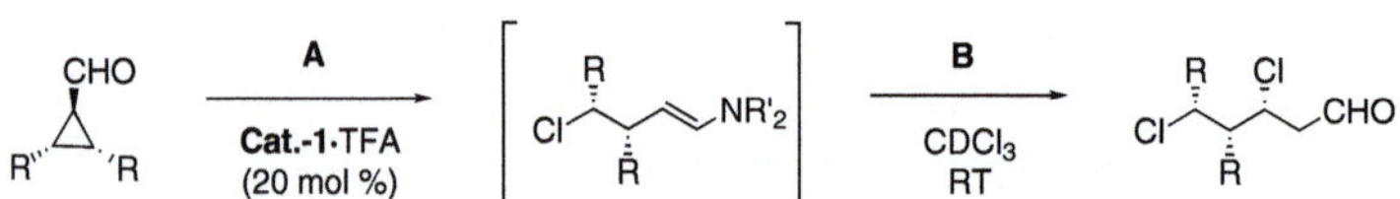

21 → 22

Entry	Substrate	Product[b]	Yield[%][c]	e.r.	d.r.
1	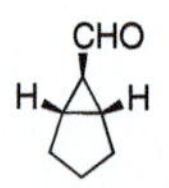	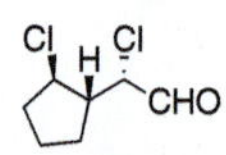	70	91 : 9	94 : 6
2	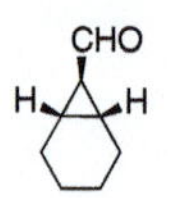	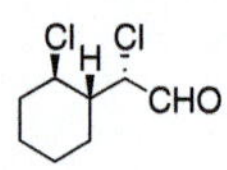	68	86 : 14	92 : 8
3	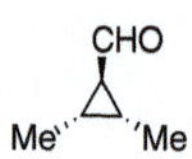	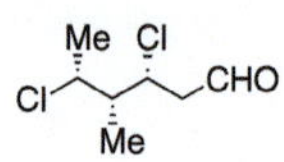	70	89 : 11	86 : 14
4	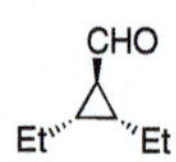	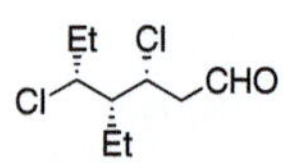	72	84 : 16	95 : 5
5	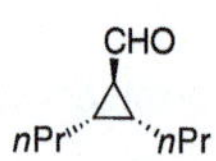	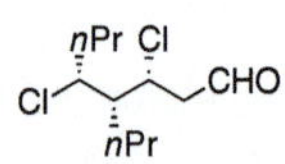	68	91 : 9	93 : 7
6	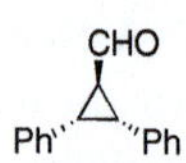	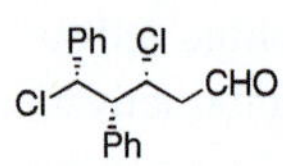	67	96 : 4	> 95 : 5

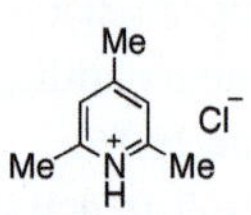

A

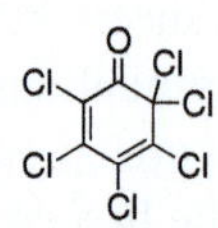

B

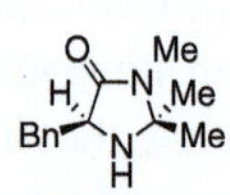

Cat.-1

Scheme 10.16 Iminium activation of D–A cyclopropanes for the asymmetric synthesis of 1,3-dichlorides

In 2016, Werz and coworkers developed another cyclopropyl iminium activation reaction of meso-cyclopropyl carbaldehydes. Sulfenyl and selenyl chlorides were employed as both the nucleophilic and the electrophilic reagents to give the corresponding 1,3-bisfunctionalization products (**23**) bearing three adjacent stereocenters in 42–84% yields with up to 93:7 e.r. and 15:1 d.r. in complete regioselectivity (Scheme 10.17) [23].

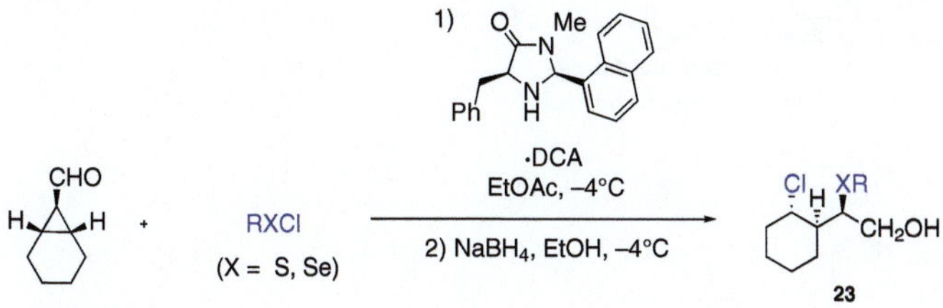

23

Selected examples

Product	Yield [%]	d.r.	e.r.
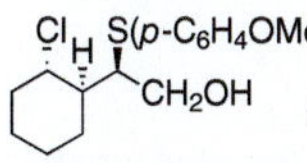 (**23a**)	63	5.7:1	87:13
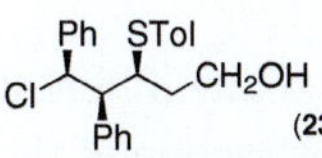 (**23b**)	70	5.2:1	81:19
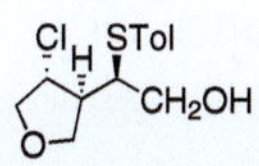 (**23c**)	65	2.1:1	71:29

Scheme 10.17 Iminium activation reaction for the asymmetric synthesis of 1,3-bisfunctionalization products.

When a nitro group was introduced in the D–A cyclopropane as an acceptor group, the urea catalysis based on hydrogen bonding activation of D–A cyclopropane was developed by Mattson and coworkers. In 2012, boronate urea was employed as a catalyst for the activation of nitrocyclopropane carboxylates (**24**) in the nucleophilic ring-opening reactions with various amines, generating a wide range of γ-amino nitro esters (**25**) in 58–99% yields (Scheme 10.18a) [24]. In 2013, they developed formal [3 + 3] cycloaddition reaction of nitrocyclopropane carboxylates with nitrones via the same hydrogen bonding activation strategy with a urea catalyst (Scheme 10.18b) [25]. This method provided an efficient approach to a large family of highly functionalized oxazinanes (**26**) in 25–99% yields.

10.3 Metal-Free Nucleophilic Activation of D–A Cyclopropanes

In contrast to the electrophilic activation approaches based on lowest unoccupied molecular orbital (LUMO) lowering strategies using Lewis acids or Brønsted acids, the nucleophilic approach for activation of cyclopropanes is based on a highest

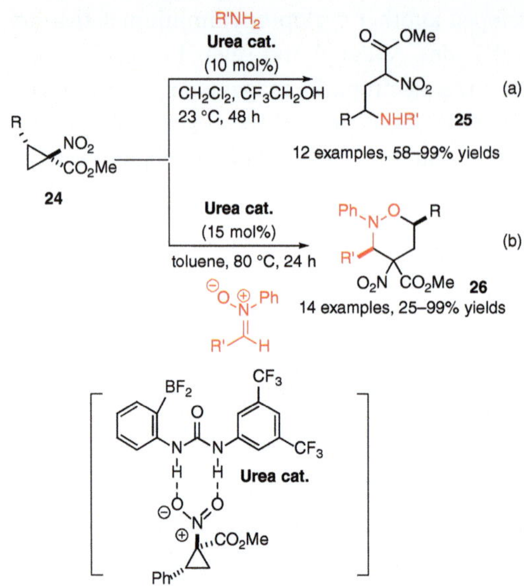

Scheme 10.18 Urea catalysis based on hydrogen bonding activation of D–A cyclopropane.

occupied molecular orbital (HOMO)-raising strategy. In this field, many different electrophilic activation strategies have emerged, including direct abstraction of the α-hydrogen of a carbonyl group with base, as well as enamine-activation for D–A cyclopropanes bearing an aldehyde unit (Scheme 10.19).

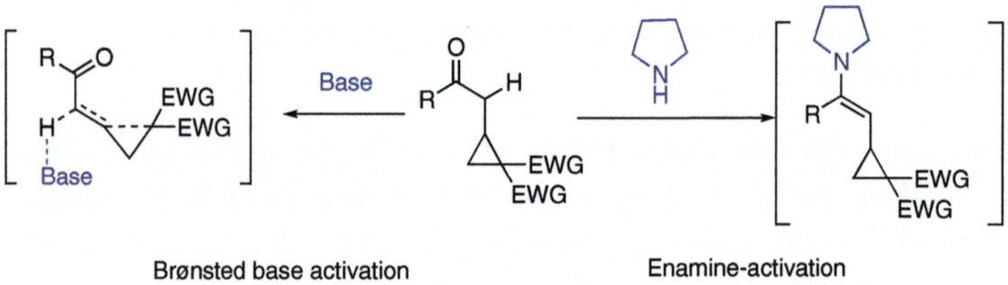

Scheme 10.19 Metal-free nucleophilic activation of D–A cyclopropanes.

In 2009, Jørgensen and coworkers reported an organocatalytic asymmetric desymmetrization of diethyl 3-oxobicyclo[3.1.0]hexane-6,6-dicarboxylate (**27**), involving a metal-free nucleophilic activation of D–A cyclopropane with quinidine-derived thiourea catalyst (**cat.-2**). Cyclopentenone (**28**) was obtained in an 84% yield with a 96% enantiomeric excess (ee) (Scheme 10.20) [26].

A direct D–A cyclopropane activation strategy based on the abstraction of α-hydrogen from ketones by a Brønsted base was developed by Jørgensen and coworkers. This type of activation was found to be more suitable for di-cyano-substituted cyclopropylketones than di-ester ones, because in the di-cyano system, an increased polarization of the labile carbon–carbon bond was determined by ^{13}C NMR experiment, as well as the more stabilized anion from the ring-opening of the di-cyano cyclopropane according to the pKa values of the related acids (Scheme 10.21) [27].

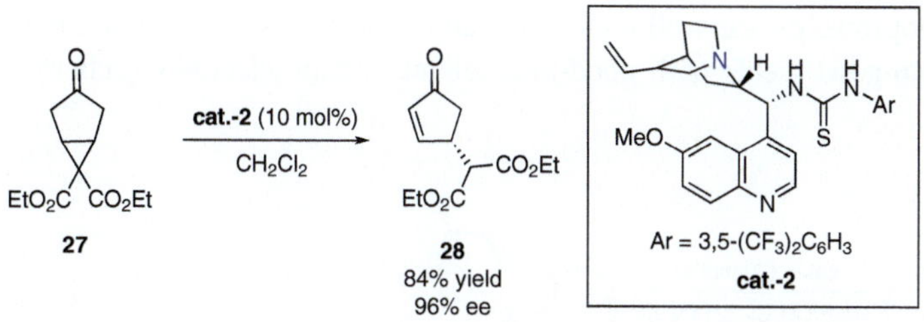

Scheme 10.20 Quinidine-derived thiourea catalyzed ring-opening of D–A cyclopropane.

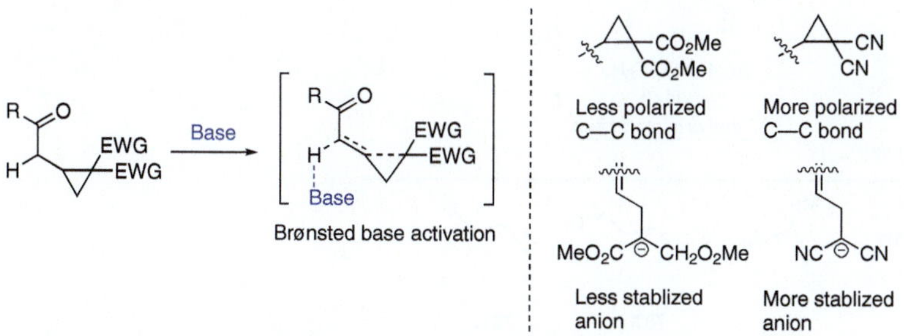

Scheme 10.21 Brønsted base activation of D–A cyclopropanes.

In 2017, Jørgensen and coworkers developed a metal-free direct activation of D–A cyclopropanes with Brønsted base catalysis (**cat.-3**) in the [3 + 2] cycloaddition reaction of di-cyano cyclopropylketones (**29**) with mono- and polysubstituted nitroolefins, providing functionalized cyclopentanes with three contiguous stereocenters (**30**) in yields up to 98% with >20 : 1 d.r. and up to 91% ee. A proposal for a mechanistic pathway that explains the high stereoselectivity, as well as chemoselective transformations of optically active cyclopentanes, was also demonstrated in their studies (Scheme 10.22) [27].

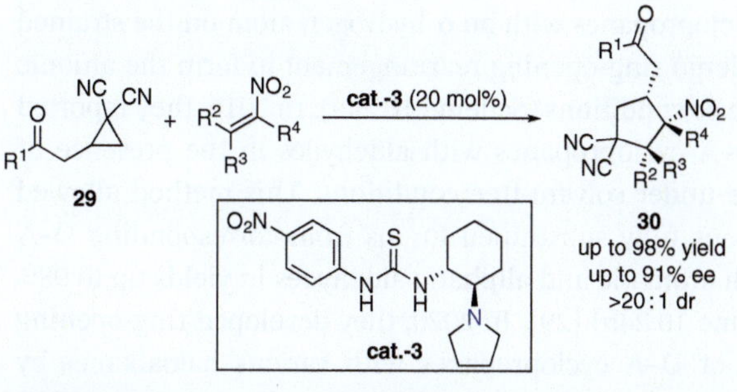

Scheme 10.22 [3+2] cycloaddition reaction of di-cyano cyclopropylketones with nitroolefins.

In 2022, Jørgensen and coworkers developed a chiral Brønsted base activated ring-opening of D–A cyclopropanes in a formal [8 + 3] cycloaddition of di-cyano cyclopropylketones (**29**) with tropone. This nucleophilic activation strategy employed a bifunctional thiourea (**cat.-4**) bearing a tertiary amine moiety as a catalyst; a broad scope of D–A cyclopropanes was well tolerated, leading to the desired cycloadducts (**31**) in moderate-to-good yields with good-to-excellent enantioselectivity (Scheme 10.23) [28].

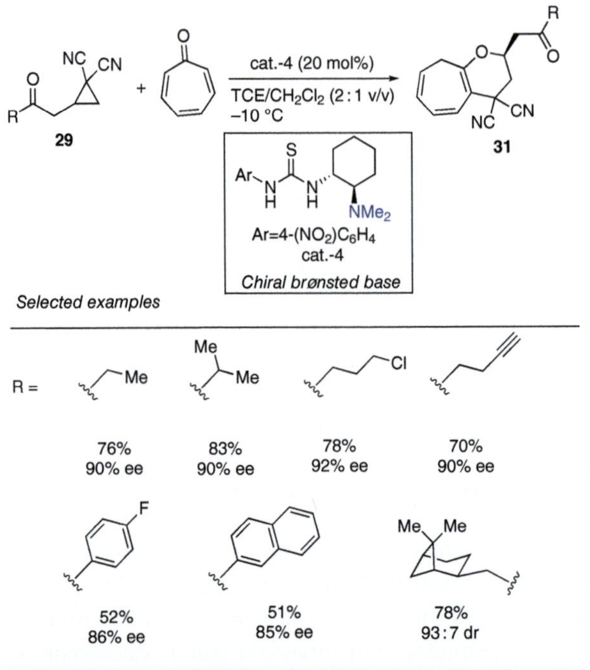

Scheme 10.23 Formal [8 + 3] cycloaddition of di-cyano cyclopropylketones with tropone.

Another type of metal-free nucleophilic activation of D–A cyclopropanes was developed by Wang and coworkers. Such type of activation is characterized by carbonyl-substituted D–A cyclopropanes with an α-hydrogen atom on the strained ring, which are likely to undergo ring-opening rearrangement to form the anionic intermediate A or B under basic conditions (Scheme 10.24a). In 2016, they reported a [3 + 2] cycloaddition of D–A cyclopropanes with aldehydes in the presence of 1.0 equiv. of DABCO as base under solvent-free conditions. This method allowed efficient preparation of various fully substituted furans from corresponding D–A cyclopropanes (**32**) with both aromatic and aliphatic aldehydes in yields up to 98% over 24 examples (**33**) (Scheme 10.24b) [29]. In 2020, they developed ring-opening reactions of the same type of D–A cyclopropanes with various haloalkanes by employing DBU as a base, leading to a wide range of functionalized chalcones (**34**) containing a quaternary carbon group in up to 94% yield (Scheme 10.24c) [30].

Jørgensen and coworkers developed a metal-free enamine-activation of cyclopropanes, in which the donor moiety of a D–A cyclopropane is *in situ* formed via condensation with a chiral secondary amine to generate an enamine intermediate. Computational studies revealed that the C1—C2 bond length of the cyclopropane was obviously elongated by the formation of the enamine intermediate.

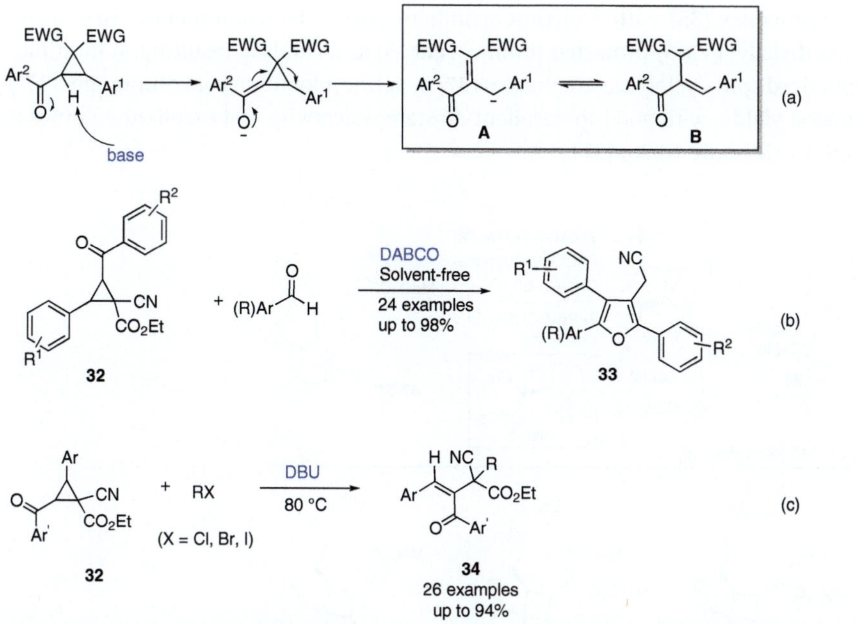

Scheme 10.24 Brønsted base activation for carbonyl-substituted D–A cyclopropanes.

Especially, when the cyclopropane bearing two acceptor groups, the C1—C2 bond length was elongated from 1.517 Å to 1.559 Å, indicating that the enamine-activation strategy was very effective (Scheme 10.25) [31].

By applying of this enamine-activation strategy, Jørgensen and coworkers demonstrated the enantioselective formal [2+2] cycloaddition reactions of D–A

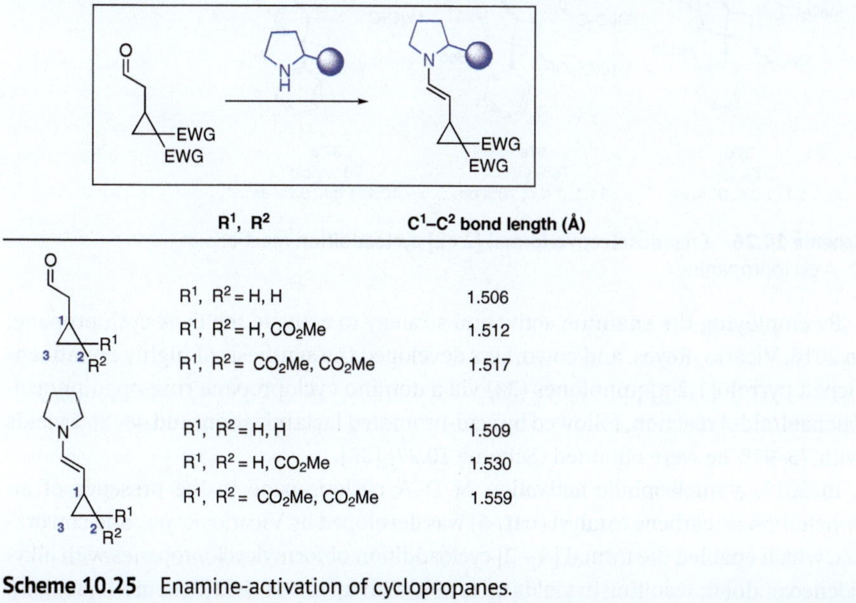

Scheme 10.25 Enamine-activation of cyclopropanes.

cyclopropanes (**35**) with 3-olefinic oxindoles (**36**) or benzofuranones (**36′**), using trimethylsilyl (TMS)-protected prolinol (**cat.-5**) as a catalyst, resulting in multifuctionalized spirocyclobutaneoxindoles (**37**) or spirocyclobutanebenzofuranones (**37′**) in good yields, with good-to-excellent diastereoselectivity and excellent enantioselectivity (Scheme 10.26) [31].

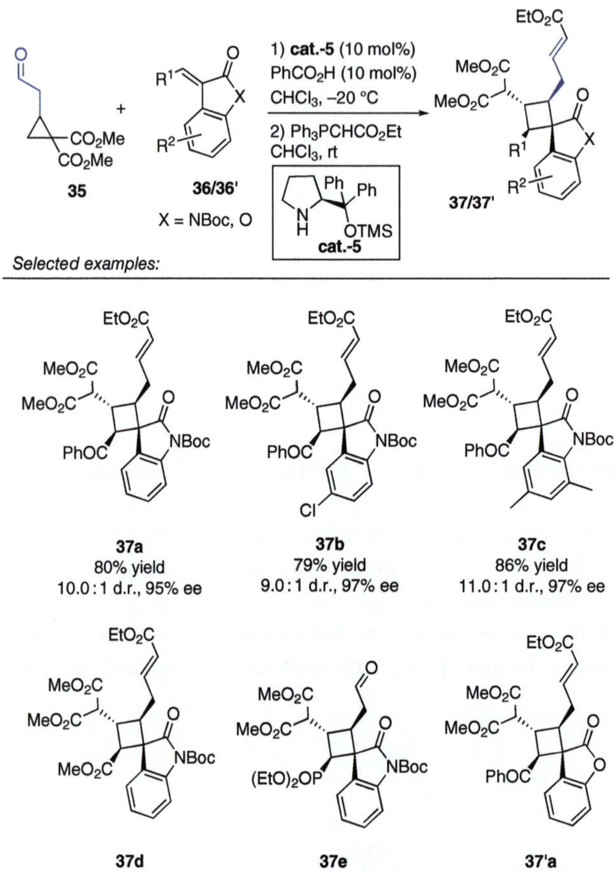

Scheme 10.26 Enantioselective formal [2 + 2] cycloaddition reactions of D–A cyclopropanes.

By employing the enamine activation strategy to activate the D–A cyclopropane, in 2016, Vicario, Reyes, and coworkers developed the synthesis of highly enantioenriched pyrrolo[1,2-a]quinolones (**38**) via a domino cyclopropane ring-opening/aza-Michael/aldol reaction, followed by acid-promoted lactamization, and 48–86% yields with 75–97% ee were obtained (Scheme 10.27) [32].

In 2017, a nucleophilic activation of D–A cyclopropane in the presence of an *N*-heterocyclic carbene catalyst (**cat.-6**) was developed by Vicario, Reyes, and coworkers, which enabled the formal [4 + 2] cycloaddition of formylcyclopropanes with alkylideneoxindoles, resulting in yields up to 99% with up to >99% ee (Scheme 10.28) [33].

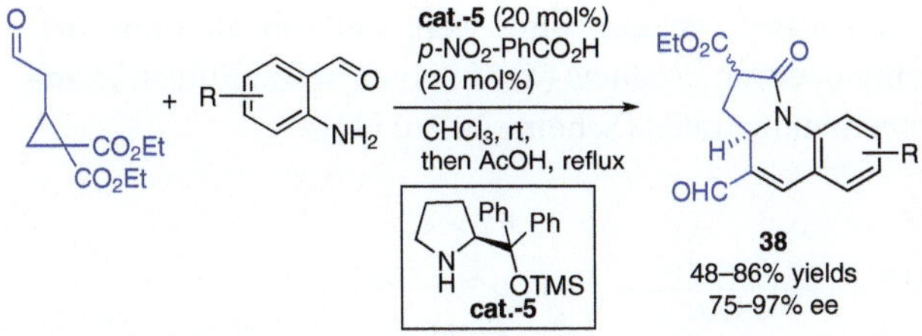

Scheme 10.27 Enamine activation of D–A cyclopropane for the synthesis of pyrrolo[1,2-a]quinolones.

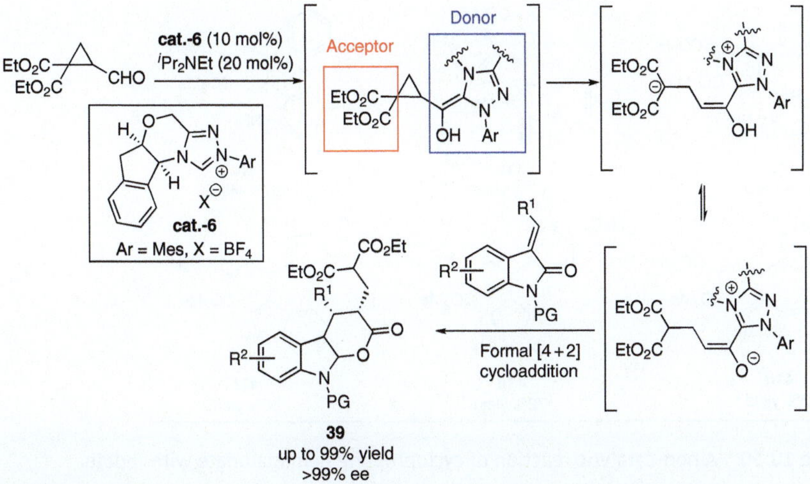

Scheme 10.28 *N*-heterocyclic carbene-promoted nucleophilic activation of D–A cyclopropane.

10.4 Catalyst-Free Activation of D–A Cyclopropanes

Kerr and coworkers developed a new mode of metal-free activation for D–A cyclopropanes. By employing cyclopropane hemimalonates as substrates, the intramolecular hydrogen bond forced coplanarity of the hemimalonate carbonyls that facilitated the nucleophilic ring-opening of the D–A cyclopropanes under high-pressure conditions (Scheme 10.29).

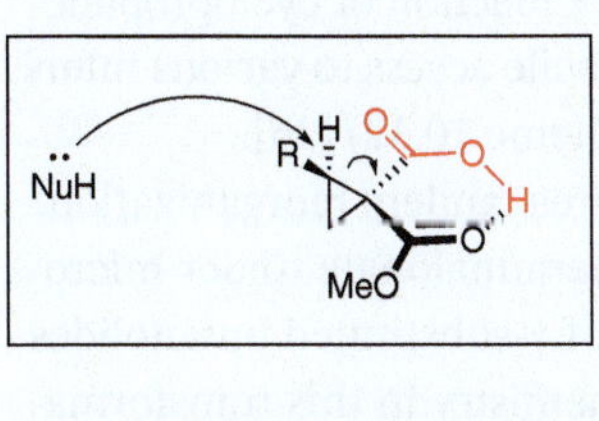

Scheme 10.29 Cyclopropane hemimalonates promoted ring-opening of D–A cyclopropanes.

In 2011, they reported a non-catalytic reaction of cyclopropane hemimalonate (**40**) with indole as a nucleophile under high-pressure (13 kbar) conditions at 50 °C. A variety of indoles with different substituents were compatible with this reaction.

Both aryl and heteroaryl-substituted cyclopropanes were well-tolerated and produced the corresponding ring-opening products (**41**) in good yields. Fifteen examples with 50–97% yields were demonstrated (Scheme 10.30) [34].

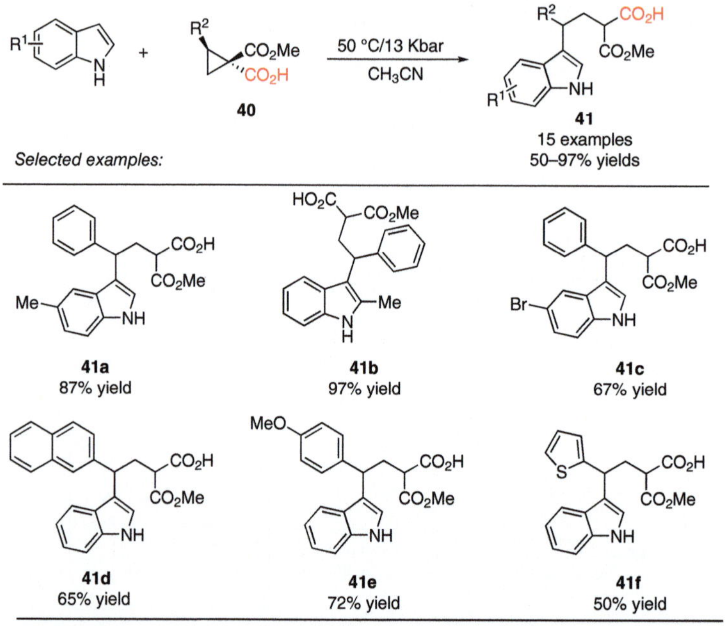

Scheme 10.30 A non-catalytic reaction of cyclopropane hemimalonate with indole.

In 2012, Kerr and coworkers developed a tandem ring-opening decarboxylation of cyclopropane hemiaminoglycate with sodium azide at 125 °C, furnishing 4-azido carboxylic acid esters (**42**) in moderate-to-excellent yields (Scheme 10.31) [35]. A possible involvement of an acyl azide mechanism, including a [3,3]-sigmatropic rearrangement to yield a ketene intermediate, was proposed. An S_N2' ring-opening process proved that an optically enriched phenyl cyclopropane (90% ee) underwent this transformation to give the ring-opening product with full retention of enantiopurity. This catalyst-free method provided a technically simple approach for the nucleophilic ring-opening of cyclopropane hemimalonates under ordinary pressure. In 2014, by delicate designing of the substrate to introduce an alkynyl moiety, they developed a catalyst-free tandem ring-opening/click reaction of cyclopropane-1,1-hemimalonate (**43**) with sodium azide, providing a facile access to various interesting linear tricyclic triazoles (**44**) in 30–80% yields (Scheme 10.32) [36].

In 2013, Kerr and coworkers demonstrated a catalyst-free tandem reorganization/dealkoxycarbonylation reaction of D–A cyclopropane hemimalonate under microwave irradiation at 150 °C, leading to a facile synthesis of γ-substituted butanolides (12 examples, 39–90% yields). The retention of stereochemistry in this transformation enabled the total synthesis of the natural product (*R*)-dodecan-4-olide in a four-step 67% overall yield from enantioenriched dimethyl ester vinyl cyclopropane (Scheme 10.33) [37].

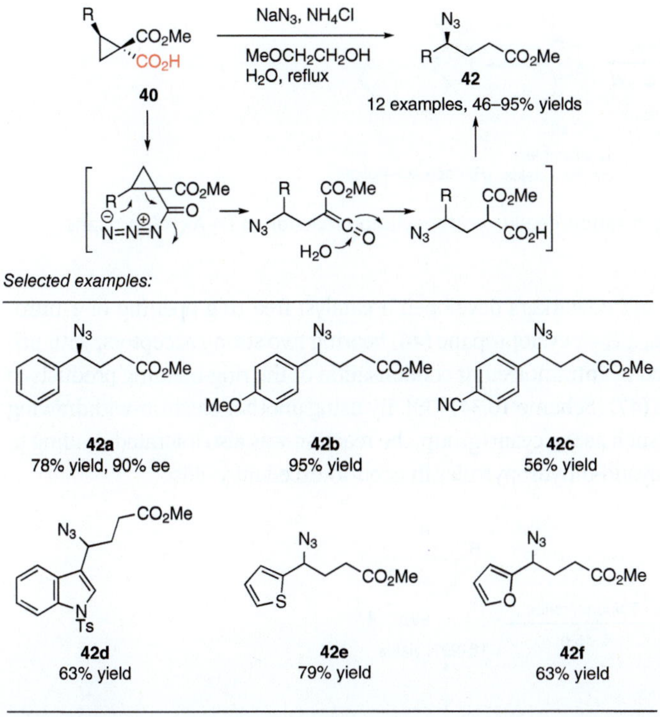

Scheme 10.31 Ring-opening decarboxylation of cyclopropane hemiaminoglycate with sodium azide.

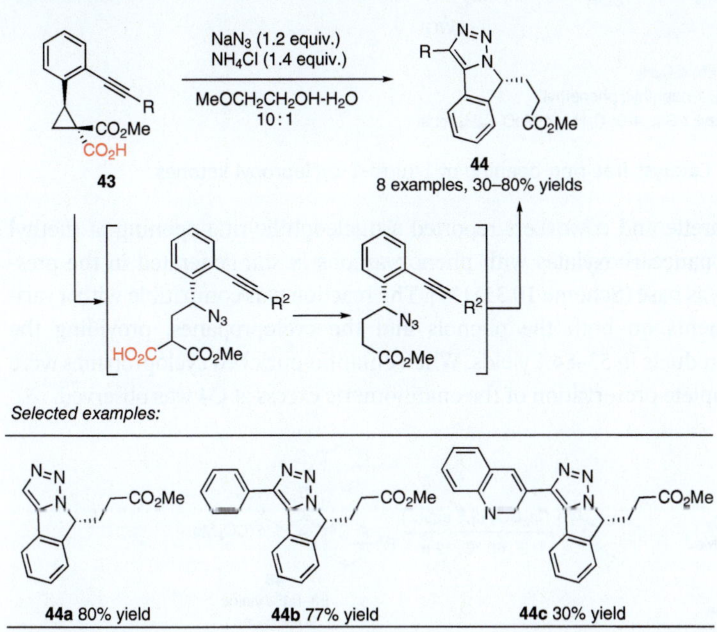

Scheme 10.32 Tandem ring-opening/click reaction of cyclopropane-1,1-hemimalonate with sodium azide.

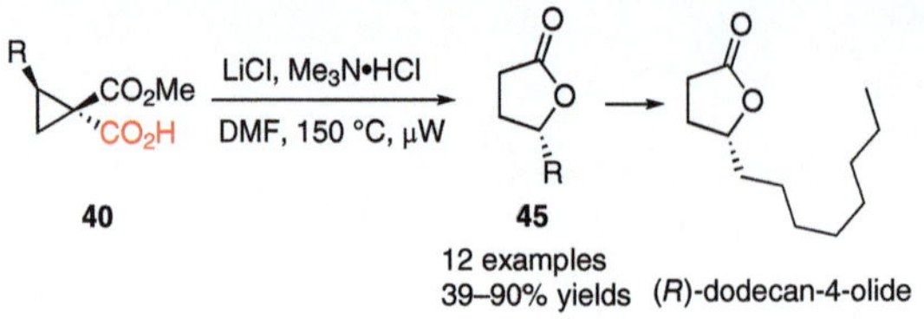

Scheme 10.33 Reorganization/dealkoxycarbonylation reaction of D–A cyclopropane hemimalonates.

In 2005, Charette and coworkers developed a catalyst-free ring-opening of 1-nitro-1-cyclopropyl ketones, a D–A cyclopropane (**46**) bearing two strong acceptors, with primary amines, followed by intramolecular condensation of the ring-opening products to yield dihydropyrroles (**47**) (Scheme 10.34) [38]. By using another electron-withdrawing group as an acceptor such as the cyano group, the reaction was also tolerated, leading to the corresponding 4-cyano-dihydropyrroles in good-to-excellent yields.

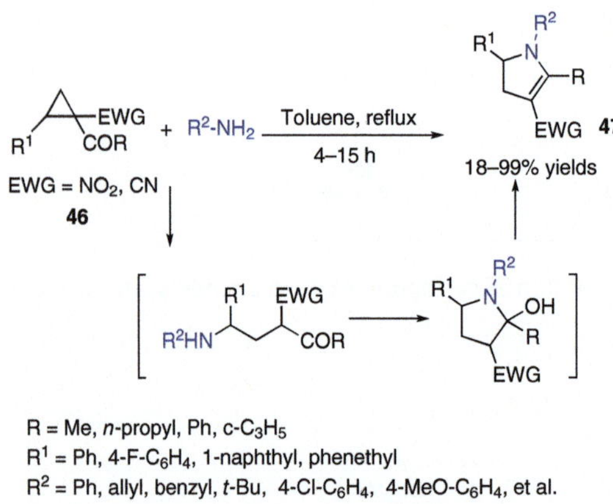

Scheme 10.34 Catalyst-free ring-opening of 1-nitro-1-cyclopropyl ketones.

In 2008, Charette and coworkers reported a nucleophilic ring-opening of methyl 1-nitrocyclopropanecarboxylates with phenoxyanions *in situ*-generated in the presence of Cs_2CO_3 as base (Scheme 10.35) [39]. The reaction was compatible with a variety of substituents on both the phenols and the cyclopropanes, providing the ring-opening products in 53–84% yields. When entantio-enriched cyclopropanes were employed, complete preservation of the enantiomeric excess at C4 was observed.

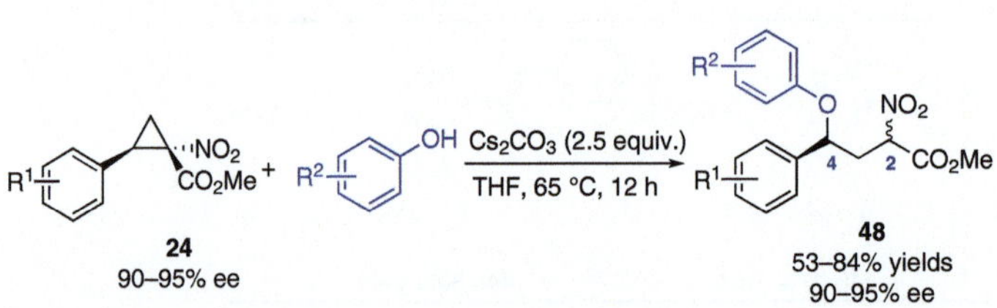

Scheme 10.35 Nucleophilic ring-opening of methyl 1-nitrocyclopropanecarboxylates with phenoxyanions.

In 2015, Budynina, Trushkov, and coworkers developed a catalyst-free ring-opening of D–A cyclopropanes with sodium azide, delivering a broad scope of polyfunctional azides (**49**) in up to 91% yield (Scheme 10.36) [40]. It is noteworthy that in this reaction for certain D–A cyclopropanes, the introduction of indolyl groups as the donor moiety or nitro group or diketone group as the acceptor moiety, high reaction temperatures were not required. DFT calculations indicate that the optimized transition state corresponds to an S_N2-like mechanism, while the nucleophilic attack on the unsubstituted C3 atom has a significantly higher energy barrier than the donor-substituted C2 atom. In the application of this method, five-, six-, and seven-membered *N*-heterocycles were synthesized with high diversity, including the total synthesis of (−)-nicotine as well as the formal total synthesis of atorvastatin.

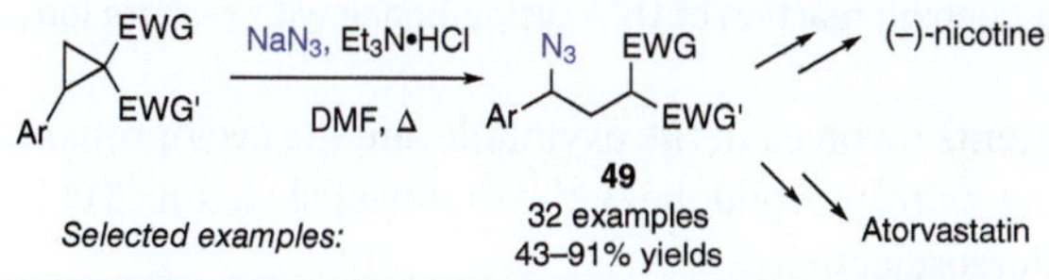

Selected examples:

Azide	*T* [°C]	*t* [h]	Yield [%]
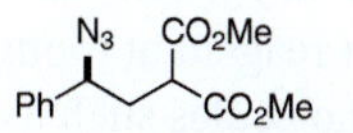 (**49a**)	100	4	88
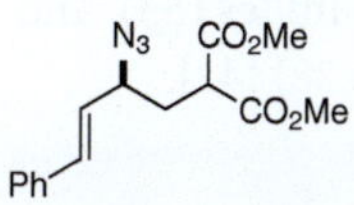 (**49b**)	100	2	80
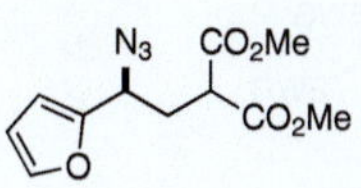 (**49c**)	100	4	75
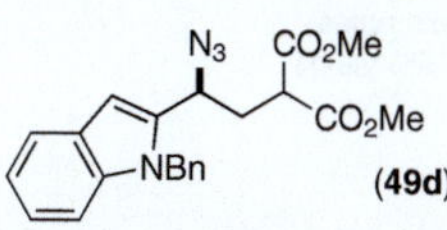 (**49d**)	50	3	91
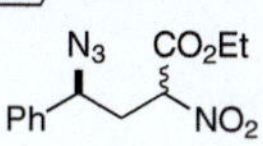 (**49e**)	55	2	76
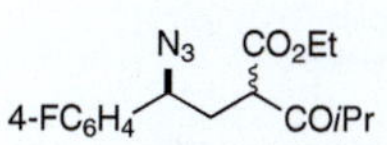 (**49f**)	90	1	90
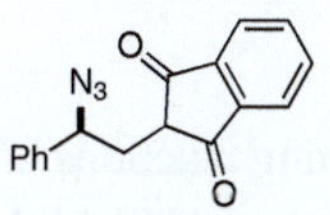 (**49g**)	25	6	85

Scheme 10.36 A catalyst-free ring-opening of D–A cyclopropanes with sodium azide.

In 2018, Budynina and coworkers developed a catalyst-free ring-opening reaction of D–A cyclopropane (**50**) with cyanate ion as a fresh *N*-nucleophile under microwave irradiation at 150 °C (Scheme 10.37) [41]. The substrate scope of this method

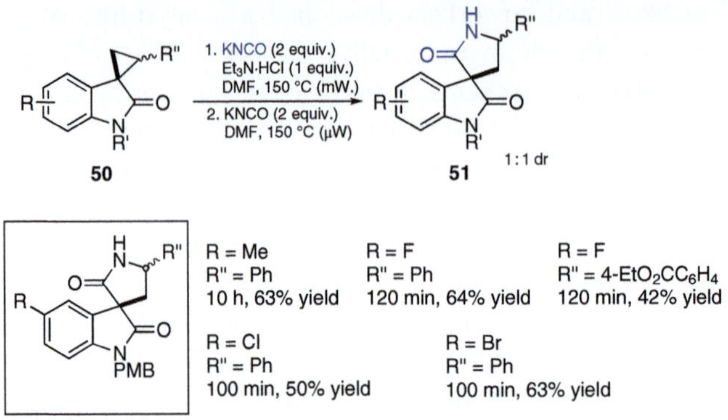

Scheme 10.37 Catalyst free ring-opening reaction of D–A cyclopropane with cyanate ion.

was broad with various substituents involved in the oxyindole and the cyclopropane units, furnishing the interesting spiro[pyrrolidone-3,3′-oxindoles] skeleton (**51**) in good yields but with poor diastereoselectivity.

In 2016, Chandrasekaran and co-workers demonstrated a series of catalyst-free ring-opening reactions of D–A cyclopropanes in the presence of benzyltriethylammonium tetrathiomolybdate {$[BnNEt_3]_2MoS_4$} as the sulfur transfer reagent at room temperature under ordinary pressure. Various sulfur-containing molecules such as symmetrical disulfides (**52**), monosulfides (**53**), unsymmetrical disulfides (**55**), and so on were synthesized under mild reaction conditions (Scheme 10.38) [42].

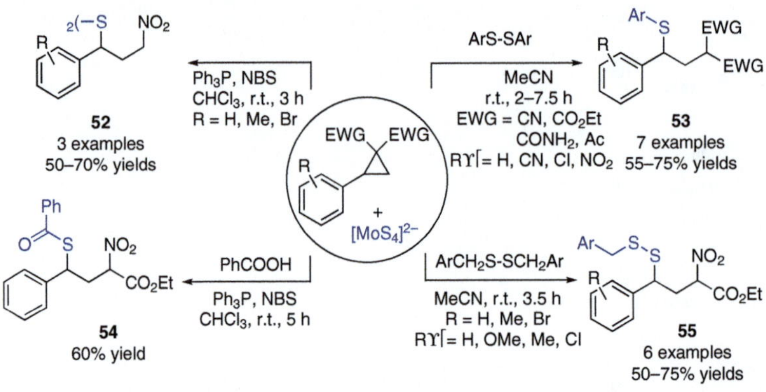

Scheme 10.38 Catalyst-free ring-opening of D–A cyclopropanes in the presence of $[BnNEt_3]_2MoS_4$.

In 2021, Ofial and coworkers studied the kinetics of the ring-opening reactions of D–A cyclopropanes with thiophenolates in DMSO at 20 °C (Scheme 10.39) [43]. Parabolic Hammett relationships were observed, verifying the electronic properties of substituents on the phenyl groups of cyclopropanes, suggesting that the inherent S_N2 reactivity of electrophilic cyclopropanes is activated by electron-rich π-systems owing to the more advanced C—C bond polarization in the transition state. Conversely, electron-poor π-systems facilitate the attack of anionic nucleophiles due to attractive electrostatic interactions, resulting in a lowering of the energetic barriers.

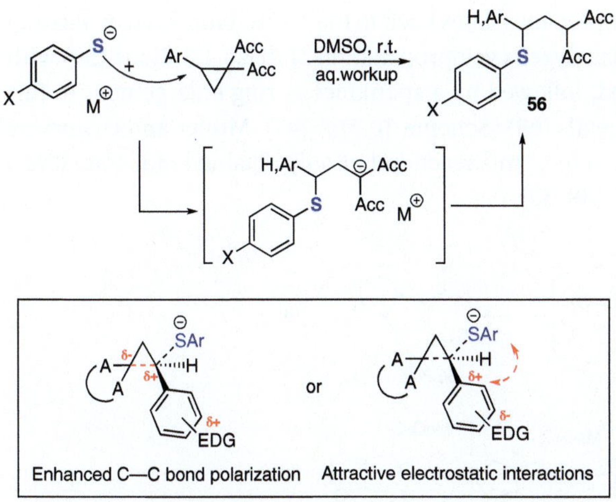

Scheme 10.39 Ring-opening reactions of D–A cyclopropanes with thiophenolates.

In 2009, Werz and coworkers developed a new approach in the synthesis of anti-fused oligoannelated tetrahydrofuran (THF) (**57**) via a synthetic sequence of cyclopropanation/reduction/oxidation/ring-opening rearrangement of D–A cyclopropanes (Scheme 10.40a) [44]. The key step is the ring-opening rearrangement of cyclopropane, which relied on the elaborate design of *in situ* formation of the cyclopropane bearing both strong electron-push and electron-pull substituents in a highly strained three-membered ring. In 2013, a similar strategy was applied to the synthesis of bisthiophenes (**58**) by Werz and coworkers (Scheme 10.40b) [45]. With Lawesson's reagents, the ketone moieties were converted into the thiocarbonyl acceptor, which could favor the immediate ring-enlargement process by heating [46].

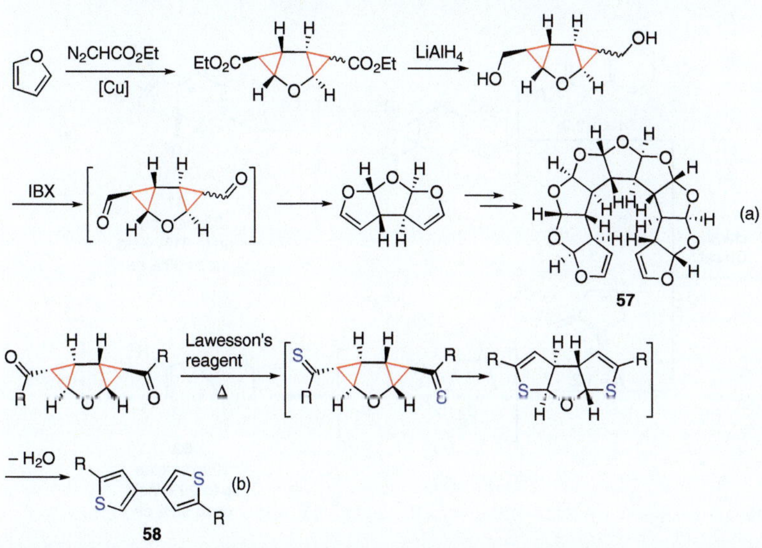

Scheme 10.40 Catalyst free ring-opening rearrangement of cyclopropane based on a THF unit.

The study on of D–A cyclopropanes dates back to the 1990s. Pioneered by Pirrung and coworkers, a rhodium-catalyzed cyclopropanation of diazo-1,3-dione (**59**) with dihydrofuran was conducted, followed by a spontaneous ring-enlargement to furnish the racemic bicyclic acetals (**60**) (Scheme 10.41a) [47]. Müller and coworkers explored the asymmetric version of this reaction but only obtained moderate levels of ee value (Scheme 10.41b) [48–50].

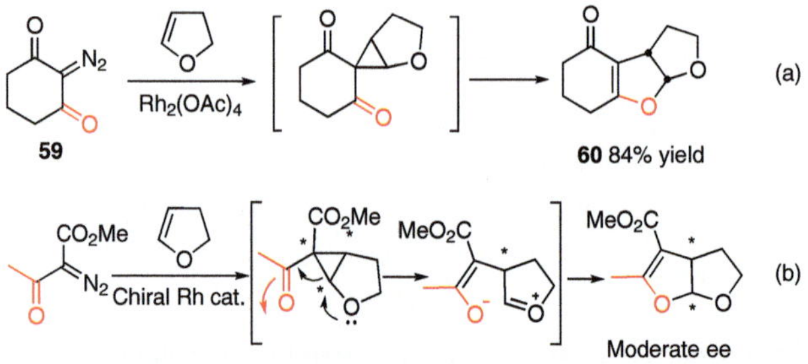

Scheme 10.41 Spontaneous ring-enlargement of D–A cyclopropanes.

In 2020, Tang, Wang, and coworkers reported a highly enantioselective synthesis of various chiral heterobicyclic molecules, including fused bicyclic acetals and spiroaminals under mild reaction conditions, involving spontaneous ring-enlargements of *in situ*-generated D–A cyclopropanes [51]. The donor unit of this type of D–A cyclopropane was expanded from oxygen-based moiety such as THF to nitrogen-based moiety such as pyrrolidine. An extremely broad substrate scope was developed, leading to 23 examples of fused bicyclic acetals (**61**) in up to 91% yield with up to 99% ee, as well as diverse chiral spiroaminals (**62**) in up to 99% yield with up to 96% ee (Scheme 10.42).

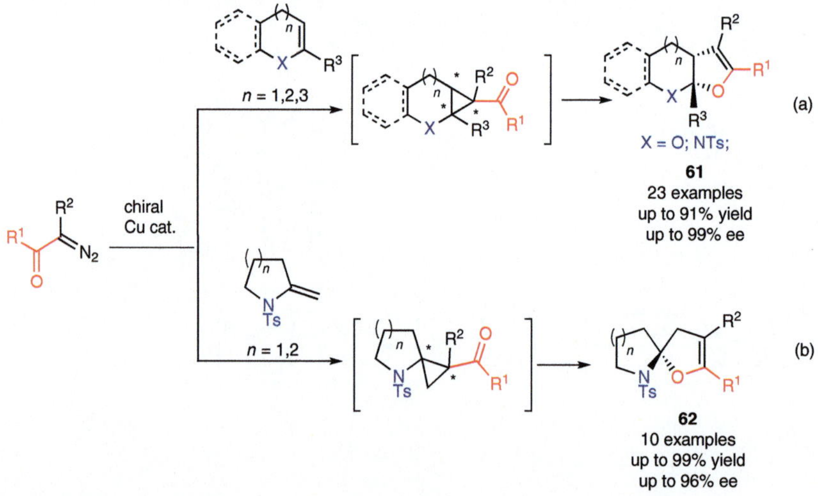

Scheme 10.42 Spontaneous ring-enlargement of D–A cyclopropanes in an enantioselective process.

For donor-substituted nitrocyclopropanes, the immediate ring-enlargement could occur in mild conditions (25 °C). In 2013, Werz and coworkers reported a rhodium-catalyzed cyclopropanation of donor-substituted alkenes with α-diazo-α-nitro ethyl acetate, and the *in situ*-generated nitrocyclopropane immediately underwent the ring-enlargement to form the cyclic nitronates (**63**) (Scheme 10.43) [52]. Interestingly, the nitro group was inserted in all cases, while the ester group, as the other acceptor unit, was not involved in the ring-opening rearrangement process.

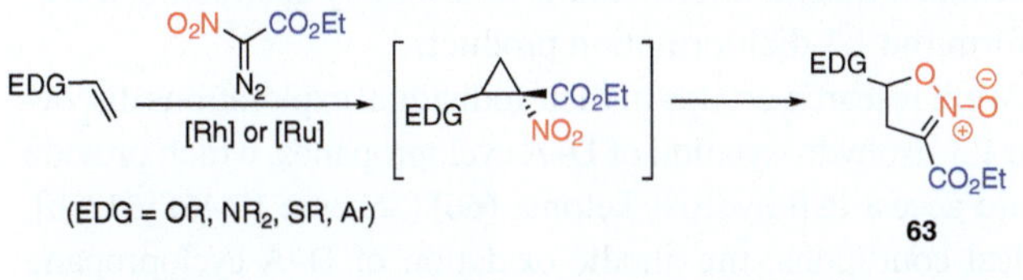

Scheme 10.43 Ring-enlargement of donor-substituted nitrocyclopropanes.

In 2022, Afonso, Candeias, and coworkers developed a formal [4+1] cycloaddition via photopromoted siloxycarbene-involved cyclopropanation/ring-opening/annulation processes, leading to a broad range of highly functionalized cyclopentene scaffolds (**64**) in up to 90% yield (Scheme 10.44) [53]. DFT calculations revealed that in this metal-free process, the energy requirements for the cyclic expansion of cyclopropane to cyclopentene are similar for both diastereomeric cyclopropanes.

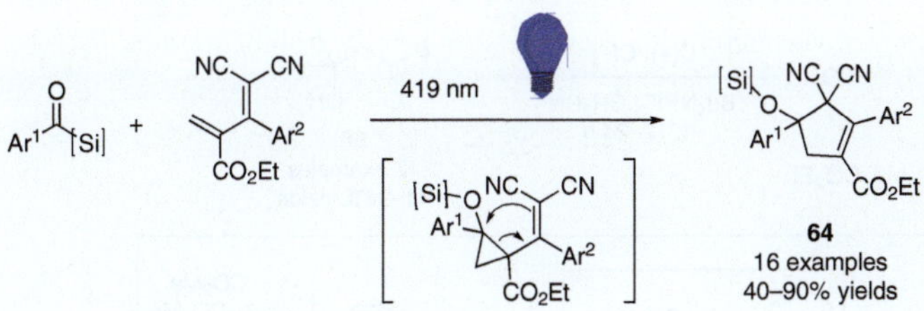

Scheme 10.44 Formal [4+1] cycloaddition via photopromoted siloxycarbene.

10.5 Metal-Free Activation of D–A Cyclopropanes via Radical, SET, and Photopromoted Process

In 2014, Werz and coworkers developed a radical 1,3-dichlorination reaction of D–A cyclopropanes with iodobenzene dichloride (Scheme 10.45) [54]. A variety of D–A cyclopropanes (22 examples) could undergo ring-opening process to furnish the corresponding 1,3-dichloro carbonyl compounds and carboxylic acid derivatives (**65**) in 53–95% yields. Different acceptor groups, such as ketone, diester, and dinitrile,

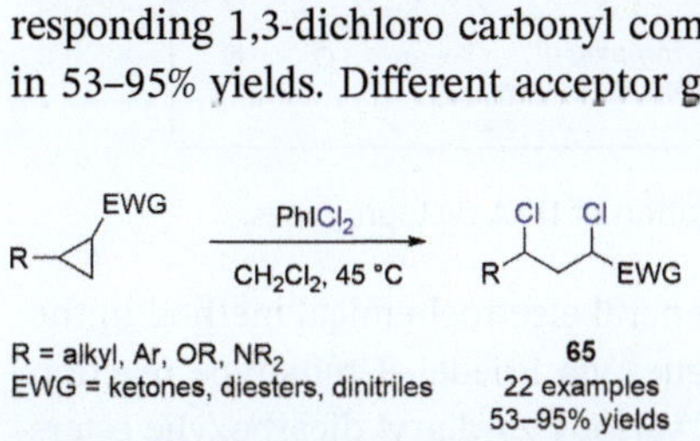

Scheme 10.45 Radical 1,3-dichlorination reaction of D–A cyclopropanes with iodobenzene dichloride.

worked well; however, the single ester group as acceptor proved insufficient to proceed with the ring-opening reaction. The scope of the donor groups for the D–A cyclopropanes are very wide, including aliphatic groups, oxygen-functional groups, nitrogen-functional groups, and the aromatic groups. A plausible mechanism was proposed that by thermolysis or photolysis, the homolytic cleavage of the I—Cl bond released the chloride radical, which attacked the C(sp^3)—C(sp^3) bond between the adjacent donor and acceptor moieties in the D–A cyclopropane, opening the strained ring and generate a delocalized radical intermediate, followed by abstracting a second chloride radical to form the 1,3-dichlorination product.

In 2021, Banerjee and Werz research groups in their individual explorations discovered an electricity-driven 1,3-oxohydroxylation of D–A cyclopropanes, which provide a mild and straightforward access to β-hydroxy ketones (**66**) (Scheme 10.46) [55 a,b]. Under the electrochemical conditions, the anodic oxidation of D–A cyclopropane occurred, generating an arene radical cation **I**, which underwent a homolytic cleavage to form a ring-opening intermediate **II**, followed by the trapping of molecular oxygen to render intermediate **III**, and then generated the endoperoxide intermediate **IV**. **IV** then accepts an electron, leading to intermediate **V**. Finally, the 1,3-oxohydroxylation product was furnished by the deprotonation of intermediate **V**. The substrate scope of this reaction was studied, and 17 examples succeeded in 41–94% yields. A strong electronic effect was observed, where the electron-rich aryl groups could benefit the reactivity. In contrast, when the NO_2 group was introduced, the reaction could not occur.

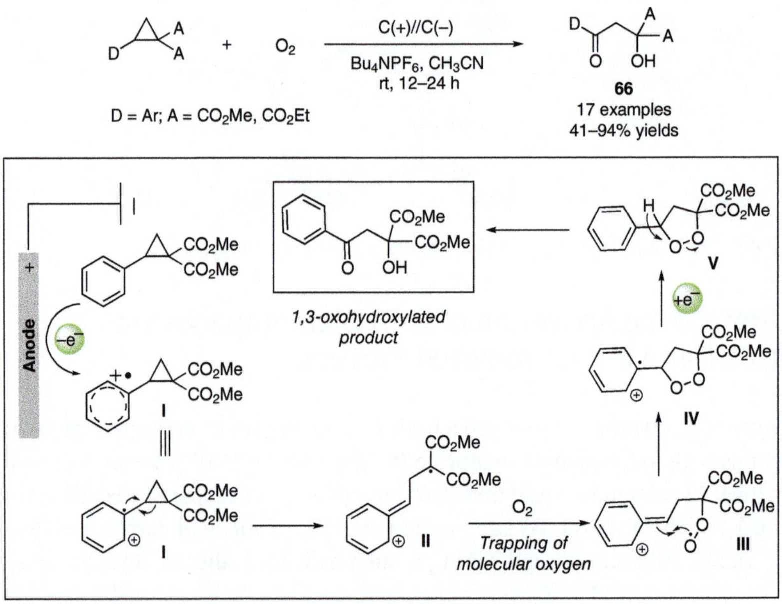

Scheme 10.46 Electricity-driven 1,3-oxohydroxylation of D–A cyclopropanes.

In 2021, Werz and coworkers described a general electrochemical method in the ring-opening of D–A cyclopropanes with arenes via Friedel–Crafts-type reaction (Scheme 10.47) [56]. This electrosynthesis of various γ,γ-diaryl dicarboxylic esters

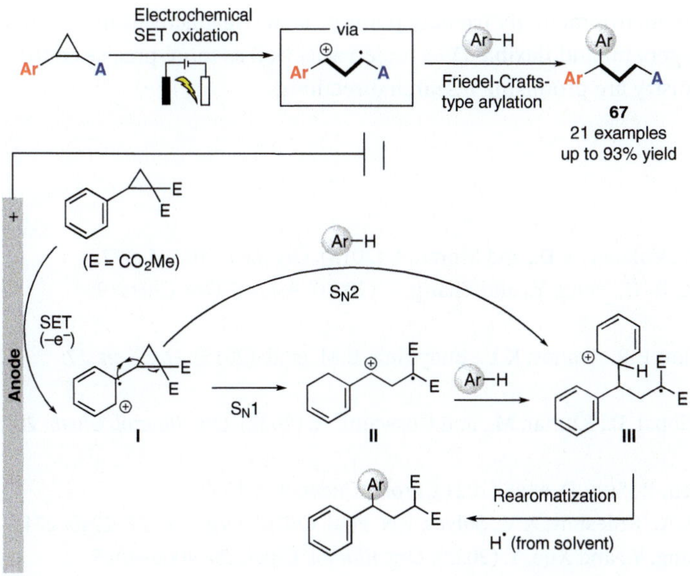

Scheme 10.47 Electrochemical-promoted ring-opening of D–A cyclopropanes with arenes.

(**67**) highlights scalability with a broad reaction scope, including 21 examples in 17–93% yields. A plausible mechanism was proposed that C(sp^3)–C(sp^3) cleavage of the D–A cyclopropanes was promoted by the direct anodic oxidation of the strained carbocycles to radical cations, followed by the Friedel–Crafts-type reaction with electron-rich arenes.

10.6 Conclusion

In summary, in the past decades, remarkable progress has been made in the field of metal-free activation of D–A cyclopropanes. Compared to the classic activation pattern of D–A cyclopropanes by means of Lewis acid catalysts, activations with protic acids result in more diverse reaction sites owing to the demand for carbocation stability. The masked D–A strategy via electrophile-induced C—C bond activation of vinylcyclopropanes provides new thoughts for the metal-free activation of D–A cyclopropanes. For D–A cyclopropanes bearing an aldehyde moiety as the acceptor unit or donor unit, the concepts of iminium activation and enamine activation bring about new opportunities in the enantioselective transformation of cyclopropanes. Studies on the catalyst-free ring-opening of D–A cyclopropanes revealed that the electric effect of the D-A system could influence the C—C bond polarization of the cyclopropane, and thus enabling the ring-opening to occur under mild reaction conditions. By elaborate substrate design, the spontaneous ring-enlargement of D–A cyclopropanes could build up interesting and complex molecular structures. The metal-free activations of D–A cyclopropanes have great vitality and promising prospects from an environmentally friendly point of view. New activation patterns concerning more general and flexible D–A moieties as well as more precise control of the stereochemistry are promising research directions.

References

1 Richmond, E., Vuković, V.D., and Moran, J. (2018). *Org. Lett. 20*: 574–577.
2 Zhou, Y., Xue, R.-C., Feng, Y., and Zhang, L. (2020). *Asian J. Org. Chem. 9*: 311–316.
3 Rakhmankulov, E.R., Ivanov, K.L., Budynina, E.M. et al. (2015). *Org. Lett. 17*: 770–773.
4 Singh, P.R., Gopal, B., Kumar, M., and Goswami, A. (2022). *Org. Biomol. Chem. 20*: 4933–4941.
5 Zhao, H., Shen, P., Sun, D. et al. (2021). *Front. Chem. 9*: 711257.
6 Mikhaylov, A.A., Kuleshov, A.V., Solyev, P.N. et al. (2020). *Org. Lett. 22*: 2740–2745.
7 Chen, Y., Zhang, Y., and Xue, Y. (2022). *Org. Biomol. Chem. 20*: 4006–4015.
8 Kaschel, J., Schneider, T.F., Kratzert, D. et al. (2012). *Angew. Chem., Int. Ed. 51*: 11153–11156.
9 Kaschel, J., Schneider, T.F., Kratzert, D. et al. (2013). *Org. Biomol. Chem. 11*: 3494–3509.
10 Kaschel, J., Schneider, T.F., Schirmer, P. et al. (2013). *Eur. J. Org. Chem. 2013*: 4539–4551.
11 Andreev, I.A., Ratmanova, N.K., Augustin, A.U. et al. (2021). *Angew. Chem., Int. Ed. 60*: 7927–7934.
12 De Simone, F., Saget, T., Benfatti, F. et al. (2011). *Chem. Eur. J. 17*: 14527.
13 De Simone, F., Andrès, J., Torosantucci, R., and Waser, J. (2009). *Org. Lett. 11*: 1023–1026.
14 Otohiko, T., Shuji, K., Tatsuya, O., and Toshiro, S. (1988). *Bull. Chem. Soc. Jpn. 61*: 2897–2908.
15 Morra, N.A., Morales, C.L., Bajtos, B. et al. (2006). *Adv. Synth. Catal. 348*: 2385–2390.
16 Chagarovskiy, A.O., Ivanova, O.A., Rakhmankulov, E.R. et al. (2010). *Adv. Synth. Catal. 352*: 3179–3184.
17 Budynina, E., Ivanova, O., Skvortsov, D. et al. (2014). *Synlett 25*: 2289–2292.
18 Ortega, V. and Csaky, A.G. (2016). *J. Org. Chem. 81*: 3917–3923.
19 Ganesh, V., Sureshkumar, D., and Chandrasekaran, S. (2011). *Angew. Chem. Int. Ed. 50*: 5878–5881.
20 Ganesh, V., Sureshkumar, D., Chanda, D., and Chandrasekaran, S. (2012). *Chem. Eur. J. 18*: 12498–12511.
21 Ganesh, V., Kundu, T., and Chandrasekaran, S. (2013). *J. Org. Chem. 78*: 380–399.
22 Sparr, C. and Gilmour, R. (2011). *Angew. Chem. Int. Ed. 50*: 8391–8395.
23 Wallbaum, J., Garve, L.K.B., Jones, P.G., and Werz, D.B. (2016). *Chem. Eur. J. 22*: 18756–18759.

24 So, S.S., Auvil, T.J., Garza, V.J., and Mattson, A.E. (2012). *Org. Lett. 14*: 444–447.
25 Hardman, A.M., So, S.S., and Mattson, A.E. (2013). *Org. Biomol. Chem. 11*: 5793–5797.
26 Dickmeiss, G., De Sio, V., Udmark, J. et al. (2009). *Angew. Chem., Int. Ed. 48*: 6650–6653.
27 Blom, J., Vidal-Albalat, A., Jørgensen, J. et al. (2017). *Angew. Chem., Int. Ed. 56*: 11831–11835.
28 McLeod, D.A., Thoegersen, M.K., Barloese, C.L. et al. (2022). *Angew. Chem., Int. Ed. 61*: e202206096.
29 Liu, J., Ye, W., Qing, X., and Wang, C. (2016). *J. Org. Chem. 81*: 7970–7976.
30 Luo, N., Liu, J., Wang, S., and Wang, C. (2020). *Org. Biomol. Chem. 18*: 9210–9215.
31 Halskov, K.S., Kniep, F., Lauridsen, V.H. et al. (2015). *J. Am. Chem. Soc. 137*: 1685–1691.
32 Sanchez-Diez, E., Vesga, D.L., Reyes, E. et al. (2016). *Org.Lett. 18*: 1270–1273.
33 Prieto, L., Sánchez-Díez, E., Uria, U. et al. (2017). *Adv. Synth. Catal. 359*: 1678–1683.
34 Emmett, M.R. and Kerr, M.A. (2011). *Org. Lett. 13*: 4180–4183.
35 Emmett, M.R., Grover, H.K., and Kerr, M.A. (2012). *J. Org. Chem. 77*: 6634–6637.
36 Flisar, M.E., Emmett, M.R., and Kerr, M.A. (2014). *Synlett 25*: 2297–2300.
37 Grover, H.K., Emmett, M.R., and Kerr, M.A. (2013). *Org. Lett. 15*: 4838–4841.
38 Wurz, R.P. and Charette, A.B. (2005). *Org. Lett. 7*: 2313–2316.
39 Lifchits, O., Alberico, D., Zakharian, I., and Charette, A.B. (2008). *J. Org. Chem. 73*: 6838–6840.
40 Ivanov, K.L., Villemson, E.V., Budynina, E.M. et al. (2015). *Chem. Eur. J. 21*: 4975–4987.
41 Zaytsev, S.V., Ivanov, K.L., Skvortsov, D.A. et al. (2018). *J. Org. Chem. 83*: 8695–8709.
42 Gopinath, P. and Chandrasekaran, S. (2016). *Synthesis 48*: 3087–3096.
43 Juestel, P.M., Stan, A., Pignot, C.D., and Ofial, A.R. (2021). *Chem. Eur. J. 27*: 15928–15935.
44 Schneider, T.F., Kaschel, J., Dittrich, B., and Werz, D.B. (2009). *Org. Lett. 11*: 2317–2320.
45 Kaschel, J., Schmidt, C.D., Mumby, M. et al. (2013). *Chem. Commun. 49*: 4403–4405.
46 Schneider, T.F. and Werz, D.B. (2011). *Org. Lett. 13*: 1848–1851.
47 Pirrung, M.C., Zhang, J., and McPhail, A.T. (1991). *J. Org. Chem. 56*: 6269–6271.
48 Müller, P., Allenbach, Y.F., and Bernardinelli, G. (2003). *Helv. Chim. Acta 86*: 3164–3178.
49 Müller, P., Allenbach, Y.F., Ferri, M., and Bernardinelli, G. (2003). *ARKIVOC 2003*: 80–95.
50 Müller, P. and Chappellet, S. (2005). *Helv. Chim. Acta 88*: 1010–1021.
51 Zhou, L., Yan, W.-G., Sun, X.-L. et al. (2020). *Angew. Chem. Int. Ed. 59*: 18964–18969.
52 Schmidt, C.D., Kaschel, J., Schneider, T.F. et al. (2013). *Org. Lett. 15*: 6098–6101.

53 Vale, J.R., Gomes, R.F., Afonso, C.A.M., and Candeias, N.R. (2022). *J. Org. Chem. 87*: 8910–8920.

54 Garve, L.K.B., Barkawitz, P., Jones, P.G., and Werz, D.B. (2014). *Org. Lett. 16*: 5804–5807.

55 (a) Saha, D., Maajid Taily, I., and Banerjee, P. (2021). *Eur. J. Org. Chem. 2021*: 5053–5057; (b) Kolb, S., Petzold, M., Brandt, F. et al. (2021). *Angew. Chem. Int. Ed.* 60: 15928–15934.

56 Kolb, S., Ahlburg, N.L., and Werz, D.B. (2021). *Org. Lett. 23*: 5549–5553.

11

Asymmetric Catalytic Activation of Donor–Acceptor Cyclopropanes

Yong Xia, Xiaohua Liu, and Xiaoming Feng

Sichuan University, Key Laboratory of Green Chemistry & Technology, Ministry of Education, College of Chemistry, Chengdu, 610064, China

CHAPTER MENU

11.1 Introduction

Due to the ease of synthesis and high reactivity, donor–acceptor (D–A) cyclopropanes have proven to be powerful synthetic building blocks in organic synthesis. In particular, the enantioselective transformations of D–A cyclopropanes provide easy access to a myriad of optically active carbo-, heterocycles, as well as γ-functionalized carbon scaffolds, and have garnered extensive attention from the synthetic chemistry community. Due to the breaking C—C bond the three-membered ring via various pathways, the stereo-arrangement in D–A cyclopropanes might be preserved or destroyed. Thus, stereospecific transformations from optically enriched cyclopropanes, or enantioselective reactions from racemic or prochiral cyclopropanes, are two general approaches for accessing chiral targets.

Efforts toward asymmetric catalytic transformations can be traced back to Scheffold and Troxler's desymmetrization-fragmentation of *meso*-cyclopropanes in 1994 [1]. A decade later, an important breakthrough in chiral Ni(II)/BOX complex-catalyzed asymmetric [3 + 3] cycloadditions of nitrones and cyclopropane-1, 1-diesters, was achieved by Sibi et al., in which tetrahydro-1,2-oxazines were synthesized in good enantioselectivity but low diastereoselectivity [2]. Afterwards,

Donor–Acceptor Cyclopropanes in Organic Synthesis, First Edition.
Edited by Prabal Banerjee and Akkattu T. Biju.

Tang's group improved the diastereoselectivity of such a reaction by employing a chiral Ni(II) catalyst with an arm-side trioxazoline ligand [3], which was proven to be useful for enantioselective ring-opening of D–A cyclopropanes with a number of nucleophiles. The low-valent transition metal-catalyzed asymmetric reactions of vinylcyclopropanes (VCPs) via π-allyl-metal intermediates were introduced by the groups of Johnson, Krische, and Trost in 2011 [4, 5]. Since then, tremendous strategies of enantioselective transformations of D–A cyclopropanes have been developed and also been applied to the construction of complex molecules. Related reviews have been dedicated to this topic [6]. In this chapter, we aim to give a picture of recent advances in asymmetric catalytic transformations of D–A cyclopropanes. The developed reactions will be classified by a chiral catalyst in connection with the activation modes: Lewis acid activation (Section 11.2); low-valent transition metal catalyst activation (Section 11.3); organocatalyst activation and miscellaneous (Section 11.4).

11.2 Chiral Lewis Acid-Catalyzed Reactions of D–A Cyclopropanes

Chiral Lewis acid activation occupies a preeminent position in enantioselective reactions of D–A cyclopropanes. The LUMO energy of cyclopropane can be lowered by a Lewis acid catalyst through coordination with the acceptor moiety and the chiral information is thus loaded meanwhile. Due to the fact that D–A cyclopropanes are often chiral themselves, a dynamic kinetic asymmetric transformation (DYKAT) or a kinetic resolution is usually involved in chiral Lewis acid-promoted reactions [7]. A DYKAT has an interconversion between two Lewis acid-activated diastereomeric intermediates of cyclopropanes, while there is no or much slow conversion between these two diastereomers in a kinetic resolution. Stereospecific S_N2-like attack by a nucleophile to the activated intermediates leads to the reversed stereocenter on the donor site of cyclopropanes.

11.2.1 Asymmetric Reactions of Two-Substituted Cyclopropane-1,1-Dicarboxylates

2-Substituted cyclopropyl diesters, as one of the most common types of D–A cyclopropane, have been widely used in enantioselective ring-opening and annulation reactions. The ring-opening of such a three-membered ring with heteroatom and electron-rich arene nucleophiles is one of the simplest and most efficient synthetic approaches to homoconjugated addition products. D–A cyclopropanes also serve as formal 1,3-zwitterions in asymmetric [3+n] annulations, leading to enantioenriched carbocycles and heterocycles.

11.2.1.1 Ring-Opening Reactions

Taking the advantage of side-arm strategy [8], Tang and coworkers developed a Ni(II)/**L1**-catalyzed asymmetric reaction of cyclopropyl diesters with secondary aliphatic amines, yielding γ-aminobutyric acid (GABA) derivatives. The introduction of an extra oxazoline group as the side arm of the chiral ligand is crucial for regulating the rate and stereocontrol of the reaction, which underwent the ring-opening process in a kinetic resolution manner. Due to the steric repulsion between the donor group of the (*R*)-enantiomer of cyclopropane **1** and the ligand **L1** (Scheme 11.1, **TS2**), (*S*)-cyclopropane was preferentially activated to perform the attack of secondary amine to deliver (*R*)-product **2**, leaving the less-matched (*R*)-cyclopropane in high optical purity (Scheme 11.1) [9].

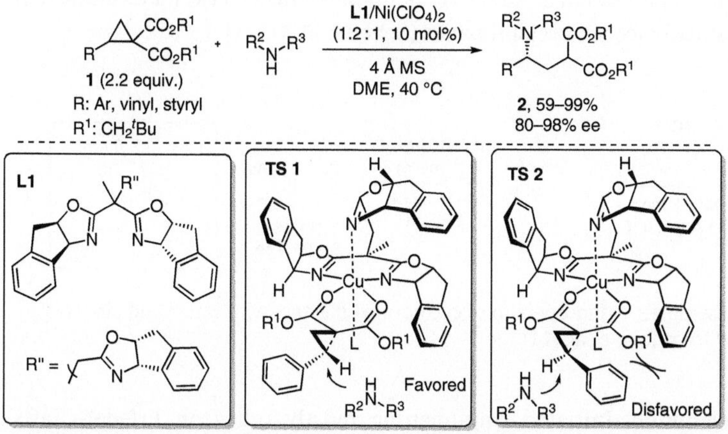

Scheme 11.1 Asymmetric ring-opening of cyclopropyl diesters with secondary amines. Source: Modified from Tang et al. [9].

Direct ring-opening of cyclopropane with primary amines might be accompanied by the overreaction raised from the competitive nucleophilic attack of the formed secondary amine product. Using chiral bimetallic Lewis acid catalysis, Wang and coworkers realized an asymmetric ring-opening of cyclopropanes with primary anilines (Scheme 11.2) [10]. Optically active γ-amino acids **4** were produced in good yield and enantioselectivity by treatment with excessive racemic cyclopropanes. The combination of $Y(OTf)_3$ and (*R*)-Yb[**P**]$_3$ greatly enhanced both activity and enantioselectivity in comparison with employing each metal salt. This reaction was

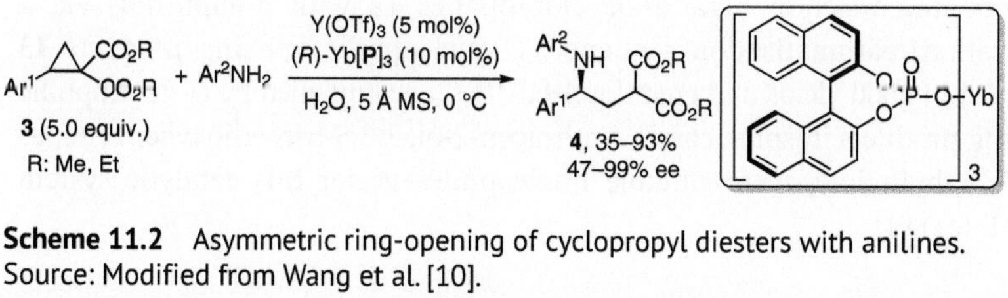

Scheme 11.2 Asymmetric ring-opening of cyclopropyl diesters with anilines. Source: Modified from Wang et al. [10].

proposed to proceed via a two-step S_N2/proton transfer mechanism, and the initial step was the rate- and enantio-determining step.

Asymmetric ring-opening with oxygen-containing nucleophiles yields enantioenriched γ-hydroxybutyric acid (GHB) derivatives. Tang and coworkers successfully developed a chiral **L2**/Cu(II)-catalyzed ring-opening of D–A cyclopropanes with aliphatic alcohols, providing optically active GHB products **6** in high enantioselectivity. They then investigated the reaction with water nucleophiles, which is particularly challenging, as its nucleophilicity is quite weak and the obtained alcohol products can react further. Here, the $Cu(ClO_4)_2{\cdot}6H_2O$ acted as both a Lewis acid and a buffer for the water nucleophile source. A DYKAT dominated this process for electron-rich aromatic substituted racemic cyclopropanes (Scheme 11.3) [11].

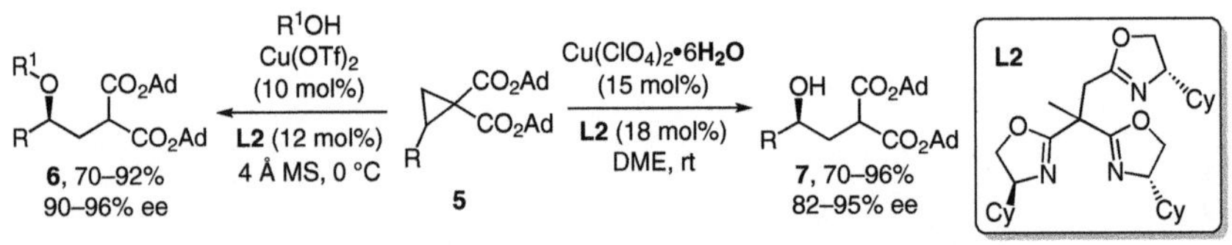

Scheme 11.3 Asymmetric ring-opening of cyclopropyl diesters with water and alcohols. Source: Modified from Tang et al. [11].

Carbon nucleophiles-initiated ring-opening mostly relies on Friedel-Crafts alkylations with electron-rich arenes. In 2013, Johnson et al. developed an **L3**/MgI_2-promoted Friedel-Crafts alkylation of cyclopropyl diesters with indoles. The use of sterically hindered and/or electronically deactivating TBS-protected indoles was important to reach high stereocontrol. Electron-rich aromatically substituted racemic cyclopropanes underwent the reaction via a DYKAT process, in which each isomer could be transformed into the product **9** through the chiral Lewis acid-mediated enantiomer interconversion. However, for less active phenylcyclopropane, the reaction was sluggish and underwent a kinetic resolution process, in which the (*S*)-cyclopropane consumed faster than its enantiomer (Scheme 11.4a) [12].

A desymmetrization process was involved by using *meso*-diaminocyclopropanes. With a modified bis(oxazoline) ligand **L4**/Cu(II) catalyst, chiral urea products **11** were harvested in high enantio- and diastereoselectivities by using indole or pyrrole nucleophiles (Scheme 11.4b) [13]. Wang and Guo also developed an enantioselective ring-opening reaction of aminocyclopropanes **12** with β-naphthols via a dearomatization/rearomatization sequence. *C*-nucleophilic opening products **13** were obtained in good yields and enantioselectivities, accompanied by *O*-nucleophilic ring-opening products in some cases. Anthracen-2-ol, 1,3,5-trimethoxybenzene, as well as *N*-methylindole, were suitable nucleophiles under this catalytic system (Scheme 11.4c) [14].

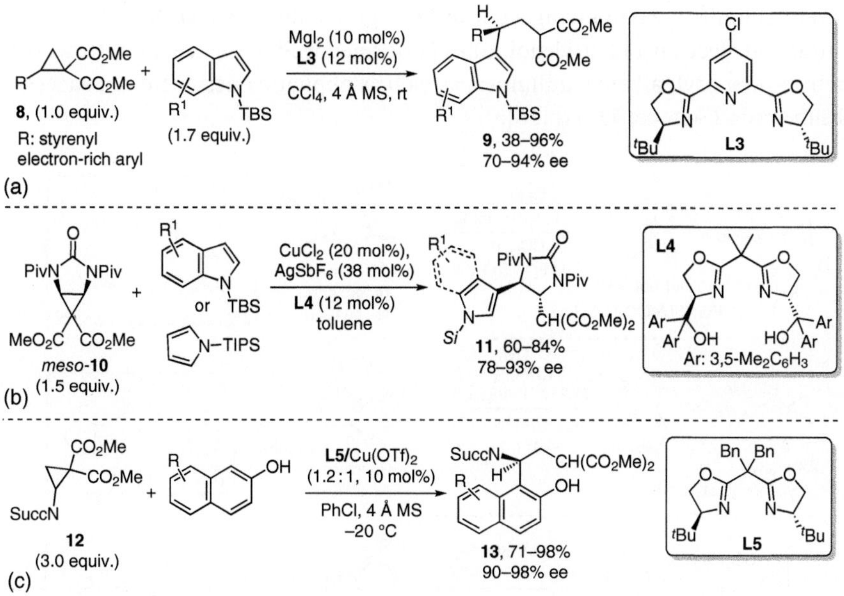

Scheme 11.4 Asymmetric alkylation of cyclopropyl diesters with arenes.

11.2.1.2 [3 + n] Annulations

One of the most important synthetic applications of D–A cyclopropanes has been serving as 1,3-zwitterionic synthons in [3 + n] annulations, resulting in a variety of carbo- and heterocyclic products. In 2009, Johnson and coworkers developed a chiral **L3**/MgI_2-catalyzed asymmetric [3 + 2] cycloaddition of racemic cyclopropanes **8** with aldehydes, allowing the synthesis of optically active *cis*-2,5-disubstituted tetrahydrofurans **14** (Scheme 11.5) [15]. The authors ascribed the high diastereoselectivity to an unusual S_N2 mechanism. A DYKAT dominated the reaction for electron-rich aromatic substituted cyclopropanes, while a kinetic resolution was observed when less active phenylcyclopropane was used.

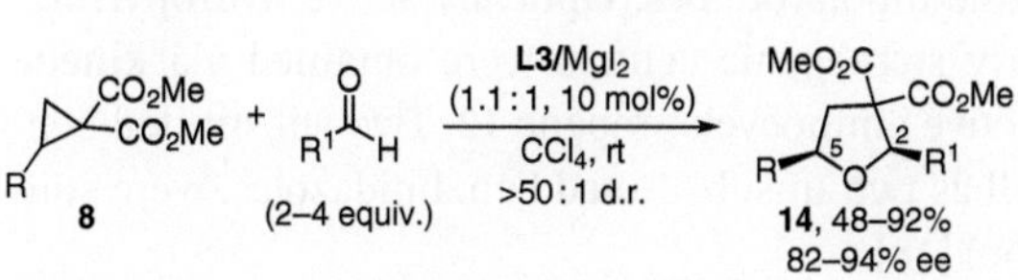

Scheme 11.5 Asymmetric [3 + 2] cycloaddition of cyclopropyl diesters with aldehydes. Source: Modified from Johnson et al. [15].

The same group later finished an enantioselective [3 + 2] cycloaddition of (*E*)-aldimines with cyclopropanes. Enantioenriched 2,5-*cis*-disubstituted pyrrolidines

15 were obtained by employing a chiral **L6**/MgI_2 catalyst via a DYKAT. Electron-neutral and electron-rich aryl aldimines performed well in the reaction. However, electron-poor aryl, alkenyl aldimines, as well as aliphatic ones, were not successful substrates (Scheme 11.6a) [16].

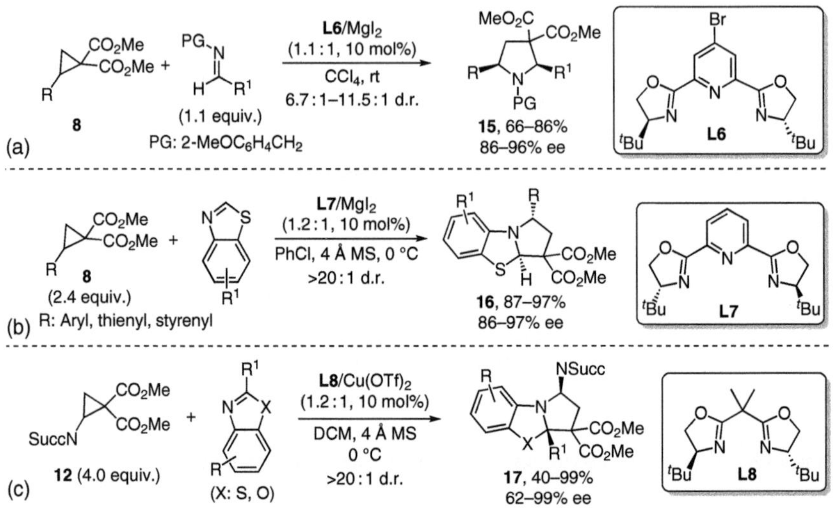

Scheme 11.6 Asymmetric [3 + 2] cycloaddition of cyclopropyl diesters with C=N bonds.

Unsaturated C=N bonds in heterocycles have successfully served as dipolarophiles in [3 + 2] cycloadditions, and the dearomatization of heterocycles has led to the generation of fused-ring systems. The groups of Guo and You jointly developed a chiral **L7**/MgI_2 complex-catalyzed dearomative [3 + 2] cycloaddition with benzothiazoles via a kinetic resolution process, yielding hydropyrrolo[2,1-*b*]thiazoles **16** in excellent enantioselectivities (Scheme 11.6b) [17]. D–A aminocyclopropane has also been used by the same groups in an **L8**/$Cu(OTf)_2$-facilitated [3 + 2] annulation with two-substituted benzothiazole dipolarophiles. Optically active hydropyrrolobenzazoles **17** bearing quaternary stereogenic centers were obtained via kinetic resolution even using highly reactive aminocyclopropane **12**. Two-substituted benzothiazoles, benzoxazoles, as well as two-unsubstituted benzimidazoles, were suitable dipolarophiles (Scheme 11.6c) [18].

Five-membered carbocycles were accessed by [3 + 2] annulations of cyclopropanes with strongly polarized C=C double bonds. Tang and coworkers developed a chiral **L9**/Cu(II) complex-catalyzed [3 + 2] annulation with cyclic silyl enol ethers, allowing the formation of enantioenriched [n.3.0] carbocycles **18**. The key to success was the use of a steric bulky chiral ligand and large adamantyl ester groups on the cyclopropanes. The employment of TBDPS-protecting group suppressed the single ring-opening side reaction, to finally yield the desired annulated products. This [3 + 2] annulation was indeed a stepwise nucleophilic ring-opening and intramolecular cyclization sequence (Scheme 11.7a) [19].

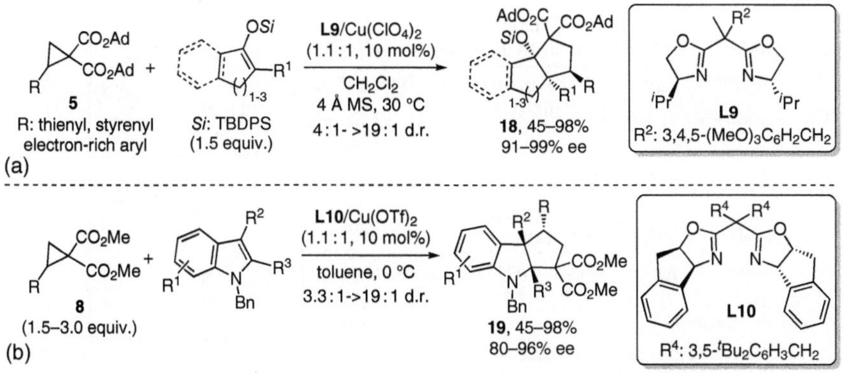

Scheme 11.7 Asymmetric [3+2] annulation of cyclopropyl diesters with enol ethers.

As illustrated in Section 11.2.1.1, acyclic ring-opened products could be obtained from three-nonsubstituted indoles, whereas the reaction of three-substituted indoles with D–A cyclopropanes led to the annulated products. Tang and coworkers described the first asymmetric variant to the synthesis of enantioenriched C2, C3-fused indolines **19** by using an **L10**/Cu(II) catalyst. The cagelike BOX ligand **L10** with two sterically hindered *tert*-butyl groups at the *meta*-positions of the pendant benzyl groups is essential to achieve high diastereo- and enantioselectivity. The reaction was also a DYKAT with electron-rich aromatically substituted cyclopropanes and a kinetic resolution with less active cyclopropanes (Scheme 11.7b) [20].

Aminocyclopropane has been used by Waser and coworkers in copper-catalyzed asymmetric annulations with enol ethers or aldehydes through a DYKAT process, leading to enantioenriched cyclopentanes or tetrahydrofurans (**20** or **21**). High levels of enantio- and diastereoselectivity were observed with a broad range of dipolarophiles. Aromatic aldehydes, cinnamaldehyde, as well as aliphatic aldehydes, were well-tolerated (Scheme 11.8) [21].

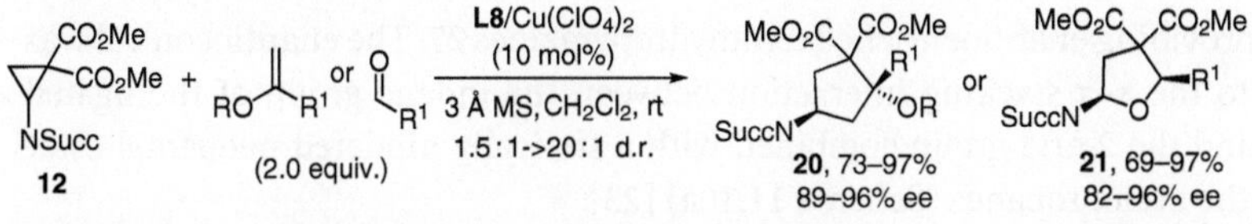

Scheme 11.8 Asymmetric [3+2] cycloaddition of aminocyclopropanes with enol ethers and aldehydes. Source: Modified from Waser et al. [21].

The asymmetric [3+3] annulations of D–A cyclopropanes with 1,3-dipoles are valuable tools for the synthesis of six-membered-ring systems. As early as 2005, Sibi and coworkers disclosed a chiral **L11**/Ni(II)-catalyzed [3+3] cycloaddition of nitrones with cyclopropyl diesters, resulting in enantioenriched tetrahydro-1,2-oxazines **23** (Scheme 11.9a). Excellent yield and enantioselectivity were observed with activated cyclopropanes **22**. However, the use of two-substituted **22** as substrates led to low diastereoselectivity [2]. Two years later, Tang and coworkers

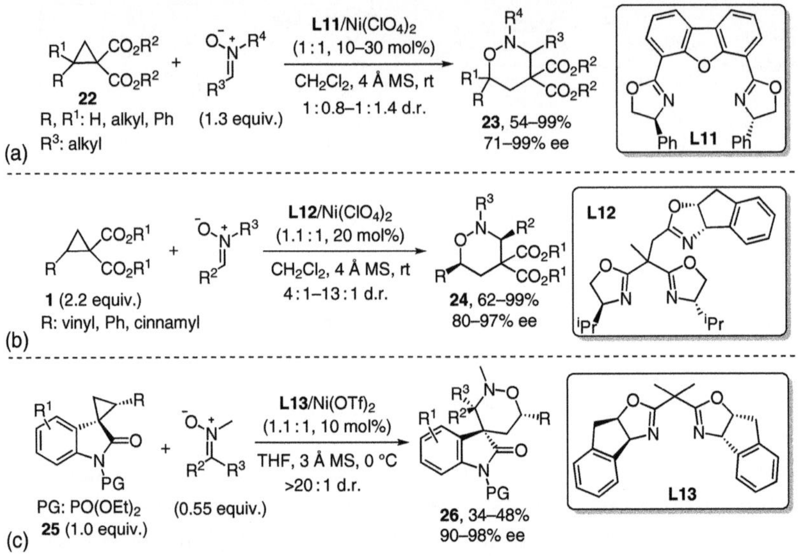

Scheme 11.9 Asymmetric [3 + 3] cycloaddition of cyclopropyl diesters with nitrones.

greatly improved the diastereoselectivity by using the chiral trisoxazoline ligand **L12**, which contained an extra pendant oxazolinyl group at the bridging carbon atom of the BOX backbone. This trisoxazoline ligand/Ni(II) catalyst was also used in the kinetic resolution of racemic **1**, providing access to optically active cyclopropanes (Scheme 11.9b) [3]. Spirocyclic tetrahydro-1,2-oxazines **26** could be accessed by an asymmetric nickel(II)-catalyzed cycloaddition of spirocyclopropyl oxindoles **25** with nitrones, as reported by Zhou and coworkers (Scheme 11.9c) [22]. Both aldonitrones and ketonitrones engaged in the reaction to form **26** in good enantioselectivity via the kinetic resolution process.

Utilizing the side-arm strategy, Tang and coworkers developed a chiral **L1**/Ni(II) complex-catalyzed [3 + 3] cycloaddition of azomethine imines with cyclopropyl diesters, providing enantioenriched tetrahydropyrazines **27**. The enantiocontrol was ascribed to the π–π stacking interaction between the indane group of the ligated sidearm and the 2-aryl group combined with a sterically hindered neopentyl ester group of the cyclopropanes (Scheme 11.10a) [23].

The Tang group also described a tandem ring-opening/cyclization sequence for the one-pot construction of chiral tetrahydrocarbazoles **28** from cyclopropyl diesters and 2-alkynyl indoles. A chiral Cu(II)/**L14** catalyst promoted the nucleophilic ring-opening of electron-rich aryl cyclopropanes with indoles to provide enantioenriched acyclic **29**, which underwent subsequent conia-ene cyclization by $InCl_3$ and DBU catalyst, leading to the formation of optically active carbazoles **28** (Scheme 11.10b) [24].

[4 + 3] Annulation of D–A cyclopropanes with common dienolsilyl ethers meets the competitive pathways, such as [3 + 2] annulation and ring-opening. Tang and coworkers successfully achieved such a transformation of formal 4π-component

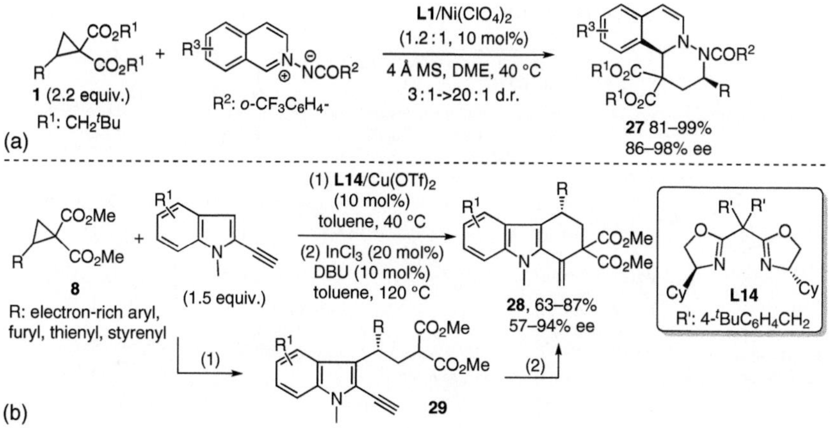

Scheme 11.10 Asymmetric [3 + 3] annulation of cyclopropyl diesters with azomethine imines and 2-alkynyl indoles.

silyl ethers by using a chiral Cu(II)/**L2** catalyst. Seven-membered cyclic products **30** were obtained in good diastereoselectivity and enantioselectivity. Experimental studies suggested this reaction proceeded through a kinetically controlled [3 + 2] cycloaddition, followed by an unusual rearrangement of cyclopentane intermediate **31** to yield the thermodynamically favored [4 + 3] annulated product **30** (Scheme 11.11) [25].

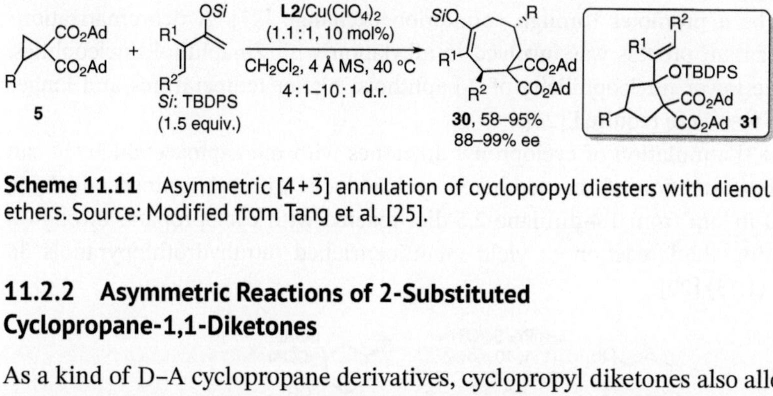

Scheme 11.11 Asymmetric [4 + 3] annulation of cyclopropyl diesters with dienol silyl ethers. Source: Modified from Tang et al. [25].

11.2.2 Asymmetric Reactions of 2-Substituted Cyclopropane-1,1-Diketones

As a kind of D–A cyclopropane derivatives, cyclopropyl diketones also allow enantioselective ring-opening to synthesize γ-functionalized carbonyls. However, more complex domino sequence might occur with the ketone units, to construct complex and interesting heterocycles in a straightforward fashion. Chiral *N,N′*-dioxide/Sc(III) complexes have been found to be efficient in enabling these transformations. The enantioselection raised from kinetic resolution of the racemic substrates, often requires excessive cyclopropyl diketones.

An enantioselective ring-opening between cyclopropyl diketones and various nucleophiles catalyzed by chiral *N,N′*-dioxide **L-PiPr₃**/Sc(OTf)₃ complex was developed by Liu and Feng (Scheme 11.12a) [26]. Thioethers were obtained in good yields

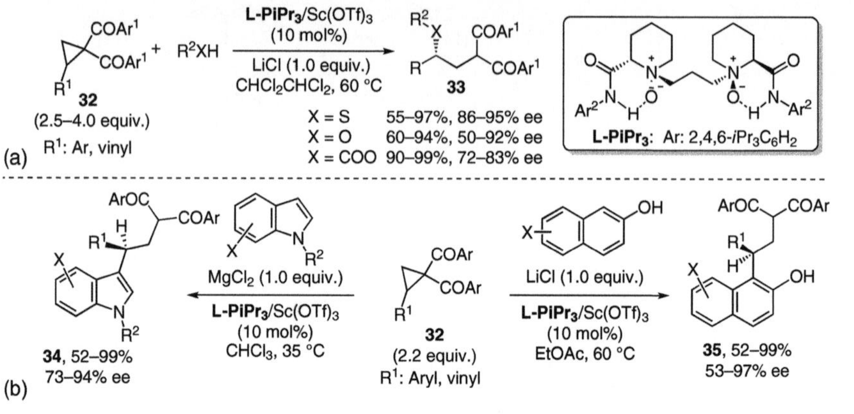

Scheme 11.12 Asymmetric ring-opening of cyclopropyl diketones with hetero- and carbon-nucleophiles.

and generally high enantioselectivity by using thiol nucleophiles. Both aromatic thiophenols and aliphatic structures were well-tolerated. In addition, alcohol nucleophiles also worked well to give optical active γ-oxy diketones **33**. Relative lower enantioselectivity was obtained when carboxylic acid was used as the nucleophile.

This catalytic system was then applied to asymmetric Friedel–Crafts alkylation of indoles with cyclopropyl diketones, resulting in enantioenriched alkylated products (Scheme 11.12b). *N*–H, *N*–Me, and *N*–TBS-protected indoles were suitable substrates. The $MgCl_2$ greatly increased both reactivity and enantioselectivity, which was proposed to be a promoter through counterion exchange [27]. A dearomatization/rearomatization process was involved in alkylation with 2-naphthol nucleophiles. Due to the lower nucleophilicity of 2-naphthols, higher temperatures and longer reaction times were required [28].

The [3+3] annulation of cyclopropyl diketones with mercaptoacetaldehyde can lead to the formation of six-membered cycloadducts. The mercaptoacetaldehyde generated *in situ* from 1,4-dithiane-2,5-diol reacted with cyclopropane through a ring-opening/aldol reaction to yield enantioenriched tetrahydrothiopyranols **36** (Scheme 11.13) [29].

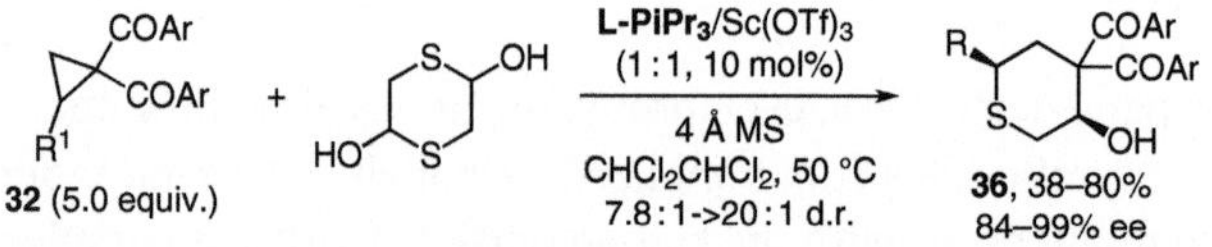

Scheme 11.13 Asymmetric [3+3] annulation of cyclopropyl diketones with mercaptoacetaldehyde. Source: Modified from Feng et al. [29].

Optically active 2,3-dihydropyrroles could be accessed by a domino ring-opening/cyclization/dehydration reaction of cyclopropyl ketones and primary amines (Scheme 11.14a) [30]. Chiral *N,N'*-dioxide **L-PiPr**$_3$/Sc(OTf)$_3$ selectively activated

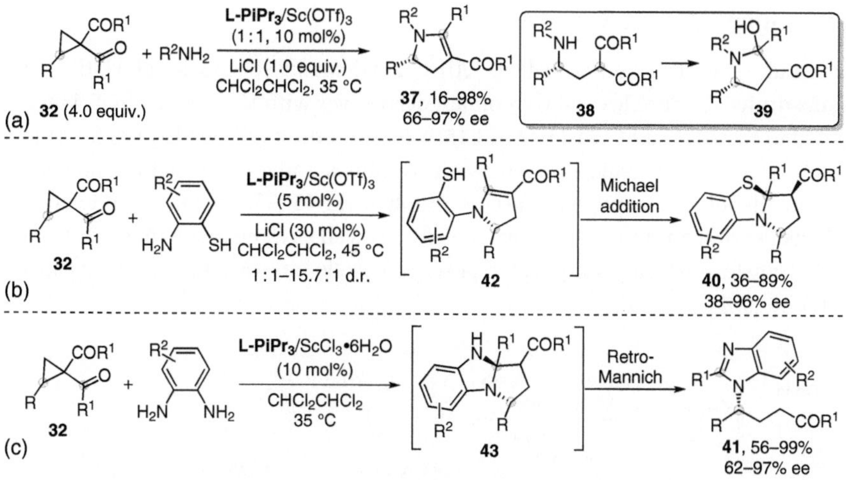

Scheme 11.14 Asymmetric ring-opening of cyclopropyl diketones with amines.

the (*S*)-enantiomer of cyclopropane **32** to perform the attack of amines in an S_N2 fashion, to generate (*R*)-**38**. Subsequent cyclization gave hemiaminal **39** and further dehydration delivered 2,3-dihydropyrroles **37** as the final products.

An asymmetric ring-opening/cyclization/thio-Michael cascade reaction of cyclopropyl diketones with 2-aminothiophenol was also realized by the same chiral catalyst. Upon the formation of 2,3-dihydropyrrole **42**, a thio-Michael reaction occurred to produce enantioenriched benzothiazole derivative **40** containing quaternary stereogenic centers (Scheme 11.14b) [31]. However, in the case of aryl-1,2-diamine nucleophile, once the tricyclic intermediate **43** was produced, a retro-Mannich process occurred to provide the thermodynamically favorable benzimidazole **41**. Here, various aryl-, vinyl-, as well as aliphatic-substituted cyclopropanes were well-tolerated (Scheme 11.14c) [32].

11.3 Chiral Low-Valent Transition Metal Promoted Reactions of Vinyl Cyclopropanes

Low-valent transition metal complexes were able to facilitate the ring-opening of VCPs to form transient π-allyl metal intermediates, which could react with electrophiles or Michael acceptors to give acyclic or cyclic targets. Chiral palladium and iridium catalysts have proven to be efficient in rendering these transformations enantioselectively. Since reactions usually begin with low-valent metals complex-mediated ring cleavage of VCPs, these transformations generally underwent a dynamic kinetic process. Although a few significant accomplishments have been made in this area, the most frequent transformation by far is the [3+2] annulation. Less common ring-opening and [4+3] annulation have been reported in some instances.

11.3.1 Ring-Opening Reactions

Trost and coworkers described in 2018 a Pd-catalyzed allylic alkylation of C3-substituted 1*H*-indoles and tryptophan derivatives with VCP. When VCP **44** was exposed to $Pd_2(dba)_3$ and Trost ligand **L15**, a zwitterionic π-allyl-palladium intermediate **47** was formed, which served as electrophiles and reacted with 3-alkylated indole nucleophiles to provide various indolenines **45**. The obtained indolenine products bearing an imine, an internal olefin, as well as a malonate motif could be readily transformed into complicated polycyclic skeletons **46** via a tandem cyclization (Scheme 11.15) [33].

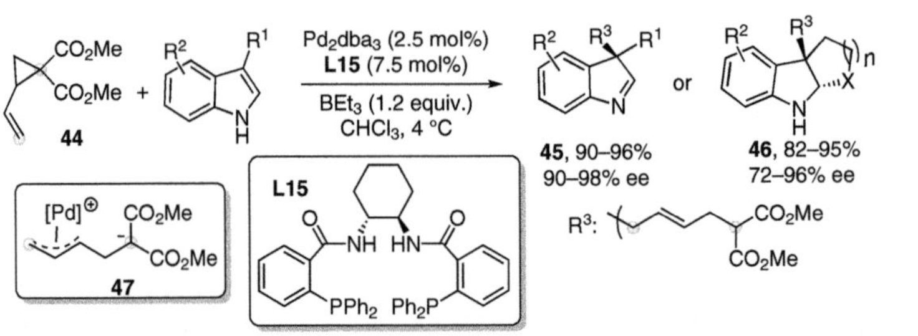

Scheme 11.15 Asymmetric ring-opening of VCP with indoles. Source: Modified from Trost et al. [33].

In 2011, the groups of Krische and Johnson jointly developed an asymmetric reductive C—C coupling of VCP with aldehydes or *in situ*-generated aldehydes by dehydrogenation of alcohols (Scheme 11.16) [4]. In the presence of chiral cyclometalated iridium catalyst **16**, diester-substituted cyclopropane **44** underwent ring-opening to form a nucleophilic π-allyl-iridium intermediate **49**. Subsequent reductive coupling with aldehydes produced homoallylic products **48**. Here, the polarity at the donor site of VCP was inversed, and **44** served as a 1,3-bisnucleophilic synthon.

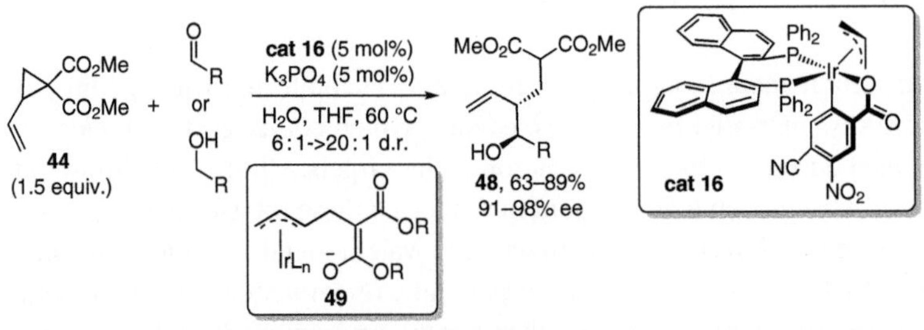

Scheme 11.16 Asymmetric electrophilic ring-opening of VCP with aldehydes. Source: Modified from Krische et al. [4].

11.3.2 [3 + n] Annulations

An asymmetric [3+2] annulation of VCP with alkylidene azlactones was disclosed to construct spiro-annulated vinyl cyclopentanes **51**. The π-allyl-palladium complex generated from the ring-opening of the bis(2,2,2-trifluoroethyl)malonate VCP **50** engaged in Michael addition with alkylidene azlactone acceptors, and further followed by ring closure to give **51** (Scheme 11.17a) [5]. Later, the same group extended the scope to Meldrum's acid-substituted VCP and alkylidenes. Enantioenriched spiro-cyclopentanes **53** were produced with an excellent level of stereoselectivity (Scheme 11.17b) [34]. The remote stereoinduction was ascribed to the reversible conjugate addition of the π-allyl-palladium intermediate to Michael acceptors.

A palladium(0)-catalyzed [3 + 2] annulation of VCPs proceeding via conjugate addition of acceptor-stabilized π-allyl-palladium intermediate was then extended to other Michael acceptors. Utilizing chiral imidazoline–phosphine ligand **L17** and Pd(0) catalyst, Shi and coworkers described an enantioselective [3 + 2] annulation with β,γ-unsaturated α-keto esters and α,β-unsaturated *N*-acyl pyrrole, providing optically active cyclopentanes **55** containing multiple stereocenters in good diastereo- and enantioselectivity (Scheme 11.17c) [35]. The use of

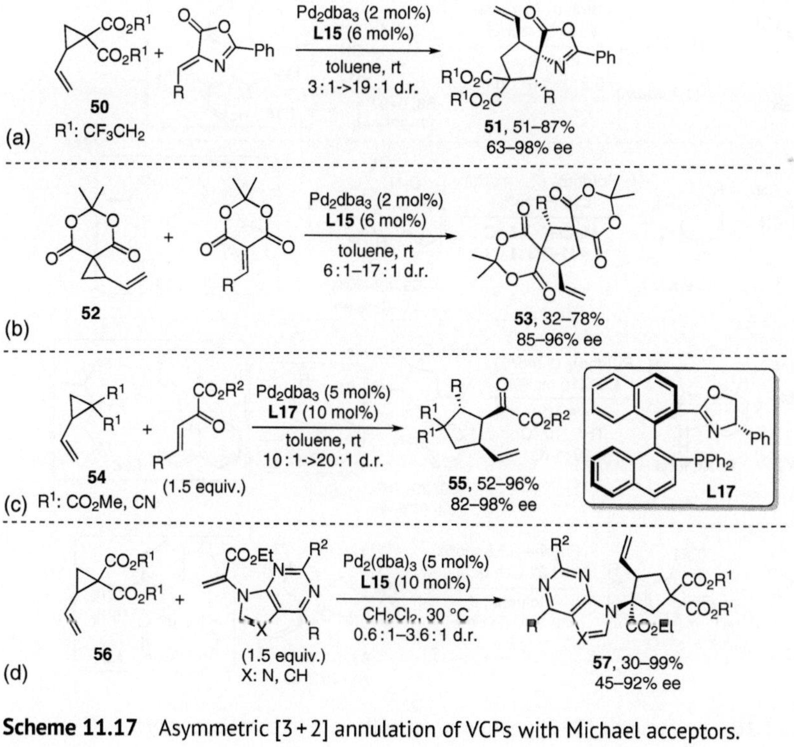

Scheme 11.17 Asymmetric [3 + 2] annulation of VCPs with Michael acceptors.

α-nucleobase-substituted acrylates as dipolarophiles for an asymmetric [3 + 2] annulation with VCPs was also shown by Guo and coworkers (Scheme 12.17d). Various purine-, pyrimidine-, as well as α-heteroaryl-substituted acrylates engaged in such transformation to give enantioenriched carbocyclic nucleoside analogs **57** in good-to-high enantioselectivities and yields, albeit with low distereoselectivities [36].

In 2015, Liu and coworkers described a chiral MeObiphep ligand **L18**/Pd(0)-catalyzed [3 + 2] annulation of nitroolefins with VCP, allowing the synthesis of chiral nitrocyclopentane derivatives **58** with three consecutive stereocenters in moderate enantioselectivities but with low diastereoselectivities (Scheme 11.18a) [37a]. A higher level of diastereocontrol was observed when indane-1,3-dione-derived vinyl cyclopropane was employed [37b]. An enantioselective dearomative [3+2] annulation occurred by using 3-nitroindoles as the Michael acceptors (Scheme 11.18b). While the transformation tolerated different substituted 3-nitroindoles, only the dicyano VCP **54** was investigated in the transformation [38].

Enantioenriched spirocyclopentane-1,3′-indolenines **60** were accessed by a palladium(0)-catalyzed [3 + 2] annulation with aryl sulfonyl indoles, as reported by Liu and coworkers (Scheme 11.18c). A zwitterionic π-allyl palladium complex intermediate was proposed to act as both a base to generate α,β-unsaturated imines **61**

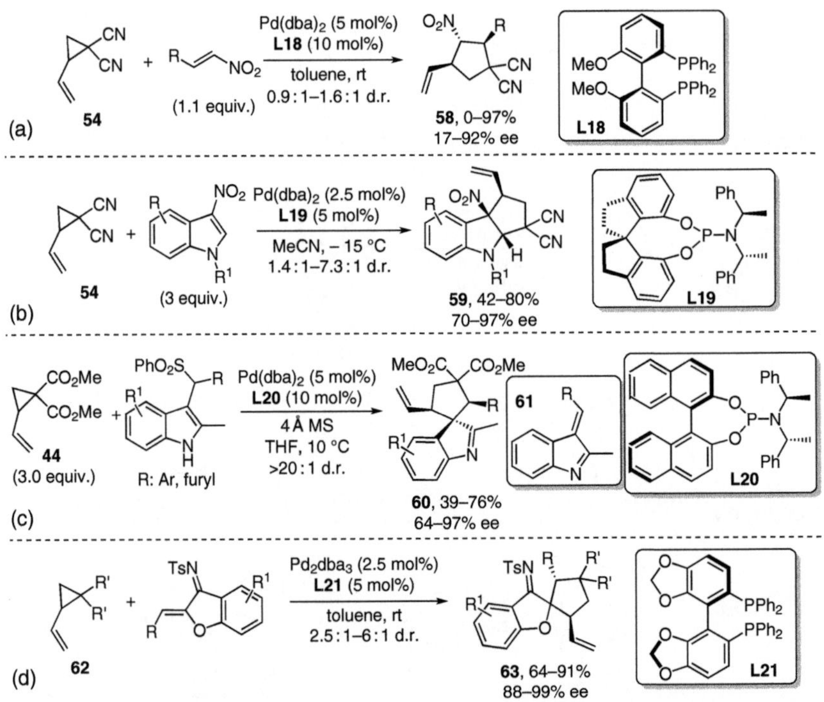

Scheme 11.18 Asymmetric [3 + 2] annulation of VCPs with Michael acceptors.

from sulfonyl indoles and as a reaction partner [39]. Cyclic 1-azadienes was also applied to a similar transformation afterward, leading to [5] spirocyclic carbocycles **63** (Scheme 12.18d) [40, 41]. Meldrum's acid-, indane-1,3-dione-, dicyano-, as well as barbituric acid-derived VCPs, reacted smoothly to give annulated products in excellent regio- and enantioselectivities.

A Pd(0)-catalyzed [3+2] annulation of *para*-quinone methides with VCPs was uncovered by Zhao et al. Optically active spiro[4.5]decanes **64** were harvested via a 1,6-conjugate addition/annulation process. Different ester groups on the VCPs were well-tolerated (Scheme 11.19a) [42]. Enantioenriched cyclopentenes **65** could also be synthesized by using ketoesters- or 1,2-diones-activated alkynes as Michael acceptors. Here, the introduction of an additional carbonyl group at the α-position

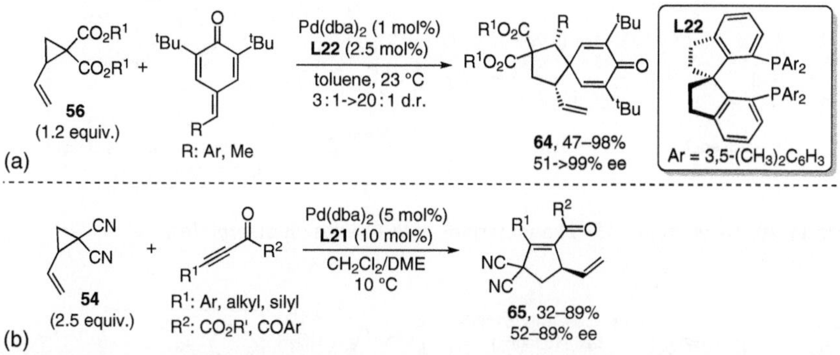

Scheme 11.19 Asymmetric [3+2] annulation of VCPs with Michael acceptors

of alkynyl esters was essential to activate the C—C triple bond to engage in the transformation (Scheme 11.19b) [43].

This type of annulation was successfully extended to heteroatom-containing dipolarophiles. Chiral tetrahydrofuran derivatives **66** were yielded via the reaction of VCP with isatin dipolarophiles (Scheme 11.20a). The diastereoselectivity could be improved by using a LiCl additive [44]. As heteroanalogous carbonyl systems, aldimines or isatin-derived ketimines underwent [3+2] annulation with cyclopropanes, allowing for the formation of tetrahydropyrrole derivatives **67**. Chiral phosphoramidite ligand **L24**/$Pd_2(dba)_3$ catalyst promoted the reaction of various isatin partners and different diester-substituted VCPs efficiently (Scheme 11.20b) [45]. Other dipolarophiles, such as 3-diazooxindoles, engaged in the palladium-catalyzed [3+2] annulation with VCPs, resulting in 1,3-indanedione- and oxindole-fused spiropyrazolidines **69** in high yields and good enantioselectivities (Scheme 11.20c) [46].

Besides extensive studies in [3+2] annulation, VCPs also participated in [4+3] annulation to construct seven-membered rings. You and coworkers investigated a

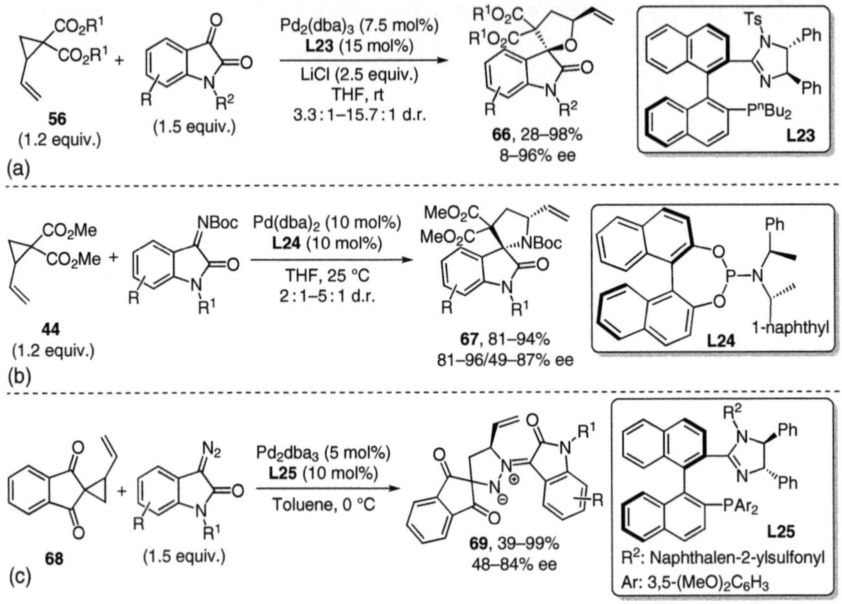

Scheme 11.20 Asymmetric [3 + 2] annulations of VCPs with dipolarophiles.

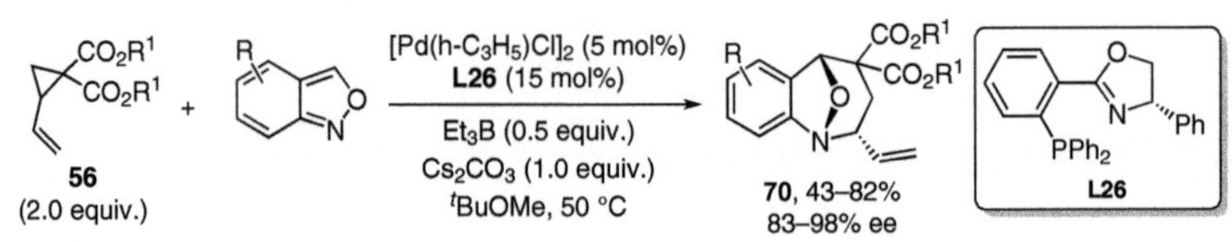

Scheme 11.21 Asymmetric [4 + 3] annulation of VCPs with anthranils. Source: Modified from You et al. [47].

Pd(II)-catalyzed dearomative [4 + 3] annulations with anthranils. With PHOX ligand **26** and triethyl borane, enantioenriched bridged cyclic compounds **70** could be harvested in excellent enantioselectivity. The use of borane was an activator for the anthranils to engage in this transformation (Scheme 11.21) [47].

Based on the synergistic merging of iminum/enamine organocatalysis with palladium(0) catalysis, Vitale and coworkers disclosed an asymmetric [3 + 2] annulation between VCPs and enals. In this reaction, highly activated α,β-unsaturated iminium ion dipolarophile **72** was generated in the presence of Hayashi–Jørgensen catalyst **27** and Brønsted acid additive, which further reacted with zwitterionic π-allyl-palladium species **51**, in a stereoselective manner, to give cyclopentane products **71**. Various VCPs and aromatic enals were successful partners for the transformation (Scheme 11.22) [48]. This dual-activation strategy was also extended to indane-1,3-dione-, azlactone-, dicyano-, and cyanoester-substituted VCPs [49].

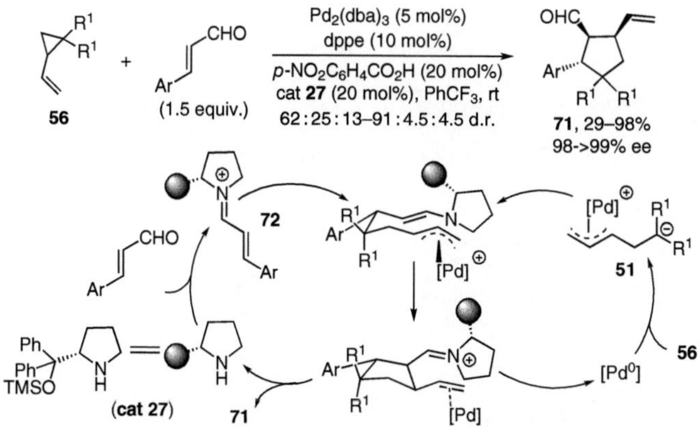

Scheme 11.22 Asymmetric [3 + 2] annulation of VCP with enals. Source: Modified from Vitale et al. [48].

11.4 Chiral Organocatalytic Reactions of D–A Cyclopropanes and Miscellaneous

Chiral organocatalysts, bearing acidic or basic functionalities, could activate D–A cyclopropanes through catalyst–substrate interactions, such as nucleophilic/electrophilic activation, hydrogen bonding, and ion pairing. The activation mode highly depended on the structure of cyclopropanes. Upon activation, the ring-cleavage of cyclopropane resulted in reactive intermediates, which reacted with a nucleophile, an electrophile, or a radical acceptor to deliver functionalized molecules.

11.4.1 Enamine/Iminium Catalysis Activation

Utilizing the HOMO-raising approach, cyclopropyl-substituted aldehydes were activated by chiral secondary aminocatalyst through the formation of covalently bonded reactive intermediates. Jørgensen and coworkers developed a stereoselective [2 + 2] annulation between cyclopropyl monoacetaldehydes **73** and 3-olefinic oxindoles or benzofuranone for the synthesis of chiral cyclobutanes **74** (Scheme 11.23a). Chiral aminocatalyst **27** activated **73** to form an enamine intermediate **76**, which underwent a ring-opening and protonation process to generate an electrophilic α,β-unsaturated iminium ion **77** in the presence of Brønsted acid. Subsequent deprotonation led to nucleophilic dienamine intermediate **78**, which further reacted with 3-olefinic oxindoles or benzofuranones to give **74** [50]. Utilizing electrophilic intermediate **77**, Vicario and coworkers described a domino cyclopropane ring-opening/aza-Michael/aldol condensation sequence with *o*-aminobenzaldehydes (Scheme 11.23b). The obtained chiral dihydroquinolines **75** could be converted into pyrroloquinolines in one pot by acid-promoted lactamization [51].

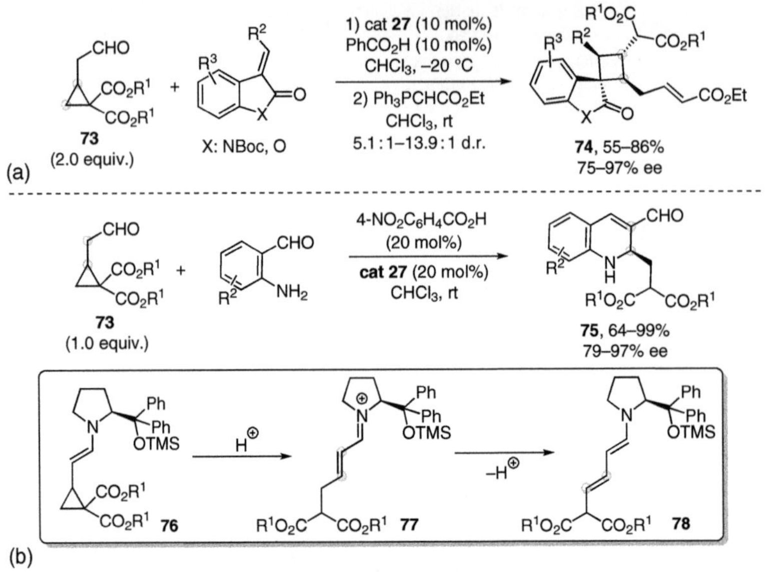

Scheme 11.23 Asymmetric cascade reactions of cyclopropyl monoacetaldehydes.

Gilmour and coworker developed a formal 1,3-addition of Cl_2 to *meso*-cyclopropane carbaldehydes **79** through a merged iminium–enamine activation by using MacMillan-type imidazolidinone catalyst **28**. Nucleophilic ring-opening of amine catalyst-activated cyclopropyl iminium species with collidine hydrochloride gave enamine **80**, which further reacted with electrophilic "Cl^+" reagent perchlorinated quinone to form 1,3-dichlorinated **81** (Scheme 11.24a) [52]. Utilizing a similar strategy, Werz and coworkers reported an asymmetric 1,3-chlorochalcogenation using sulfenyl or selenyl chlorides as 1,3-difunctionalization reagents (Scheme 11.24b) [53]. The nucleophilic ring-opening of secondary amine-activated cyclopropyl carbaldehydes with carboxylic acid was also reported by Vicario and coworkers (Scheme 11.24c) [54]. Enantioenriched γ-acyloxy-substituted aldehydes **83** were provided in good yields and enantioselectivities with benzoic or aliphatic carboxylic acids.

11.4.2 Brønsted Base Catalyst Activation

Based on hydrogen bonding/nucleophilic activation, in 2009, Jørgensen and coworkers described an enantioselective fragmentation of *meso*-cyclopropane using a bifunctional urea–amine catalyst **31**. The *meso*-cyclopropane **84** was activated by the basic quinidine moiety of catalyst **31**, whereas the acidity of the α-H was increased through hydrogen bonding between the carbonyl and the thiourea (Scheme 11.25a) [55].

Employing urea–amine catalyst **32**, the same group also developed an organocatalytic asymmetric [3+2] annulation of dicyano-substituted cyclopropylpropan-2-one **86** with nitroolefins to synthesize cyclopentanes **87**. Here, the deprotonation

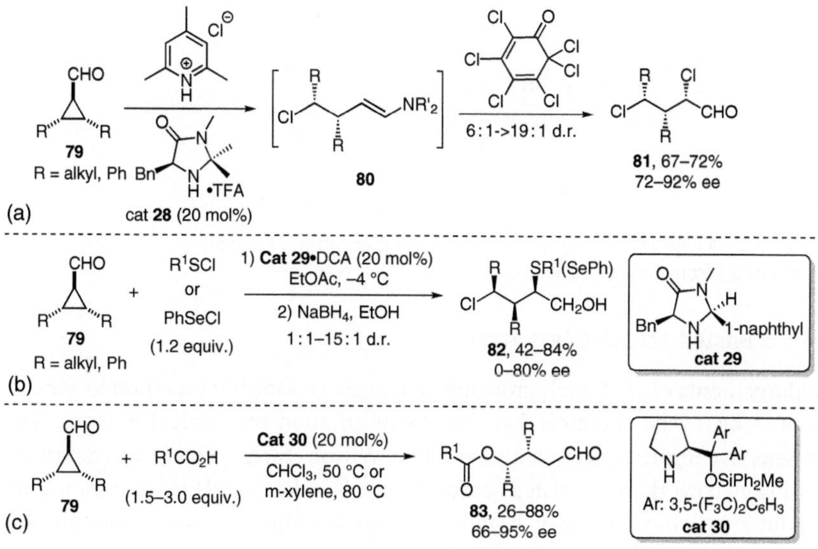

Scheme 11.24 Asymmetric ring-opening of cyclopropane carbaldehydes with electrophiles and nucleophiles.

of the cyclopropyl ketone by the basic pyrrolidine in **32** led to a reactive D–A cyclopropane, which reacted with nitroolefins to give chiral annulated products (Scheme 11.25b) [56].

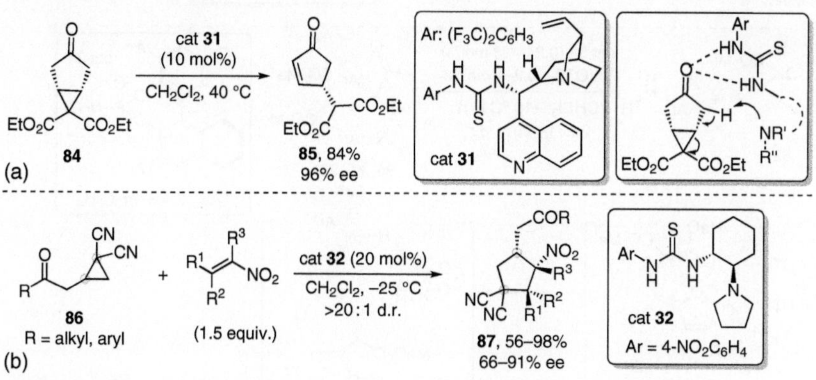

Scheme 11.25 Brønsted base-facilitated transformations of cyclopropanes.

11.4.3 Nucleophilic Catalyst Activation

Troxler and Scheffold developed a vitamin B_{12}-catalyzed enantioselective desymmetrization rearrangement of *meso*-cyclopropanes **88**. The Cob(I)alamin catalyst, served as the nucleophile, and engaged in the ring-opening of spiroactivated cyclopropanes. The resulting Co(III) intermediate **89** underwent reductive elimination to form cycloalk-2-enyls **90** (Scheme 11.26) [1].

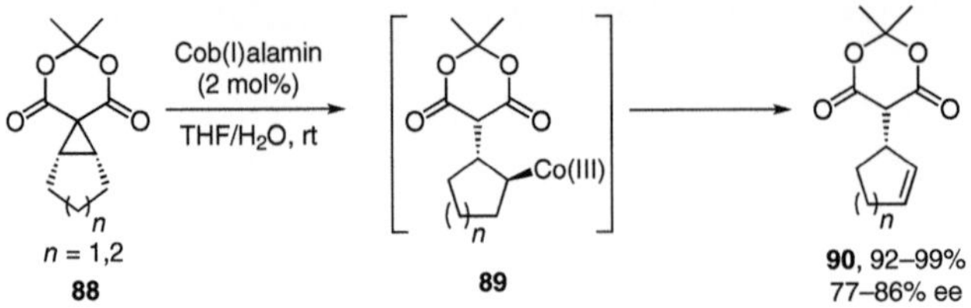

Scheme 11.26 Asymmetric desymmetrization-fragmentation of *meso*-cyclopropanes. Source: Modified from Troxler et al. [1].

11.4.4 Brønsted Acid Catalyst Activation

Ring enlargements of D–A cyclopropanes represent an expedient method to generate more stable cyclic chemical frameworks in an atom-economical manner. An enantioselective Brønsted acid-catalyzed Cloke–Wilson rearrangement was reported by Vicario and coworkers. Brønsted acid protonated the ketones **91** to form transient carbocationic intermediates, which underwent cyclization to give dihydrofurans **92** through a DYKAT (Scheme 11.27a) [57].

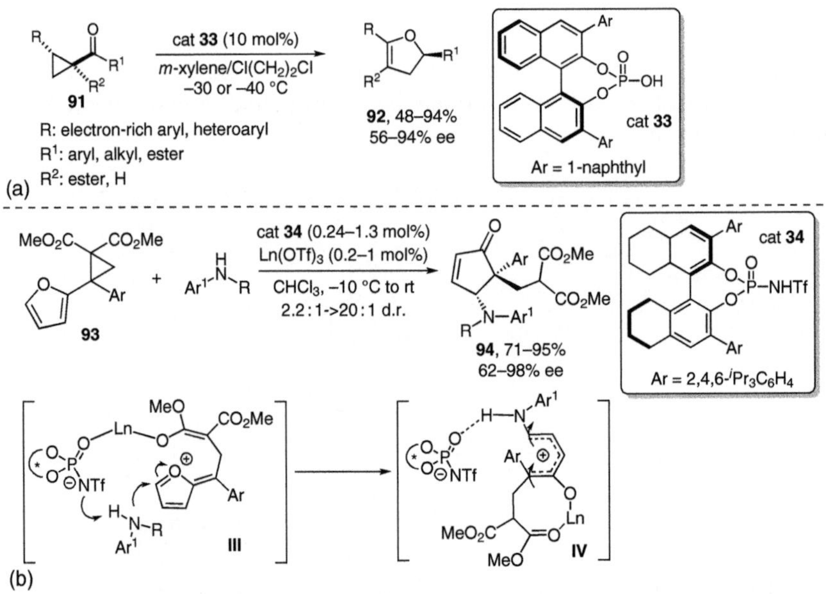

Scheme 11.27 Brønsted acid-facilitated transformations of D–A cyclopropanes.

An achiral rare-earth metal Lewis acid and chiral Brønsted acid cooperatively catalyzed-cascade ring-opening/aza-Piancatelli rearrangement of furyl-substituted D–A cyclopropanes **93** with anilines or *N*-alkyl anilines has been reported by Tang and Cai (Scheme 11.27b) [58]. The authors proposed that a chiral anion-stabilized oxocarbenium enolate–Ln(III) intermediate **III** was generated via Ln(III)-facilitated ring-opening of the cyclopropane. Subsequent *N*-triflylphosphoramide anion promoted nucleophilic attack by the amine and further Ln(III)-assisted ring-opening of the furan led to a pentadienyl carbocation intermediate **IV**. Stereoselective 4π-electrocyclization resulted in functionalized aminocyclopentenone **94** bearing an α-quaternary carbon stereocenter in good enantioselectivity.

11.4.5 Radical Pathway

Besides the ionic pathway, the enantioselective radical reactions of D–A cyclopropanes also provide efficient access to enantioenriched molecules. In 2014, Maruoka and coworkers designed a stereoselective [3+2] annulation of VCPs with electron-rich alkenes by using thiyl radical precatalyst **35** (Scheme 11.28) [59]. The thiyl radical generated by oxidation was added to VCPs, resulting in carbon-centered radical **97**, which then reacted with alkenes to form radical **98**. Subsequent cyclization gave cyclopentanes **96**, and the thiyl radical was released.

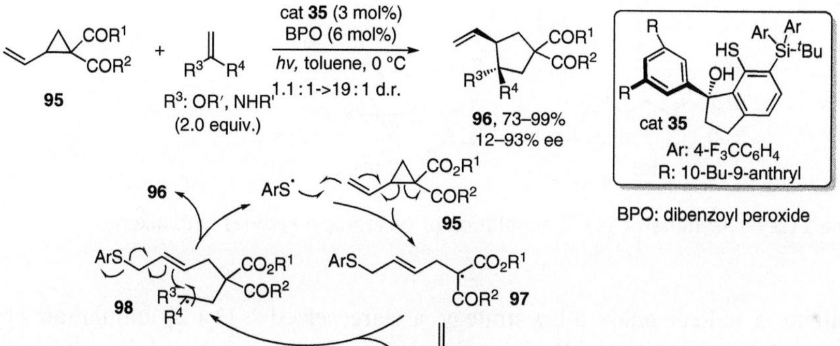

Scheme 11.28 Asymmetric [3+2] annulation of VCPs with alkenes. Source: Modified from Maruoka et al. [59].

The combination of a chiral Gd(III)/pybox and $[Ru(bpy)_3](PF_6)_2$ photoredox catalyst system was developed to facilitate a [3+2] photocycloaddition between aryl cyclopropyl ketones and alkenes (Scheme 11.29a). Photoredox catalyst **36** promoted the one-electron reduction of Lewis acid-activated aryl cyclopropyl ketone to form a ketyl radical. The latter underwent fragmentation and stereocontrolled radical cyclization to give a chiral acyl-substituted cyclopentane derivative **100** [60].

Meggers and coworkers also described a rhodium catalyst Δ-RhS **37**-catalyzed [3+2] cycloaddition of cyclopropyl imidazolyl ketones with alkenes (Scheme 11.29b). Here, the rhodium catalyst acted as both a Lewis acid catalyst and a photoredox catalyst. The direct photoexcitation of cyclopropane/catalyst complex **103** generated a strong photooxidant **103***, which could be reduced to provide a ketyl radical **104**. The followed fragmentation and enantioselective radical addition delivered optically active cyclopentanes [61].

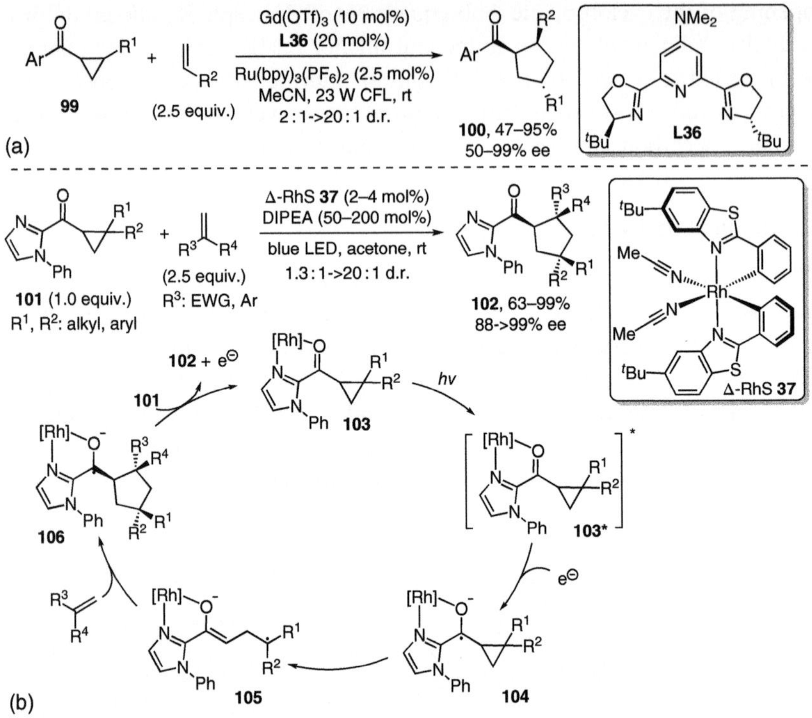

Scheme 11.29 Asymmetric [3 + 2] annulation of cyclopropyl ketones with alkenes.

Utilizing a radical redox relay strategy, a stereoselective [3 + 2] annulation of cyclopropyl ketones with radical acceptor alkenes was developed by Lin and coworkers (Scheme 11.30). A chiral Ti^{IV}(salen) complex was reduced by Mn to Ti^{III}(salen), which coordinated with cyclopropyl ketone to form redox-active species. The followed single-electron transfer/ring cleavage/radical cyclization process gave annulated cyclopentanes **108**. Such Ti^{IV}/Ti^{III} redox catalytic system was quite efficient for styrene, and a high level of enantiocontrol was obtained. However, lower enantioselectivity was observed by using electron-deficient alkenes [62].

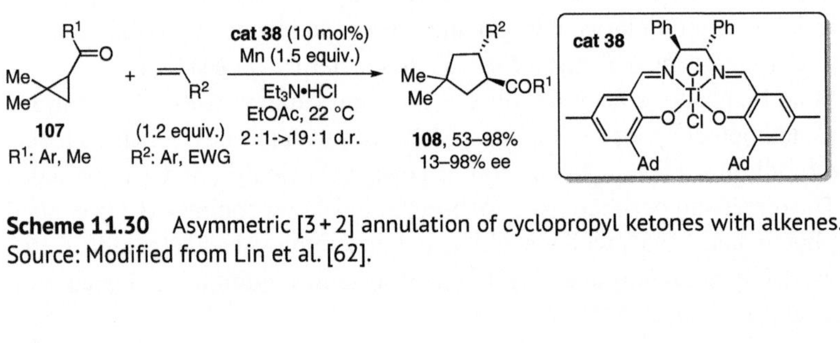

Scheme 11.30 Asymmetric [3 + 2] annulation of cyclopropyl ketones with alkenes. Source: Modified from Lin et al. [62].

11.5 Conclusion

Over these past two decades, catalytic asymmetric transformations of D–A cyclopropanes have benefited greatly from the rapid development of organic synthesis methodology. A number of chiral Lewis/Brønsted acids, Brønsted bases, organocatalysts, as well as radical-involved catalysts have been successfully used to render transformations enantioselectively, providing efficient approaches to optically active acyclic and cyclic compounds. Their applications have also been demonstrated by the synthesis of biologically active and related products. However, there are still some limitations and unresolved problems. The activation mode between cyclopropanes and chiral catalysts is highly dependent on the structure of the cyclopropanes. The catalytic asymmetric transformations of D–A cyclopropanes are still dominated by chiral Lewis acid catalysts. Furthermore, the donor group of cyclopropanes is mostly confined to aryl/vinyl/heteroatom substituents, and less reactive D–A cyclopropanes, such as an alkyl donor, remain elusive. Enantioselective radical transformations of D–A cyclopropanes are still largely undeveloped. Thus, the design of novel activation model, new catalyst systems, as well as their applications in organic synthesis are still highly demanding.

References

1 Troxler, T. and Scheffold, R. (1994). *Helv. Chim. Acta* 77: 1193–1202.

2 Sibi, M.P., Ma, Z., and Jasperse, C.P. (2005). *J. Am. Chem. Soc.* 127: 5764–5765.

3 Kang, Y.B., Sun, X.L., and Tang, Y. (2007). *Angew. Chem. Int. Ed.* 46: 3918–3921.

4 Moran, J., Smith, A.G., Carris, R.M. et al. (2011). *J. Am. Chem. Soc.* 133: 18618–18621.

5 Trost, B.M. and Morris, P.J. (2011). *Angew. Chem. Int. Ed.* 50: 6167–6170.

6 (a) Wang, L. and Tang, Y. (2016). *Isr. J. Chem.* 56: 463–475. (b) Pirenne, V., Muriel, B., and Waser, J. (2021). *Chem. Rev.* 121: 227–263. (c) Xia, Y., Liu, X.H., and Feng, X.M. (2021). *Angew. Chem. Int. Ed.* 60: 9192–9204.

7 Steinreiber, J., Faber, K., and Griengl, H. (2008). *Chem. Eur. J.* 14: 8060–8072.

8 Liao, S., Sun, X.L., and Tang, Y. (2014). *Acc. Chem. Res.* 47: 2260–2272.

9 (a) Zhou, Y., Wang, L., Li, J. et al. (2012). *J. Am. Chem. Soc.* 134: 9066–9069. (b) Kang, Q., Wang, L., Zheng, Z. et al. (2014). *Chin. J. Chem.* 32: 669–672.

10 Luo, W., Sun, Z., Fernando, E.H.N. et al. (2019). *ACS Catal.* 9: 8285–8293.

11 Kang, Q.K., Wang, L., Liu, Q.J. et al. (2015). *J. Am. Chem. Soc.* 137: 14594–14597.

12 Wales, S.M., Walker, M.M., and Johnson, J.S. (2013). *Org. Lett.* 15: 2558–2561.

13 Perrotta, D., Wang, M., and Waser, J. (2018). *Angew. Chem. Int. Ed.* 57: 5120–5123.

14 Zhu, M., Wang, D.C., Xie, M.S. et al. (2018). *Chem. Eur. J.* 24: 15512–15516.

15 Parsons, A.T. and Johnson, J.S. (2009). *J. Am. Chem. Soc.* 131: 3122–3123.

16 Parsons, A.T., Smith, A.G., Neel, A.J., and Johnson, J.S. (2010). *J. Am. Chem. Soc.* 132: 9688–9692.

17 Wang, D.C., Xie, M.S., Guo, H.M. et al. (2016). *Angew. Chem. Int. Ed.* 55: 14111–14115.

18 Zhang, M.C., Wang, D.C., Xie, M.S. et al. (2019). *Chem* 5: 156–167.

19 Xu, H., Qu, J.P., Liao, S. et al. (2013). *Angew. Chem. Int. Ed.* 52: 4004–4007.

20 (a) Xiong, H., Xu, H., Liao, S. et al. (2013). *J. Am. Chem. Soc.* 135: 7851–7854. (b) Yan, W.G., Wang, P., Wang, L. et al. (2017). *Acta Chim. Sinica* 75: 783–787.

21 de Nanteuil, F., Serrano, E., Perrotta, D., and Waser, J. (2014). *J. Am. Chem. Soc.* 136: 6239–6242.

22 Xu, P., Liu, J., Shen, L. et al. (2017). *Nat. Commun.* 8: 1619.

23 Zhou, Y., Li, J., Ling, L. et al. (2013). *Angew. Chem. Int. Ed.* 52: 1452–1456.

24 Liu, Q.J., Yan, W.G., Wang, L. et al. (2015). *Org. Lett.* 17: 4014–4017.

25 Xu, H., Hu, J., Wang, L. et al. (2015). *J. Am. Chem. Soc.* 137: 8006–8009.

26 Xia, Y., Lin, L.L., Chang, F.Z. et al. (2015). *Angew. Chem. Int. Ed.* 54: 13748–13752.

27 Chang, F.Z., Lin, L.L., Xia, Y. et al. (2018). *Adv. Synth. Catal.* 360: 2608–2612.

28 Xia, Y., Chang, F.Z., Lin, L.L. et al. (2018). *Org. Chem. Front.* 5: 1293–1296.

29 Fu, X., Lin, L.L., Xia, Y. et al. (2016). *Org. Biomol. Chem.* 14: 5914–5917.

30 Xia, Y., Liu, X.H., Zheng, H.F. et al. (2015). *Angew. Chem. Int. Ed.* 54: 227–230.

31 Chang, F.Z., Shen, B., Wang, S. et al. (2020). *Chem. Commun.* 56: 13429–13432.

32 Xia, Y., Lin, L.L., Chang, F.Z. et al. (2016). *Angew. Chem. Int. Ed.* 55: 12228–12232.

33 Trost, B.M., Bai, W.J., Hohn, C. et al. (2018). *J. Am. Chem. Soc.* 140: 6710–6717.

34 Trost, B.M., Morris, P.J., and Sprague, S.J. (2012). *J. Am. Chem. Soc.* 134: 17823–17831.

35 Mei, L., Wei, Y., Xu, Q., and Shi, M. (2012). *Organometallics* 31: 7591–7599.

36 Xie, M.S., Wang, Y., Li, J.P. et al. (2015). *Chem. Commun.* 51: 12451–12454.

37 (a) Li, W.K., Liu, Z.S., He, L. et al. (2015). *Asian. J. Org. Chem.* 4: 28–32. (b) Wei, F., Ren, C., Wang, D., and Liu, L. (2015). *Chem. Eur. J.* 21: 2335–2338.

38 Sun, M., Zhu, Z.Q., Gu, L. et al. (2018). *J. Org. Chem.* 83: 2341–2348.

39 Liu, Z.S., Li, W.K., Kang, T.R. et al. (2015). *Org. Lett.* 17: 150–153.

40 Zhou, Q., Chen, B., Huang, X.B. et al. (2019). *Org. Chem. Front.* 6: 1891–1894.

41 (a) Trost, B.M. and Zuo, Z. (2021). *Angew. Chem. Int. Ed.* 2021 (60): 5806–5810. (b) Liu, K., Yang, J., and Li, X. (2021). *Org. Lett.* 2021 (23): 826–831.

42 Ma, C., Huang, Y., and Zhao, Y. (2016). *ACS Catal.* 6: 6408–6412.

43 Ding, W.P., Zhang, G.P., Jiang, Y.J. et al. (2019). *Org. Lett.* 21: 6805–6810.

44 Mei, L., Wei, Y., Xu, Q., and Shi, M. (2013). *Organometallics* 32: 3544–3556.

45 Huang, X.B., Li, X.J., Li, T.T. et al. (2019). *Org. Lett.* 21: 1713–1716.

46 Cao, B., Mei, L.Y., Li, X.G., and Shi, M. (2015). *RSC Adv.* 5: 92545–92548.

47 Cheng, Q., Xie, J.H., Weng, Y.C., and You, S.L. (2019). *Angew. Chem. Int. Ed.* 58: 5739–5743.

48 Laugeois, M., Ponra, S., Ratovelomanana-Vidal, V. et al. (2016). *Chem. Commun.* 52: 5332–5335.

49 (a) Zhu, H., Du, P., Li, J. et al. (2016). *Beilstein J. Org. Chem.* 12: 1340–1347. (b) Meazza, M. and Rios, R. (2016). *Chem. Eur. J.* 22: 9923–9928. (c) Zhang, K.,

Meazza, M., Izaga, A. et al. (2016). *Synthesis* 49: 167–174. (d) Kamlar, M., Franc, M., Císařová, I. et al. (2019). *Chem. Commun.* 55: 3829–3832.

50 Halskov, K.S., Kniep, F., Lauridsen, V.H. et al. (2015). *J. Am. Chem. Soc.* 137: 1685–1691.

51 Sanchez-Diez, E., Vesga, D.L., Reyes, E. et al. (2016). *Org. Lett.* 18: 1270–1273.

52 Sparr, C. and Gilmour, R. (2011). *Angew. Chem. Int. Ed.* 50: 8391–8395.

53 Wallbaum, J., Garve, L.K.B., Jones, P., and Werz, D.B. (2016). *Chem. Eur. J.* 22: 18756–18759.

54 Díaz, E., Reyes, E., Uria, U. et al. (2018). *Chem. Eur. J.* 24: 8764–8768.

55 Dickmeiss, G., De Sio, V., Udmark, J. et al. (2009). *Angew. Chem. Int. Ed.* 48: 6650–6653.

56 Blom, J., Vidal-Albalat, A., Jørgensen, J. et al. (2017). *Angew. Chem. Int. Ed.* 56: 11831–11835.

57 Ortega, A., Manzano, R., Uria, U. et al. (2018). *Angew. Chem. Int. Ed.* 57: 8225–8229.

58 Xu, L., Yang, Q., Zhong, S. et al. (2020). *Org. Lett.* 22: 9016–9021.

59 Hashimoto, T., Kawamata, Y., and Maruoka, K. (2014). *Nat. Chem.* 6: 702–705.

60 Amador, A.G., Sherbrook, E.M., and Yoon, T.P. (2016). *J. Am. Chem. Soc.* 138: 4722–4725.

61 Huang, X., Lin, J., Shen, T. et al. (2018). *Angew. Chem. Int. Ed.* 57: 5454–5458.

62 Hao, W., Harenberg, J.H., Wu, X. et al. (2018). *J. Am. Chem. Soc.* 140: 3514–3517.

12

Application of Donor–Acceptor Cyclopropanes in Total Synthesis of Natural Products

Amrita Saha, Karuna Mahato, Satysen Yadav, and Manas K. Ghorai

Indian Institute of Technology Kanpur, Department of Chemistry, Kanpur, Uttar Pradesh, 208016, India

CHAPTER MENU

12.1 Introduction

Cyclopropanes are an important class of compounds in organic chemistry. Despite high strain (~115 kJ/mol), cyclopropane does not lose its cyclic structure, and its C—C bond also remains kinetically inert. Enhancement of the reactivity of these strained three-membered rings could be possible by introducing electron-donating and electron-withdrawing groups in the vicinal carbons of the rings, and these reactive cyclopropanes are well known as donor–acceptor (DA) cyclopropanes [1, 2]. The developed potential general methods for preparation as well as high reactivity make DA-cyclopropane a very popular and very useful synthetic building block for the construction of various important structural cores and also in the total synthesis of many natural products. Cyclopropanation followed by ring opening, or cycloaddition and cyclopropanation-ring opening-cyclization are popular strategies widely used in the total synthesis of complex natural products. Over the years, DA-cyclopropanes have been utilized as a reactive subunit for the synthesis of many alkaloids, terpenes/terpenoids, and also many other natural products. However, here we demonstrate the importance of DA-cyclopropanes in the total synthesis of bioactive natural products from 2001 to 2021.

Donor–Acceptor Cyclopropanes in Organic Synthesis, First Edition.
Edited by Prabal Banerjee and Akkattu T. Biju.

12.2 Synthesis of Alkaloids

Alkaloids are pharmacologically valuable natural products that mostly contain basic nitrogen atoms. To date, more than 20 000 alkaloids have been documented in the literature. Although the major sources of alkaloids are plants, apart from plants, alkaloids can also be produced by bacteria, fungi, insects, and animals. In this chapter, first, we highlight and summarize the recent advancements in the total synthesis of various alkaloids of different families utilizing DA-cyclopropane chemistry.

The simplest naturally occurring alkaloid, (±)-horsfiline (**4**), is isolated from the leaves of *Horsfieldia superba*, a tree native to Malaysia [3]. The Carreira group developed a versatile route to the alkaloid (±)-horsfiline from commercially available isatin **1** in 41% overall yield. The synthesis involves the MgI_2-catalyzed cyclopropane ring expansion of **3** leading to the formation of the spiro[pyrrolidine-3,3'-oxindole] scaffold, as shown in Scheme 12.1 [4].

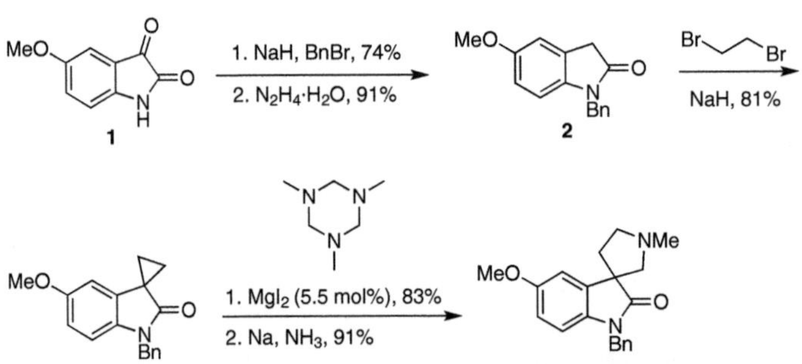

Scheme 12.1 Total synthesis of (±)-Horsfiline.

The Carreira group modified their earlier [5a] report for the total synthesis of the anticancer alkaloid (±)-strychnofoline (**12**) [5b], isolated from the leaves of *Strychnos usambarensis* [6]. The synthesis involved the ring expansion of spiro[cyclopropan-1,3′-oxindole] **8** with a cyclic disubstituted aldimine **10** in the presence of MgI_2 to provide the corresponding ring-expansion product **11** in 55% yield as a single diastereomer. The ring-expansion product **11** possessed the same stereochemical pattern as that of the (±)-strychnofoline, and it was converted to the targeted natural product (±)-strychnofoline (**12**) in eight additional linear steps (Scheme 12.2).

In 1996, Osada and co-workers discovered the bioactive indole alkaloid spirotryprostatin B (**17**) from the fermentation broth of *Aspergillus fumigatus* BM939 [7]. The synthesis of spirotryprostatin B was performed by the MgI_2-catalyzed annulation reaction of a spiro[oxindole-3,1′-vinylcyclopropane] and an alkynyl imine, as shown in Scheme 12.3 [8]. The synthesis began with a known diazoketone **13**, prepared in two steps from readily available isatin. Rhodium-catalyzed cyclopropanation of diazoketone **13** with piperylene resulted in the formation of spiro[cyclopropane-1,3′-oxindole] **14** in 71% yield as an inseparable mixture of diastereomers. The reaction of cyclopropane **14** with imine **15** in the presence of MgI_2 afforded **16** in a combined

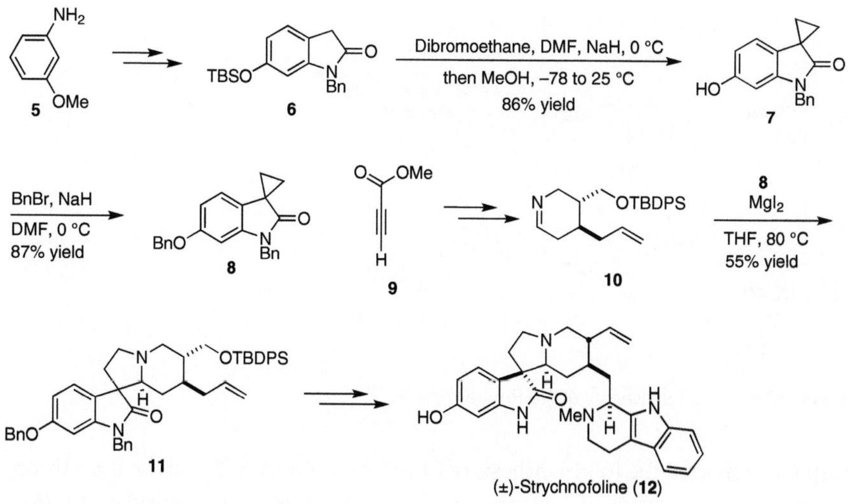

Scheme 12.2 Total synthesis of (±)-Strychnofoline.

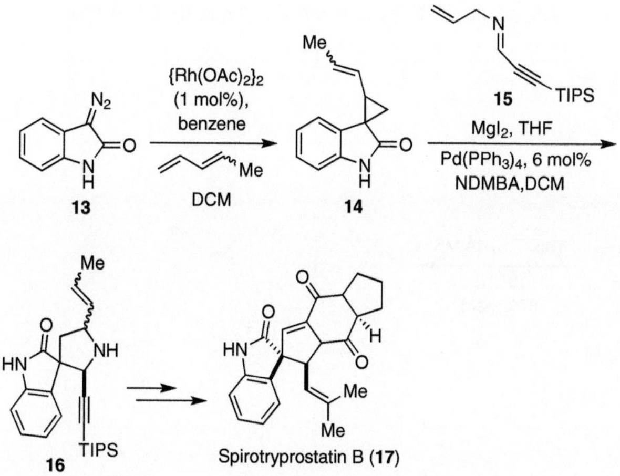

Scheme 12.3 Total synthesis of Spirotryprostatin B.

68% yield, as a mixture of diastereomers (dr 6:1). It was further converted to spirotryprostatin B (**17**) after a number of steps.

(+)-Phyllantidine (**22**) is a securinega alkaloid, and it was isolated from *Phyllanthus discoides* and *Seurinega suffruticosa* [9]. Kerr group reported the total synthesis of (+)-phyllantidine (**22**) alkaloid, where the first step involved the three-component coupling of vinyl cyclopropane **18**, aldehyde **19**, and hydroxylamine **20** in the presence of ytterbium triflate hydrate as the Lewis acid to produce tetrahydo-1,2-oxazine **21** as a mixture of diastereomers (dr 12:1) in 86% yield. The compound **21** was subsequently converted in several steps into (+)-phyllantidine (**22**) (Scheme 12.4) [10].

The marine alkaloid (+)-nakadomarin A (**27**) was isolated from an Okinawan sea sponge and shows anticancer, antifungal, and antibacterial activities [11]. The Kerr

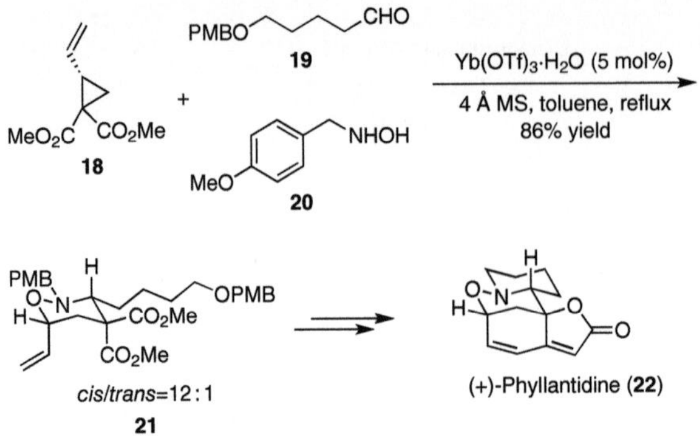

Scheme 12.4 Total synthesis of (+)-Phyllantidine.

group next reported the total synthesis of (+)-nakadomarin A (**27**) involving a three-component cycloaddition of cyclopropane **23**, aldehyde **24**, and hydroxylamine **25** to generate highly functionalized tetrahydro-1,2-oxazine **26** as a single diastereomer. The chirality of cyclopropane **23** dictated the absolute configuration of the resulting oxazine **26**, from which (+)-nakadomarin A (**27**) was synthesized in 22 linear steps (Scheme 12.5) [12].

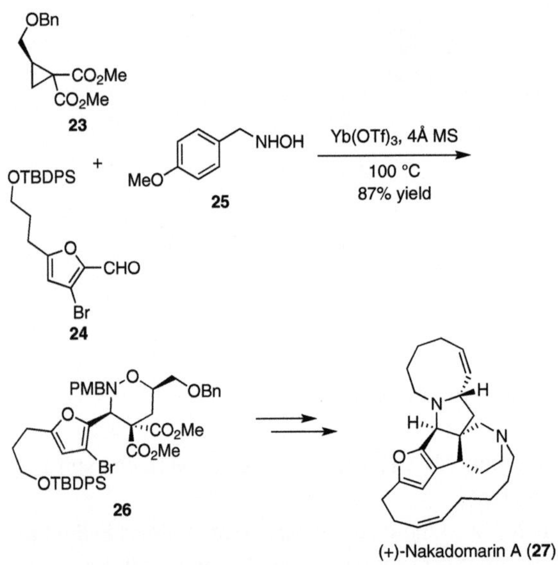

Scheme 12.5 Total synthesis of (+)-Nakadomarin A.

The Qin group developed the first total synthesis of (±)-communesin F (**33**) that can act as an insecticide [13]. The synthesis commenced with the condensation of indole **28** and acid **29** followed by ketone-to-diazo functional transformation to produce **30**. Next, CuOTf-mediated intramolecular cyclopropanation led to the formation of the stable cyclopropane intermediate **31** as diastereomeric mixtures (dr 1.6:1).

Treatment of **31** with PBu_3 in aqueous THF led to cascade ring opening and ring-closing reactions of cyclopropane with in-situ generated aniline to furnish the pentacyclic compound **32** in diastereomerically pure form. (±)-Communesin F (**33**) was synthesized from **32** in several steps with about 3% overall yield (Scheme 12.6) [14].

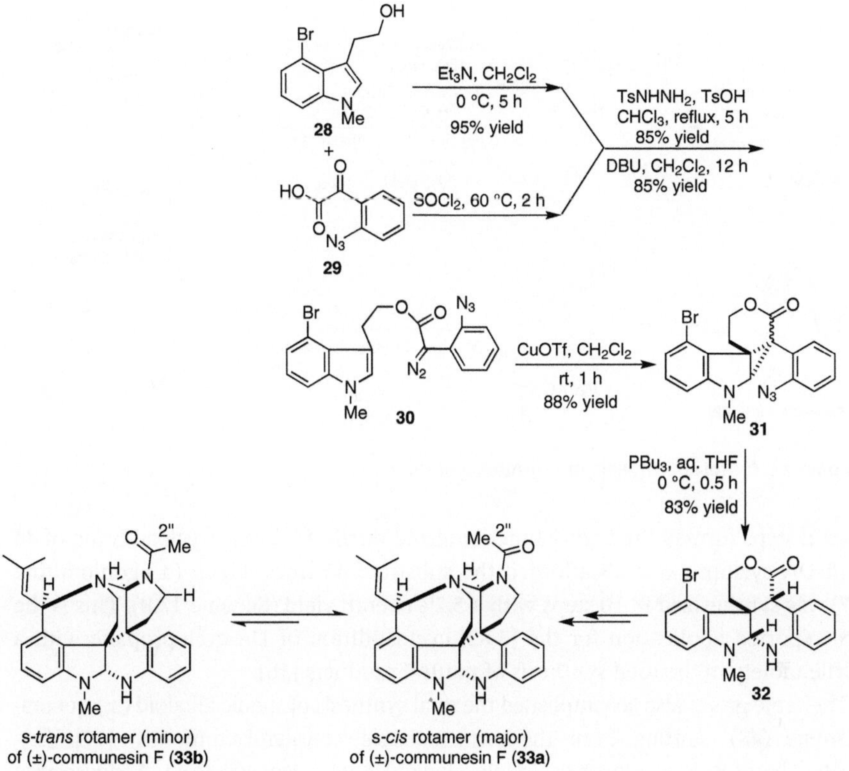

Scheme 12.6 Total synthesis of (±)-Communesin F.

The Kerr group also demonstrated the total synthesis of *Securinega* alkaloid (−)-allosecurinine (**42**) that was isolated from the plants of the Euphorbiaceae family [15]. Enantiopure cyclopropane **37** was the crucial substrate for this synthesis. Homoallylic alcohol **35** upon protection as *p*-methoxyphenyl ether via Mitsunobu reaction followed by asymmetric dihydroxylation and dimesylation provided the compound **36**. A double displacement reaction of **36** with dimethyl malonate produced cyclopropane **37**, which upon deprotection followed by tosylation and displacement with *N*-hydroxyphthalimide furnished the cyclopropane **38**. Deprotection of **38** with ethanolic hydrazine formed the free cyclopropane **39**, which underwent $Yb(OTf)_3$-catalyzed cyclization reaction with protected aldehyde **40** to provide the substrate **41**. (−)-Allosecurinine (**42**) was obtained from **41** in 13 steps with a 5% overall yield (Scheme 12.7).

Pagenkopf and coworkers developed the shortest total synthesis of indole alkaloid (±)-goniomitine (**47**) that was isolated from the root bark of *Gonioma malagasy*. The synthesis began with commercially available δ-valerolactum, which upon

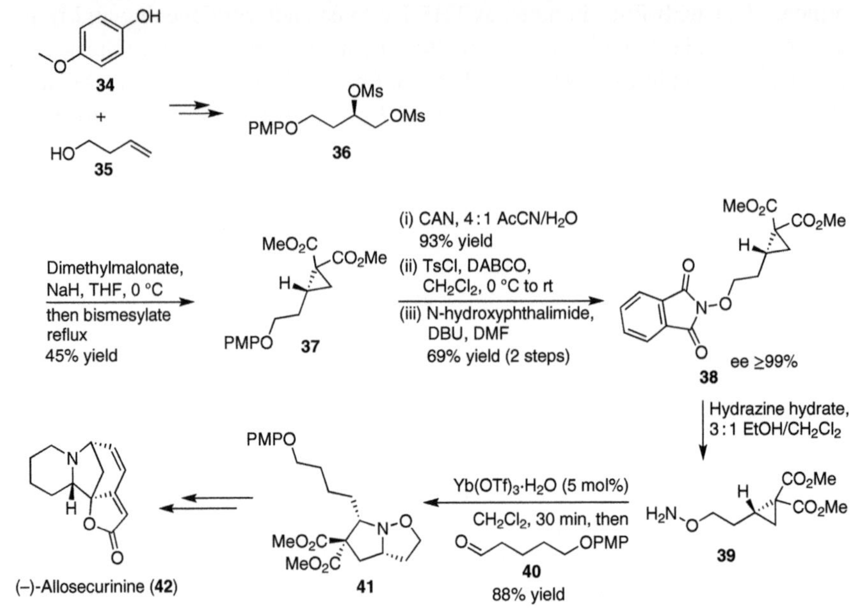

Scheme 12.7 Total synthesis of (−)-Allosecurinine.

several steps formed the highly functionalized nitrile **44**. [3+2] Cycloaddition of **44** with DA-cyclopropane **45** afforded the substrate **46** from which (±)-goniomitine (**47**) was synthesized in 10 steps with a 5.2% overall yield (Scheme 12.8). This is the first reported application for the [3+2] cycloaddition of DA-cyclopropane with a nitrile moiety in the total synthesis of natural products [16].

The same group also accomplished the total synthesis of indole alkaloid (±)-quebrachamine (**48**), starting from the same substrate δ-valerolactum (**43**) [17]. (±)-Quebrachamine is a tetracyclic indole alkaloid and is isolated from *Aspidosperma quebracho* tree bark. Once again, the key step was the cycloaddition of DA-cyclopropane

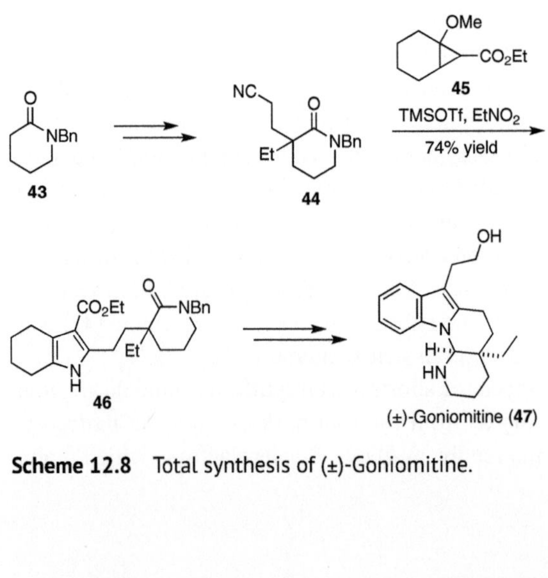

Scheme 12.8 Total synthesis of (±)-Goniomitine.

45 with functionalized nitrile moiety **44** to furnish the compound **46**, as shown in Scheme 12.9. Another crucial step involved in the synthesis was the formation of the nine-membered ring by photo-induced cyclization strategy. Starting from δ-valerolactum (**43**), (±)-quebrachamine (**48**) was obtained with an overall yield of 17.8% over 13 steps (Scheme 12.9).

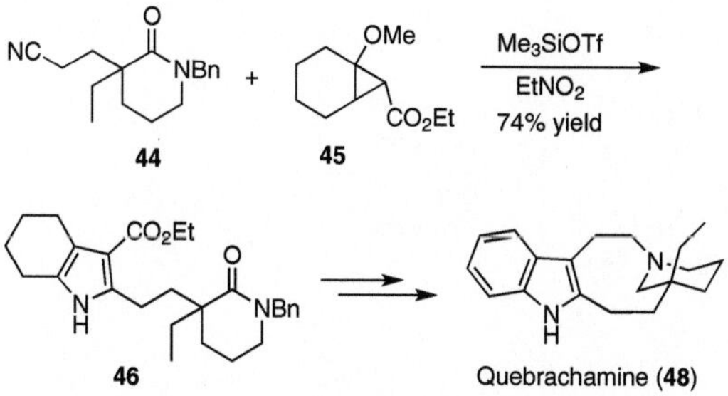

Scheme 12.9 Total synthesis of (±)-Quebrachamine.

The Qin group also demonstrated the total synthesis of anti-multi-drug-resistant indole alkaloid (−)-ardeemin (**55**), that was isolated from the fermentation of a strain of *Aspergillus fischeri* [18]. The key step in the synthetic route involved a three-step, one-pot cascade reaction of **50** with diazoester **51** to construct the non-racemic 3-substituted hexahydropyrrolo[2,3-*b*]indole **54** via intermolecular cyclopropanation, ring opening, and ring closure. The total synthesis has been completed in 20 steps with an overall yield of about 2% (Scheme 12.10) [19].

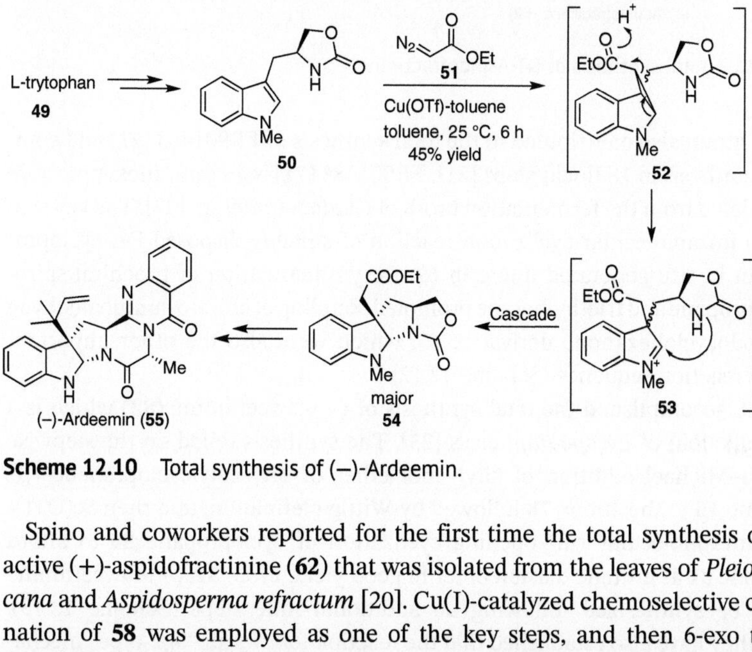

Scheme 12.10 Total synthesis of (−)-Ardeemin.

Spino and coworkers reported for the first time the total synthesis of optically active (+)-aspidofractinine (**62**) that was isolated from the leaves of *Pleiocarpa tubicana* and *Aspidosperma refractum* [20]. Cu(I)-catalyzed chemoselective cyclopropanation of **58** was employed as one of the key steps, and then 6-exo trig radical cyclization of **59b** afforded the compound **60**. Treatment of **60** with sodium anthracene radical anion resulted in the removal of the *N*-protecting group, followed by ring-opening of cyclopropane to synthesize the imine **61**, which in some additional steps produced (+)-aspidofractinine (**62**). Total synthesis of (+)-aspidofractinine (**62**) has been accomplished from cheap indole moiety **56** in 21 steps with a 2.1% overall yield (Scheme 12.11).

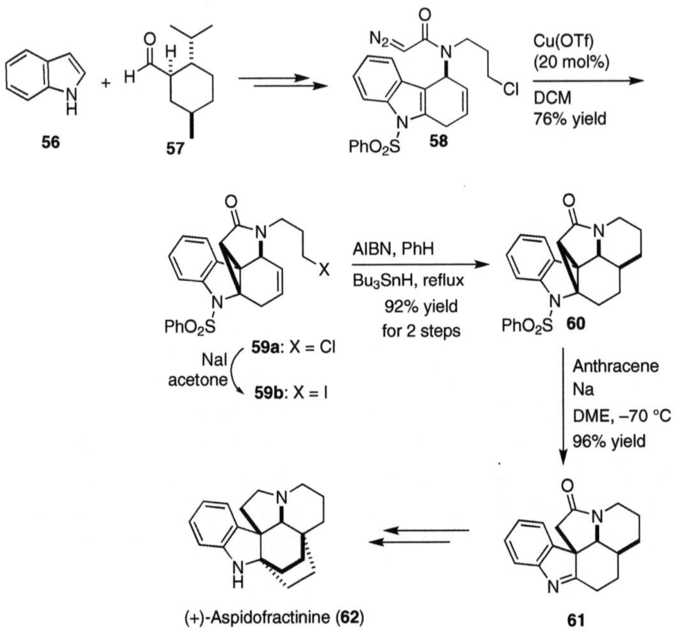

Scheme 12.11 Total synthesis of (+)-Aspidofractinine.

The Kerr group also contributed to the total synthesis of FR901483 (**71**) and completed the synthesis in 18 linear steps [21]. FR901483 (**71**) is an immunosuppressive alkaloid isolated from the fermentation broth of *Cladobotryum* sp. [22]. The key step involved an intramolecular cyclization reaction of suitably disposed DA-cyclopropane and an in situ-generated imine in **69**. Desymmetrization of prochiral spiro-fused cyclopropane and triethylborane promoted coupling of chiral *α*-aminoaldehyde **66** and *α*-iodocyclohexanone derivative **65**, which were also the other important steps in the reaction sequence (Scheme 12.12).

Jung et al. accomplished the total synthesis of (+)-fawcettimine (**82**) which is a tetracyclic alkaloid of *Lycopodium* class [23]. The synthesis relied on the stepwise Mukaiyama–Michael addition of silyl enol ether of an acetylcyclopropane-1,1-dicarboxylate **76** to the enone **78** followed by Wittig olefination, and then $Sc(OTf)_3$ promoted intramolecular ring-opening cyclization of cyclopropane **80** to afford hydrindanone **81** as a single diastereomer in good yield. From **81**, (+)-fawcettimine (**82**) could be synthesized following an additional nine steps (Scheme 12.13). Moreover, they have also established that the reaction follows an "S_N2-type" mechanism and is a diastereospecific reaction proceeding with complete retention of stereochemistry at the cyclopropyl center.

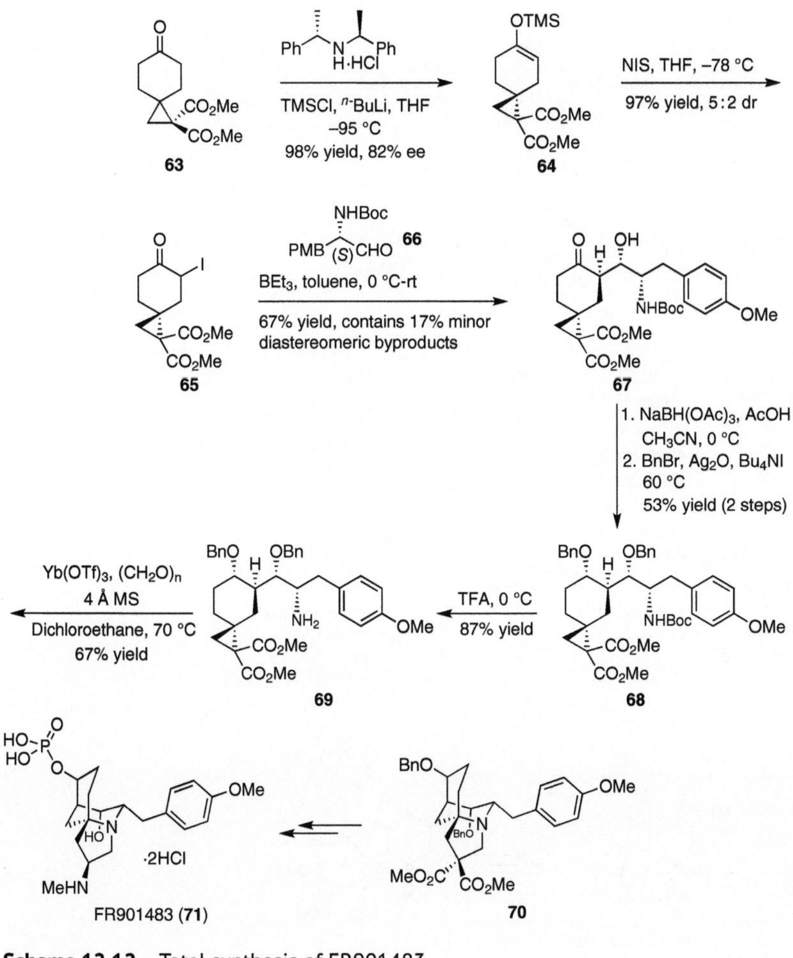

Scheme 12.12 Total synthesis of FR901483.

The Shair group completed the total synthesis of Lycopodium alkaloid (+)-fastigiatine (**87**) in 15 steps from cyclopropane **83** with an overall yield of 30% [24a]. (+)-Fastigiatine belongs to the lycodine structural class and was isolated from the alpine club moss *Lycopodium fastigatum* [24b]. The synthesis commenced with cyclopropane **83**, which was prepared from (*S*)-epichlorohydrin in four steps. **83** was then transformed to cyclopropane **84** in 83% yield upon transesterification with 2-(trimethylsilyl)ethanol followed by treatment with Boc-anhydride. Cyclopropane **84** underwent regioselective ring opening with mixed cuprate **85** to afford the key synthetic intermediate **86** in 93% yield (Scheme 12.14). Other important key steps involved in this total synthesis were diastereoselctive [3+3] cycloaddition and the transannular Mannich reaction to build the core of (+)-fastigiatine (**87**).

The Fukuyama group accomplished the first total synthesis of gelsemium alkaloid gelsemoxonine (**100**), which was isolated from the leaves of *Gelsemium elegans* [25]. The embedded spiro quaternary carbon center connected to bicyclic seven-membered core **99** in this natural product was constructed via divinylcyclopropane-cycloheptadiene rearrangement of highly functionalized substrate **97** (Scheme 12.15).

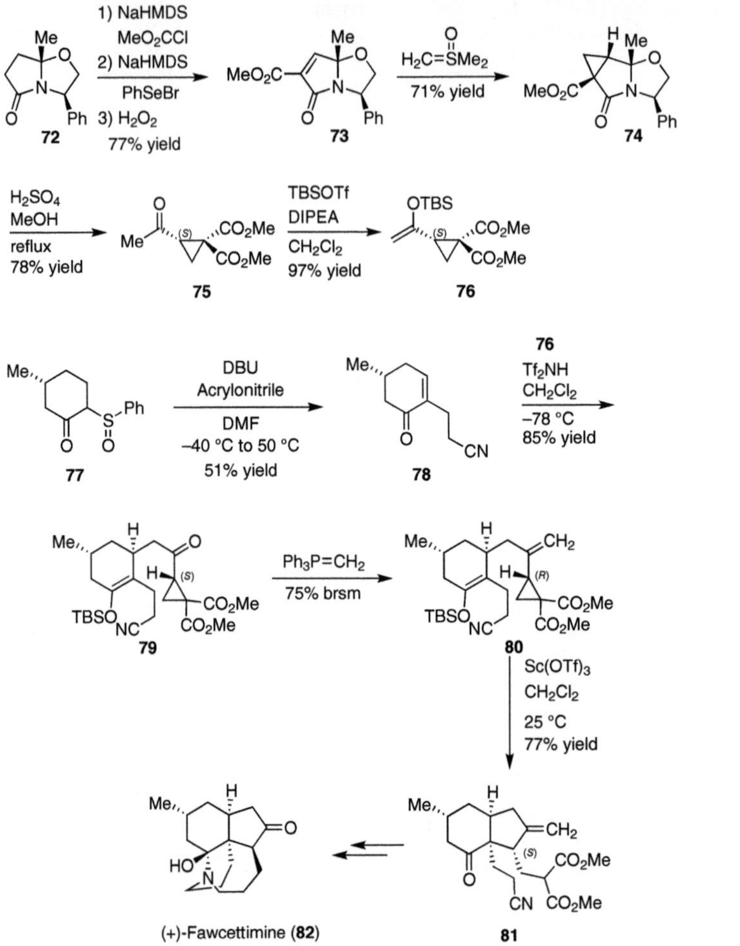

Scheme 12.13 Total synthesis of (+)-Fawcettimine.

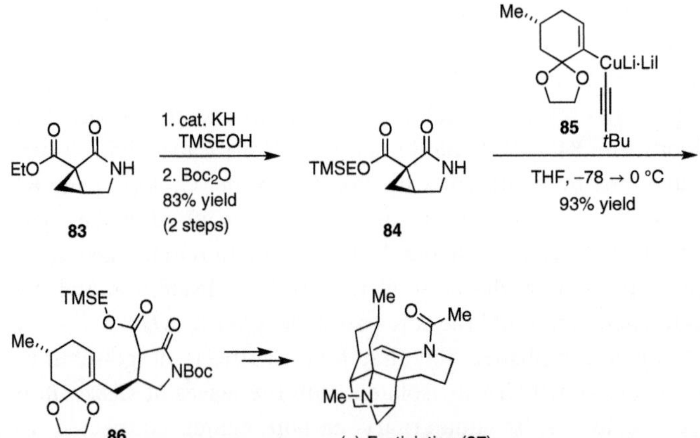

Scheme 12.14 Total synthesis of (+)-Fastigiatine.

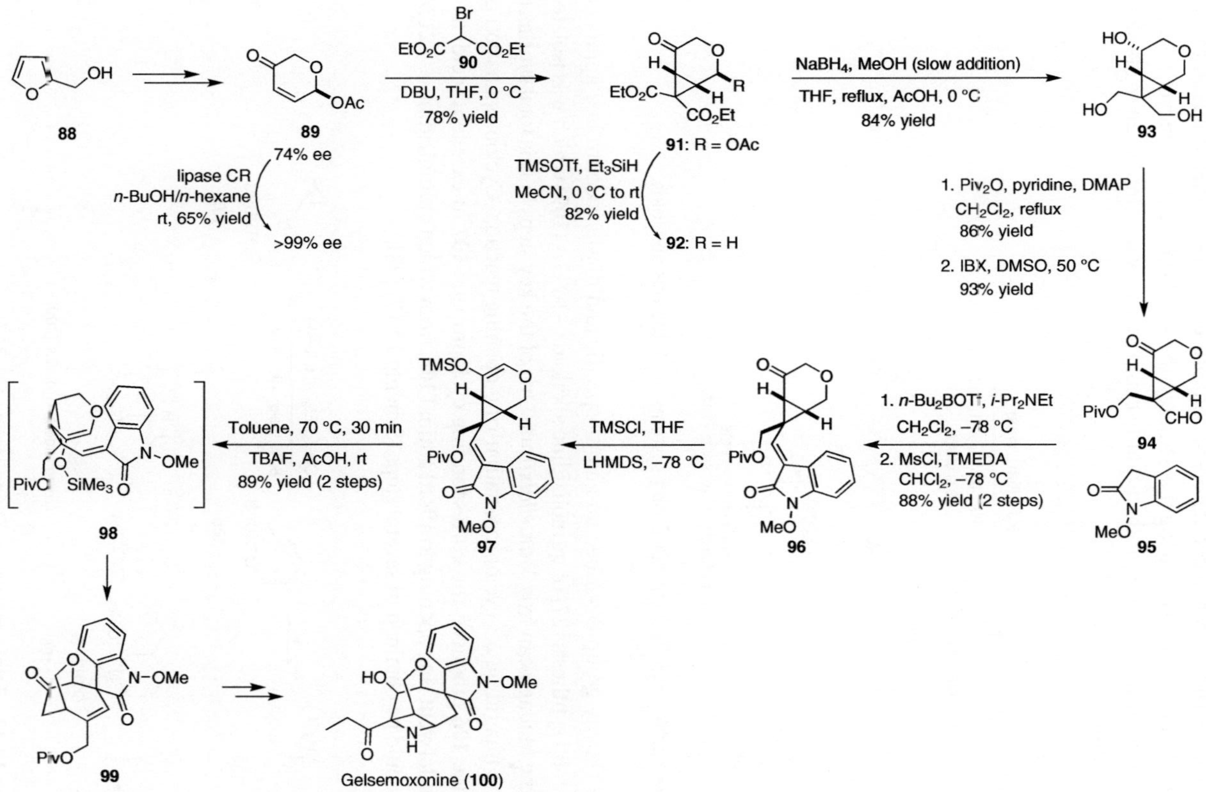

Scheme 12.15 Total synthesis of Gelsemoxonine.

In 2011, the Stoltz group reported an efficient approach for the synthesis of melodinus alkaloids. These alkaloids belong to the dihydroquinolinone family of natural products and were isolated from *Melodinus scandens* Forst. They performed palladium-catalyzed [3+2] cycloaddition reaction of β-nitrostyrene **101** with vinylcycloprpane **102** to construct the cyclopentane core **103** of melodinus alkaloids (**104**) (Scheme 12.16). Following this strategy, the synthesis of the tetracyclic ABCD ring system of this natural product family could be easily accomplished in six steps from commercial sources [26].

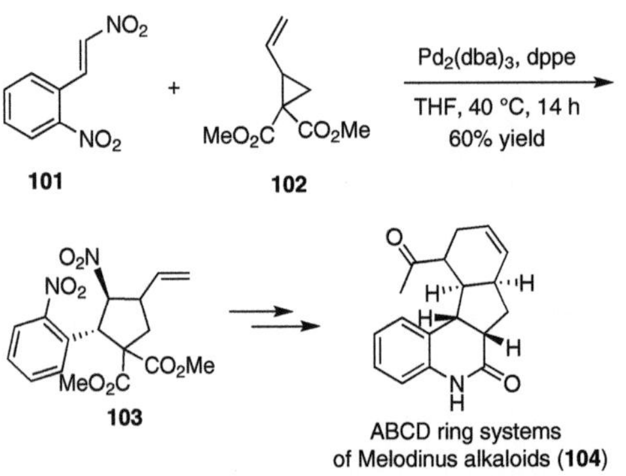

Scheme 12.16 Synthesis of ABCD ring systems of Melodinus alkaloids.

The Harrity group described an enantiospecific and diastereocontrolled approach to the total synthesis [27] of quinolizidine alkaloid (−)-217A (**109**) which was isolated from the Madagascan frog *Mantellabaroni*. One of the key steps in this total synthesis was the piperidine 2,3-cyclopropanation-ring-opening reaction. Cyclopropanation of enamide **106** produced the corresponding cyclopropane **107** in excellent yield, and **107** underwent ring opening to afford aminal **108**, from which quinolizidine (−)-217A (**109**) could be obtained in several steps (Scheme 12.17) [28].

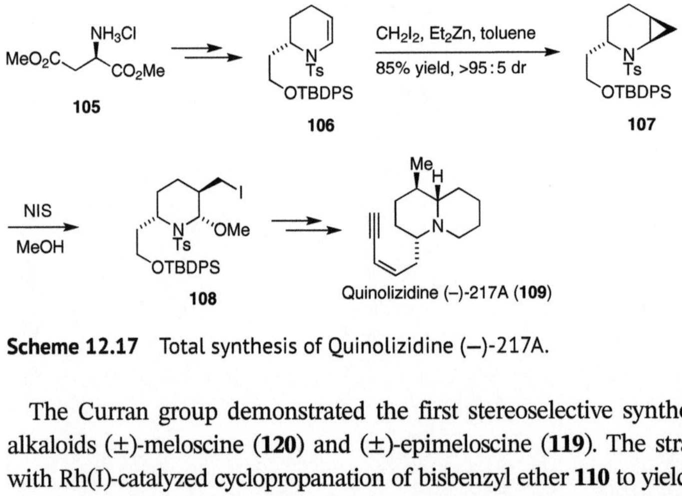

Scheme 12.17 Total synthesis of Quinolizidine (−)-217A.

The Curran group demonstrated the first stereoselective synthesis of Melodinus alkaloids (±)-meloscine (**120**) and (±)-epimeloscine (**119**). The strategy commenced with Rh(I)-catalyzed cyclopropanation of bisbenzyl ether **110** to yield the trisubstituted cyclopropane **111**. One key chemical step in the reaction sequence was the stereoselective cascade radical annulation of divinylcyclopropane **116** for the generation of two rings in compound **117**. The total synthesis of (±)-epimeloscine (**119**) has been completed with an overall yield of 6%, and (±)-meloscine (**120**) was synthesized from (±)-epimeloscine (**119**) by epimerization (Scheme 12.18) [29].

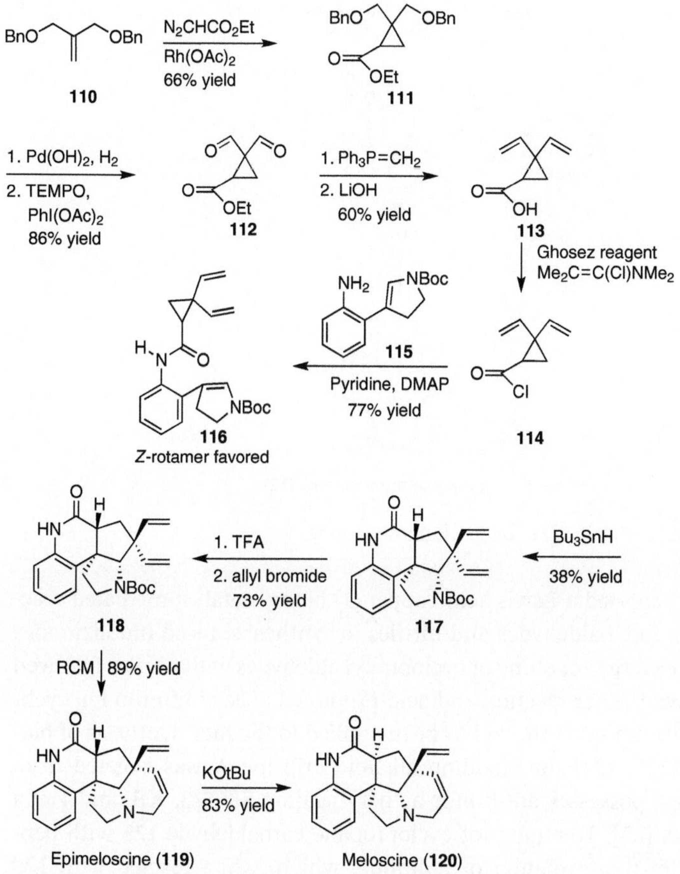

Scheme 12.18 Total synthesis of (±)-Epimeloscine and (±)-Meloscine.

The France group completed a diastereoselective synthesis of (±)-deethyleburnamonine (**127**) in six steps with about 18% overall yield. (±)-Deethyleburnamonine represents the simplest example of both *eburnan* and *tacaman* classes of alkaloids (no ethyl group present), which exhibit interesting pharmacological properties. Generation of ABDE portion of the target **126** via tandem ring-opening/Friedel–Crafts alkylation of donor–acceptor–acceptor amino cyclopropane **125** under the influence of an indium(III) catalyst served as the key step for the synthesis (Scheme 12.19). Apart from that, this total synthesis exploited the following key reactions: (i) Generation of C ring via TFA-mediated *N*-Boc deprotection/*N*-alkylation and (ii) a Krapcho decarboxylation to produce the target compound [30].

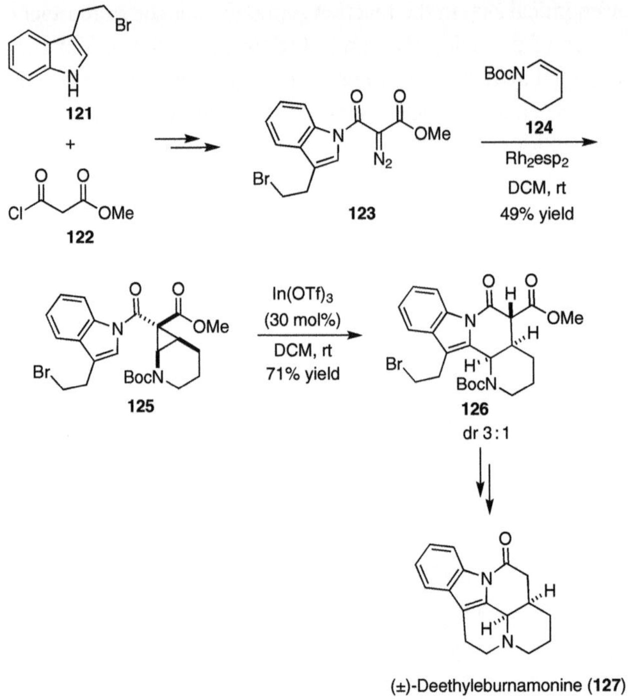

Scheme 12.19 Total synthesis of (±)-Deethyleburnamonine.

The Jiang group reported a Lewis acid-copper (I) binary catalyst-mediated reaction between cyclopropyl aldehydes and nitriles to synthesize fused indolizinones via Lewis acid-induced ring-opening of cyclopropyl aldehydes with nitriles followed by copper(I)-catalyzed Ritter reaction and acid-promoted *N*-acyliminium ion cyclization sequence. This unique process has been applied to the total synthesis of bioactive crispine A (**130**) [31]. Indolizidine alkaloid cripsine A was isolated from *Carduus crispus* and possesses antitumor activity against SKOV3, KB, and HeLa human cancer lines [32]. Treatment of cyclopropane carbaldehyde **128** with benzylamine resulted in the formation of aldimine, which, when reacted with **129** under standard indolizinone formation reaction conditions led to 3,4-dimethoxyphenyl indolizinone. It underwent lithium aluminum hydride-induced reduction to produce crispine A (**130**) (Scheme 12.20).

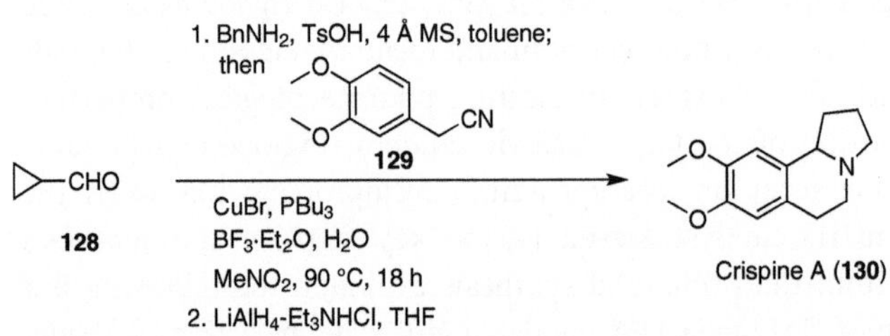

Scheme 12.20 Synthesis of Crispine A.

The Waser group accomplished the first total synthesis of *Aspidosperma* indole alkaloid (±)-jeratinine E (**136**), which was isolated from a leaf extract of the Malayan plant *Tabernaemontanan corymbosa* and was found to exhibit cytotoxic activity against human KB cells. δ-Valerolactum served as the starting material for the synthesis. The key synthetic step was the selective formal homo-Nazarov cyclization of aminocyclopropane to form four of the five rings contained in the final product. The synthesis commenced with the preparation of Weinreb amide **132** from δ-valerolactum **131** in seven steps, followed by treatment of **132** with the organolithium reagent generated from *N*-carboxy indole **133** to afford the aminocyclopropane **134**, which underwent homo-Nazarov cyclization to produce *cis*-diastereomer of **135**. Other key reactions involved in the reaction sequence were the installation of an ester group onto the core and a late-stage selective demethylation. Utilizing this protocol (±)-jeratinine E (**136**) was obtained in 16% overall yield over 17 steps starting from δ-valerolactum **131** (Scheme 12.21) [33].

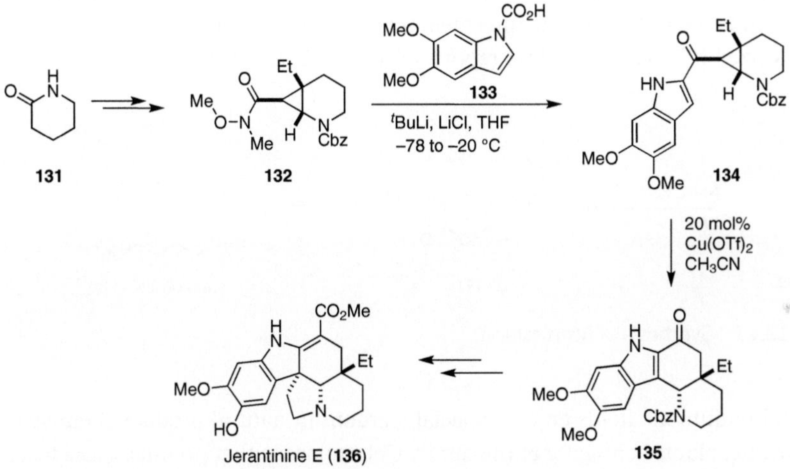

Scheme 12.21 Total synthesis of (±)-Jerantinine E.

Ivanov et al. developed a new method for S_N2-type ring-opening of various DA cyclopropanes with azide ions, and the reaction proceeds in a highly regioselective and stereoselective manner where the nucleophilic attack takes place at the more substituted C2 atom of the cyclopropane with complete inversion of configuration at C2. This developed methodology was successfully applied to the synthesis of natural products and medicines like (−)-nicotine (**139**) and atorvastatin (Lipitor) (**142**). (−)-Nicotine belongs to the tobacco alkaloid family and is well known for its effect on the central nervous system. It may also be used as an insecticide. The synthesis began with the ring opening of optically active 3-pyridyl-derived cyclopropane (*R*)-**137** with azide ions under the standard conditions to afford the ring-opening product (*S*)-**138**. Krapcho dealkoxycarbonylation of the ring-opening product (*S*)-**138** followed by treatment with Ph_3P furnished the corresponding lactum, which upon reduction with $LiAlH_4$ and then reductive methylation produced (−)-nicotine (**139**) (Scheme 12.22).

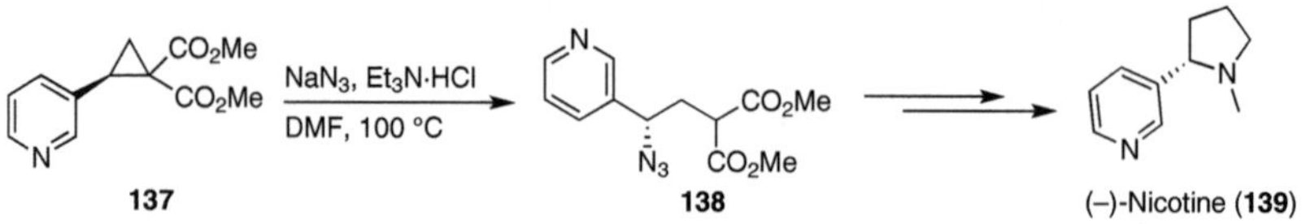

Scheme 12.22 Total synthesis of (−)-Nicotine.

This methodology was further used for the preparation of synthetic drug atorvastatin (**142**), the calcium salt of which is being used as the cholesterol-lowering medication Lipitor. The synthesis commenced with ring opening of 4-F-phenylcyclopropane-1,1-diester **140** with azide ion under the developed reaction condition to form the ring-opening product **141**, which underwent Staudinger/aza-Wittig reaction, dehydrogenation, bromination, and Suzuki cross-coupling to produce tetrasubstituted pyrrole intermediate from which atorvastatin (**142**) could be synthesized following literature reports (Scheme 12.23) [34].

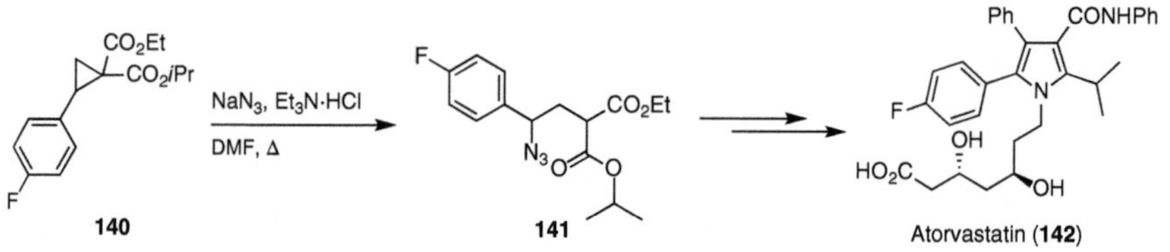

Scheme 12.23 Synthesis of Atorvastatin.

The Gelsemium alkaloids enjoy a special stardom in natural product chemistry. The flowering plants belonging to the family Gelsemiaceae that produce these toxic alkaloids have been well documented in the literature. They find application in traditional Asian medicine for the treatment of various ailments [35]. Ferreira and co-workers have reported the first total synthesis of (±)-gelsenicine featuring cyclopropanation in an important step. From the known aldehyde **143**, the diyne precursor *EE*-**144** was synthesized in three steps. Au-catalyzed cycloisomerization/Cope rearrangement of *EE*-**144** gave bicycle **146** via the formation of the cyclopropane intermediate ***E*-145**. This compound **146** after a series of transformations converted to (±)-gelsenicine (**147**) in 66% yield (Scheme 12.24) [36].

The steroidal alkaloid (−)-batrachotoxin (**153**) is a lethal component of the Colombian poison dart frogs and acts as a powerful agonist of voltage-gated sodium channels (NaVs). In 1963, the toxin (−)-batrachotoxin was first isolated by Witkop and co-workers from the membrane extracts of *Phyllobates bicolor* [37]. Thus, on account of its compelling biological activity and exhilarating chemical structure, Du Bois and co-workers accomplished the total synthesis of both (−)-batrachotoxin and its enantiomer in 2016, as shown in Scheme 12.25 [38]. Ketalization and enol ether

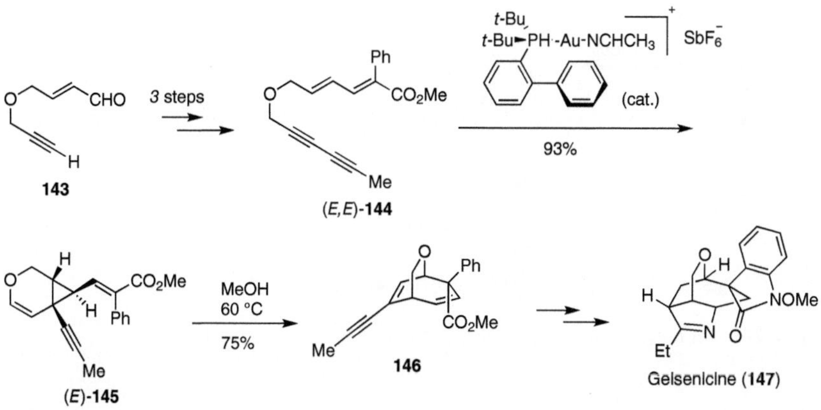

Scheme 12.24 Total synthesis of (±)-Gelsenicine.

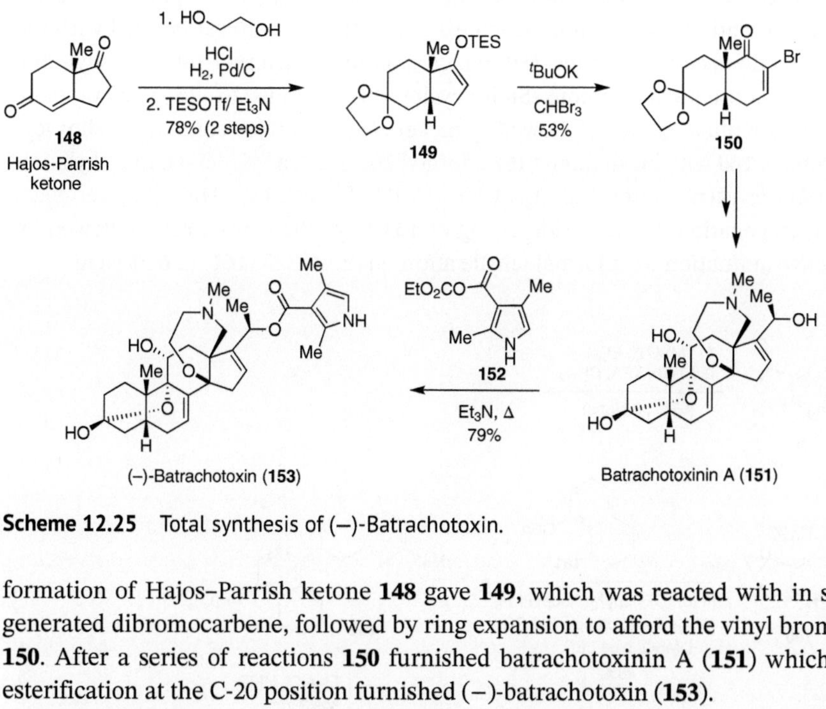

Scheme 12.25 Total synthesis of (−)-Batrachotoxin.

formation of Hajos–Parrish ketone **148** gave **149**, which was reacted with in situ-generated dibromocarbene, followed by ring expansion to afford the vinyl bromide **150**. After a series of reactions **150** furnished batrachotoxinin A (**151**) which on esterification at the C-20 position furnished (−)-batrachotoxin (**153**).

The kainoid amino acids have stimulated great attention on account of their numerous biological properties, such as neuroexcitatory and excitotoxic activities [39]. *β*-Allokainic acid (**158**) belongs to such kainoid alkaloid family. Kerr and coworkers have developed a simple route to substituted 2,5-dihydrooxepine rings **157** through a vinylogous Cloke–Wilson rearrangement [40] of DA cyclopropane **156**. The compound **157** was converted to (±)-*β*-allokainic acid (**158**) after a few steps (Scheme 12.26) [41].

GB alkaloids are found in the bark of *Galbulimima belgraveana* and *G. baccata* native to Papua New Guinea, Malaysia, and Northern Australia. It leads to

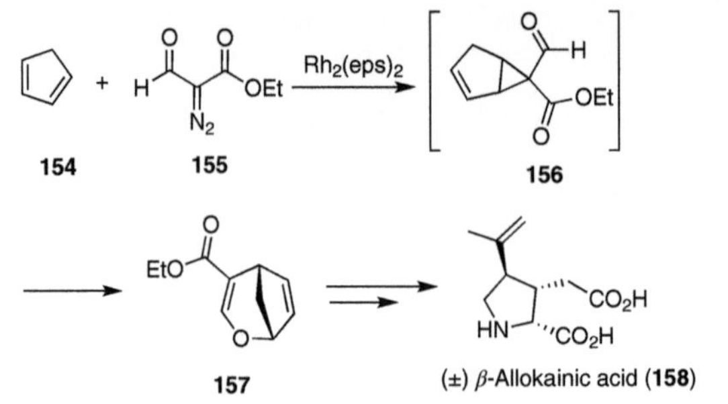

Scheme 12.26 Total synthesis of (±)-β-Allokainic acid.

psychedelic effects when the barks are munched, and in addition, it also finds application in traditional medicine [42]. In 2019, Shenvi and co-workers accomplished the synthesis of GB22 (**164**) via Ir/Ni dual-catalytic siloxycyclopropane C—C bond cleavage/arylation, as shown in Scheme 12.27 [43]. The synthesis started with the commercially available ketone **159**, which was first converted into the silyl-enol ether, and subsequent Simmons–Smith cyclopropanation led to siloxycyclopropane **160**. Iridium photocatalysis followed by nickel-catalyzed cross-coupling of siloxycyclopropane **160** with bromoarene **161** afforded **162**. Next, a Friedel–Crafts alkylation reaction was carried out using Et_2AlCl in HFIP to furnish **163**. After diastereoselective hydrogenation of the pyridine ring of **163** from the convex face, followed by reductive amination with formaldehyde afforded (±)-GB22 (**164**) in 64% yield.

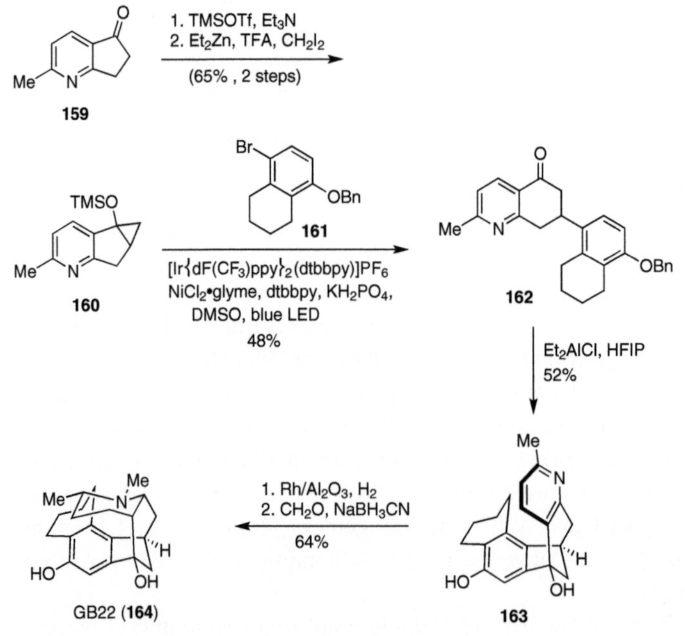

Scheme 12.27 Total synthesis of GB22.

The indole alkaloid Melokhanine E (**171**) was isolated in 2016 from *Melodinus khasianus*, a subtropical plant used in traditional Chinese medicine for the treatment of rheumatic heart disease and meningitis [44]. Recently, the Pierce group unveiled a 12-step convergent synthesis of (±)-melokhanine E (**171**) with an 11% overall yield, as shown in Scheme 12.28 [45]. The approach employed the application of two cyclopropane moieties **166** and **169** as reactive precursors to a 1,3-dipole and imine species to assist stereoselective creation of the core scaffold through MgI_2-mediated formal [3+2] cycloaddition. Piperidine derivative **169** and spirocyclic cyclopropane **166** were prepared by the reported literature procedure from anthranilic acid **167** and delta-valerolactam **165** in 6 and 4 steps, respectively. MgI_2 promoted ring-opening/cyclization of coupled biscyclopropane **170** under microwave conditions, resulting in the spirocycle opening and the thermolysis of the Boc-carbamate to form the reactive imine intermediate, which subsequently underwent the formal [3 + 2] cycloaddition, affording the (±)-melokhanine (**171**) in 88% yield as a single diastereomer.

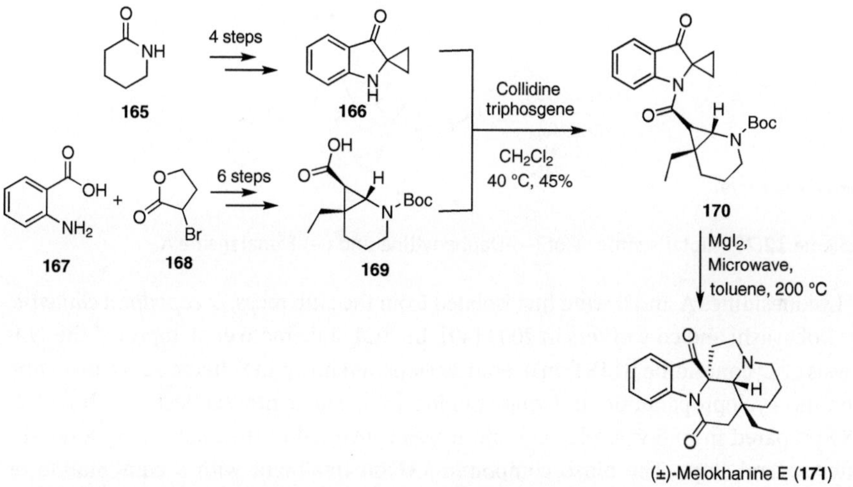

Scheme 12.28 Racemic total synthesis of (±)-Melokhanine E.

The Daphniphyllum alkaloids show a wide range of biological activities spanning from anticancer, anti-HIV, and antioxidation, to vasorelaxant activities [46]. Qiu and coworkers developed an efficient general strategy for the synthesis of the total synthesis of (−)-daphenylline (**177**) and (−)-himalensine A (**179**) (Scheme 12.29) [47]. The amide atropisomers **173** were generated in several steps from (*S*)-carvone **172**. The addition of diazo acetoacetamide into a solution of $[Cu(tbs)_2]$ in toluene afforded cyclopropyllactone **174**, which underwent a formal Cope rearrangement in the presence of tri-tert-butylphosphine to afford cycloheptenone **175** in an 82% yield [48]. Next, the diene **176** was generated from **175** in an overall 81% yield. The diene **176** acted as a common intermediate for the synthesis of calyciphylline A-type alkaloids, i.e. (−)-daphenylline (**177**) and (−)-himalensine A (**179**).

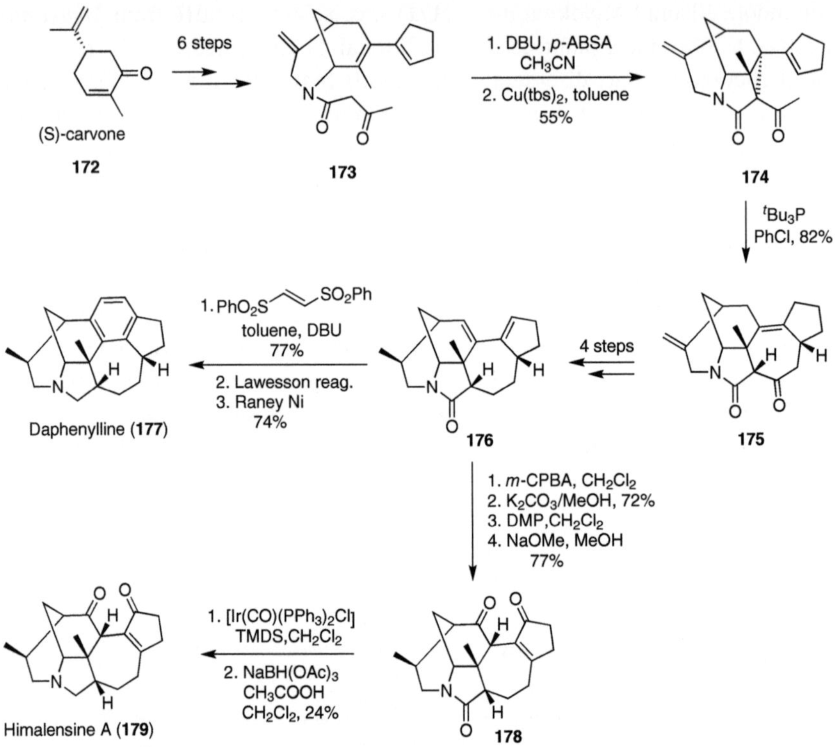

Scheme 12.29 Total synthesis of (−)-Daphenylline and (−)-Himalensine A.

Lyconesidines A and B were first isolated from the club moss, *Lycopodium chinense*, by Kobayashi and co-workers in 2002 [49]. In 2021, Takemoto et al. reported the synthesis of Lyconesidine B (**183**) in overall 30 steps, as shown in Scheme 12.30, involving domino cyclopropanation and ring-opening [50]. The synthesis began with triflate **180** prepared from δ-valerolactam and it was converted to the diazo compound **181** after several steps. The diazo compound **181** on treatment with a combination of

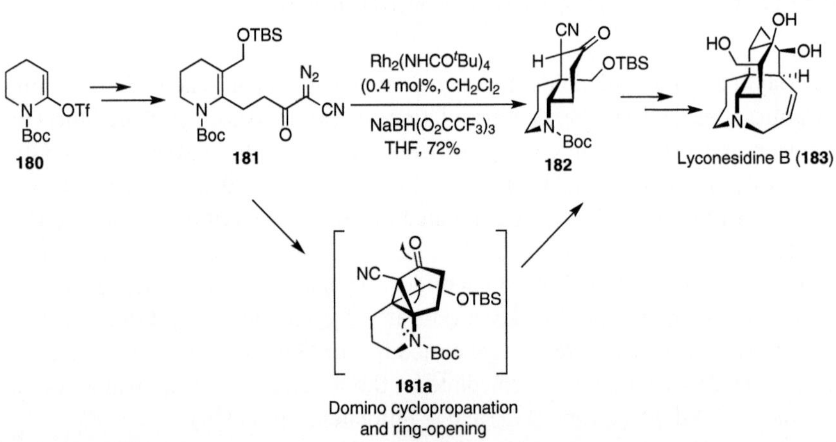

Scheme 12.30 Total synthesis of Lyconesidine B.

$Rh_2(NHCOtBu)_4$ and $NaBH(O_2CCF_3)_3$, prepared from $NaBH_4$ and TFA (1:3), gave **182** in 72% yield as the single product. The reaction advanced through cyclopropanation, ring-opening of the unstable intermediate (**181a**), followed by the reduction of an iminium ion to generate decahydroquinoline **182**. Lyconesidine B (**183**) was obtained from **182** after several steps with a 0.52% overall yield.

12.3 Synthesis of Terpene/Terpenoids

Terpene or terpenoids are a prominent class of natural products found in various fungi, algae, plants, and sponges. They have achieved significant pharmaceutical value because of their broad spectrum of medical applications. Apart from that, they also provided humans with hormones, flavors, fragrances, and commercial products. This section of this chapter highlights the recent total syntheses of terpenes/terpenoids employing DA-cyclopropane chemistry.

The sesquiterpenes (+)-(*R*, *S*)-isoclavukerin (**188**) and (−)-(*S*, *S*)-clavukerin A (**189**) were isolated from the Okinawan soft coral *Clavularia koellikeri* [51]. Alexakis et al. reported a tandem asymmetric formal total synthesis of (+)-(*R*, *S*)-isoclavukerin (**188**) and (−)-(*S*, *S*)-clavukerin A (**189**), as shown in Scheme 12.31 [52]. Silylation of the conjugated adduct zinc enolate **185** of **184** was carried out with TMSOTf at room temperature. This silyl enol ether was then subjected to cyclopropanation employing diiodomethane to furnish the cyclopropanol **186** with excellent yield (95%) and enantiomeric excess (98%). $FeCl_3$ promoted radical cleavage of the cyclopropane ring followed by the elimination of the corresponding α-chloro ketone to give the γ-methylenone **187** in an excellent yield (90%). **187** acted as an intermediate for the synthesis of (+)-(*R*, *S*)-isoclavukerin (**188**) and (−)-(*S*, *S*)-clavukerin A (**189**).

The Narasaka group reported for the first time the total synthesis of (−)-sordarin (**198**). It is a potent antifungal agent, and its structure contains a diterpene aglycon. One of the key steps in this synthesis involves oxidative radical cyclization of

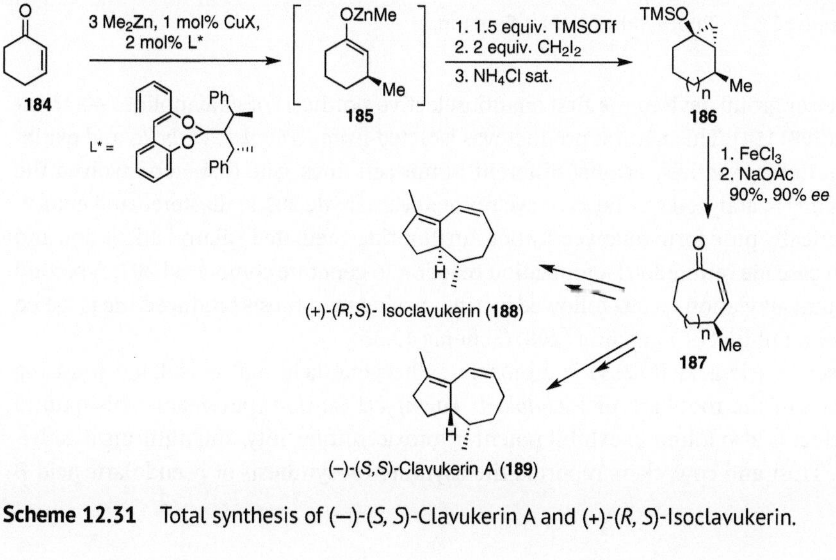

Scheme 12.31 Total synthesis of (−)-(*S*, *S*)-Clavukerin A and (+)-(*R*, *S*)-Isoclavukerin.

cyclopropanol derivative **192** in the presence of an Ag(I) catalyst. The reaction of **192** with $AgNO_3$-$(NH_4)_2S_2O_8$-pyridine system in the presence of 1,4-cyclohexadiene in DMF leads to the stereoselective formation of optically active bicyclo[5.3.0]decan-3-one skeleton **194** via cyclization of β-keto radical intermediate **193**. Bicyclic compound **194** was further converted to (−)-sordarin (**198**) in several steps (Scheme 12.32) [53].

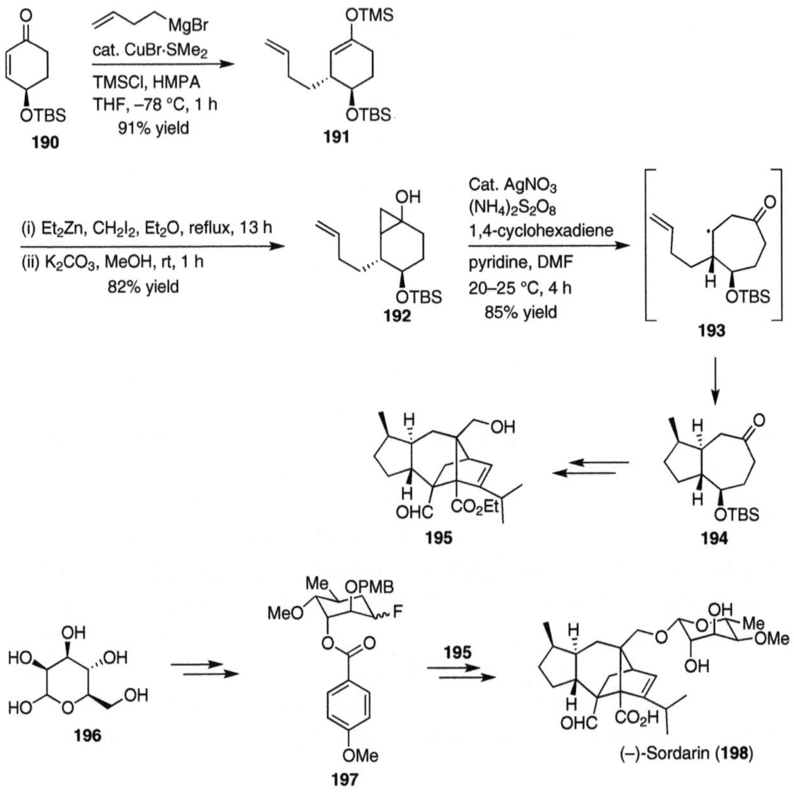

Scheme 12.32 Total synthesis of (–)-Sordarin.

Reiser group devised the first enantioselective synthesis of guaianolide (+)-arglabin (**208**) [54]. This natural product was isolated from *Artemisia glabella* and exhibits antitumor activity against different tumor cell lines. The first step involved the copper(I)-catalyzed synthesis of cyclopropanealdehyde **201** in diastereo and enantiomerically pure form followed by borontrifluoride-mediated Sakurai alkylation and then cascade retroaldol/lactonization reaction to generate compound **207**. A second Sakurai alkylation on **207** followed by ring-closing metathesis produced the targeted natural product (+)-arglabin (**208**) (Scheme 12.33).

Pseudolaric acid B (**215**) is a bioactive diterpene acid and is isolated from the extract of the root bark of *Pseudolarix kaempferi Gordon* (pinaceae). This natural product is also found to exhibit potent cytotoxic, antifertility, and antifungal activities. Trost and coworkers reported the asymmetric synthesis of pseudolaric acid B

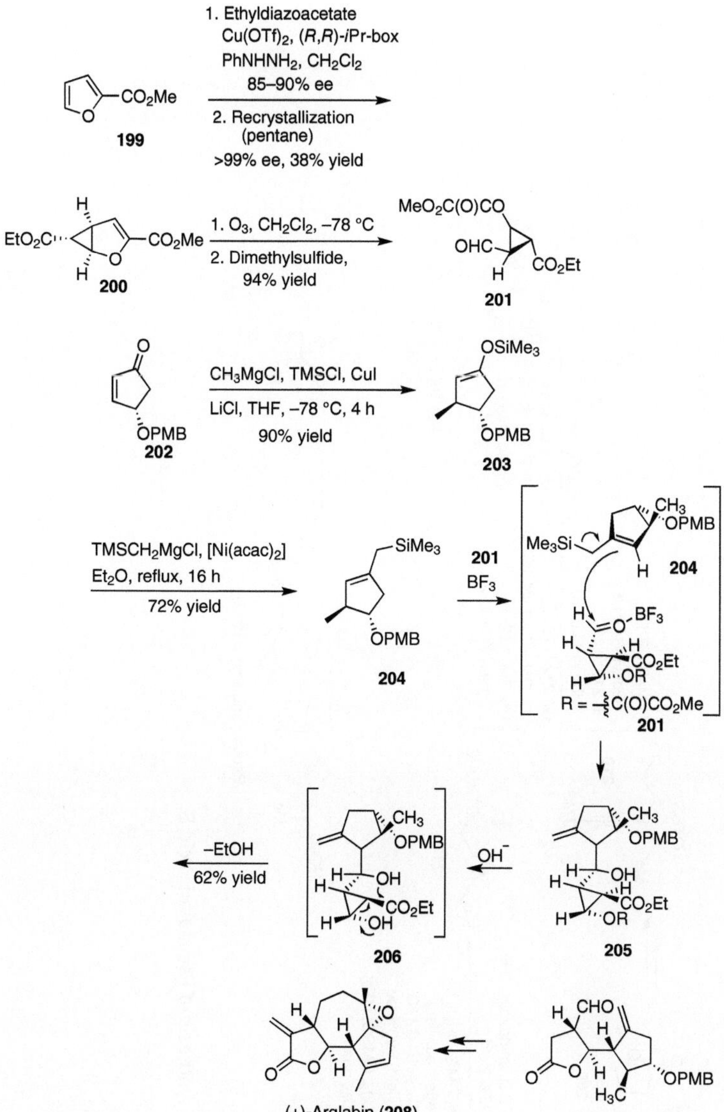

Scheme 12.33 Total synthesis of (+)-Arglabin.

starting from two simple precursors, 2-acetylbutyrolactone **209** and *cis*-butenediol **211** [55]. Cyclopropanaldehyde **212** was synthesized via Charette cyclopropanation from **211**, and subsequent reactions furnished the vinylcyclopropane **213** with 10:1 *E*/*Z* selectivity. The key step involved the Ru-catalyzed [5+2] cycloaddition reaction in **213** to give polyhydroazulene core **214** with 15:1 diastereosectivity (Scheme 12.34). Other key steps involved alkoxycarbonyl radical cyclization and selective cerium acetylide addition to form the tricyclic core of pseudolaric acid B (**215**) stereoselectively.

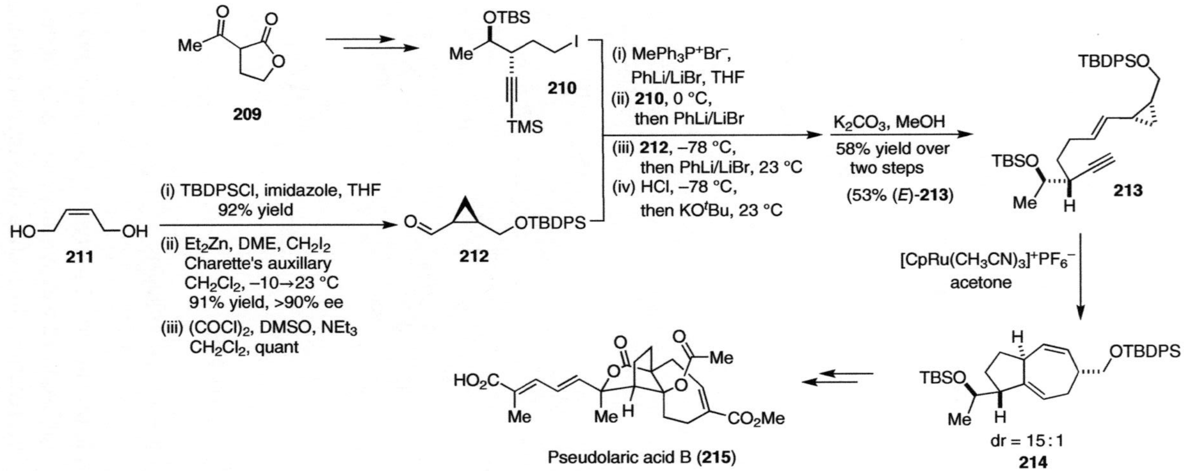

Scheme 12.34 Total synthesis of Pseudolaric acid B.

Yu and coworkers demonstrated a general method to build up triquinane skeleton via tandem Rh(I)-catalyzed two-component [(5+2)+1] cycloaddition/aldol condensation strategy [56]. This devised method was employed for the racemic synthesis of two representative linear triquinanes, (±)-Hirsutene (**220**) and (±)-1-Desoxyhypnophilin (**221**), which are a subset of polyquinanes representing an important class of sesquiterpenoids. The syntheses commenced with Horner–Wadsworth–Emmons olefination of dimethylhexenal **216**, followed by treatment with TBSOTf and Et_3N to form silylenol ether **217**. Chemoselective cyclopropanation of **217** provided pure (*Z*)-ene-vinylcycloprpane **218**, which underwent [(5+2)+1] cycloaddition to afford the desired tricyclic ketone **219** in diastereomerically pure form. This was the common key intermediate for the synthesis of two targeted natural products: (±)-hirsutene (**220**) and (±)-1-desoxyhypnophilin (**221**). From **219**, these two natural products were synthesized following some additional steps (Scheme 12.35).

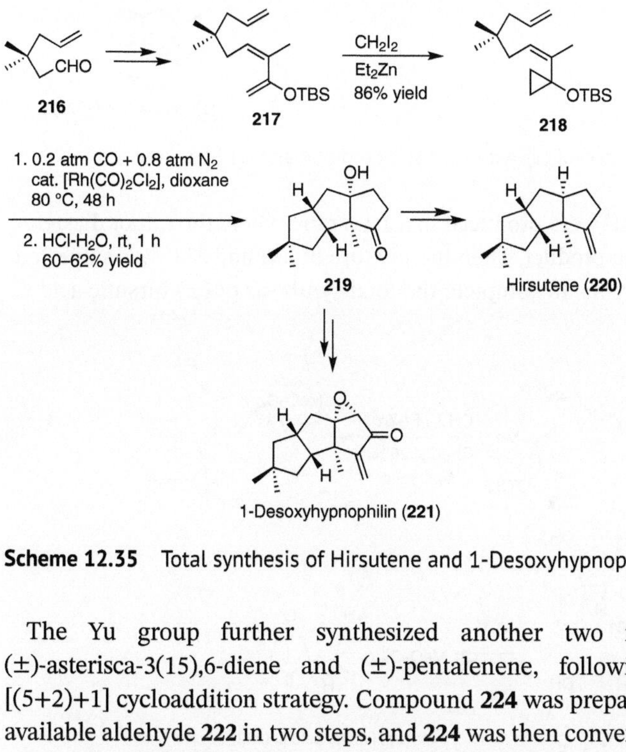

Scheme 12.35 Total synthesis of Hirsutene and 1-Desoxyhypnophilin.

The Yu group further synthesized another two important terpenoids, (±)-asterisca-3(15),6-diene and (±)-pentalenene, following the Rh-catalyzed [(5+2)+1] cycloaddition strategy. Compound **224** was prepared from commercially available aldehyde **222** in two steps, and **224** was then converted to bicyclic cyclooctenone **225** under Rh-catalyzed [(5+2)+1] cycloaddition reaction. **225** was the common intermediate for the synthesis of two natural products: (±)-asterisca-3(15),6-diene (**226**) [57] and (±)-pentalenene (**227**) (Scheme 12.36) [58].

The same analogy, tandem Rh(I)-catalyzed [(5+2)+1] cycloaddition/aldol condensation had been again used by the Yu group to synthesize linear triquinane natural product (±)-hirsutic acid C (**232**) [59], which belongs to the tricyclic sesquiterpene class [60]. The *β*-ene-vinylcyclopropane substrate **230** was prepared first starting from ester **228** and then **230** was subjected to the tandem [(5+2)+1]/aldol reaction conditions. This reaction showed poor diastereoselectivity, and two inseparable

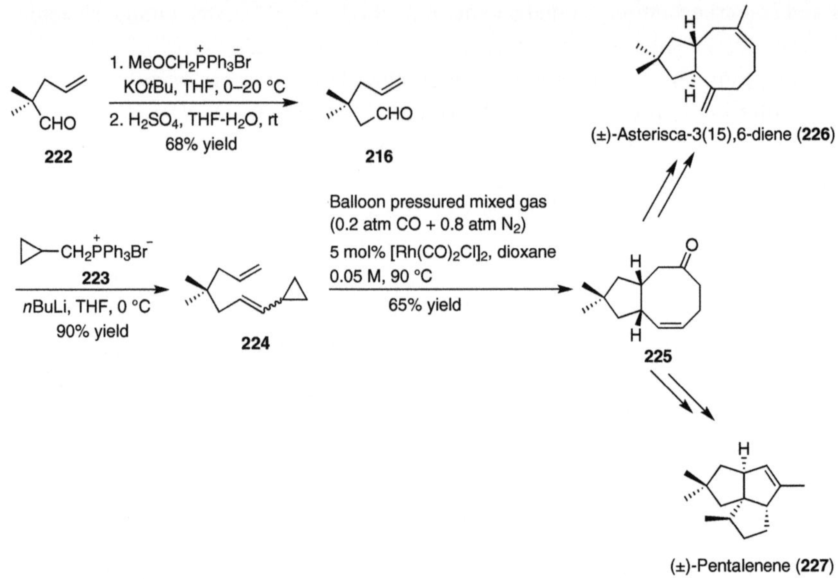

Scheme 12.36 Total synthesis of (±)-Asterisca-3(15),6-diene and (±)-Pentalenene.

cycloadducts **231** and **231′** were produced in a 1:1.5 ratio where the minor diastereomer **231** was the desired product. Then the mixture of **231** and **231′** was subjected to some sequential reactions to complete the total synthesis of (±)-hirsutic acid C (**232**) (Scheme 12.37).

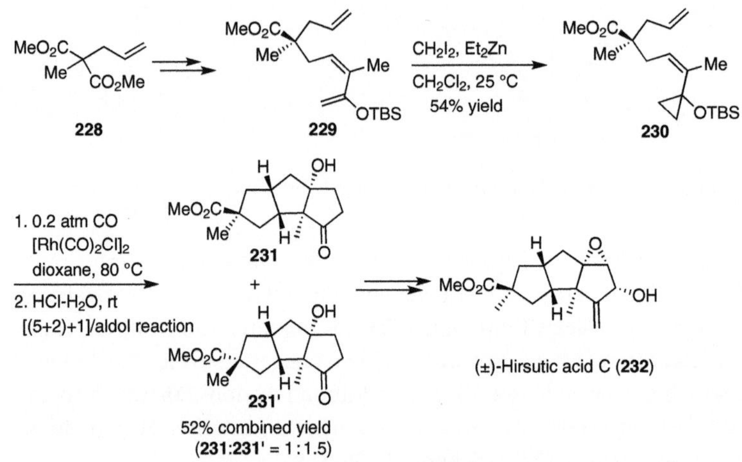

Scheme 12.37 Total synthesis of (±)-Hirsutic acid C.

The Yu group further developed a new strategy for the synthesis of bicyclic cyclohexenones and cyclohexanones by Rh(I)-catalyzed [(3+2)+1] cycloaddition reaction of 1-yne/ene-vinylcyclopropanes and CO [61]. This reaction can be considered as a homologous Pauson–Khand reaction. This analogy has been used for the synthesis of (±)-α-agarofuran, a furanoid sesquitepenoid natural product. The key step for the synthesis was [(3+2)+1] reaction of 1-yne-VCP **238** and CO to furnish bicyclic skeleton **239**. The targeted natural product (±)-α-agarofuran (**240**) was synthesized from **239** following some additional steps (Scheme 12.38).

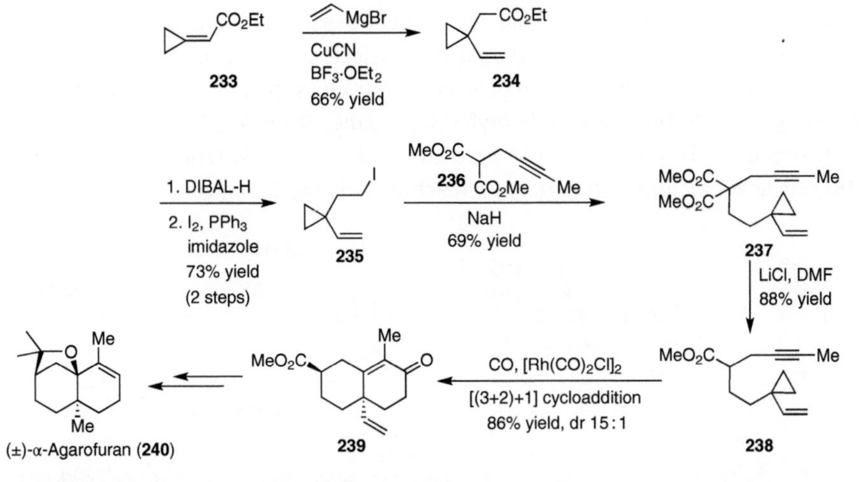

Scheme 12.38 Total synthesis of (±)-α-Agarofuran.

The Wang group first time reported the construction of bridged oxa- and aza-[n.2.1] (n = 2, 3, 4) skeletons via Lewis acid-catalyzed type II intramolecular [3+2] cycloaddition reaction of cyclopropane-1,1-diester with carbonyls and imines. The devised analogy was further employed for the total synthesis of platensimycin (**246**) [62], a promising antibacterial agent [63]. The synthesis commenced with the preparation of cyclopropane-1,1-diester **243** from **241**. The key step was Sc(OTf)$_3$-catalyzed [3+2] intramolecular cycloaddition of **244** to afford compound **245**, which, following some additional steps [64, 65] produced platensimycin (**246**) (Scheme 12.39).

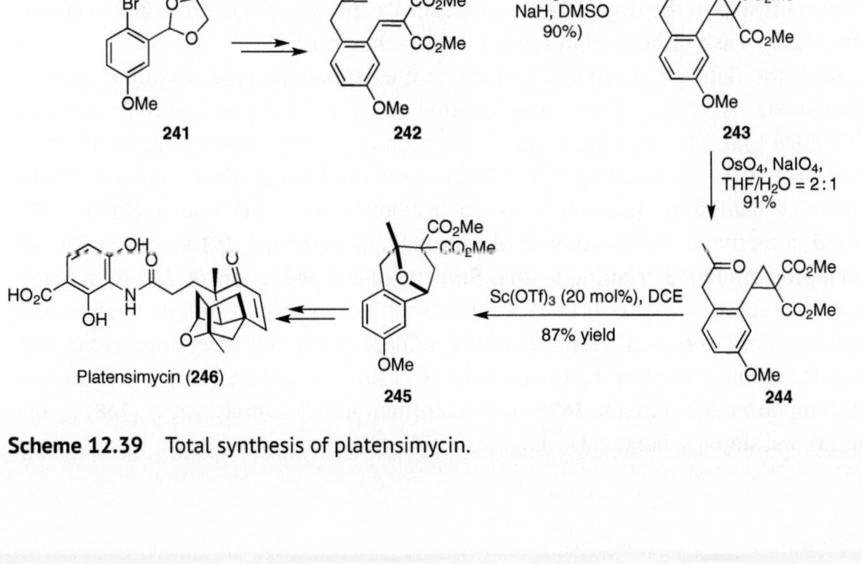

Scheme 12.39 Total synthesis of platensimycin.

Davies and Sarpong groups synthesized [66] two important diterpenes, (+)-barekoxide (**250**) and (−)-barekol (**251**) that were isolated from the Kenyan sponge *Raspailia* sp. [67]. The synthesis was based on dirhodium catalyst Rh_2(R-PTAD)$_4$-catalyzed tandem cyclopropanation/Cope rearrangement of bicyclic diene **247** and siloxyvinyldiazoacetate **248** to furnish the tricyclic compounds **249** and **249′** with a 6:1 diastereomeric ratio from which **249** could be separated in pure form with a 47% yield via recrystallization. Here, the diastereoselectivity was completely controlled by the chiral catalyst. The reaction proceeded by cyclopropanation, followed by cope rearrangement of the resulting divinylcyclopropane. From **249**, (+)-barekoxide was synthesized in six steps. Now (+)-barekoxide (**250**) was converted to (−)-barekol (**251**) via acid-catalyzed isomerization reaction (Scheme 12.40) [68].

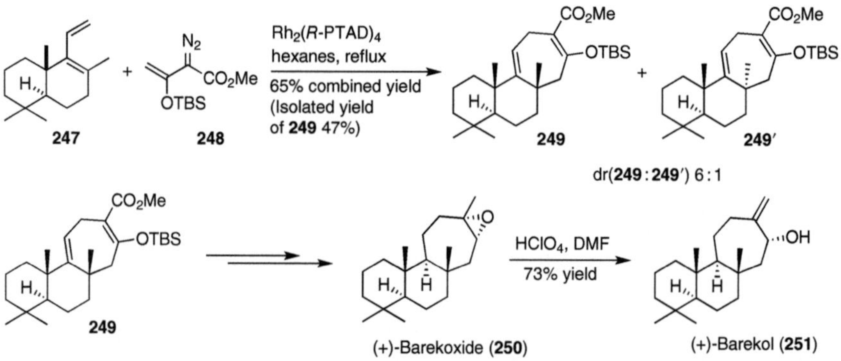

Scheme 12.40 Total synthesis of (+)-Barekoxide and (+)-Barekol.

Banwell and coworkers accomplished the first total synthesis of hirsutene-type sesquiterpenoid (+)-connatusin B (**261**), which was isolated from the culture broth of the fungus *Lentinus connatus* [69]. One of the key steps for this synthesis was the oxa-di-π-methane rearrangement of **254** to generate cyclopropane ring-fused linear triquinane **255**. The cleavage of a three-membered ring in **259** upon treatment with tri-*n*-bytyltin hydride and AIBN to construct the linear triquinane **260** was another important step in the way of (+)-connatusin B (**261**) synthesis. From **260**, connatusin B (**261**) was synthesized in several steps (Scheme 12.41).

Next, the Banwell group synthesized another hirsutane-type sesquiterpene (−)-connatusin A (**268**), which was isolated from the fungus *Lentinus connatus* BCC8996 [48]. The reaction commenced with enantiomerically pure *cis*-1,2-dihydrocatechol **262**. Noteworthy transformations of this total synthesis include a Diels–Alder cycloaddition reaction between cyclopentenone **252** and acetonide **253**, oxa-di-π-methane rearrangement of bicyclo[2.2.2]octenone **263** to synthesize the cyclopropannulated triquinane **264**, SmI_2-mediated reduction of **264** to generate **265**, treatment of **265** with LiHMDS followed by trapping of kinetic enolate with MeI to afford **266**, and nBu_3SnH-AIBN-induced reduction of cyclopropane ring-fused triquinane **266** leading to the cleavage of three-membered ring and formation of compound **267**. Finally, **267** was transformed to (−)-connatusin A (**268**) in five additional steps (Scheme 12.42).

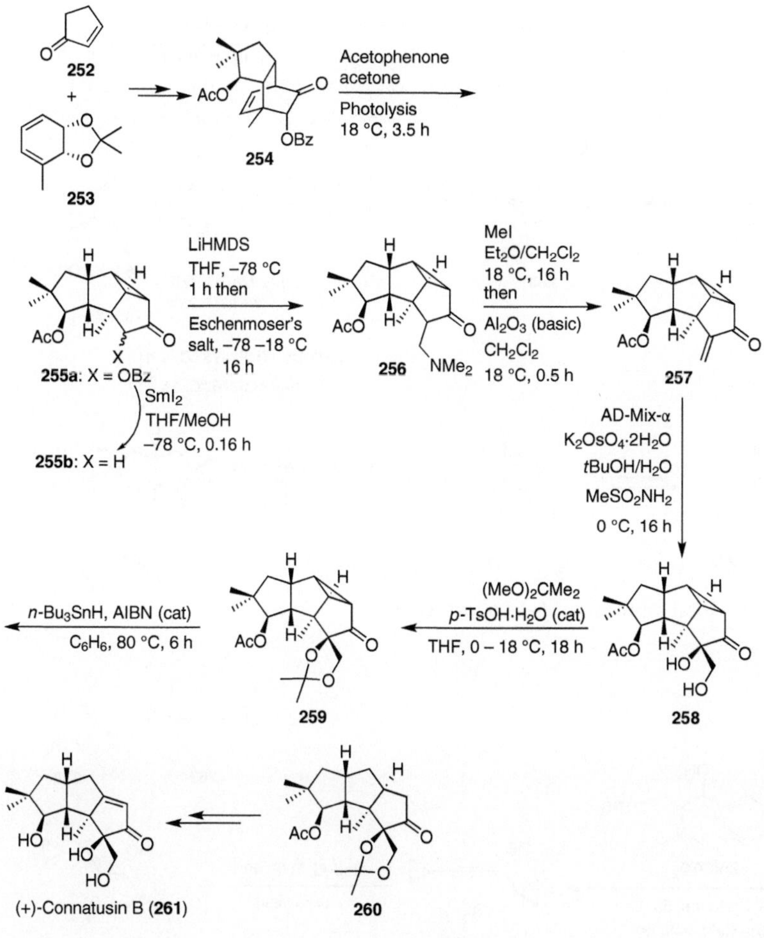

Scheme 12.41 Total synthesis of (+)-Connatusin B.

After successful synthesis of (±)-asterisca-3(15),6-diene and (±)-pentalenene terpenoids utilizing Rh-catalyzed [(5+2)+1] cycloaddition strategy, the Yu group further explored this reaction in the enantioselective total synthesis of sesquiterpenoid (+)-asteriscanolide (**275**) [70]. The synthesis proceeded by the formation of chiral ene-vinylcyclopropane substrate utilizing catalytic asymmetric alkynylation of aldehyde. Again, the construction of [6.3.0] carbocyclic core **274** via a chiral substrate-induced Rh(I)-catalyzed [(5+2)+1] cycloaddition was the key step for this synthesis (Scheme 12.43). The total synthesis was completed in 19 steps, with an overall yield of 3.8%.

Xiao et al. accomplished the diastereoselective total synthesis of nortriterpenoid (±)-schindilactone A (**281**) [71]. This natural product was isolated from the plants of *Schisandraceae* and has been used for rheumatic lumbago and related disease treatment. Key step associated with the synthesis includes an intermolecular Diels–Alder reaction of diene **276** and dienophile **277** and cyclopropanation with dibromocarbene of vinyl silyl ether **278** derived from the initial adduct, followed by $AgClO_4$-mediated rearrangement of cyclopropane **279** to afford vinyl bromide **280**, which

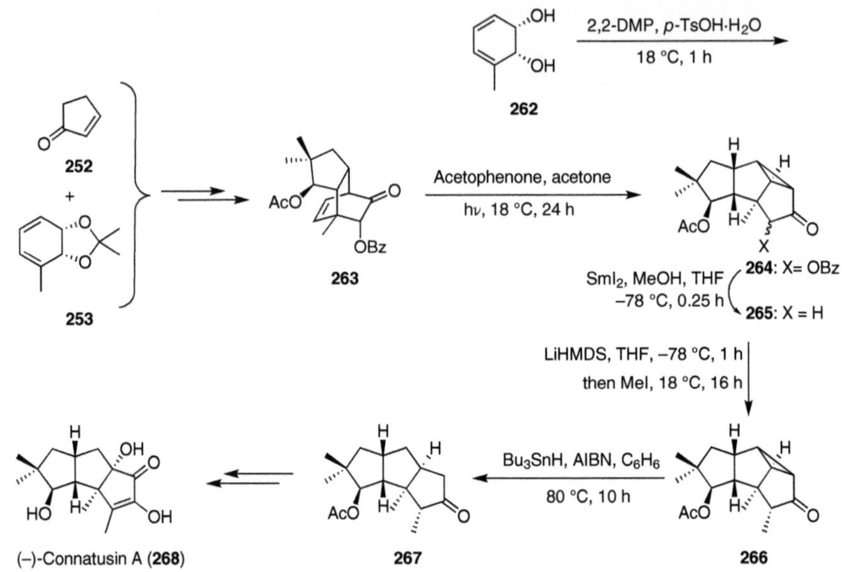

Scheme 12.42 Total synthesis of (−)-Connatusin A.

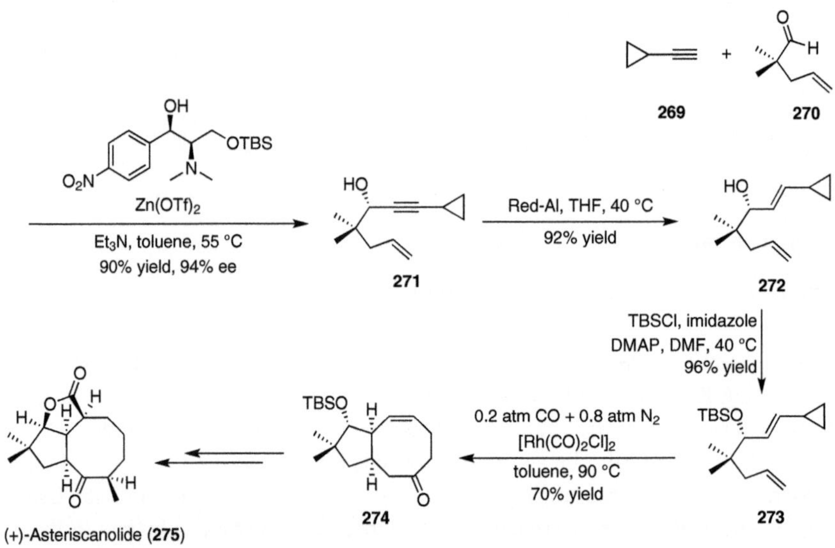

Scheme 12.43 Total synthesis of (+)-Asteriscanolide.

upon subsequent reactions generated the desired natural product (Scheme 12.44). This is the first reported total synthesis of (±)-schindilactone A (**281**).

The Reiser group also completed the total synthesis of monoterpene (−)-paeonilide (**288**) in 2012 [72]. Furan-3-carboxylic acid **282** served as the starting material for the reaction. The reaction proceeded by the asymmetric cyclopropanation of **283** to generate **284** followed by selective hydrolysis to synthesize **285**, which was selectively

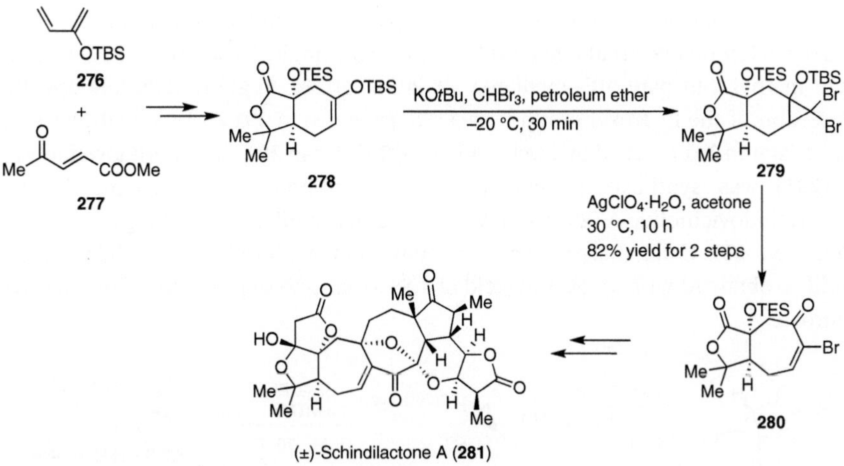

Scheme 12.44 Total synthesis of (±)-Schindilactone A.

transformed to **286** in which the carboxylic acid group was placed in the concave face of the bicycle **286**. Acid-induced cyclopropane ring-opening/lactonization converted **286** to **287** from which (−)-paeonilide (**288**) was obtained in seven steps with a 4.4% overall yield (Scheme 12.45).

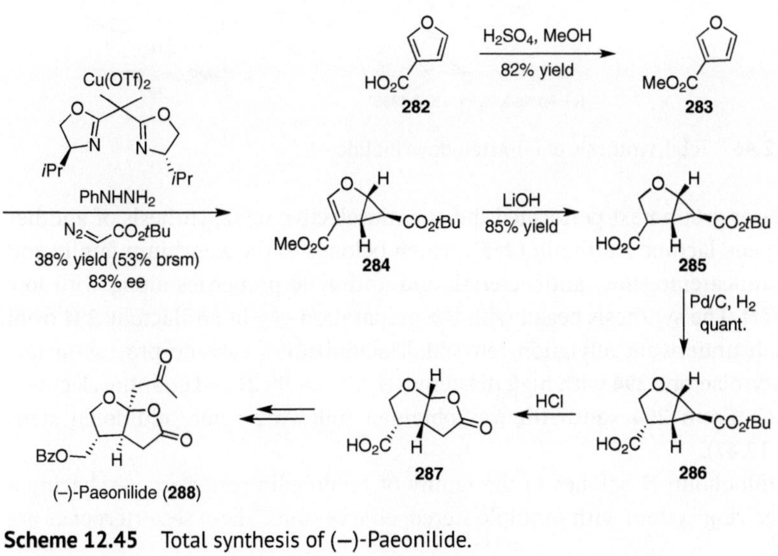

Scheme 12.45 Total synthesis of (−)-Paeonilide.

Further, the Reiser group accomplished the first enantioselective total synthesis of (+)- and (−)-arteludovicinolide A (**293**) [73]. The sesquiterpene lactone (+)-arteludovicinolide A (**293**) was discovered in the aerial parts of *Artemisia ludoviciana* and was found to have anti-inflammatory activity (NO inhibition via iNOS pathway). Moreover, (+)-arteludovicinolide A (**293**) has 15 times higher anti-inflammatory activity at noncyctotoxic concentrations (≤10 μM) than previously reported for concentrations of ≥45 μM. The synthesis commenced with the preparation of cyclopropanecarbaldehyde **201** (90% ee) as a single diastereomer in two steps from methyl 2-furoate. BF_3-mediated allylation of **201** to afford **290**, followed by protection of the hydroxyl group with TIPS, gave rise to **291** and the hydrolysis of oxalic ester in **291** resulted in acyclic aldehyde **292**. From **292**, (+)-arteludovicinolide A (**293**) was synthesized following some additional steps (Scheme 12.46). (−)-Arteludovicinolide A (**293**) was synthesized in a similar way starting from enantiopure *ent*-**201**. Using this protocol, either enantiomer of arteludovicinolide A (**293**) could be obtained with an overall yield of 4.8% over nine steps starting from methyl 2-furoate.

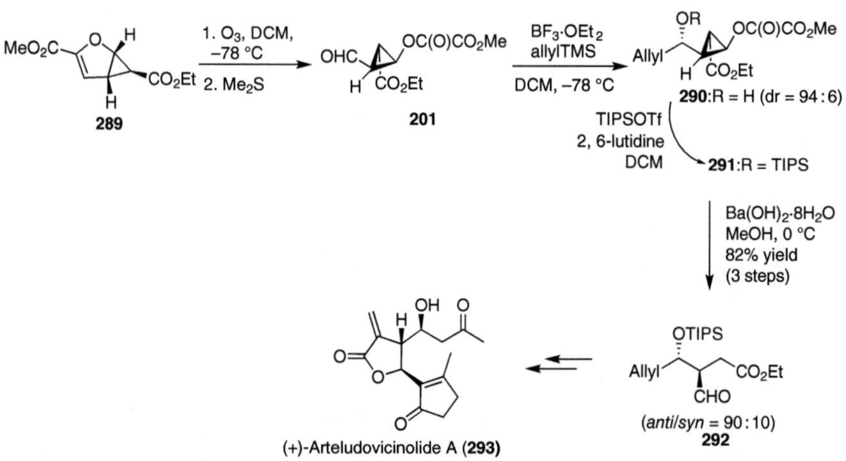

Scheme 12.46 Total synthesis of (+)-Arteludovicinolide A.

The Reiser group next performed the enantioselective total synthesis of another sesquiterpene lactone xanthatin (**295**), which belongs to the Xanthium family and displays anticarcinogenic, antibacterial, and antimytic properties along with low toxicity [74]. The synthesis began with the preparation of γ-butyrolactone **294** from **201**, which underwent allylation/retroaldol/lactonization cascade process to furnish γ-butyrolactone **294** with high diastereo- (*trans*/*cis* 98:2) and enantioselectivity (>99% ee). From **294** xanthatin was obtained following some additional steps (Scheme 12.47).

6-*epi*-Ophiobolin N belongs to the family of ophiobolin sesterterpene, having a 5-8-5 fused ring system with multiple stereogenic centers. These sesterterpenes are

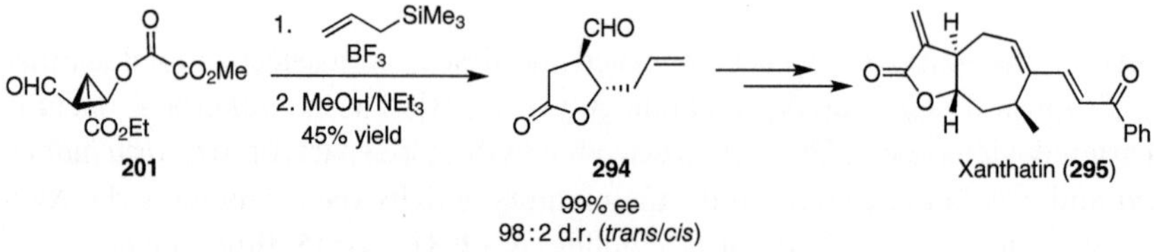

Scheme 12.47 Total synthesis of Xanthatin.

reported to exhibit significant bioactivities [75], including cytotoxicity against various cancer cell lines [76]. Their gifted biological potential and synthetically exciting chemical skeleton have drawn synthetic attention [77]. In 2016, Maimone et al. reported a nine-step, asymmetric synthesis of (-)-6-epi-Ophiobolin N (**303**), featuring cyclopropanation (Scheme 12.48) [78]. Inexpensive (−)-linalool was used as a chiral pool building block for the synthesis of **298**. In a separate sequence, *trans, trans* farnesol **296** when subjected to Charette's asymmetric cyclopropanation followed by an Appel-type reaction led to the formation of iodide **297**. For the crucial C—C bond-forming step, iodide **297** was treated with *tert*-butyllithium to induce anionic cyclopropane ring opening. Next, this presumed lithiated intermediate on reaction with copper iodide, enone **298**, and trichloroacetyl chloride furnished **299** in 60% yield (dr = 3:1) via a 1,4-addition/trapping sequence. Further synthetic manipulation, as shown in Scheme 12.48, afforded (−)-6-*epi*-ophiobolin N (**303**).

Marine cembranoids are a group of macrocycle-derived diterpenes produced by several species of coral [79]. The soft coral *Sinularia pavida* produces (−)-pavidolide B (**313**), a tetracyclic 7,5,5,6-fused cembranoid that displays inhibitory activity against various human promyelocytic leukemia cell lines [80]. In 2017, Gong and coworkers developed a novel route to (−)-pavidolide B based on a radical cascade employing a vinyl cyclopropane cleavage/formal annulation process [81]. Domino Michael's addition/alkylation of dimethyl bromomalonate **304** and dieneal **305** in the presence of proline-derived catalyst gave the key enantioenriched cyclopropane fragment **306** (Scheme 12.49) [82]. Acetal formation and subsequent mild ester monohydrolysis of the aldehyde group of cyclopropane **306** gave carboxylic acid **307** an inseparable pair of diastereomers. Hydrogenation of the isopropenyl group in **309** employing Wilkinson's catalyst afforded the alcohol **310**. The amalgamation of acid **307** and alcohol **310** obtained from (+) carvone under Mitsunobu conditions gave an intermediate **311** in high yield. **311** underwent radical cascade reaction in the presence of visible light, Ir(III) complex, *p*-toluidine, and thiophenol via the formation of phenylthiyl radical, and the concurrent cyclopropane fragmentation furnished the annulated product **312** as a single diastereomer in 50% yield [83–85]. **312** was transformed to (−)-pavidolide B (**313**) after several steps.

Yu and co-workers reported the asymmetric total synthesis of the clovane-type sesquiterpenes (-)-clovan-2,9-dione (**324**) with a [6.3.1.0^{1,5}]dodecane skeleton. The synthesis involves an Rh(I)-catalyzed [3+2+1] cycloaddition of 1-yne-vinylcyclopropane (1-yne-VCP) **322** with CO to obtain the framework of the target natural product, as shown in Scheme 12.50 [86a]. The aldehyde **314** [86b] and ester **315** [86c] were prepared from the known starting materials. The intermolecular aldol reaction of **314** with **315** followed by LAH reduction and subsequent selective oxidation of the primary hydroxyl gave the aldehyde (±)-**317** in the presence of TEMPO, NCS, and TBAC. (±)-**317** was converted to **322** after several steps, as shown in Scheme 12.50. Rh(I)-catalyzed [3+2+1] cycloaddition of (+)-**322** with CO under a 0.2 atm CO atmosphere afforded the desired product (−)-*trans*-**323** and (−)-*cis*-**323** in 83% total combined yield with a 3:1 diastereomeric ratio. These two diastereomers were separated by column chromatography. (−)-*trans*-**323** was transformed into the sesquiterpenes (-)-Clovan-2,9-dione (**324**) after several reaction steps.

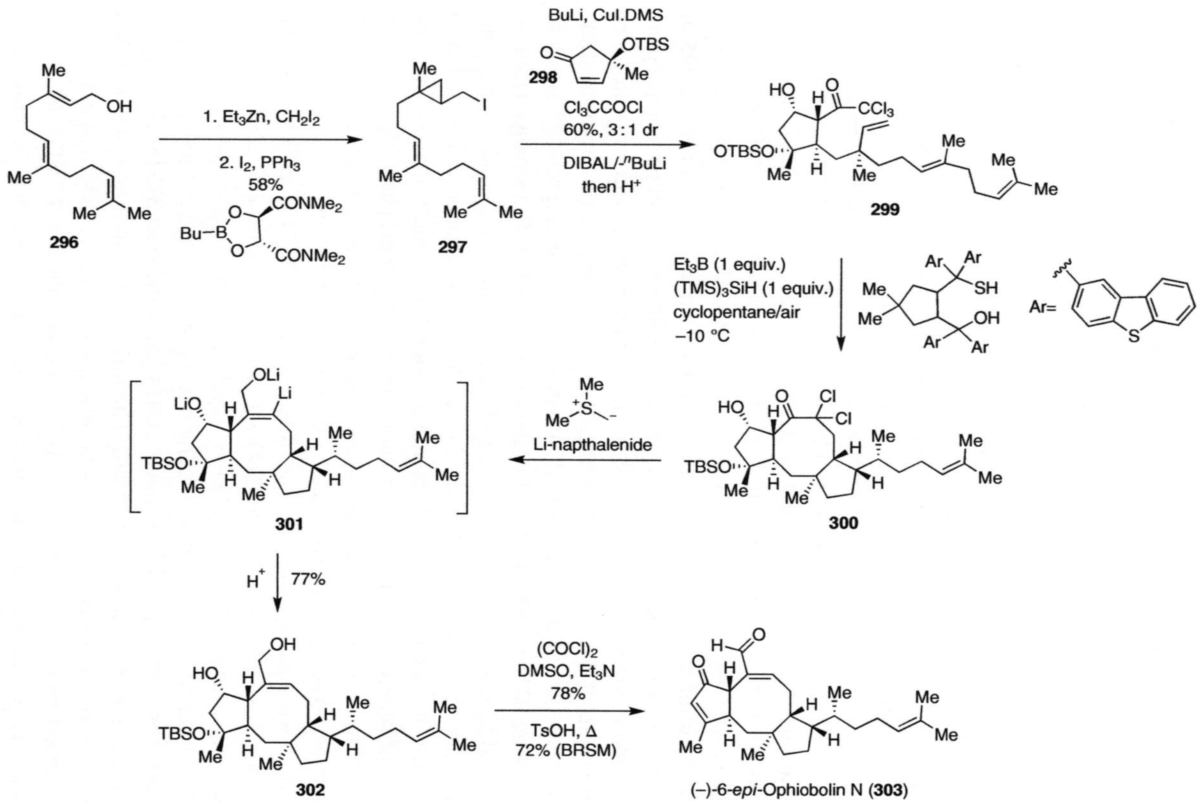

Scheme 12.48 Total synthesis of (−)-6-*epi*-Ophiobolin N.

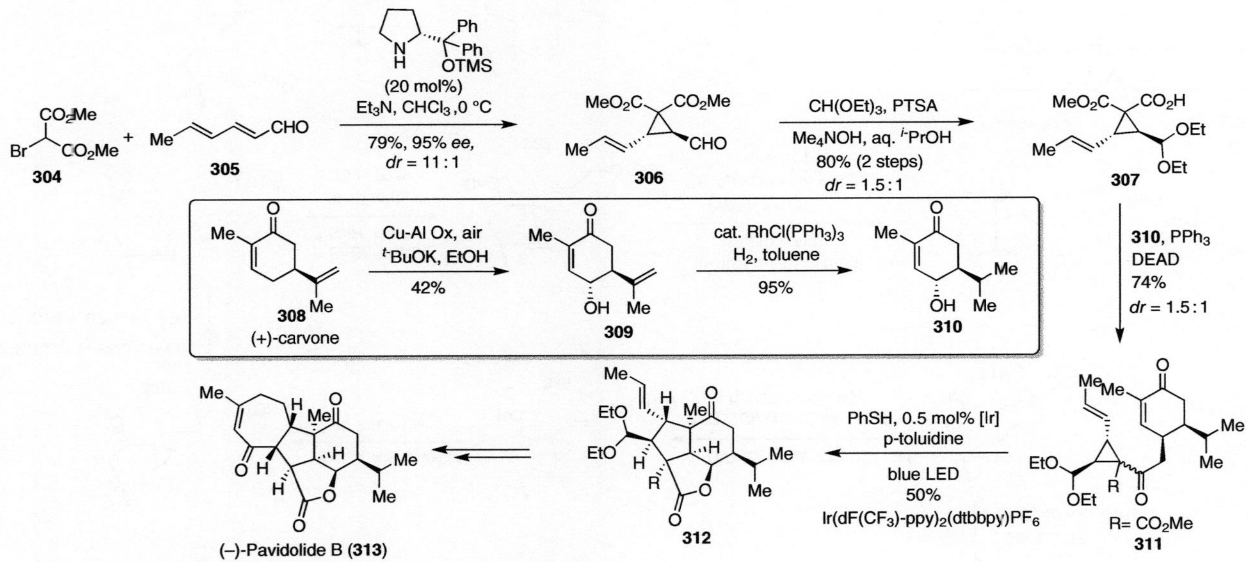

Scheme 12.49 Total synthesis of (−)-pavidolide B.

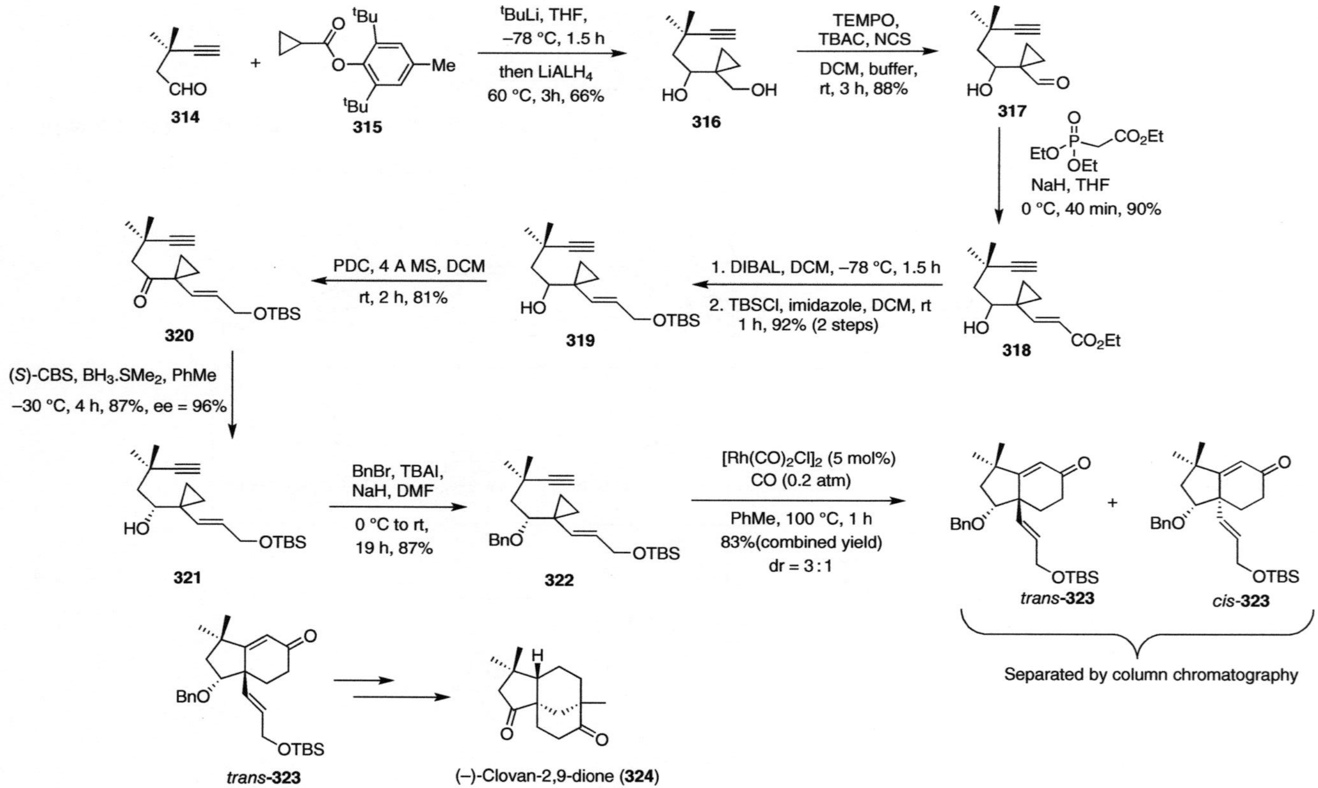

Scheme 12.50 Total synthesis of (–)-Clovan-2,9-dione.

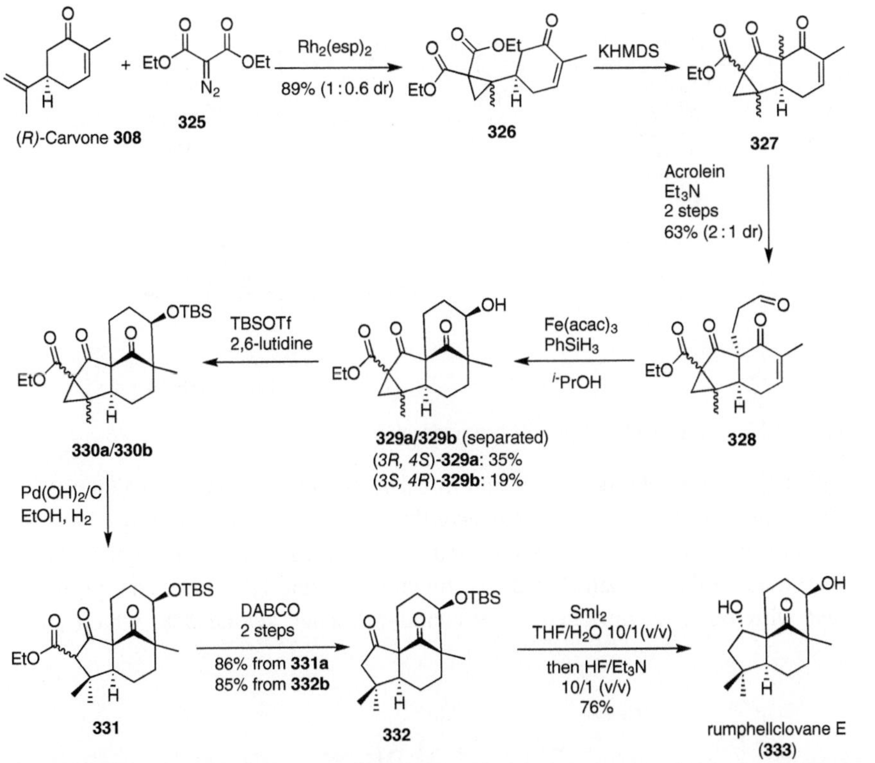

Scheme 12.51 Total synthesis of Rumphellclovane E.

Rumphellclovanes E (**333**) is isolated from the gorgonian coral Rumphella antipathies. It belongs to the family of clovane-type sesquiterpenoids [87]. In 2021, Liu reported the synthesis of Rumphellclovane E (**333**), starting from (*R*)-carvone, as shown in Scheme 12.51 [88]. Rh-catalyzed cyclopropanation of (*R*)-carvone **308** with diazo compound **325** gave the corresponding DA cyclopropane **326**. Intramolecular acylation of **326** in the presence of KHMDS afforded **327**, which on treatment with acrolein and NEt_3 in acetonitrile gave compound **328** as a mixture of inseparable diastereomers (2:1 dr). Treatment of **328** with $Fe(acac)_3$ catalyst, delivered **329a** and **329b** in 35% and 19% isolated yield, respectively. The TBS ether **330a** upon hydrogenolysis, hydrolysis, and decarboxylation produced the compound **332** with an 86% yield over two steps via the intermediate **331**. Likewise, **330b** was also transformed to **332** with an 85% overall yield. **332** on treatment with SmI_2 in THF/H_2O and subsequent treatment with $HF{\cdot}Et_3N$ furnished rumphellclovane E (**333**) in 86% yield.

The chromodorolides are diterpenoid natural products isolated from nudibranchs in the genus *Chromodoris* [89]. In 2018, Overman and co-workers reported the synthesis of Chromodorolide B (**337**), as shown in Scheme 12.51 [90]. The synthesis began with enantioenriched enedione **334**; a five-step sequence of reactions was used to obtain the *trans*-hydrindanone cyclopropane **335**. Employing Simmons–Smith cyclopropanation as the key step, reductive cleavage of the strained cyclopropane **335** was carried out via hydrogenation with Adam's catalyst, followed by PCC oxidation, giving the ketone **336**, which was further transformed to (−)-chromodorolide B (**337**).

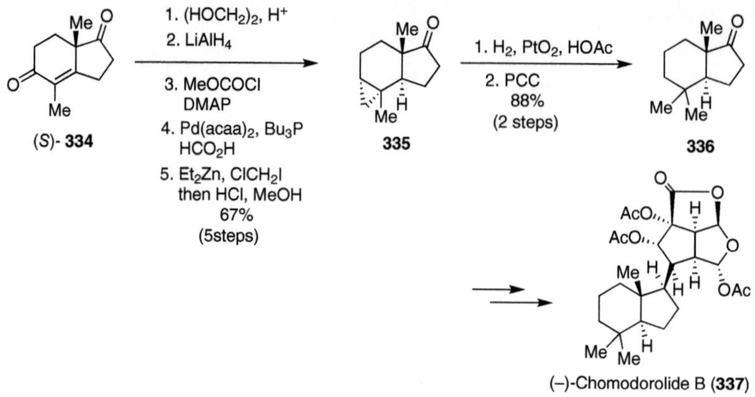

Scheme 12.52 Total synthesis of (−)-Chromodorolide B.

JBIR-03 is an indole diterpene found in *Dichotomomyces cejpii* var. cejpii NBRC 103559, a marine-derived strain of *Aspergillus* oryzae [91] and of *D. cejpii* [92]. In 2018, Kuwahara and co-workers reported the enantioselective total synthesis of JBIR-03 (**343**) and asporyzin C (**342**), as shown in Scheme 12.53 [93]. The synthesis started with optically enriched (+)-Wieland–Miescher ketone **338**, which gave

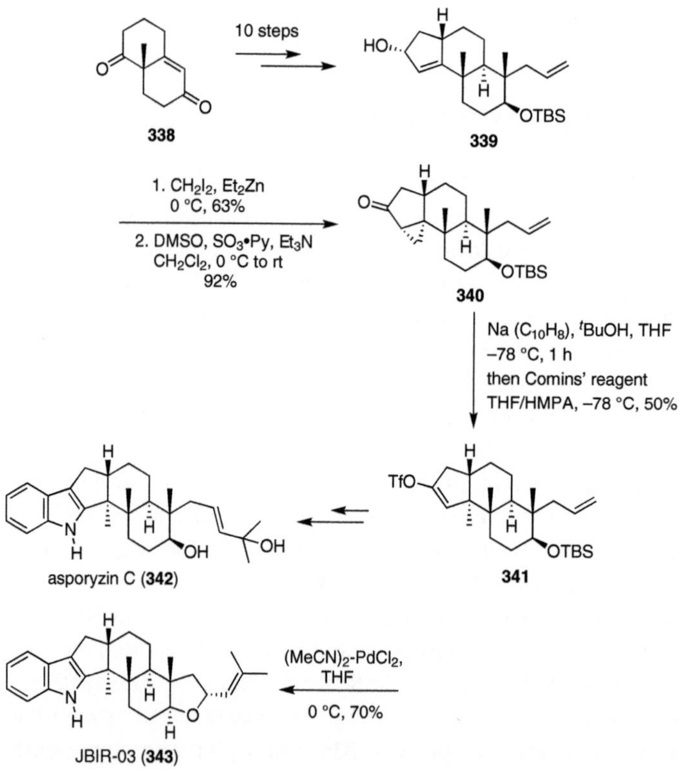

Scheme 12.53 Total synthesis of JBIR-03 and asporyzin C.

access to allylic alcohol **339** after several steps. The hydroxy-directed Simmons–Smith cyclopropanation of **339** resulted in the formation of its corresponding cyclopropyl alcohol, which underwent subsequent Parikh–Doering oxidation to furnish the cyclopropyl ketone **340** in diastereomerically pure form. The reaction of sodium naphthalenide in the presence of ^{t}BuOH acting as a proton source led to reductive cleavage of the cyclopropane ring of **340**, generating an enolate intermediate, which was then trapped with Comin's reagent to accomplish the synthesis of enol triflate **341** with the anti-stereochemistry between the contiguous C3/C4 quaternary stereocenters. The compound **341** acted as an intermediate for the synthesis of asporyzin C (**342**) and JBIR-03 (**343**) in an absolutely diastereoselective manner [94].

The asteraceae plant *Dichrocephala benthamii* is an annual herb found in China and India. Traditionally, in folk medicine, the whole plant is used for the treatment of cold and fever in children, indigestion, hepatitis, and pneumonia [95]. A detailed investigation of *D. benthamii* led to the isolation of two new modhephane sesquiterpenoids analogs dichrocephones A and B (**350–351**) [96]. In 2018, the Christmann group reported an enantioselective synthesis of dichrocephones A and B (**350–351**), as shown in Scheme 12.54 [97]. Alkyne **344** was synthesized from commercially available cyclopentane-1,3-dione. **344** afforded pseudosymmetric dienone **345** after a few steps. The spirocycle **346** was synthesized in 63% yield by the monocyclopropanation of **345** at the sterically less hindered alkene. Hydroxy-directed epoxidation of the

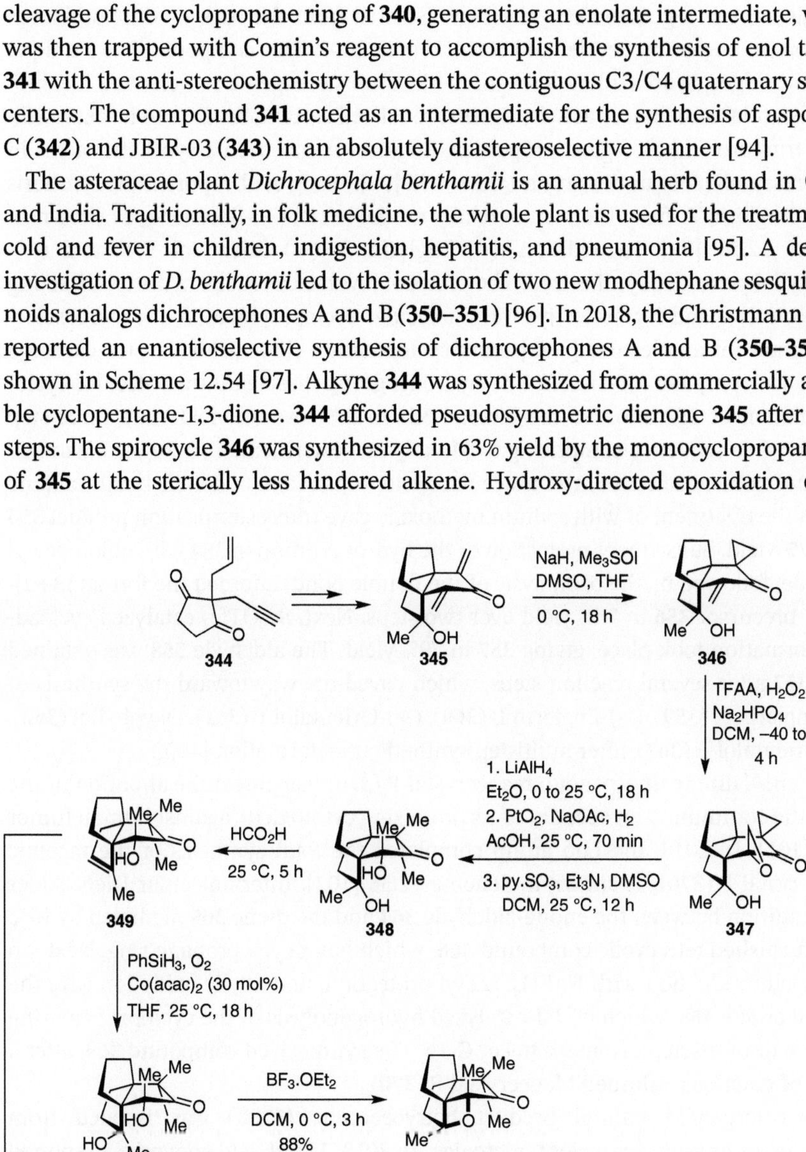

Scheme 12.54 Total synthesis of Dichrocephones A and B.

enone with trifluoroperacetic acid gave access to the crucial bis-spirocyclic intermediate **347**. Next, the simultaneous ketone reduction and epoxide opening using $LiAlH_4$ followed by regioselective hydrogenolytic cyclopropane opening with Adams Q catalyst in acetic acid generated geminal dimethyl group in **348**. **348** on simple synthetic manipulation was converted to Dichrocephones A and B (**350–351**).

Englerin A is a guaiane sesquiterpene isolated from the bark of Euphorbiaceae *Phyllanthus engleri* by Beutler and coworkers in 2009. This compound has a unique oxygen-bridged hydroazulene skeleton with seven stereogenic centers, and it exhibits substantial anti-cancer activity against renal cell carcinoma. Due to its extraordinary structure and biological properties, Englerin A has been a synthetic target in many research laboratories around the world [98]. Wang et al. reported the synthesis of (−)-Englerin A from the commercially available (R)-(+)-limonene, as shown in Scheme 12.55 [99]. In addition, (−)-Englerin B, (+)-Orientalol E/F, and (−)-Oxyphyllol were also synthesized. The protocol involved a crucial hydroxyl-directing stereoselective and regioselective intramolecular cyclopropanation and stereoselective formal intramolecular [3+2] cross cycloaddition ([3+2]-IMCC) of a carbonyl with a cyclopropane 1,1-diester. The hydroxyl group directed intramolecular cyclopropanation of **352** afforded the lactone **353** in 58% overall yield via a three-step isolation-free sequence comprising of Mukaiyama esterification, a Regitz diazotization, and a chemoselective intramolecular cyclopropanation of double bond. Lactone **353** on the treatment of with sodium methoxide gave transesterification product **354** in a 87% yield. Subsequent protection of the hydroxyl group in **354** with chloroacetyl chloride, followed by the ozonolysis of the double bond, afforded the formal [3+2]-IMCC precursor **356** in 75% yield over two steps. Next, $Sc(OTf)_3$-catalyzed cycloadduct formation took place, giving **357** in 79% yield. The aldehyde **358** was obtained from **357** after several reaction steps, which paved the way toward the synthesis of (−)-Englerin A (**359**), (−)-Englerin B (**360**), (+)-Orientalol F (**361**), Oxyphyllol (**362**), and Orientalol E (**363**), after multistep synthetic transformation [100].

The ent-kaurene diterpenoid, maoecrystal P (**370**) has drawn the attention of the scientific community on account of its promising cytotoxicity against human tumor cells [101]. In 2018, the Luo group completed the total synthesis of the racemic maoecrystal P (**370**), as shown in Scheme 12.56 [102]. Intermolecular Diels–Alder cycloaddition between the enone-aldehyde **364** and the diene **365** mediated by $BF_3\cdot OEt_2$ furnished tetracyclic compound **366**, which has a cyclopropane ring. Next, on concurrent reduction with $NaBH_4$, acetyl protection, and allylic oxidation gave the desired enone **368**, which on Pd-catalyzed hydrogenolysis of the cyclopropane ring gave a pair of diastereoisomers **369** at C-16. The synthesized compound **369**, after a series of reactions, afforded Maoecrystal P (**370**).

The triterpenoid natural product botryococcene (**378b**) was isolated from *Botryococcus braunii*, an ancient microalga. In 2018, Marek and co-workers reported an elegant strategy for the synthesis of botryococcene, as shown in Scheme 12.57 [103]. The synthesis involves rhodium-catalyzed (2+1) cycloaddition reaction of alkyne **372** with diazo ethylacetate to generate cyclopropene **373**. The latter underwent a one-pot copper-catalyzed carbometallation, transmetallation, and palladium-catalyzed crosscoupling with alkenyl iodide **374** to produce alkenylcyclopropane ester. It was reduced in situ to alcohol, and subsequent phosphonation afforded the

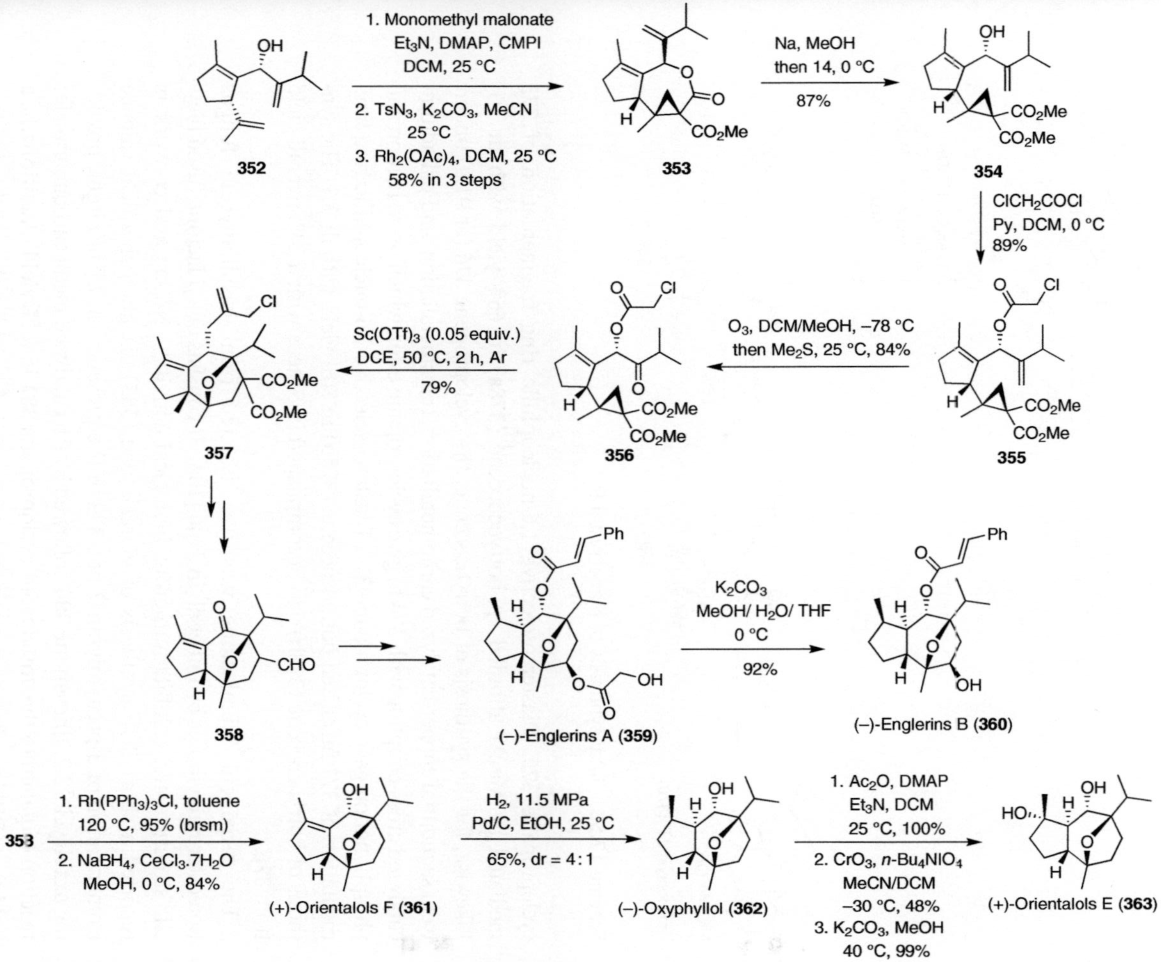

Scheme 12.55 Total synthesis of Englerin A-B, (+)-Orientalol F, (–)-Oxyphyllol, and (+)-Orientalol E.

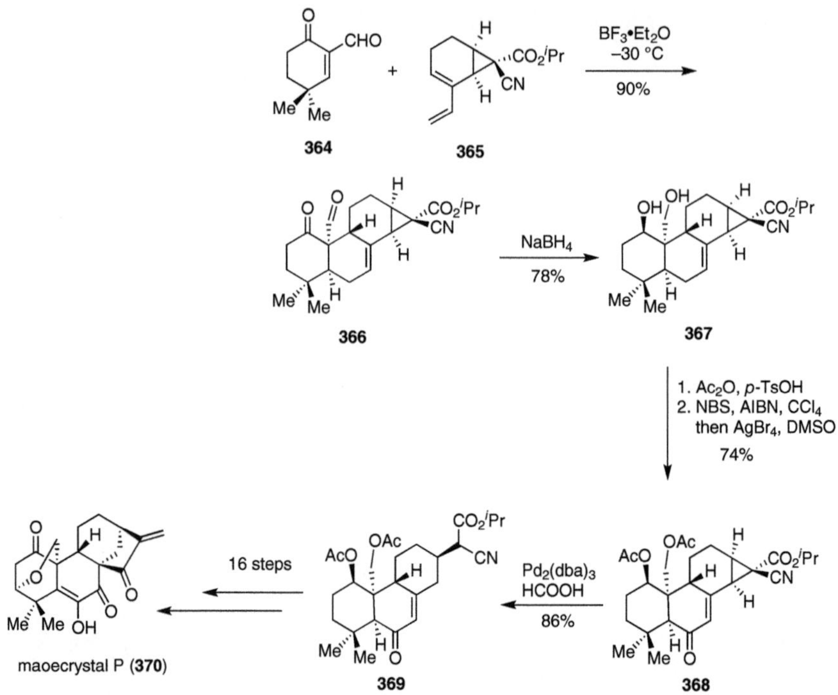

Scheme 12.56 Total synthesis of Maoecrystal P.

cyclopropane **375**. Diastereoselective 1,6-nucleophilic ring fragmentation of **375** mediated by Me_3Al afforded *epi*-botryococcene (**378a**) in 66% yield with dr 3:1. Likewise, for the synthesis of botryococcene, the cyclopropene **376** (er 91:9) underwent sequential copper-catalyzed carbometallation, transmetallation, and palladium-catalyzed cross-coupling with **374** to give cyclopropane ester, which on reduction and phosphonation gave cyclopropane **377**. Diastereoselective 1,6-nucleophilic ring fragmentation of **377** furnished botryococcene (**378b**) in 80% yield with dr 3:1. The synthesis of botryococcene (**378b**) was accomplished in 5 steps with a 24% overall yield from **376**.

The terpenoid Vibralactone was first isolated from basidiomycete fungus Boreostereum vibrans by Liu et al. in 2006 [104]. It consists of a β-lactone fused bicyclic structure and exhibits numerous biological activities. Nelson and co-workers reported a short PGF synthesis of vibralactone (**383**) via photochemical valence isomerization of 3-prenyl-pyran-2-one **379** in the presence of 300 nm light, providing oxabicyclo[2.2.0]hexenone **380**, followed by Rh catalyzed metal carbene transfer reaction to generate the anticipated cyclopropane **381** in 22% yield. Additionally, a mild base DBU-mediated ring expansion occurred, furnishing the single allyl isomer **382** with a 69% yield. The latter, on DIBAL-H reduction, furnished racemic vibralactone (**383**). This four-step synthesis was accomplished with a 4.3% overall yield (Scheme 12.58) [105].

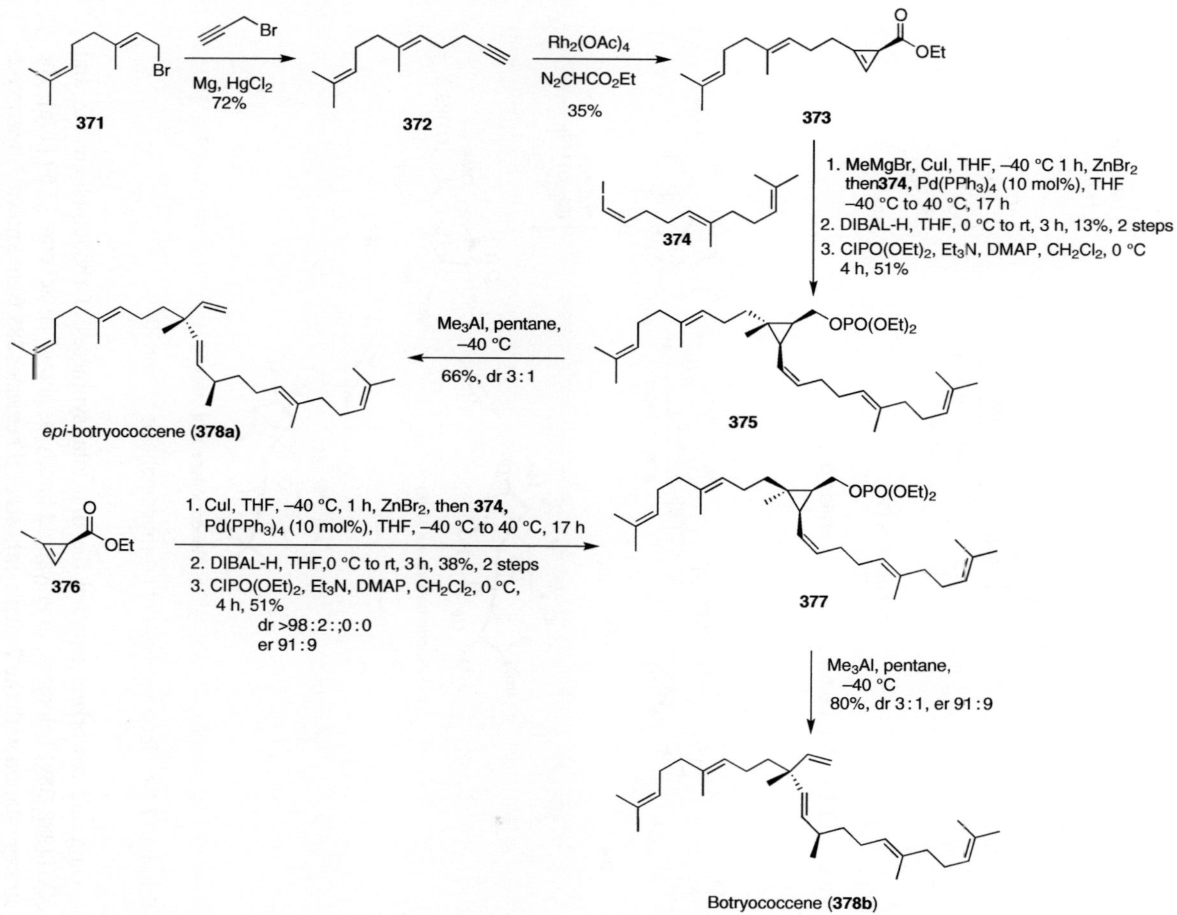

Scheme 12.57 Total synthesis of Botryococcene.

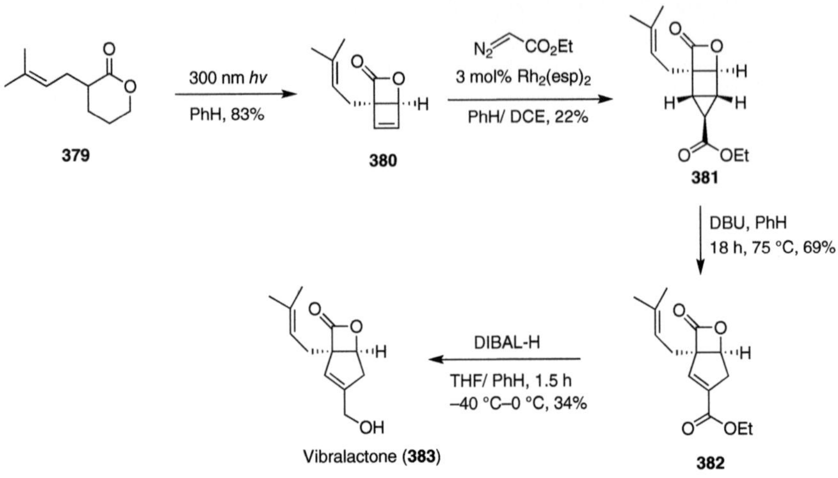

Scheme 12.58 Total synthesis of (±) Vibralactone.

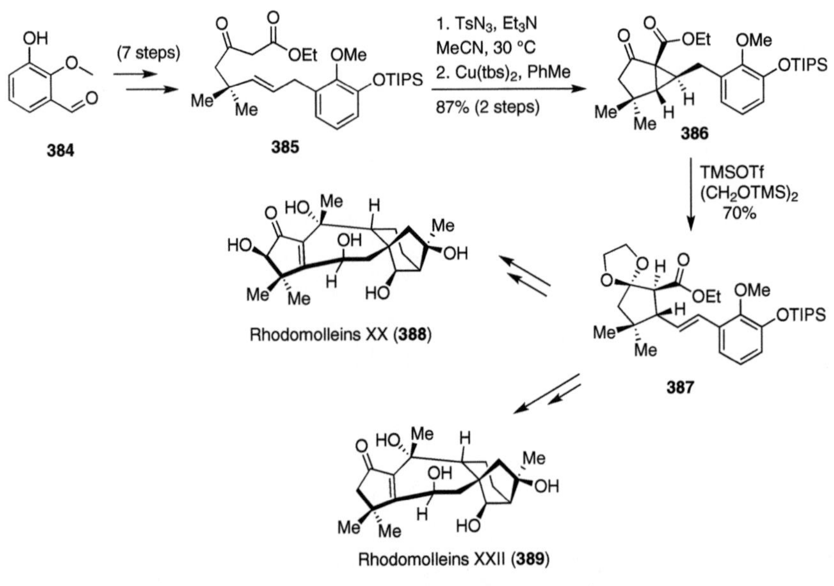

Scheme 12.59 Total synthesis of (−)-rhodomolleins XX and XXII.

Ding and coworkers reported the first total syntheses of rhodomolleins XX and XXII (**388–389**) belonging to grayanoid family, as shown in Scheme 12.59 [106]. The strategy started with the 7-step synthesis of β-ketoester **385** from known 3-hydroxy-2-methoxybenzaldehyde **384**. The compound **385** on Regitz diazo transfer delivered an a-diazo-β-ketoester, followed by the treatment with Cu(II) salen complex, which generated the cyclopropane **386**. Treatment of **386** with TMSOTf and $(TMSOCH_2)_2$ resulted in the protection of the C-2 ketone as well as simultaneous ring opening of the cyclopropane furnishing the styrene **387**, which could be converted to (−)-rhodomollein XX (**388**) and (−)-rhodomollein XXII (**389**) after multistep conversation.

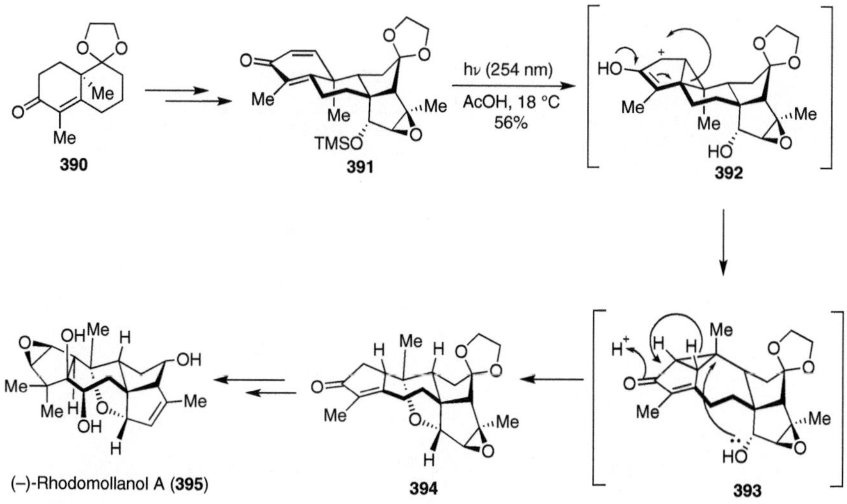

Scheme 12.60 Total synthesis of (−)-Rhodomollanol A.

(−)-Rhodomollanol A (**395**) is an oxygenated grayanane-type diterpenoid with an uncommon [3,5,7,5,5,5] hexacyclic framework and displays a moderate PTP1B inhibitory activity. Ding and co-workers recently reported the first total synthesis of (−)-rhodomollanol A, characterized by a series of synchronized domino reactions and skeletal rearrangements, as shown in Scheme 12.60 [106, 107]. The divinyl ketone **391**, obtained from bicyclic enone **390**, was subjected to Photo-Nazarov cyclization to generate allylic carbocation **392**, which triggered the ring expansion of the central cyclohexane ring to afford **393** having a cyclopropane ring via 1,4-alkyl migration. Under acidic conditions (AcOH), the trimethylsilyl group was removed, and intramolecular ring-opening of the cyclopropane by the tethered hydroxyl group afforded **394** in a 56% yield. The latter was transformed to (−)-rhodomollanol A (**395**) in 12 steps.

12.4 Synthesis of Miscellaneous Natural Products

Apart from alkaloids and terpenes/terpenoids, the total synthesis of many other natural products involved DA-cyclopropane synthesis followed by its synthetic transformations. This part of the chapter highlights the syntheses of different other natural products where a few key steps involve the DA-cyclopropane reagent.

(−)-Doliculide (**401**) is a unique 16-membered depsipeptide obtained from the Japanese sea hare *Dolabella auricularia* (Aplysiidae) [108]. Doliculide (**401**) is known to show potent cytotoxicity against HeLa-S3 cells. In 2001, Ghosh et al. reported the synthesis of (−)-Doliculide (**401**), as shown in Scheme 12.61 [109]. The synthesis began with the Charette asymmetric cyclopropanation of allyl alcohol **396** using the amphoteric chiral dioxaborolane ligand and $Zn(CH_2I)_2$•DME complex to furnish the cyclopropane derivative **397** in quantitative yield and with high diastereoselectivity (91% de). Selective ring opening of the cyclopropane **397** with PPh_3,

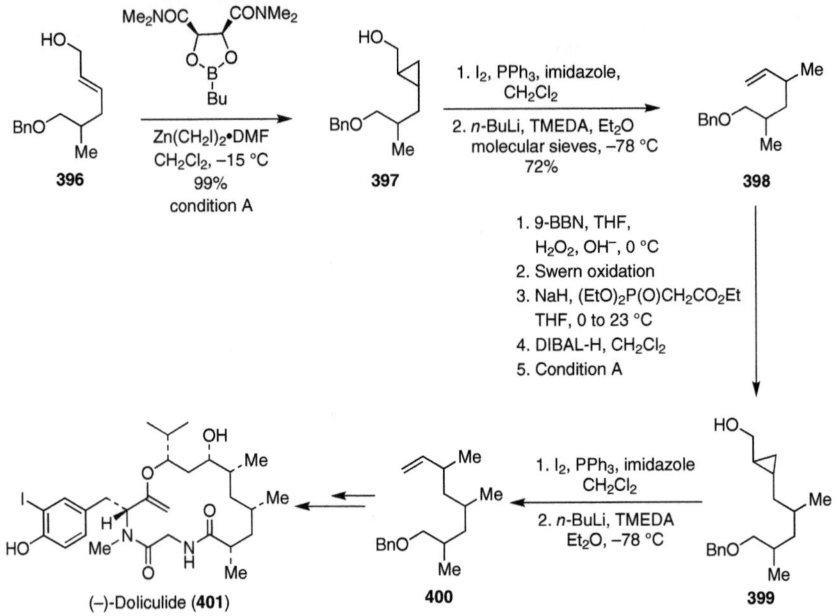

Scheme 12.61 Total synthesis of (–)-Doliculide.

iodine, and imidazole provided the iodide, followed by the treatment with *n*-BuLi in the presence of TMEDA and molecular sieves, which afforded the alkene **398** in 72% yield over two steps [110]. Hydroboration of alkene **398** with 9-BBN, followed by Swern oxidation, delivered the analogous aldehyde. Next, Horner–Emmons homologation of the aldehyde, DIBAL-H reduction, followed by Charette asymmetric cyclopropanation of the subsequent allylic alcohol, furnished the cyclopropane **399** in 55% overall yield starting from **398**. Cyclopropane derivative **399** was transformed to olefin **400** in 72% yield. This olefin **400** was transformed to (−)-Doliculide (**401**) after several steps.

(−)-Roccellaric acid (**404**) and (−)-Nephrosteranic acid (**406**) are trisubstituted γ-butyrolactones belonging to the family of paraconic acids, a class of γ-butyrolactone natural products isolated from various species of lichens, mosses, and fungus. They possess antifungal, antibiotic, antitumor, and antibacterial properties [111]. Because of their pharmacological profile, they have attracted considerable interest from organic chemists. The total synthesis of four paraconic acids, i.e. (−)-roccellaric acid (**404**), (−)-protolichesterinic acid (**405**), (−)-nephrosteranic acid (**406**), and (−)-protopraesorediosic acid (**407**), was carried out by Reiser et al. by taking the advantage of a retro-aldol/lactonization sequence as vital steps (Scheme 12.62) [112]. Enantioselective cyclopropanation of methyl ester of furan-2-carboxylic acid (**199**) gave the DA cyclopropane **200a**, which underwent furan ring cleavage by ozonolysis to furnish the cyclopropane **201**. The strategic intermediate **201** was subjected to allylation in the presence of BF_3•OEt_2 followed by the reaction with base affording the lactone **402** via a retro-aldol/lactonization sequence. The lactone **403** was used in a divergent approach to synthesize various paraconic acids (**404–407**).

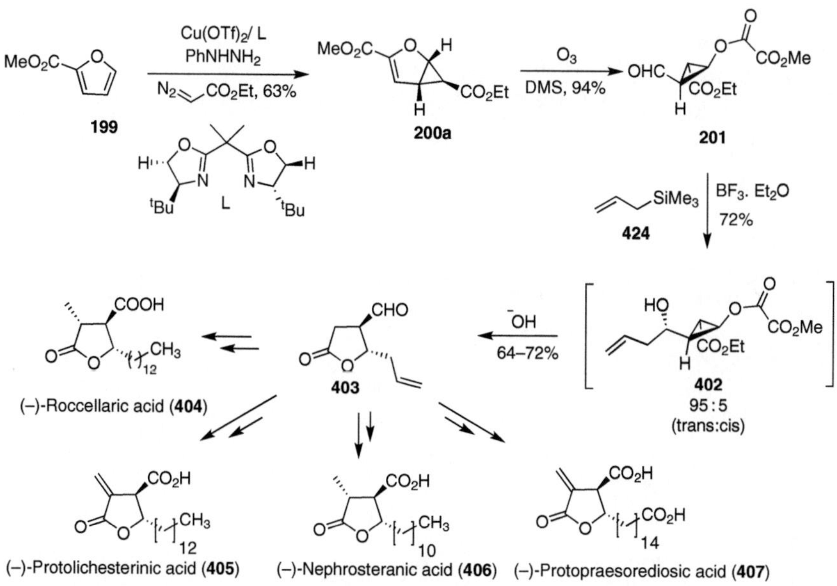

Scheme 12.62 Total synthesis of Paraconic Acids.

Borschberg reported the discovery of β-araneosene, a metabolite of the mold *Sordaria araneosa* [113]. Corey and co-workers reported the synthesis of isoedunol (**411**) and β-araneosene (**412**), as shown in Scheme 12.63 [114]. Treatment of **408** with EtMgBr, ClTi(O^{i}Pr)$_3$ furnished cyclopropyl carbinol **409**, which when reacted with Me$_3$Al furnished cyclobutanone **410** in 90% yield. **410** was converted to isoedunol (**411**) after several steps. The reaction of chloromethyl methylether in the presence of ^{i}Pr$_2$NEt, TBAI gave methoxymethyl (MOM) ether of isoedunol, which on dissolving metal reduction with excess Li in NH$_3$ afforded β-araneosene (**412**).

Apart from the terpenoids, the Reiser group also synthesized (*S*)-vigabatrin (**418**), starting from the inexpensive pyrrole moiety **413** [115]. (*S*)-Vigabatrin is an irreversible inhibitor of GABA-T and is being used for adjunctive therapy in patients suffering from epilepsy. First, 3,4-didehydropyrohomoglutamate **417** was successfully synthesized with 91% *ee* from pyrrole moiety **413** via cyclopropanation followed by radical ring opening, and then **417** was further converted to (*S*)-vigabatrin (**418**) following several additional steps (Scheme 12.64).

The Inaba group describes the synthesis of protein kinase C-β (PKCβ)-selective inhibitor JTT-010 (**423**) [116]. Synthesis of five-membered heteroaromatic fused pyrrolidine **421** via ring opening of cyclopropane **420** with five-membered heteroaromatic **419** with a leaving group in C(2) position served as the foundation for the synthesis. Compound **421** was synthesized with a 76% yield in optically pure form, and it had all the necessary components, including the absolute configuration needed to be the precursor for JTT-010 (**423**). Compound **422** was also found to be optically active, indicating that ring opening of cyclopropane and consequent decarboxylation proceeded with retention of configuration. Further JTT-010 (**423**) was synthesized from **422** following some sequential additional steps (Scheme 12.65).

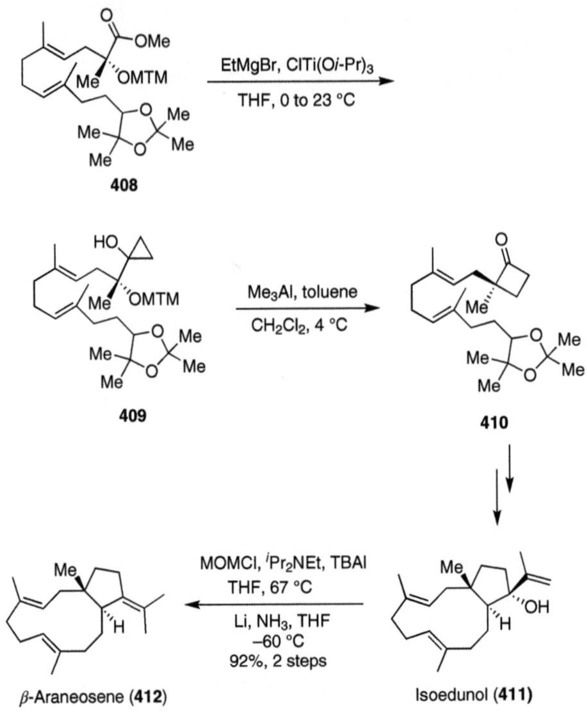

Scheme 12.63 Total synthesis of Isoedunol and β-Araneosene.

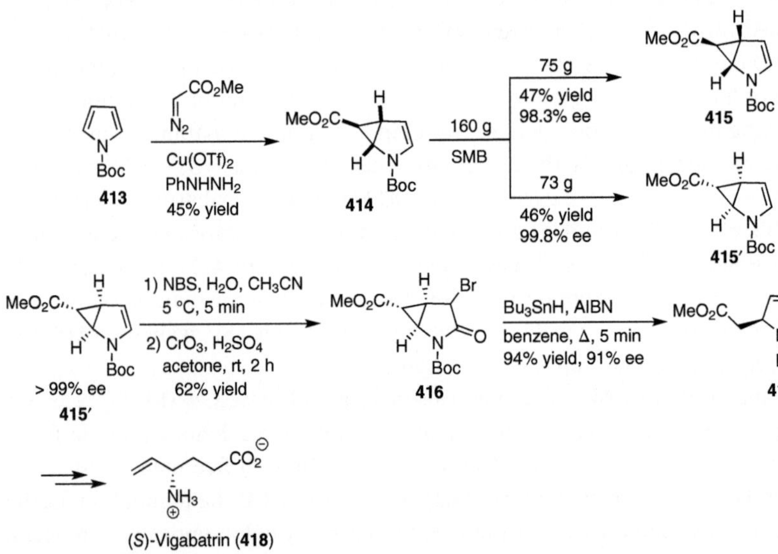

Scheme 12.64 Total synthesis of (*S*)-Vigabatrin.

The Johnson group utilized their developed [3+2] cyclopropane-aldehyde cycloaddition strategy for the synthesis of natural product (+)-virgatusin (**431**), which is identified as a potent antibacterial and antifungal agent [117]. The synthesis commenced with the preparation of enantioenriched cyclopropane **427**, which

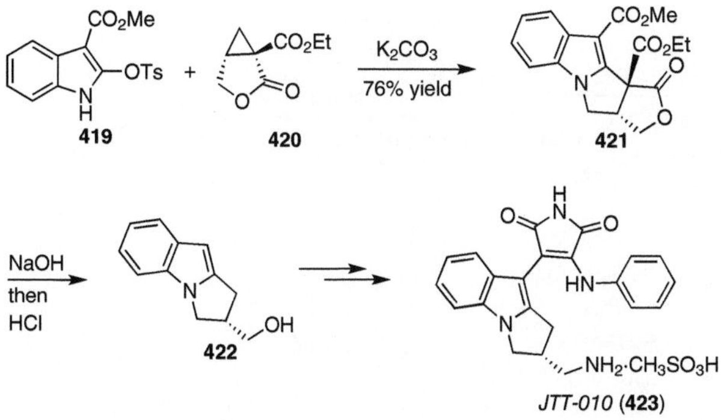

Scheme 12.65 Total synthesis of JTT-010.

underwent $AlCl_3$-catalyzed formal [3+2] cycloaddition with piperonal **428** to afford tetrahydrofuran moiety **429** as a single diastereomer. Compound **429** was further converted to (+)-virgatusin core **430** with >20:1 diastereoselectivity via benzyl ester hydrogenolysis followed by decarboxylation. Finally, **430** upon $LiAlH_4$ reduction and methylation afforded the targeted natural product (+)-virgatusin (**431**) (Scheme 12.66).

The Johnson group also accomplished the total synthesis of (+)-Polyanthellin A (**437**), a reported antimalarial agent [118]. The synthesis relied on the $MADNTf_2$-catalyzed [3+2] cycloaddition of DA-cyclopropane **434** and β-silyloxy aldehyde **435** to afford **436** in 76% yield with good diastereoselectivity. (+)-Polyanthellin A (**437**) was synthesized from **436** following some additional steps (Scheme 12.67).

The Wang group also completed the total synthesis of the natural product (±)-bruguierol A (**443**) from commercially available 3-bromoanisole **438** [119]. (±)-Bruguierol A (**443**) was isolated from the stem of *Bruguiera gymnorrhiza* tree and exhibits potent biological activity. The key strategy for the synthesis was $Sc(OTf)_3$-catalyzed intramolecular [3+2] cycloaddition of cyclopropane **441** to construct 8-oxabicyclo[3.2.1]octane skeleton **442** (Scheme 12.68). The total synthesis has been completed in 10 steps from 3-bromoanisole **438** with a 16.8% overall yield.

The Kerr group also synthesized (+)-isatisine A (**447**), a natural product present in *Isatis indigotica* Fort plant species [120]. The first step of the synthesis followed Johnson tetrahydrofuran synthesis from (*S*)-vinylcyclopropane diester **445** and indole-2-carboxaldehyde **444** in the presence of $Sn(OTf)_2$ catalyst to afford **446** in 89% yield as a mixture of diastereomers. The diastereomeric ratio between 2,5-*cis*:2,5-*trans* (furan numbering) isomers of **446** was found to be 11:1 (Scheme 12.69). The total synthesis was completed in 14 steps with a 5.8% overall yield.

The Nakada group reported an enantioselective approach to the natural product (−)-platencin (**453**), which contains a tricyclic framework of only carbon atoms [121]. It was isolated from *Streptomyces platensis* MA 7339 and acts as a moderate inhibitor of both FabF and FabH, an enzyme required for catalyzing the initial condensation in the synthesis of bacterial fatty acid. The reaction proceeded by preparation of tricycle[4.4.0.0]decene derivative **450** via enantioselective catalytic asymmetric intramolecular cyclopropanation (CAIMCP) followed by conversion of **450** to chiral intermediate **451** containing α,β-unsaturated sulfone functionality, which can act as

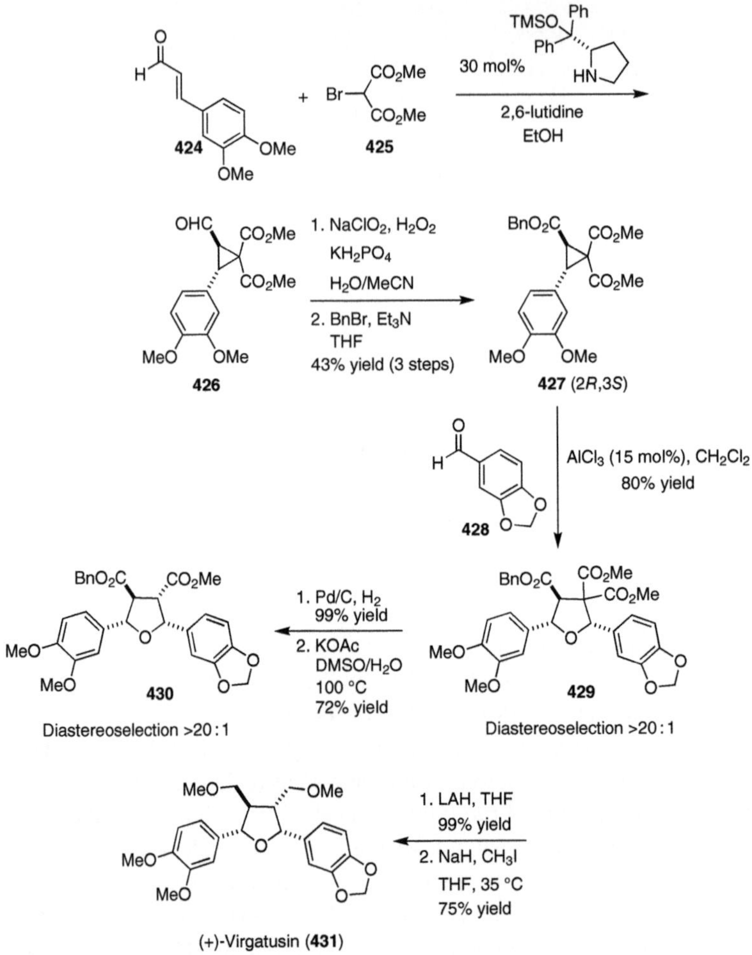

Scheme 12.66 Total synthesis of (+)-Virgatusin.

a potential radical acceptor. **451** could be successfully transformed in four steps to Nicolaou's intermediate [122] **452** for his total synthesis of (−)-platencin (**453**) (Scheme 12.70).

The Gharpure group reported the enantioselective and diastereoselective total synthesis of *cis* isomers of (+)-Hagen's gland lactones (**460**) in 2011 [123]. Hagen's glands are located near the abdominal tips of braconid wasps, *Diachasmimorpha lingicaudata* (Ashmead), *Diachasmimorpha tryoni* (Cameron), and *Fopius arisanus* (Bioesters), and contain fragrance-rich lactones [124, 125]. The total synthesis proceeded by intramolecular cyclopropanation of vinylogous carbonates **455a–b** in a regio- and stereoselective manner in presence of $Cu(acac)_2$ to furnish the DA-substituted cyclopropanes **456a–b** and then regioselective radical ring opening of cyclopropanes **458a–b** using nBu$_3$SnH and AIBN (Scheme 12.71). The total synthesis was completed in 12 steps from readily available aldehydes **454a–b**.

The Snapper group reported the total synthesis of marine natural product norrisolide (**467**), which was first isolated from the nudibranch mollusc *Chromodoris*

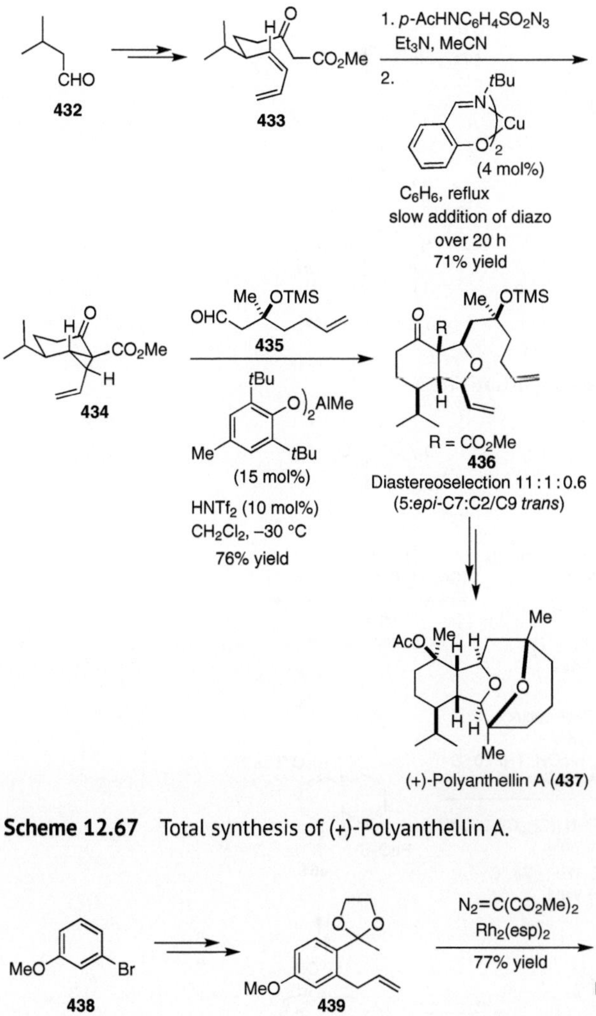

Scheme 12.67 Total synthesis of (+)-Polyanthellin A.

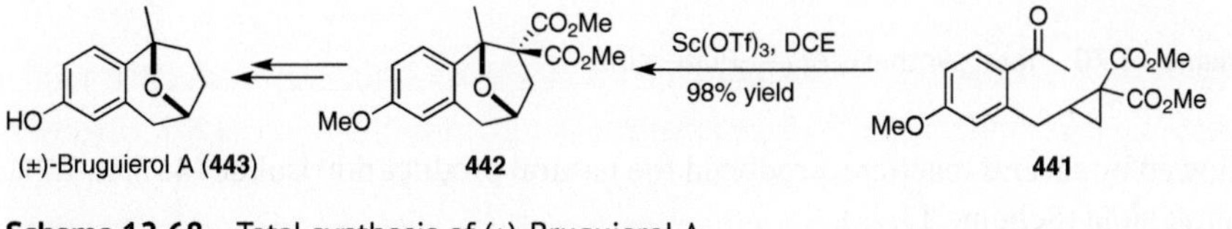

Scheme 12.68 Total synthesis of (±)-Bruguierol A.

norrisi [126]. It has also been found in sponges from the same area in minor amounts. They divided the synthesis into two parts. The first part involved the synthesis of hydrindane portion **465** of the molecule, and the second part involved the preparation of side chain **464**. The enantioselective cyclopropanation of furan-2-one **461** synthesized **462**, followed by rearrangement of **462** generated **463**, which furnished side chain **464** in four sequential steps. The coupling of two parts **464** and **465**

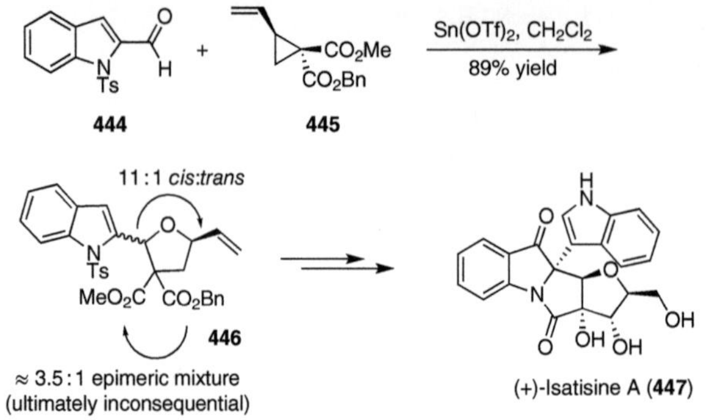

Scheme 12.69 Total synthesis of (+)-Isatisine A.

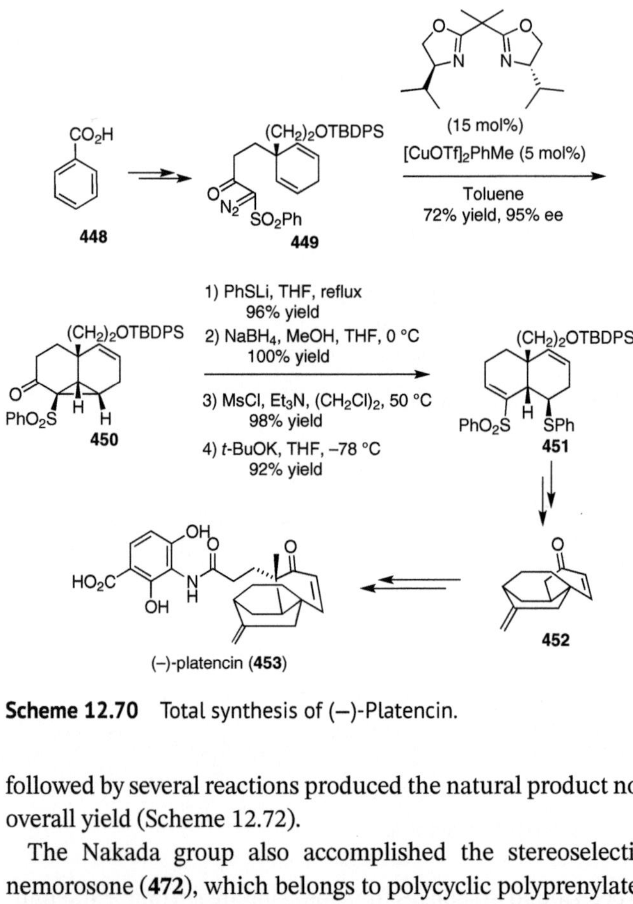

Scheme 12.70 Total synthesis of (−)-Platencin.

followed by several reactions produced the natural product norrisolide (**467**) in 1.7% overall yield (Scheme 12.72).

The Nakada group also accomplished the stereoselective total synthesis of nemorosone (**472**), which belongs to polycyclic polyprenylated acylphloroglucinols (PPAPs) and exhibits anti-HIV and antitumor activities [127]. In the way, of this natural product synthesis, the authors developed a new approach for the synthesis of bicyclo[3.3.1]nonane-2,4,9-trione core via intramolecular cyclopropanation of **469** to generate **470**, which underwent stereoselective alkylation followed by regioselective ring opening of cyclopropane to form diketone **471** as a single isomer. **471** produced the targeted natural product nemorosone (**472**) following some additional steps (Scheme 12.73).

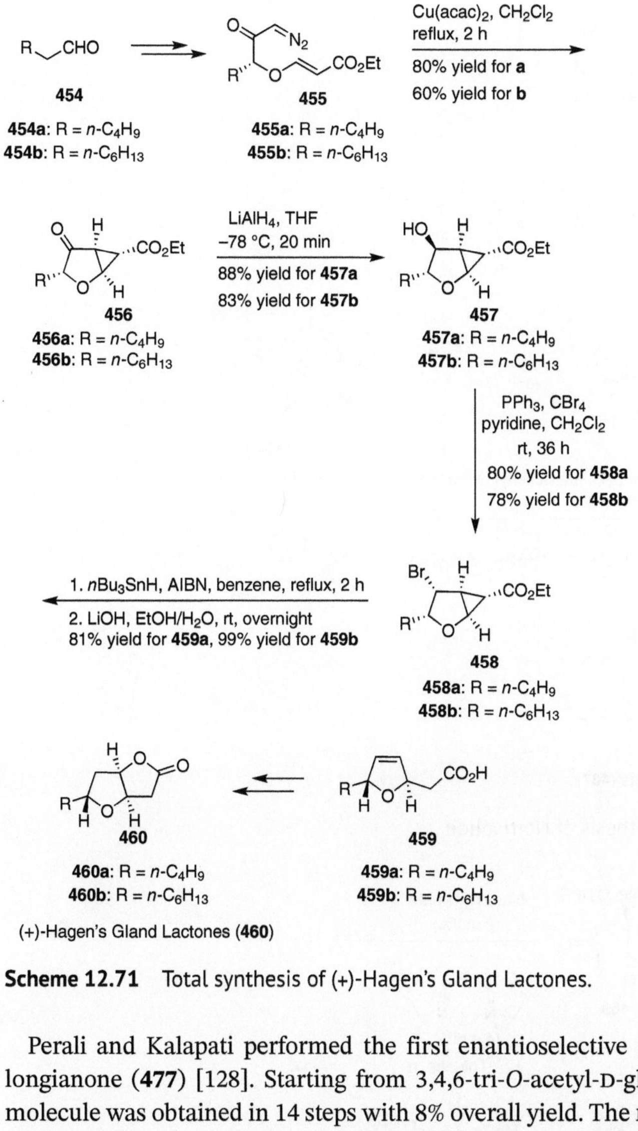

(+)-Hagen's Gland Lactones (**460**)

Scheme 12.71 Total synthesis of (+)-Hagen's Gland Lactones.

Perali and Kalapati performed the first enantioselective total synthesis of (*S*)-longianone (**477**) [128]. Starting from 3,4,6-tri-*O*-acetyl-D-glucal **473** the targeted molecule was obtained in 14 steps with 8% overall yield. The reaction proceeded via Rh-catalyzed stereoselective cyclopropanation of glycal **474** to afford enantiomerically pure 1,2-cyclopropanecarboxylate **475**, which upon treatment with *N*-bromosuccinimide resulted in bromonium ion-induced electrophilic ring opening and produced bromohydrin **476** (Scheme 12.74). Other notable transformations in this total synthesis included one-pot dehydrohalogenation, stereoselective intramolecular hetero-Michael addition, and ester hydrolysis to furnish the final product.

In 2013, the Kerr group reported that cyclopropane hemimalonates could be readily converted to γ-substituted butanolides in the presence of inorganic salts under microwave irradiations with retention of stereochemistry [129]. This reaction was successfully applied to the synthesis of a small natural product (*R*)-(+)-dodecan-4-olide (**482**). (*R*)-(+)-Dodecan-4-olide (**482**) is a naturally occurring butanolide, and it was isolated from various natural sources like the pyridial glands of rove beetles,

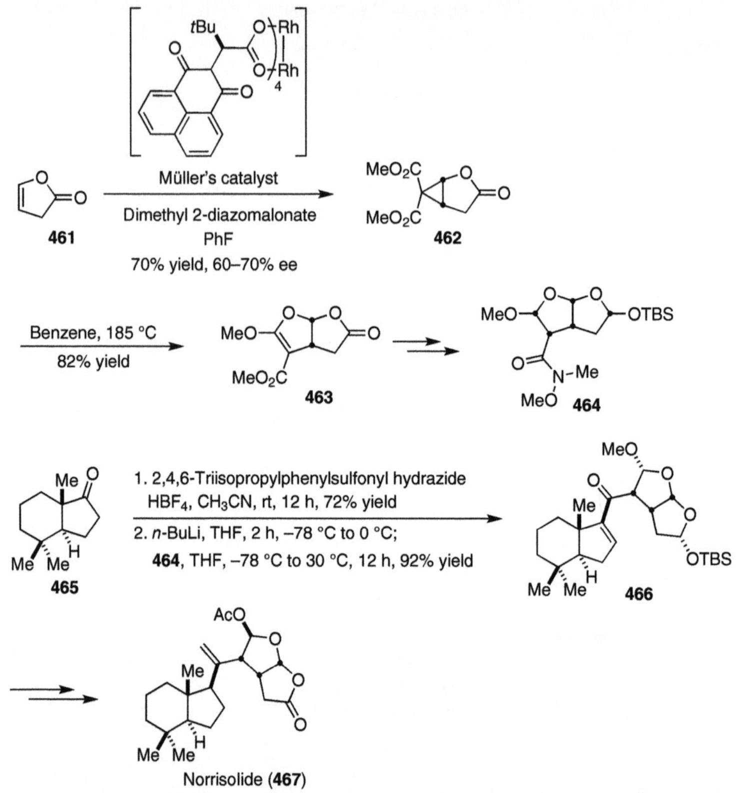

Scheme 12.72 Total synthesis of Norrisolide.

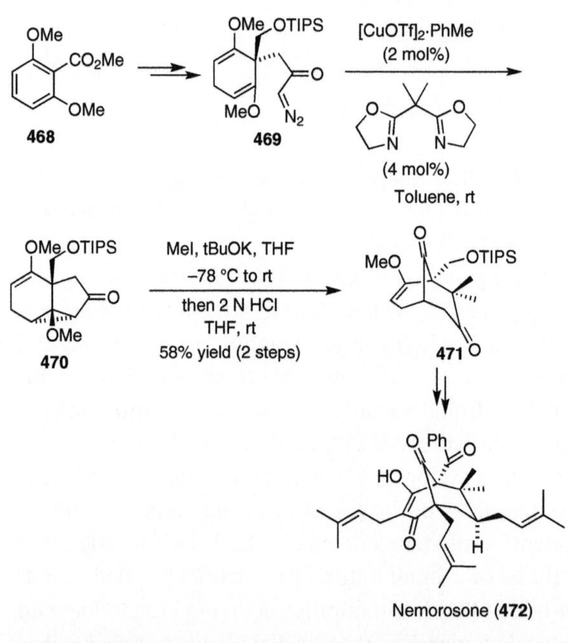

Scheme 12.73 Total synthesis of Nemorosone.

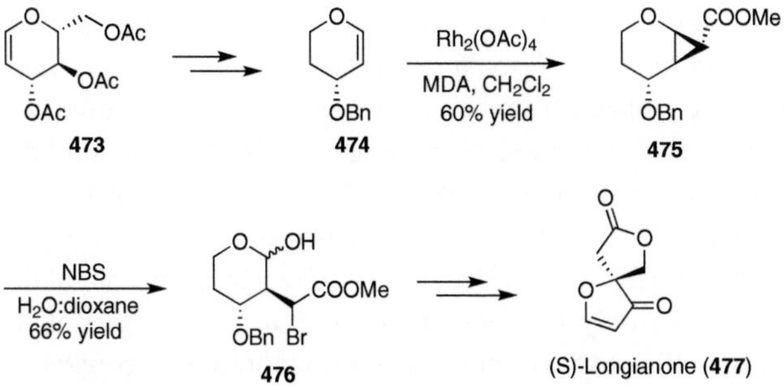

Scheme 12.74 Total synthesis of (S)-Longianone.

butterfat, fruits, and territorial marking fluid of the Bengal tiger. Grubbs second-generation ruthenium catalyst mediated the cross-metathesis of dimethyl ester vinyl cyclopropane **478** and oct-1-ene **479**, followed by monosaponification generating cyclopropane hemimalonate **480**, which under standard butanolide synthesis conditions produced alkenyl butanolide **481**. Finally, reduction of **481** with tosylhydrazide afforded (R)-(+)-Dodecan-4-olide (**482**) in 67% overall yield with 94% ee (Scheme 12.75).

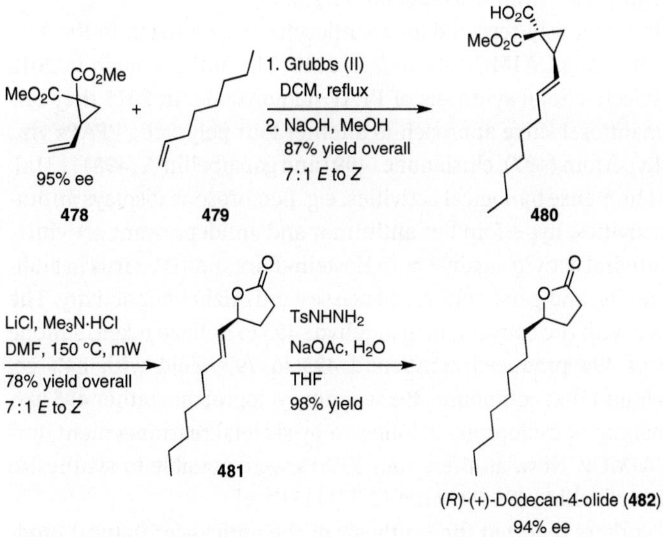

Scheme 12.75 Total synthesis of (R)-(+)-dodecan-4-olide.

In 2014, Gharpure et al. accomplished the enantiospecific collective total synthesis of butanolide and butenolide natural products utilizing oxygen-substituted DA cyclopropane as a common precursor [130]. The authors performed the first total synthesis of (+)-juruenolide D (**487b**), butanolide (**489a-b**), butenolide (**490**), (+)-hydroxyancepenolide (**492a**), (+)-hydroxyancepenolide acetate (**492b**), and also did the total synthesis of (+)-juruenolide C (**487a**), (+)-blastmycinone (**488a**), (+)-antimycinone (**488b**), and (+)-ancepsenolide (**491**). Among these natural products,

(+)-juruenolide C and D (**478a–478b**) were isolated from *Virola surinamensis,* and butanolide (**489a**) was found in *Iryanthera* sp. (+)-Blastmycinone (**488a**) and (+)-antimycinone (**488b**) were isolated from *Streptomyces* sp. and were found to possess antitumor and antifungal activities. Marine natural product butanolide (**489b**) and its dehydration product butenolide (**490**), (+)-ancepsenolide (**491**), and (+)-hydroxyancepenolide (**492**), were isolated from Caribbean gorgonian *Pterogorgia anceps* and displays the defensive mechanism of the species.(+)-Hydroxyancepenolide acetate (**492b**) was isolated from only *Pterogorgia citrima* but (+)-ancepsenolide (**491**) and (+)-hydroxyancepenolide (**492a**) were isolated from both *Pterogorgia citrima* and *Pterogorgia guadalupensis.* (+)-Ancepsenolide (**491**) exhibits pesticidal activities as well as antitumoral, immunosuppressive, and antimalarial properties. The synthesis commenced with the conversion of DA cyclopropane **483** to lactol **486**, which acted as the common intermediate for the synthesis of all butanolide and butenolide-based natural products (**487–492**). $LiAlH_4$-induced chemo- and stereoselective reduction of ketone moiety in **483** generated alcohol **484**, which upon treatment with cat. amount H_2SO_4 in MeOH furnished lactone **485** via tandem regioselective cyclopropane ring-opening/trapping of oxonium ion by MeOH/intramolecular lactonization sequence. Lactone **485** underwent reduction with DIBAL-H to afford the desired lactol **486**. All the aforementioned natural products were successfully synthesized from **486** (Scheme 12.76).

The Nakada group previously reported an enantioselective approach to the synthesis of (−)-platencin using CAIMCP strategy in 2010. The authors again in 2012 reported the stereoselective total synthesis of PPAP nemorosone. In 2015 they further proposed the enantioselective approach to another four polycyclic PPAPs viz., nemorosone (**496**), hyperfoin (**497**), clusianone (**499**) and garsubellin A (**498**) [131a]. These PPAPs exhibit immense biological activities, e.g. nemorosone displays antitumor and anti-HIV activities, hyperfoin has antitumor and antidepressant activities, clusianone shows antiviral activity against both Epstein–Barr and HIV virus in addition to antitumor activity, and garsubellin A possesses anti-Alzheimer activity. The synthesis commenced with the conversion of aldehyde **493** to α-diazo β-keto sulfone **494**, and CAIMCP of **494** produced compound **495** in 79% yield with 84% ee. Interestingly, it was found that compound **495** was not cyclopropane rather **495** was obtained via ring opening of cyclopropane followed by skeletal rearrangement during the course of CAIMCP. Now, all these four PPAPs were possible to synthesize from the common intermediate **495** (Scheme 12.77) [131b, e].

Magauer and co-workers reported the synthesis of the anticancer natural product chartarin (**506**) from commercially available indanones, as shown in Scheme 12.78. The protocol proceeded through a two-step process involving the thermally induced fragmentation of a cyclopropane indanone with a simultaneous 1,2-chloride shift [132]. The synthesis began with the coupling of Indanon **500** obtained from commercially available 7-methoxy-1-indanone with para-quinone **501** in the presence of catalytic amounts of hydroquinidine and subsequent decarboxylation of the *tert*butyl ester in the presence of trifluoroacetic acid to afford the 2-arylated indanone **502** in good overall yield on gram scale [133]. Next, **502** was oxidized employing StahlÏs palladium-catalyzed aerobic dehydrogenation conditions [134] and subsequent treatment with KHMDS in the presence of 18-crown-6

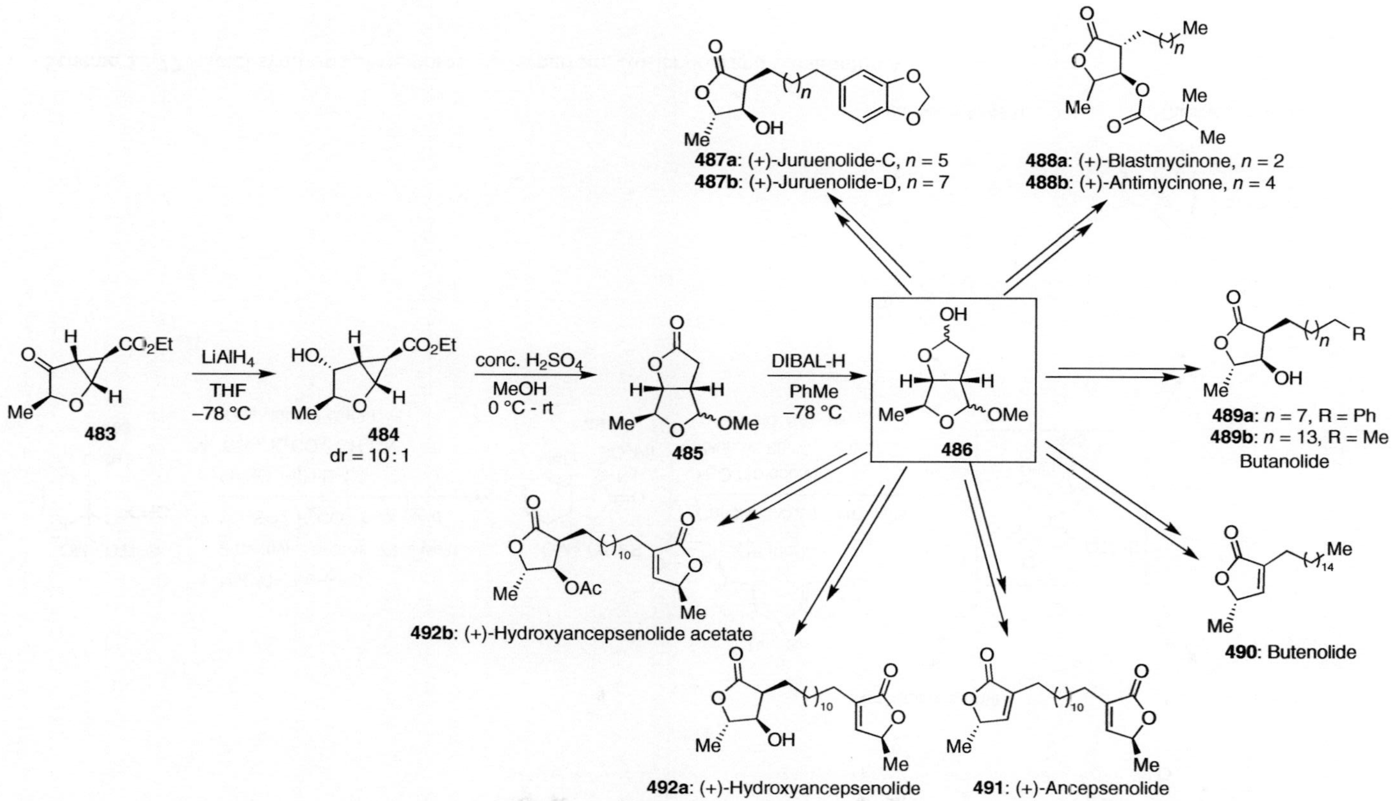

Scheme 12.76 Total synthesis of Butanolide and Butenolide natural products.

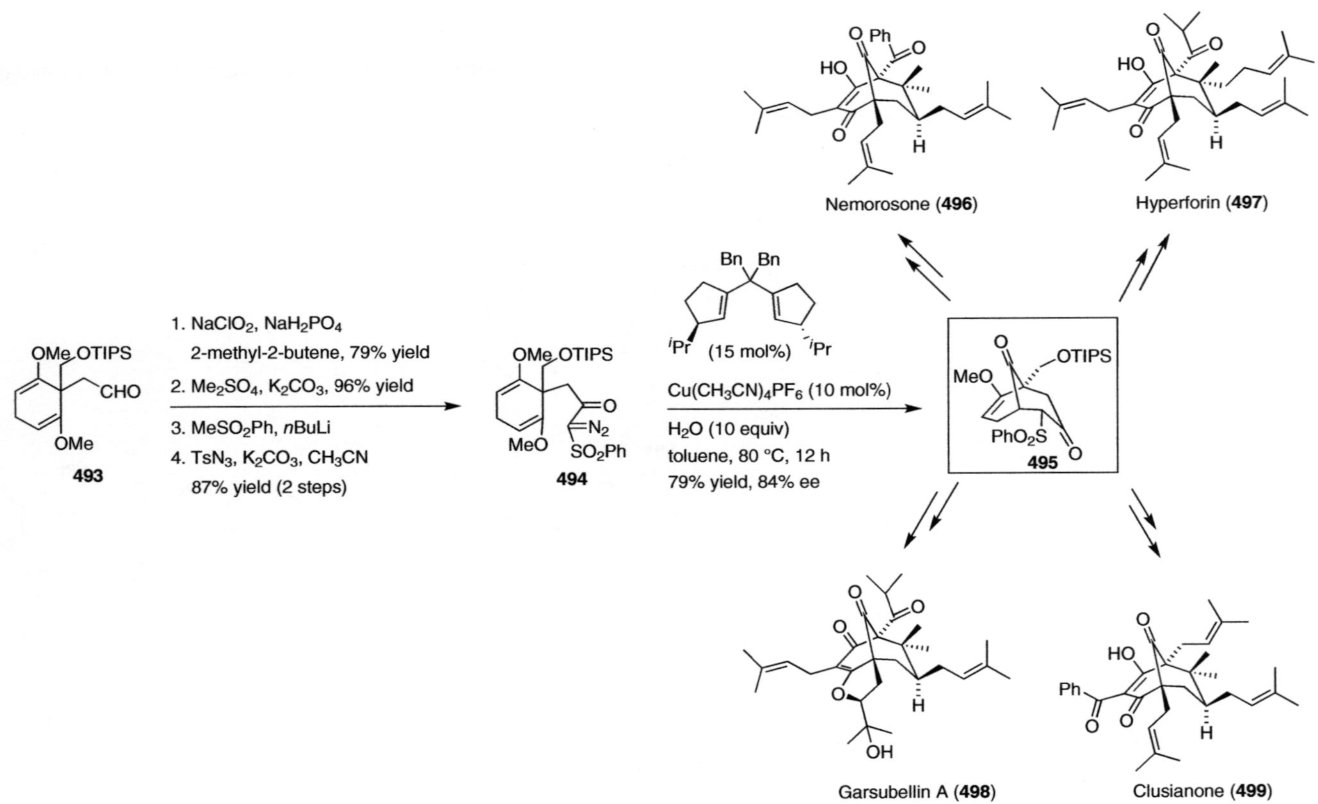

Scheme 12.77 Total synthesis of Nemorosone, Hyperfoin, Clusianone, and Garsubellin A.

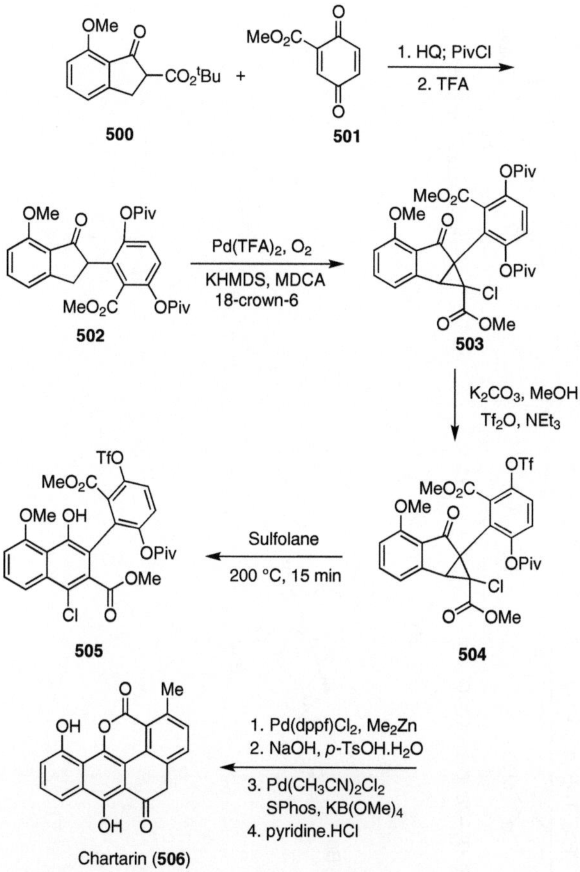

Scheme 12.78 Total synthesis of chartarin.

to give the cyclopropane **503** in 75% yield. The sequential reaction of **503** with methanolic potassium carbonate and then triflic anhydride provided cyclopropane **504**. The ring-opening reaction of the cyclopropane **504** was carried out by heating its solution in sulfolane to generate the biaryl intermediate **505** in 75% yield. Next **505** on heating with dimethyl zinc in the presence of Pd(dppf)Cl_2 and subsequent treatment with NaOH/*p*-TsOH·H_2O, followed by the treatment with potassium tetramethoxyborate, bis(acetonitrile)dichloropalladium(II), SPhos, pyridine hydrochloride gave chartarin (**506**).

Piperarborenine B (**512**) is a cyclobutane-containing natural product isolated from *Piper arborescens* [135]. It exhibits activity against P-388, A-549, and HT-29 cancer cell lines. Fox and co-workers reported a highly efficient, enantioselective total synthesis (8% overall yield) of the natural product, piperarborenine B (**512**), as shown in Scheme 12.79 [136]. The strategy involves the synthesis of diazoester **508** from commercially available veratraldehyde **507** using a Tsuji–Trost allylation, followed by treatment with *p*-ABSA and NaOH in a mixture of acetonitrile/H_2O in 47% yield over three steps. The diazoester **508** on reaction with $Rh_2(SNTTL)_3(dCPA)$ followed by

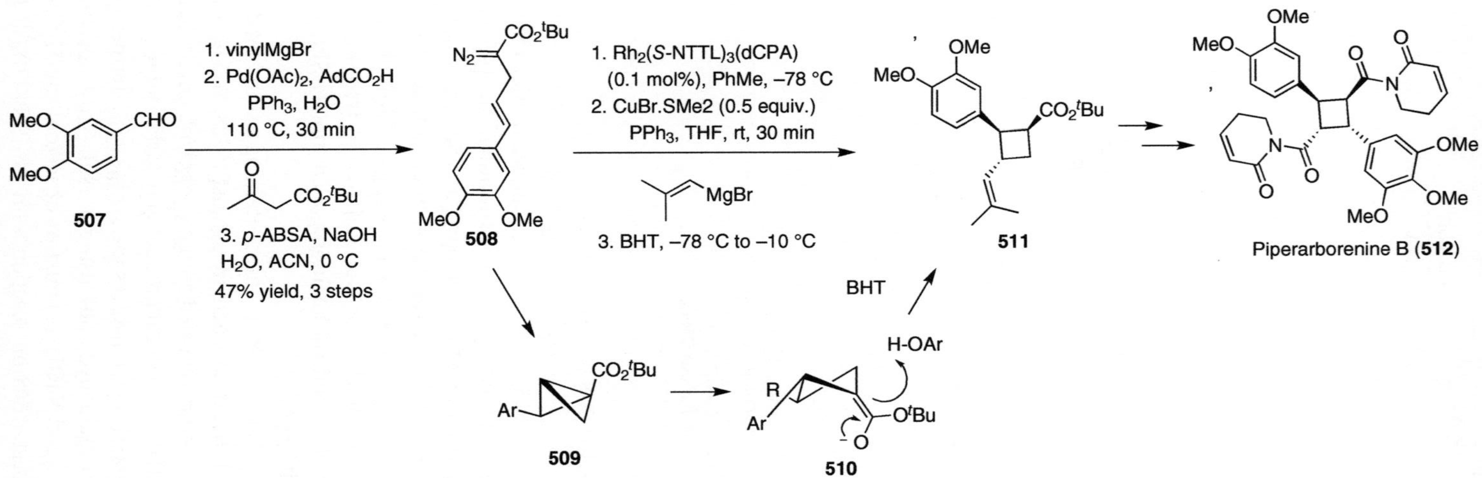

Scheme 12.79 Total synthesis of Piperarborenine B.

cuprate addition using CuBr·SMe_2, PPh_3, and 2-methyl-1-propenylmagnesium bromide provided the desired trisubstituted cyclobutane **510** via the formation of bicyclo-propane **509** in 69% yield, 92% ee, and 4:1 *dr* after kinetic protonation with BHT. This cyclobutane **511** furnishes piperarborenine B (**512**) after several steps.

Diarylheptanoids are a class of secondary plant metabolites belonging to diarylheptanoid natural products. It was isolated in 2003 from the rhizomes of *Dioscorea spongiosa* and is reported to have anti-osteoporotic activity [137]. Gharpure and co-workers reported the synthesis of diospongins B (**517**) via the cleavage of the DA-substituted cyclopropapyranones **515**, which was harvested by stereoselective intramolecular cyclopropanation of vinylogous carbonates **514** with carbenes using a copper catalyst, as shown in Scheme 12.80. Regioselective ring-opening reaction of **515** under radical conditions provided by $^{n}Bu_3SnH$ and AIBN in refluxing benzene afforded the pyranone **516**, which led to the rac-diospongin B (**517**) in 68% yield [138].

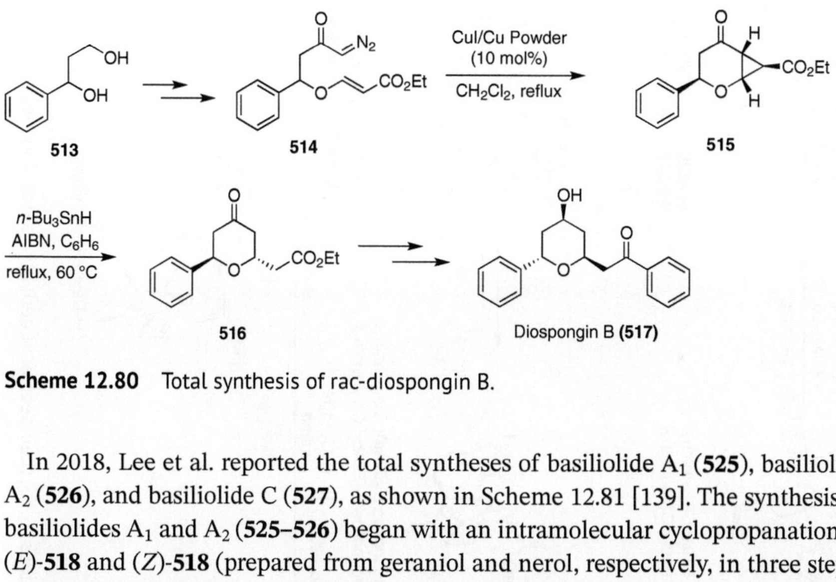

Scheme 12.80 Total synthesis of rac-diospongin B.

In 2018, Lee et al. reported the total syntheses of basiliolide A_1 (**525**), basiliolide A_2 (**526**), and basiliolide C (**527**), as shown in Scheme 12.81 [139]. The synthesis of basiliolides A_1 and A_2 (**525–526**) began with an intramolecular cyclopropanation of (*E*)-**518** and (*Z*)-**518** (prepared from geraniol and nerol, respectively, in three steps) to obtain (±)-**519a** and (±)-**519b**, respectively. The ketone moiety was reduced with $NaBH_4$ to afford β-hydroxy esters (±)-**520** as single diastereoisomers, where the attack took place from the less hindered α face to escape a steric interaction with the γ-lactone moiety. Subsequent hydrolysis of the γ-lactones under basic conditions, followed by a reaction with allyl bromide, generated the allyl esters. The free alcohols were transformed into iodides (±)-**521** using PPh_3/I_2/imidazole. Lithium-iodine interchange took place with ^{n}BuLi leading to the ring-opening of the cyclopropane furnishing esters (±)-**522** as single diastereoisomers. (±)-**522** on treatment with $VO(acac)_2$/TBHP and Ag_2O/MeI gave (±)-**523**, which was transformed to (±)-**524** after several steps. (±)-**523** gave rise to seco acid derivatives (±)-transtaganolides E and F **524** after a few steps. O-acylation of (±)-**524** with Tf_2O in toluene to led (±)-basiliolides A_1 (**525**) and A_2 (**526**). Likewise, (±)-**524** was also converted to (±)-basiliolide C (**527**).

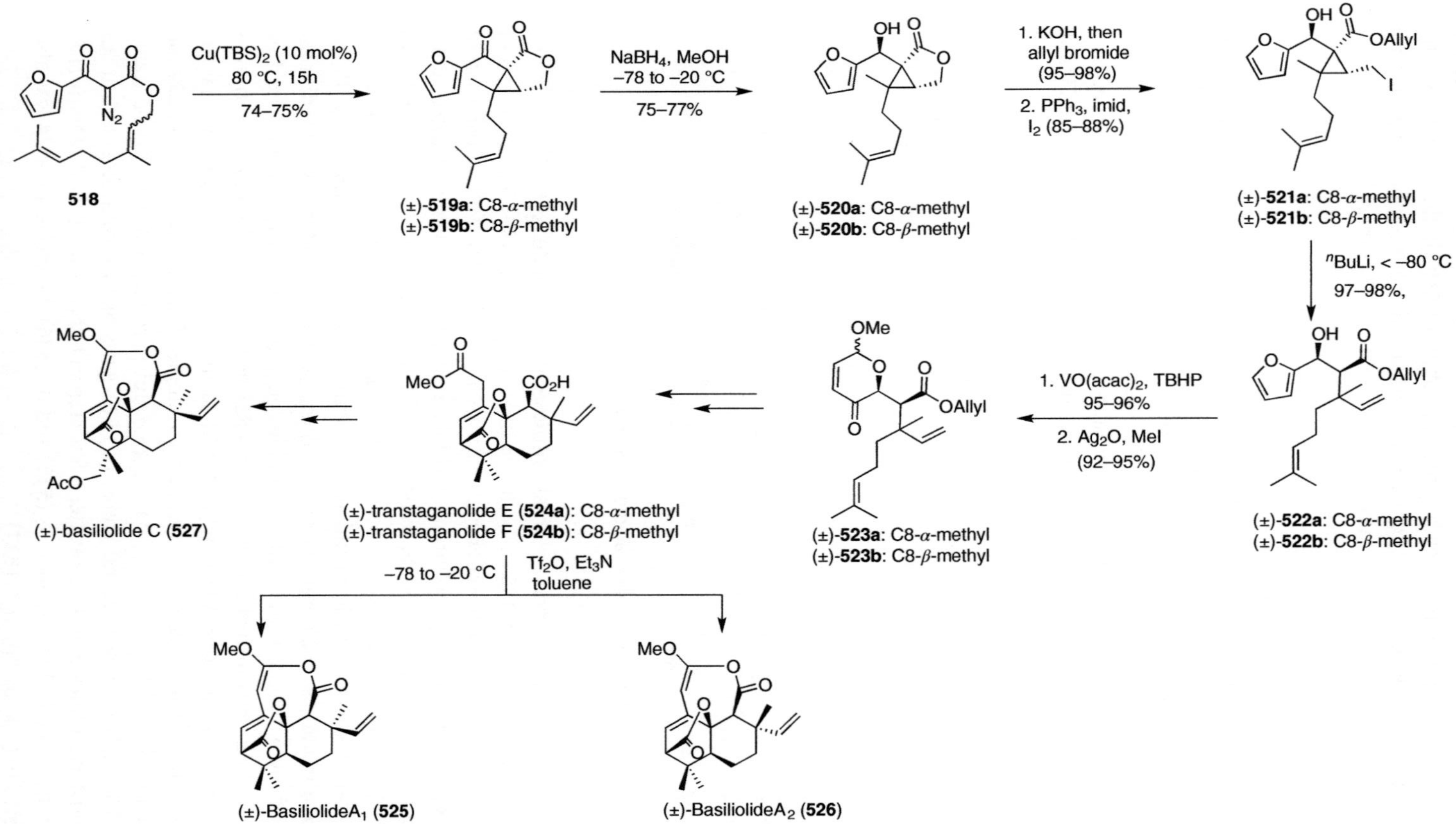

Scheme 12.81 Total syntheses of basiliolides A_1, A_2, their seco acid derivatives transtaganolides E and F and basiliolide C.

From the stem bark of *lophira lanceolate*, Bodo and Martin et al. first isolated Lophirone F a hexaphenolic biflavonoid natural product [140]. Lophirone F shows inhibitory activities against Epstein–Barr virus (EBV) and early antigen (EA) induction by a tumor promoter, teleocidin B-4 [141]. In 2019, Do and coworkers reported the total synthesis of Lophirone F hexamethyl ether (**531**), as shown in Scheme 12.82 [142]. Corey–Chaykovsky cyclopropanation of benzylidene **528** with the dimethylsulfonium bromide salt led to *trans*-cyclopropane (±)-**529** (80% yield, >20:1 dr). Next, aluminum trichloride-catalyzed formal [3+2] cycloaddition of the cyclopropane **529** with 4-methoxybenzaldehyde resulted in the formation of tetrahydrofuran **530** in very good yield and excellent diastereoselectivity (80% yield, >20:1 dr). The *tert*-butyl ester **530** transformed to lophirone F hexamethyl ether (**531**) in an excellent dr (>20:1 dr).

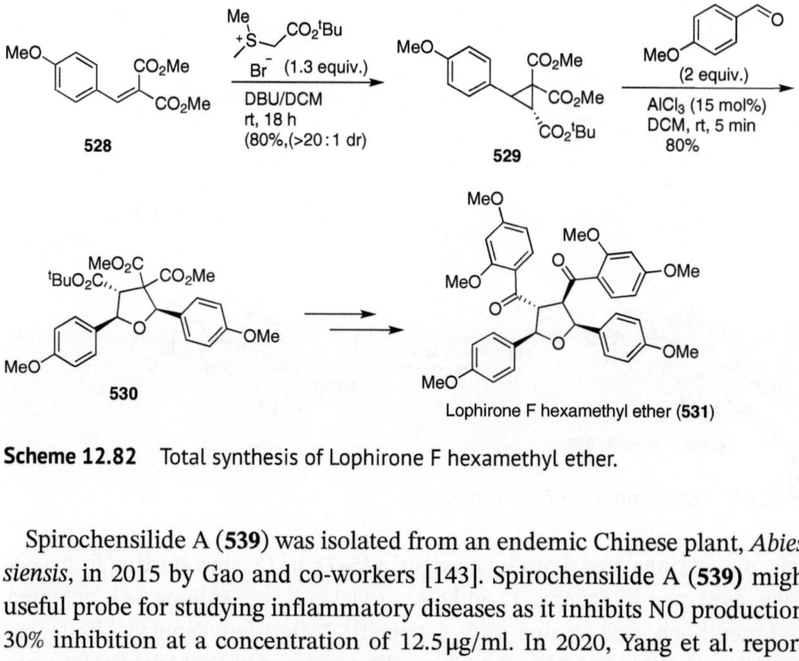

Scheme 12.82 Total synthesis of Lophirone F hexamethyl ether.

Spirochensilide A (**539**) was isolated from an endemic Chinese plant, *Abies chensiensis*, in 2015 by Gao and co-workers [143]. Spirochensilide A (**539)** might be a useful probe for studying inflammatory diseases as it inhibits NO production, with 30% inhibition at a concentration of 12.5 μg/ml. In 2020, Yang et al. reported an asymmetric total synthesis of Spirochensilide A (**539**) in 22 steps with a total yield of up to 2.2% via a tungsten-mediated cyclopropene-based PK reaction, as shown in Scheme 12.83 [144]. The synthesized compound **532** was alkylated with lithium reagent **533** in the presence of $CeCl_3$, and the resultant secondary alcohol was protected with TES to furnish cyclopropene **534** in 98% overall yield. **534** was converted to **535b** along with its diastereoisomer **535a**. The cyclopropane **535b** upon treatment with ^{t}BuOK gave **536**; regioselective hydrogenation of **536** using Pd/C followed by subsequent reductive ring-opening reaction of the product **537** with Li/NH_3 gave **538** in 76% yield over three steps. The compound **538** was then transformed into the natural product Spirochensilide A (**539**) after several steps.

In 2003, Ohta isolated Ikegami Exigurin (**547**) from the marine sponge Geodia exigua found on Oshima Island [145]. Ichikawa accomplished the first total

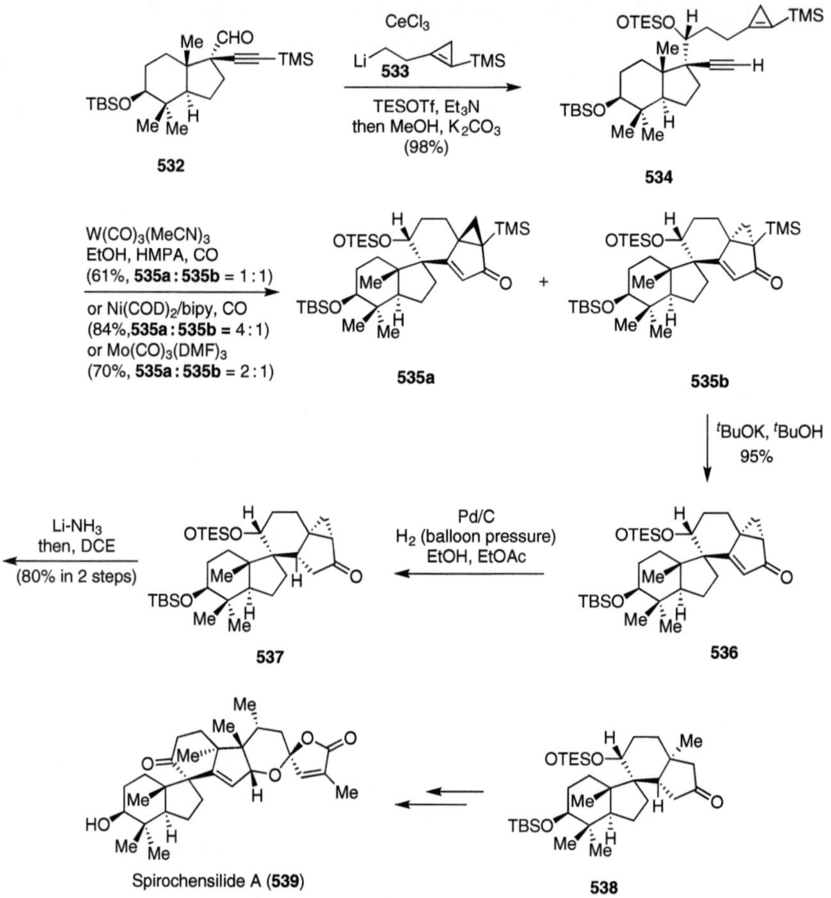

Scheme 12.83 Total synthesis of Spirochensilide.

synthesis of a marine natural product, exigurin (**547**), in 13 steps starting from (+)-menthone, as shown in Scheme 12.84 [146]. α-Diazo-β-keto sulfone **541**, obtained from (+)-menthone, was treated with copper(II) N-(tert-butyl)salicylaldimine to furnish cyclopropyl ketone **542** as a single diastereomer in 75% yield. The ring-opening reaction of the cyclopropane **542** with the azide anion (NaN_3) using a combination of magnesium perchlorate and the phase transfer catalyst ($^{n}Bu_4N{\cdot}HSO_4$) gave **543** in 66% yield. **543** was converted into **544** involving the triflation of ketosulfone with triflic anhydride and sodium hydride, followed by the palladium-catalyzed Suzuki–Miyaura coupling employing trimethylboroxine. The sulfonyl group of **544** was removed by using samarium diiodide, and simultaneous azide reduction generated the corresponding amine, which underwent formylation with acetic formic anhydride to give exiguamide **545**. Dehydration of **545** with triphosgene in the presence of triethylamine gave rise to (−)-10-epi-axisonitrile-3 **546** in 88% yield. Finally, a bioinspired Ugi five-centre four-component reaction (U-5C-4CR) using a solution of formaldehyde and sarcosine methanol with (−)-10-*epi*-axisonitrile-3 **546** gave exigurin (**547**) in 53% yield.

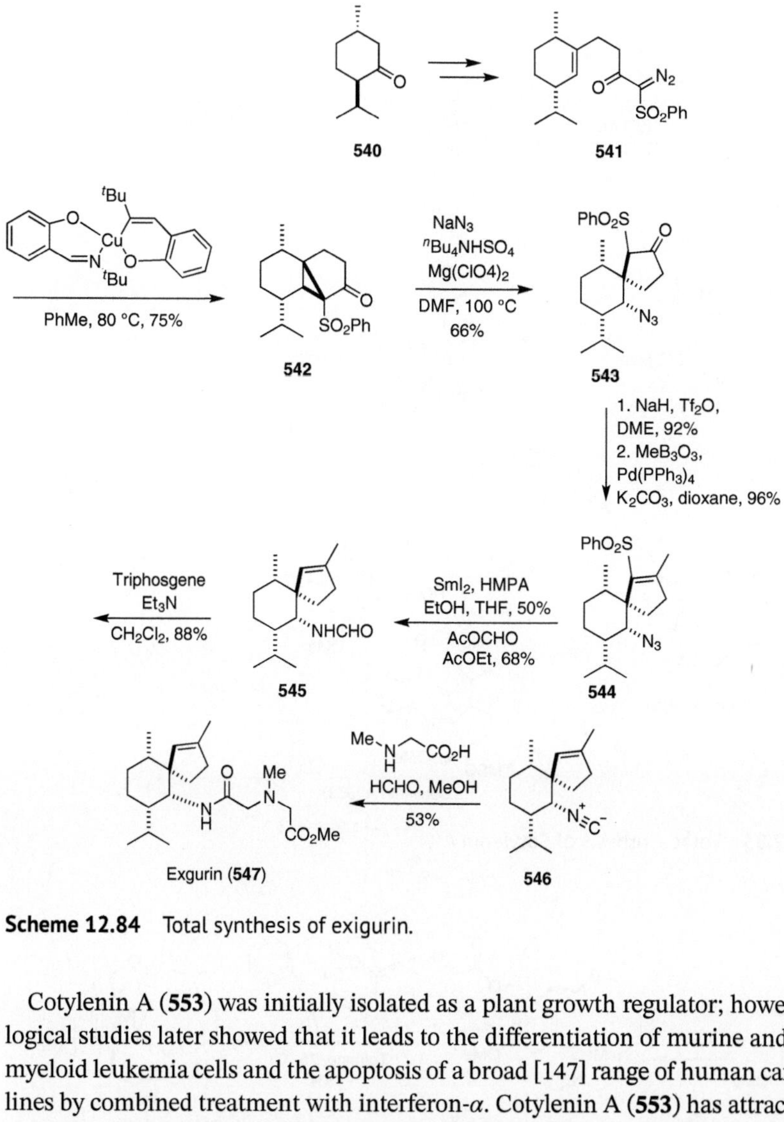

Scheme 12.84 Total synthesis of exigurin.

Cotylenin A (**553**) was initially isolated as a plant growth regulator; however, biological studies later showed that it leads to the differentiation of murine and human myeloid leukemia cells and the apoptosis of a broad [147] range of human cancer cell lines by combined treatment with interferon-α. Cotylenin A (**553**) has attracted considerable attention from the scientific community in the past decades on account of its anti-cancer activity. In 2020, Nakada et al. described a convergent enantioselective total synthesis of cotylenin A (**553**), as shown in Scheme 12.85 [148]. The synthesized compound **548**, when subjected to the CAIMCP afforded the cyclopropane **549** in 86% yield with 86% ee. **549** underwent ring opening with sodium cyanide to give **550**, which was transformed into **551** after several steps. Finally, **551** reacted with TBAF to afford cotylenol (**552**). **551** was also converted to cotylenin A (**553**).

Haliclonin A (**559**), isolated from the marine sponge *Haliclona sp.* collected from Korean waters, is a macrocyclic diamide of a novel skeletal class. Recently, Yokoshima accomplished the total synthesis of haliclonin A (**559**), starting from 3,5-dimethoxybenzoic acid (**554**), as shown in Scheme 12.86 [149]. The symmetric compound **555**, obtained from 3,5-dimethoxybenzoic acid **554** underwent cyclopropanation with a

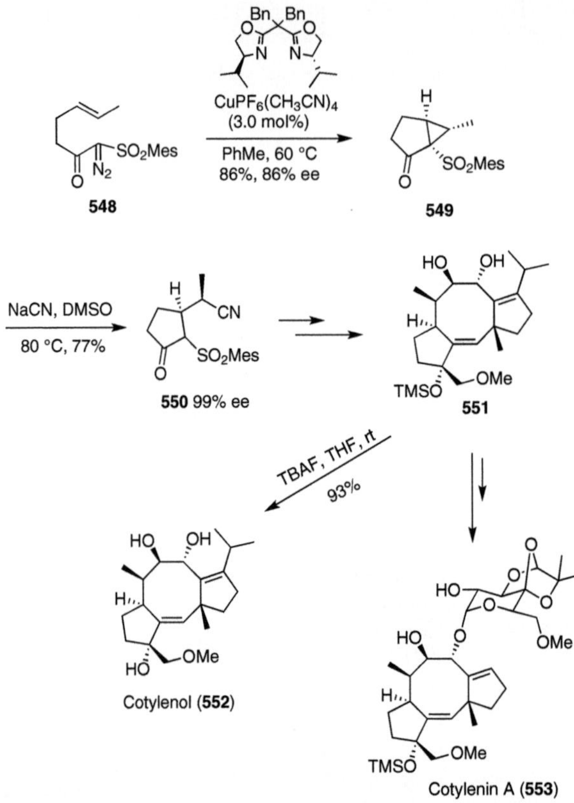

Scheme 12.85 Total synthesis of Cotylenin A.

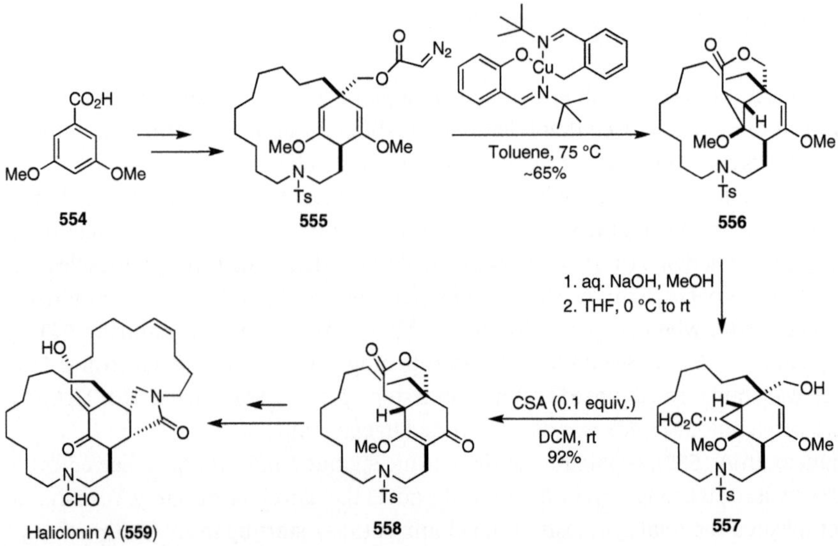

Scheme 12.86 Total synthesis of Haliclonin A.

chiral copper catalyst to afford enantioenriched cyclopropane **556**. The reaction of **556** with sodium hydroxide gave carboxylic acid **557**, which when treated with 10-camphorsulfonic acid (CSA) in dichloromethane, resulted in the formation of **558** in 92% yield over two steps. The compound **558** finally gave rise to haliclonin A (**559**) in twenty five further steps.

Dictyopterenes C' (**567**) is found in sexually mature thalli of the Japanese brown algae *Scytosiphon lomentaria* [150]. Ryu and coworkers accomplished an effective strategy for enantioselective synthesis of (−)-dictyopterene C' (**567**) on the basis of chiral oxazaborolidinium ion (COBI)-catalyzed enantioselective cyclopropanation and subsequent divinylcyclopropane-cycloheptadiene rearrangement [151]. Enantioselective cyclopropanation of α,β-unsaturated aldehyde **560** and diazoacetate **561** in the presence of COBI catalyst gave the cyclopropane **562**. Highly *trans*-selective Julia–Kocienski olefination with 1-phenyl-1H-tetrazole and potassium bis(trimethylsilyl)amide (KHMDS) furnished the alkene **563** in high yield. After the DIBAL-H reduction of *tert*-butyl ester group to aldehyde and simultaneous Wittig reaction with a methylphosphonium salt using *n*-butyllithium produced divinylcyclopropane products **564. 564** was converted to 1,4-cycloheptadienes **566** through the DVCPR endo-boat-like transition-state **565**. Finally, palladium-catalyzed reduction of vinyl bromide **566** afforded desired (−)-dictyopterene C' (**567**) in five steps and 42% yield, as shown in Scheme 12.87.

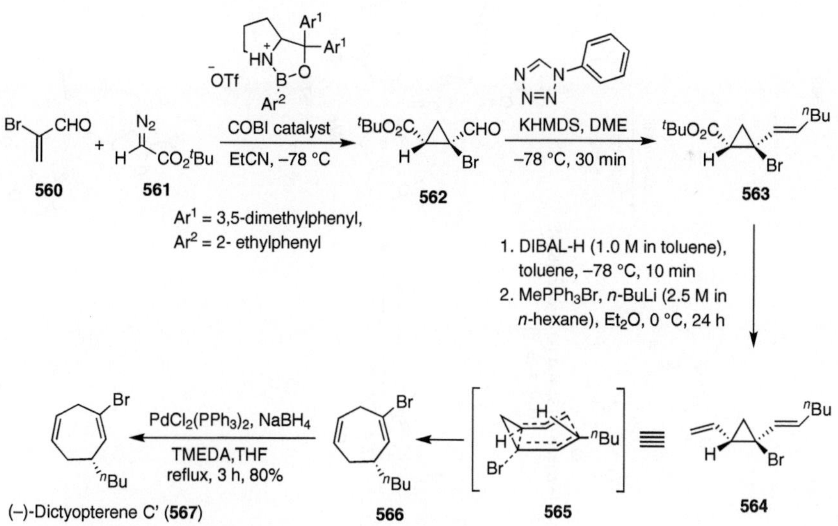

Scheme 12.87 Total synthesis of Dictyopterenes C'.

The high natural abundance and varied biological activities of α-alkylidene-γ-butyrolactones have stimulated the scientific community to dense efficient methods for their synthesis [152a]. Unsworth and co-workers reported the total syntheses of three natural products, (±)-savinin (**573**), (±)-gadain (**574**), and (±)-peperomin E (**579**) in 2016 as shown in Scheme 12.88 [152b]. α-Diazo-α-(diethoxyphosphoryl) acetate **570** was derived from allylic alcohol **568**. The compound **570** on treatment

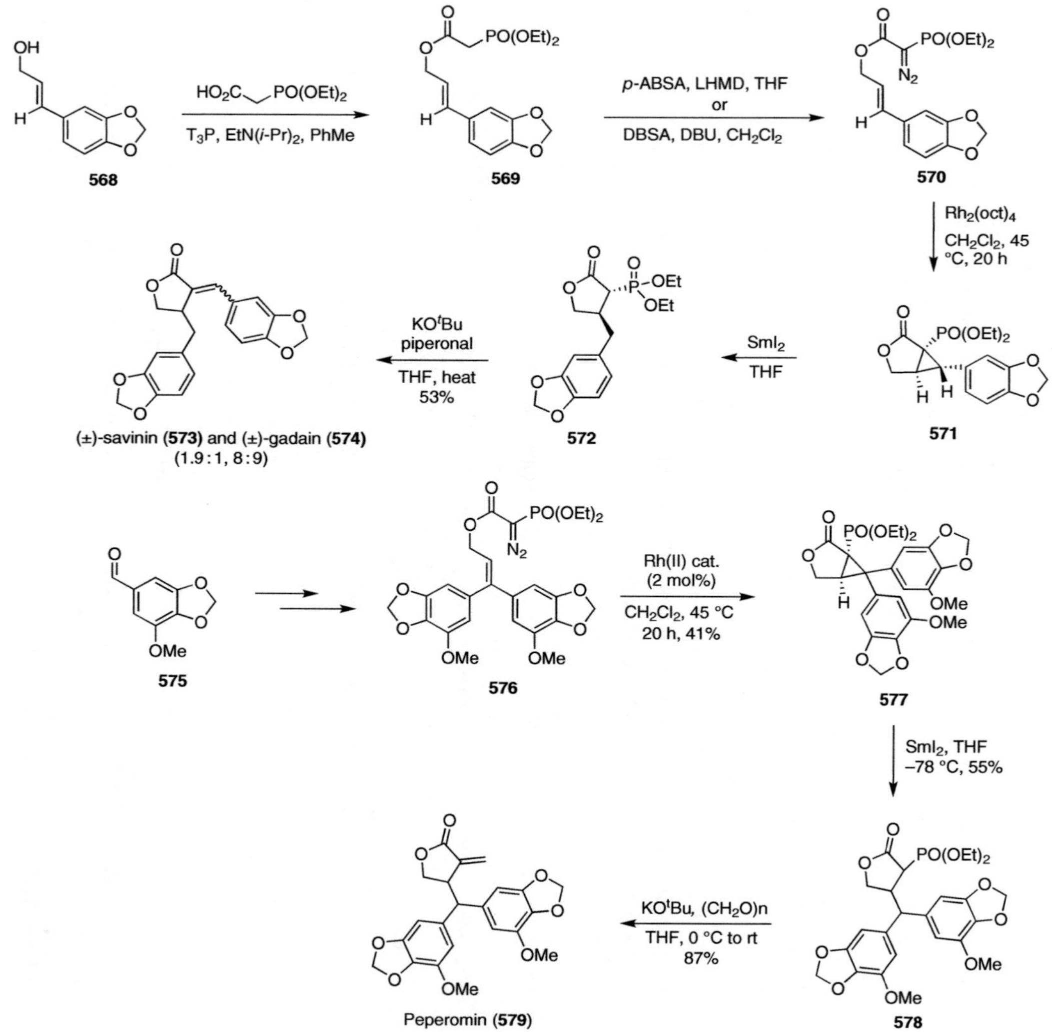

Scheme 12.88 Total synthesis of (±)-savinin, (±)-gadain and (±)-peperomin E

with 2 mol% rhodium(II) octanoate in CH_2Cl_2 under refluxing conditions, led to the efficient formation of cyclopropane **571**, which was reacted with samarium(II) to give the lactone **572** in 38% yield, as a single diastereoisomer, which on reaction with KO^tBu and piperonal, led to a 1.9:1 mixture of α-methylene-γ-butyrolactones (±)-savinin **573**, (±)-gadain **574** in 53% overall yield. These isomers were separated by column chromatography, and the spectral data for both compounds were in full agreement with the reported literature. Next, for the synthesis of (±)-peperomin E (**579**), it began with the conversion of commercially available aldehyde **575** to the key cyclopropanation precursor **576**. The rhodium-catalyzed cyclopropanation of α-diazo-α-(diethoxyphosphoryl)acetate **576** led to the formation of **577**, which underwent ring-opening with samarium(II) iodide to form the lactone **578**. Finally, Horner-Wadsworth-Emmons olefination of **578** afforded peperomin E (**579**).

12.5 Conclusion

The total synthesis of natural products has always been considered as one of the most exciting and dynamic fields in organic chemistry. This chapter summarizes the application of DA-cyclopropanes in the total synthesis of various alkaloids, terpenes/terpenoids, and other natural products of diverse biological activities. As discussed, cyclopropanation followed by synthetic transformations of the corresponding cyclopropanes were considered as some of the important key steps in each of the above total syntheses. More interestingly, in some cases, the reactions were also able to show high selectivity. Hence, good diastereocontrol as well as enantiocontrol even led to the total synthesis of natural products in an enantioselective manner. Since DA-cyclopropane chemistry is in constant growth, the syntheses described above are only some limited examples of the expanding literature on the application of DA-cyclopropanes in the total synthesis of natural products.

Acknowledgments

M.K.G. is grateful to IIT Kanpur and SERB (DST), India, for financial support. A. S. and K. M. thank IIT-Kanpur for Institute Research Fellowships and S.Y. thanks UGC for a Junior Research Fellowship.

References

1 Reissig, H.-U. and Hirsch, E. (1980). *Angew. Chem. Int. Ed. Eng.* 19: 813–814.
2 Schneider, T.F., Kaschel, J., and Werz, D.B. (2014). *Angew. Chem. Int. Ed.* 53: 5504–5523.
3 Jossang, A., Jossang, P., Hadi, H.A. et al. (1991). *J. Org. Chem.* 56: 6527–6530.
4 Fisher, C., Meyers, C., and Carreira, E.M. (2000). *Helv. Chim. Acta* 83: 1175–1181.
5 (a) Lerchner, A. and Carreira, E.M. (2002). *J. Am. Chem. Soc.* 124: 14826–14827. (b) Lerchner, A. and Carreira, E.M. (2006). *Chem. Eur. J.* 12: 8208–8219.

6 Bassleer, R., Depauw-Gillet, M.C., Massart, B. et al. (1982). *Planta Med.* 45: 123–126.
7 (a) Cui, C.-B., Kakeya, H., and Osada, H. (1996). *J. Antibiot.* 49: 832–835. (b) Cui, C.-B., Kakeya, H., and Osada, H. (1996). *Tetrahedron* 52: 12651–12666.
8 (a) Meyers, C. and Carreira, E.M. (2003). *Angew. Chem. Int. Ed.* 42: 694–696. (b) Marti, C. and Carreira, E.M. (2005). *J. Am. Chem. Soc.* 127: 11505–11515.
9 Snieckus, V. (1973). *The Alkaloids*, vol. 14, 425–506. New York: Academic Press.
10 Carson, C.A. and Kerr, M.A. (2006). *Angew. Chem. Int. Ed.* 45: 6560–6563.
11 Young, I.S. and Kerr, M.A. (2007). *J. Am. Chem. Soc.* 129: 1465–1469.
12 Kobayashi, J., Watanabe, D., Kawasaki, N., and Tsuda, M. (1997). *J. Org. Chem.* 62: 9236–9239.
13 Yang, J., Wu, H., Shen, L., and Qin, Y. (2007). *J. Am. Chem. Soc.* 129: 13794–13795.
14 Liu, P., Seo, J.H., and Weinreb, S.M. (2010). *Angew. Chem. Int. Ed.* 49: 2000–2003.
15 Leduc, A.B. and Kerr, M.A. (2008). *Angew. Chem. Int. Ed.* 47: 7945–7948.
16 Morales, C.L. and Pagenkopf, B.L. (2008). *Org. Lett.* 10: 157–159.
17 Bajtos, B. and Pagenkopf, B.L. (2009). *Eur. J. Org. Chem.* 1072–1077.
18 Hochlowski, J.E.M., Mullally, M., Spanton, S.G. et al. (1993). *J. Antibiot.* 46: 380–386.
19 He, B., Song, H., Du, Y., and Qin, Y. (2009). *J. Org. Chem.* 74: 298–304.
20 Gagnon, D. and Spino, C. (2009). *J. Org. Chem.* 74: 6035–6041.
21 Carson, C.A. and Kerr, M.A. (2009). *Org. Lett.* 11: 777–779.
22 Sakamoto, K., Tsujii, E., Abe, F. et al. (1996). *J. Antibiot.* 49: 37–44.
23 Jung, M.E. and Chang, J.J. (2010). *Org. Lett.* 12: 2962–2965.
24 (a) Liau, B.B. and Shair, M.D. (2010). *J. Am. Chem. Soc.* 132: 9594–9595. (b) DeForest, J.C., Samame, R.A., Suryn, G. et al. (2018). *J. Org. Chem.* 83: 8914–8925.
25 Shimokawa, J., Harada, T., Yokoshima, S., and Fukuyama, T. (2011). *J. Am. Chem. Soc.* 133: 17634–17637.
26 Goldberg, A.F.G. and Stoltz, B.M. (2011). *Org. Lett.* 13: 4474–4476.
27 Mancey, N.C., Sandon, N., Auvinet, A.-L. et al. (2011). *Chem. Commun.* 47: 9804–9806.
28 (a) Huang, H., Spande, T.F., and Panek, J.S. (2003). *J. Am. Chem. Soc.* 125: 626–627. (b) Fellah, M., Santarem, M., Lhommet, G., and Mansuy, V.M. (2010). *J. Org. Chem.* 75: 7803–7808.
29 Zhang, H. and Curran, D.P. (2011). *J. Am. Chem. Soc.* 133: 10376–10378.
30 Patil, D.V., Cavitt, M.A., and France, S. (2012). *Heterocycles* 84: 1363–1373.
31 Huang, H., Ji, X., Wu, W., and Jiang, H. (2013). *Chem. Commun.* 49: 3351–3353.
32 Kapat, A., Kumar, P.S., and Baskaran, S. (2007). *Beilstein J. Org. Chem.* 3: 49.
33 Frei, R., Staedler, D., Raja, A. et al. (2013). *Angew. Chem. Int. Ed.* 52: 13373–13376.
34 Ivanov, K.L., Villemson, E.V., Budynina, E.M. et al. (2015). *Chem. Eur. J.* 21: 4975–4987.
35 (a) Jin, G.-L., Su, Y.-P., Liu, M. et al. (2014). *J. Ethnopharmacol.* 152: 33–52. (b) Kitajima, M. (2007). *J. Nat. Med.* 61: 14–23.
36 Newcomb, E.T., Knutson, P.C., Pedersen, B.A., and Ferreira, E.M. (2016). *J. Am. Chem. Soc.* 138: 108–111.
37 Marki, F. and Witkop, B. (1963). *Experientia* 19: 329–338.

38 Logan, M.M., Toma, T., Thomas-Tran, R., and Bois, J.D. (2016). *Science* 354: 865–869.
39 For reviews on the Kainoids: (a) Parsons, A.F. (1996). *Tetrahedron* 52: 4149–4174. (b) Stathakis, C.I., Yioti, E.G., and Gallos, J.K. (2012). *Eur. J. Org. Chem.* 2012: 4661–4673. (c) Bhat, C. and Kumar, A. (2015). *Asian J. Org. Chem.* 4: 102–115.
40 Piotrowski, M.L. and Kerr, M.A. (2018). *Org. Lett.* 20: 7624–7627.
41 Piotrowski, M.L. and Kerr, M.A. (2019). *Eur. J. Org. Chem.* 2019: 3122–3126.
42 Rinner, U., Lentsch, C., and Aichinger, C. (2010). *Synthesis* 22: 3763–3784.
43 Burdge, H.E., Oguma, T., Kawajiri, T., and Shenvi, R. (2019). *ChemRxiv* .
44 Cheng, G.-G., Li, D., Hou, B. et al. (2016). *J. Nat. Prod.* 79: 2158–2166.
45 Cholewczynski, A.E., Williams, P.C., and Pierce, J.G. (2020). *Org. Lett.* 22: 714–717.
46 Chattopadhyay, A.K. and Hanessian, S. (2017). *Chem. Rev.* 117: 4104–4146.
47 Wang, B., Xu, B., Xun, W. et al. (2021). *Angew. Chem. Int. Ed.* 60: 9439–9443.
48 Bon, D.J.-Y., Banwell, M.G., Cade, I.A., and Willis, A.C. (2011). *Tetrahedron* 67: 8348–8352.
49 Hirasawa, Y., Morita, H., and Kobayashi, J. (2002). *Tetrahedron* 58: 5483–5488.
50 Kurose, T., Tsukano, C., Nanjo, T., and Takemoto, Y. (2021). *Org. Lett.* 23: 676–681.
51 For the isolation and absolute configuration determination of clavukerin A, see: (a) Kobayashi, M., Son, B.W., Kido, M. et al. (1983). *Chem. Pharm. Bull.* 31: 2160–2163.For the absolute configuration determination of isoclavukerin, see: (b) Kusumi, T., Hamada, T., Hara, M. et al. (1992). *Tetrahedron Lett.* 33: 2019–2022.
52 Alexakis, A. and March, S. (2002). *J. Org. Chem.* 67: 8753–8757.
53 Chiba, S., Kitamura, M., and Narasaka, K. (2006). *J. Am. Chem. Soc.* 128: 6931–6937.
54 Kalidindi, S., Jeong, W.B., Schall, A. et al. (2007). *Angew. Chem. Int. Ed.* 46: 6361–6363.
55 (a) Trost, B.M., Waser, J., and Meyer, A. (2007). *J. Am. Chem. Soc.* 129: 14556–14557. (b) Waser, J., Meyer, A., and Trost, B.M. (2008). *J. Am. Chem. Soc.* 130: 16424–16434.
56 Jiao, L., Yuan, C., and Yu, Z.-X. (2008). *J. Am. Chem. Soc.* 130: 4421–4430.
57 Fan, X., Zhuo, L.-G., Tu, Y.Q., and Yu, Z.-X. (2009). *Tetrahedron* 65: 4709–4713.
58 (a) Mehta, G. and Umarye, J.D. (2001). *Tetrahedron Lett.* 42: 8101–8104. (b) Pattenden, G. and Teague, S.J. (1987). *Tetrahedron* 43: 5637–5652.
59 Yuan, C., Jiao, L., and Yu, Z.-X. (2010). *Tetrahedron Lett.* 51: 5674–5676.
60 Trost, B.M., Shuey, C.D., and DiNinno, F. Jr. (1979). *J. Am. Chem. Soc.* 101: 1284–1285.
61 Jiao, L., Lin, M., Zhuo, L.-G., and Yu, Z.-X. (2010). *Org. Lett.* 12: 2528–2531.
62 Xing, S., Pan, W., Liu, C. et al. (2010). *Angew. Chem. Int. Ed.* 49: 3215–3218.
63 Nicolaou, K.C., Li, A., Edmonds, D.J. et al. (2009). *J. Am. Chem. Soc.* 131: 16905–16918.
64 Nicolaou, K.C., Li, A., and Edmonds, D.J. (2006). *Angew. Chem. Int. Ed.* 45: 7086–7090.
65 McGrath, N.A., Bartlett, E.S., Sittihan, S., and Njardarson, J.T. (2009). *Angew. Chem. Int. Ed.* 48: 8543–8546.
66 Lian, Y., Miller, L.C., Born, S. et al. (2010). *J. Am. Chem. Soc.* 132: 12422–12425.

67 Rudi, A., Shalom, H., Schleyer, M. et al. (2004). *J. Nat. Prod.* 67: 106–109.

68 Justicia, J., Oller-Lopez, J.L., Campan, A.G. et al. (2005). *J. Am. Chem. Soc.* 127: 14911–14921.

69 Bon, D.J.-Y., Banwell, M.G., and Willis, A.C. (2010). *Tetrahedron* 66: 7807–7814.

70 Liang, Y., Jiang, X., and Yu, Z.-X. (2011). *Chem. Commun.* 47: 6659–6661.

71 Xiao, Q., Ren, W.W., Chen, Z.X. et al. (2011). *Angew. Chem. Int. Ed.* 50: 7373–7377.

72 Harrar, K. and Reiser, O. (2012). *Chem. Commun.* 48: 3457–3459.

73 Kreuzer, A., Kerres, S., Ertl, T. et al. (2013). *Org. Lett.* 15: 3420–3423.

74 Bergmann, A. and Reiser, O. (2014). *Chem. Eur. J.* 20: 7613–7615.

75 (a) Au, T.K., Chick, W.S.H., and Leung, P.C. (2000). *Life Sci.* 67: 733–742. (b) Evidente, A., Andolfi, A., Cimmino, A. et al. (2006). *Phytochemistry* 67: 2281–2287.

76 Wei, H., Itoh, T., Kinoshita, M. et al. (2004). *Tetrahedron* 60: 6015–6019.

77 Dauben, W.G. and Hart, D.J. (1977). *J. Org. Chem.* 42: 922–923.

78 Brill, Z.G., Grover, H.K., and Maimone, T.J. (2016). *Science* 352: 1078–1082.

79 Coll, J.C. (1992). *Chem. Rev.* 92: 613–631.

80 Shen, S., Zhu, H., Chen, D. et al. (2012). *Tetrahedron Lett.* 53: 5759–5762.

81 Zhang, P.-P., Yan, Z.-M., Li, Y.-H. et al. (2017). *J. Am. Chem. Soc.* 139: 13989–13992.

82 Hayashi, Y., Gotoh, H., Hayashi, T., and Shoji, M. (2005). *Angew. Chem. Int. Ed.* 44: 4212–4215.

83 Narayanam, J.M.R. and Stephenson, C.R. (2011). *J. Chem. Soc. Rev.* 40: 102–113.

84 (a) Yoon, T.P. (2016). *Acc. Chem. Res.* 49: 2307–2315. (b) Prier, C.K., Rankic, A., and MacMillan, D.W.C. (2013). *Chem. Rev.* 113: 5322–5363.

85 Ravelli, D., Protti, S., and Fagnoni, M. (2016). *Chem. Rev.* 116: 9850–9913.

86 (a) Yang, J., Xu, W., Cui, Q. et al. (2017). *Org. Lett.* 19: 6040–6043. (b) Stevens, R.V., Beaulieu, N., Chan, W.H. et al. (1986). *J. Am. Chem. Soc.* 108: 1039–1049. (c) Haener, R., Laube, T., and Seebach, D. (1985). *J. Am. Chem. Soc.* 107: 5396–5403.

87 Chung, H.-M., Su, J.-H., Hwang, T.-L. et al. (2013). *Tetrahedron* 69: 2740–2744.

88 Liu, G., Zhang, Z., Fu, S., and Liu, L. (2021). *Org. Lett.* 23: 290–295.

89 (a) Dumdei, E.J., Silva, E.D.D., Andersen, R.J. et al. (1989). *J. Am. Chem. Soc. 111*: 2712–2713. (b) Morris, S.A., Silva, E.D.D., and Andersen, R.J. (1991). *Can. J. Chem.* 69: 768–771.

90 Tao, D.J., Slutskyy, Y., Muuronen, M. et al. (2018). *J. Am. Chem. Soc.* 140: 3091–3102.

91 Qiao, M.-F., Ji, N.-Y., Liu, X.-H. et al. (2010). *Bioorg. Med. Chem. Lett.* 20: 5677–5680.

92 Harms, H., Rempel, V., Kehraus, S. et al. (2014). *J. Nat. Prod.* 77: 673–677.

93 Murokawa, T., Enomoto, M., Teranishi, T. et al. (2018). *Tetrahedron Lett.* 59: 4107–4109.

94 (a) Kawai, N., Lagrange, J.-M., Ohmi, M., and Uenishi, J. (2006). *J. Org. Chem.* 71: 4530–4537. (b) Uenishi, J., Vikhe, Y.S., and Kawai, N. (2008). *Chem. Asian J.* 3: 473–484.

95 Wu, Z.Y. (2000). *Flora of Yunnan*, 13. Beijing: Science Press.

96 Tian, X., Li, L., Hu, Y. et al. (2013). *RSC Adv.* 3: 7880–7883.

97 Schmiedel, V.M., Hong, Y.J., Lentz, D. et al. (2018). *Angew. Chem. Int. Ed.* 57: 2419–2422.

98 Zhang, J., Zheng, S., Peng, W., and Shen, Z. (2014). *Tetrahedron Lett.* 55: 1339–1341.

99 Liu, P., Cui, Y., Chen, K. et al. (2018). *Org. Lett.* 20: 2517–2521.

100 Wang, J., Chen, S., Sun, B. et al. (2013). *Chem. Eur. J.* 19: 2539–2547.

101 (a) Wang, J., Lin, Z., Zhao, Q., and Sun, H. (1998). *Phytochemistry* 47: 307–309. (b) Niu, X.-M., Li, S.-H., Mei, S.-X. et al. (2002). *J. Nat. Prod.* 65: 1892–1896. (c) Wang, W.-G., Li, X.-N., Du, X. et al. (2012). *J. Nat. Prod.* 75: 1102–1107.

102 Su, F., Lu, Y.-D., Kong, L.-R. et al. (2018). *Angew. Chem. Int. Ed.* 57: 760–764.

103 Cormier, M., de la Torre, A., and Marek, I. (2018). *Angew. Chem. Int. Ed.* 57: 13237–13241.

104 Liu, D.-Z., Wang, F., Liao, T.-G. et al. (2006). *Org. Lett.* 8: 5749–5752.

105 Nistanaki, S.K., Boralsky, L.A., Pan, R.D., and Nelson, H.M. (2019). *Angew. Chem. Int. Ed.* 58: 1724–1726.

106 Yu, K., Yang, Z.-N., Liu, C.-H. et al. (2019). *Angew. Chem. Int. Ed.* 58: 8556–8560.

107 Gao, J., Rao, P., Xu, K. et al. (2020). *J. Am. Chem. Soc.* 142: 4592–4597.

108 Ishiwata, H., Nemoto, T., Ojika, M., and Yamada, K. (1994). *J. Org. Chem.* 59: 4710–4711.

109 Ghosh, A.K. and Liu, C. (2001). *Org. Lett.* 3: 635–638.

110 Charette, A.B. and Naud, J. (1998). *Tetrahedron Lett.* 39: 7259–7262.

111 Bandichhor, R., Nosse, B., and Reiser, O. (2005). *Top. Curr. Chem.* 243: 43–72.

112 Bandichhor, R., Nosse, B., Sorgel, S. et al. (2003). *Chem. Eur. J.* 9: 260–270.

113 Jenny, L. and Borschberg, H.-J. (1995). *Helv. Chim. Acta* 78: 715–731.

114 Kingsbury, J.S. and Corey, E.J. (2005). *J. Am. Chem. Soc.* 127: 13813–13815.

115 Gheorghe, A., Schulte, M., and Reiser, O. (2006). *J. Org. Chem.* 71: 2173–2176.

116 Tanaka, M., Ubukata, M., Matsuo, T. et al. (2007). *Org. Lett.* 9: 3331–3334.

117 Sanders, S.D., Olalla, A.R., and Johnson, J.S. (2009). *Chem. Commun.* 5135–5137.

118 Campbell, M.J. and Johnson, J.S. (2009). *J. Am. Chem. Soc.* 131: 10370–10371.

119 Hu, B., Xing, S., Ren, J., and Wang, Z. (2010). *Tetrahedron* 66: 5671–5674.

120 Karadeolian, A. and Kerr, M.A. (2010). *Angew. Chem. Int. Ed.* 49: 1133–1135.

121 Hirai, S. and Nakada, M. (2010). *Tetrahedron Lett.* 51: 5076–5079.

122 Nicolaou, K.C., Tria, G.S., and Edmonds, D.J. (2008). *Angew. Chem. Int. Ed.* 47: 1780–1783.

123 Nanda, L.N., Shukla, M.K., and Gharpure, S.J. (2011). *Eur. J. Org. Chem.* 2011: 6632–6635.

124 Kauloorkar, S.V., Jha, V., Jogdand, G., and Kumar, P. (2015). *RSC Adv.* 5: 61000–61005.

125 Fernandes, R.A. and Kattanguru, P. (2012). *J. Org. Chem.* 77: 9357–9360.

126 Granger, K. and Snapper, M.L. (2012). *Eur. J. Org. Chem.* 2012: 2308–2311.

127 Uwamori, M., Saito, A., and Nakada, M. (2012). *J. Org. Chem.* 77: 5098–5107.

128 Perali, R.S. and Kalapati, S. (2012). *Tetrahedron* 68: 3725–3728.

129 Grover, H.K., Emmett, M.R., and Kerr, M.A. (2013). *Org. Lett.* 15: 4838–4841.

130 Gharpure, S.J., Nanda, L.N., and Shukla, M.K. (2014). *Org. Lett.* 16: 6424–6427.

131 (a) Uetake, Y., Uwamori, M., and Nakada, M. (2015). *J. Org. Chem.* 80: 1735–1745. (b) Zhang, Q. and Porco, J.A. Jr. (2012). *Org. Lett.* 14: 1796–1799. (c) Sparling, B.A., Moebius, D.C., and Shair, M.D. (2013). *J. Am. Chem. Soc.* 135: 644–647. (d) Uwamori, M. and Nakada, M. (2013). *J. Antibiot.* 66: 141–145. (e) Uwamori, M. and Nakada, M. (2013). *Nat. Prod. Commun.* 8: 955–959.

132 Unzner, T.A., Grossmann, A.S., and Magauer, T. (2016). *Angew. Chem. Int. Ed.* 55: 9763–9767.

133 (a) Aleman, J., Richter, B., and Jørgensen, K.A. (2007). *Angew. Chem. Int. Ed.* 46: 5515–5519. (b) Aleman, J., Richter, B., and Jørgensen, K.A. (2007). *Angew. Chem.* 119: 5611–5615.

134 (a) Diao, T. and Stahl, S.S. (2011). *J. Am. Chem. Soc.* 133: 14566–14569. (b) Diao, T., Wadzinski, T.J., and Stahl, S.S. (2012). *Chem. Sci.* 3: 887–891.

135 (a) Lee, F.-P., Chen, Y.-C., Chen, J.-J. et al. (2004). *Helv. Chim. Acta* 87: 463–468. (b) Tsai, I.-L., Lee, F.-P., Wu, C.-C. et al. (2005). *Planta Med.* 71: 535–542.

136 Panish, R.A., Chintala, S.R., and Fox, J.M. (2016). *Angew. Chem. Int. Ed.* 55: 4983–4987.

137 (a) Kumaraswamy, G., Ramakrishna, G., Naresh, P. et al. (2009). *J. Org. Chem.* 74: 8468–8471. (b) Yin, J., Kouda, K., Tezuka, Y. et al. (2004). *Planta Med.* 70: 54–58.

138 Gharpure, S.J., Mane, S.P., Nanda, L.N., and Shukla, M.K. (2016). *Isr. J. Chem.* 56: 553–557.

139 Ma, W., Liang, X., Ye, W. et al. (2018). *Eur. J. Org. Chem.* 2018: 196–208.

140 Tih, R.G., Sondengam, B.L., Martin, M.T., and Bodo, B. (1990). *Phytochemistry* 29: 2289–2293.

141 Murakami, A., Tanaka, S., Ohigashi, H. et al. (1992). *Phytochemistry* 31: 2689–2693.

142 Le, H.M., Mac, H.D., Oh, C.H., and Do, D.T. (2019). *Eur. J. Org. Chem.* 2019: 2362–2367.

143 Zhao, Q.-Q., Song, Q.-Y., Jiang, K. et al. (2015). *Org. Lett.* 17: 2760–2763.

144 Liang, X.-T., Chen, J.-H., and Yang, Z. (2020). *J. Am. Chem. Soc.* 142: 8116–8121.

145 Uy, M.M., Ohta, S., Yanai, M. et al. (2003). *Tetrahedron* 59: 731–736.

146 Hosokawa, S., Nakanishi, K., Udagawa, Y. et al. (2020). *Org. Biomol. Chem.* 18: 687–693.

147 (a) Asahi, K., Honma, Y., Hazeki, K. et al. (1997). *Biochem. Biophys. Res. Commun.* 238: 758–763. (b) Honma, Y. (2002). *Leuk. Lymphoma* 43: 1169–1178. (c) Honma, Y., Ishii, Y., Yamaguchi, Y.Y. et al. (2003). *Cancer Res.* 63: 3659–3666.

148 Uwamori, M., Osada, R., Sugiyama, R. et al. (2020). *J. Am. Chem. Soc.* 142: 5556–5561.

149 Jin, Y., Orihara, K., Kawagishi, F. et al. (2021). *Angew. Chem. Int. Ed.* 60: 2–8.

150 Kajiwara, T., Hatanaka, A., Kodama, K. et al. (1991). *Phytochemistry* 30: 1805–1807.

151 Kim, T., Kim, J.Y., Park, K.Y., and Ryu, D.H. (2021). *Bull. Korean Chem. Soc.* 42: 675–678.

152 (a) Kitson, R.R.A., Millemaggi, A., and Taylor, R.J.K. (2009). *Angew. Chem. Int. Ed.* 48: 9426–9451. (b) Lloyd, M.G., Taylor, R.J.K., and Unsworth, W.P. (2016). *Org. Biomol. Chem.* 14: 8971–8988.

Index

Donor–Acceptor Cyclopropanes in Organic Synthesis, First Edition.
Edited by Prabal Banerjee and Akkattu T. Biju.

d

f

g

h

i